THE CITY & GUILDS TEXTBOOK

# LEVEL 3 DIPLOMA IN ELECTRICAL INSTALLATIONS 2365

BUILDINGS AND STRUCTURES UNITS 201, 301-305 AND 308

THE CITY & GUILDS TEXTBOOK

# LEVEL 3 DIPLOMA IN ELECTRICAL INSTALLATIONS 2365

## BUILDINGS AND STRUCTURES UNITS 201, 301-305 AND 308

HOWARD CAREY, PAUL COOK, PAUL HARRIS, ANDREW HAY-ELLIS AND TREVOR PICKARD

SERIES EDITOR: PETER TANNER

**About City & Guilds**

City & Guilds is the UK's leading provider of vocational qualifications, offering over 500 awards across a wide range of industries, and progressing from entry level to the highest levels of professional achievement. With over 8500 centres in 100 countries, City & Guilds is recognised by employers worldwide for providing qualifications that offer proof of the skills they need to get the job done.

**Equal opportunities**

City & Guilds fully supports the principle of equal opportunities and we are committed to satisfying this in all our activities and published material. A copy of our equal opportunities policy statement is available on the City & Guilds website.

**Copyright**



First edition 2014

ISBN 978-0-85193-283-5

Publisher: Charlie Evans

Development Editor: Hannah Cooper

Production Editor: Fiona Freel

Picture Research: Katherine Hodges

Project Management and Editorial Series Team: Vicky Butt, Anna Clark, Kay Coleman, Jo Kemp, Karen Hemingway, Jon Ingoldby, Caroline Low, Joan Miller, Shirley Wakley

Cover design by Select Typesetters Ltd

Text design by Design Deluxe, Bath

Indexed by Indexing Specialists (UK) Ltd

Illustrations by Saxon Graphics Ltd and Ann Paganuzzi

Typeset by Saxon Graphics Ltd, Derby

Printed in the UK by Cambrian Printers Ltd

**Publications**

For information about or to order City & Guilds support materials, contact 0844 534 0000 or centresupport@cityandguilds.com. You can find more information about the materials we have available at www.cityandguilds.com/publications.

Every effort has been made to ensure that the information contained in this publication is true and correct at the time of going to press. However, City & Guilds' products and services are subject to continuous development and improvement and the right is reserved to change products and services from time to time. City & Guilds cannot accept liability for loss or damage arising from the use of information in this publication.

City & Guilds
1 Giltspur Street
London EC1A 9DD

T 0844 543 0033
www.cityandguilds.com
publishingfeedback@cityandguilds.com

# CONTENTS

# ACKNOWLEDGEMENTS

City & Guilds would like to thank sincerely the following:

**For invaluable knowledge and expertise**
*Eur Ing* D. Locke, *BEng (Hons)*, *CEng*, *MIEE*, *MIEEE*, The Institution of Engineering and Technology, Technical Reviewer

Richard Woodcock, City & Guilds, Technical Reviewer and Contributor

**For supplying pictures for the front and back covers**
Jules Selmes and Shutterstock

**For their help with photoshoots**
Jules Selmes and Adam Giles (photographer and assistant);
Andrew Hay-Ellis, James L Deans and the staff at Trade Skills 4 U, Crawley,
and the following models: Jordan Hay-Ellis, Terry White, Katherine Hodges, Claire Owen.

**Picture credits**
Every effort has been made to acknowledge all copyright holders as below and the publishers will, if notified, correct any errors in future editions.

**Alternative Water Solutions** p134; **BSI** p7, p308, p312, p394, p398, p404, p427, p447, p449, p455, p468, p469, p471, p498, p608 (Permission to reproduce extracts from British Standards is granted by BSI Standards Limited (BSI). No other use of this material is permitted. British Standards can be obtained in PDF or hard copy formats from the BSI online shop: www.bsigroup.com/Shop or by contacting BSI Customer Services for hard copies only: Tel: +44 (0)20 8996 9001, Email: cservices@bsigroup.com); **Eaton** p467; **Ecoplay Nederland B.V.** p134; **ECS** p606; **Emily Cooper** p186; **Electricity North West** p336; **Hager** p467; **Hansgrohe/Freewater UK Ltd.** p135; **HSE** p361, p365, p370, p510, p526; **IET** p221, p252, p253, p263, p282, p310, p322, p323, p331, p344, p345, p346, p363, p372, p374, p375, p376, p377, p379, p380, p381, p382, p384, p388, p396, p399, p412, p414, p416, p417, p432, p434, p436, p438, p441, p443, p445, p447, p448, p451, p452, p457, p458, p462, p464, p467, p474, p475, p476, p477, p479, p500, p501, p548, p569, p608; **Kensa Engineering Ltd.** p81; **Kewtech** p432, p531; **Martindale Electric Company Ltd.** p543; **Megger Ltd.** p344, p408, p411, p440; ; **RSI Video Technologies** p243; **ScrewFix** p11, p15, p16, p17, p21, p22, p26, p61; **Shutterstock** p1, p11, p16, p44, p52, p63, p76, p78, p84, p96, p98, p99, p100, p101, p112, p147, p184, p185, p199, p228, p235, p236, p249, p267, p359, p497, p517, p521, p544, p574, p583, p585; **Staco Energy Products Co.** p170; **test-meter.co.uk** p363; **Titan Products Ltd.** p244; **UNECE** p18; **Worcester Bosch Thermotechnology Ltd.** p78; **Wylex Electrium** p458.

**Text permissions**
For kind permission of text extracts:

Crown copyright. Contains public sector information published by the Health and Safety Executive and licensed under the Open Government Licence v.1.0 on page 24 and includes extracts from the following HSE publications: 'Dangerous Substances and Explosive Atmosphere Regulations' pages 521, 557; 'Electricity at Work Regulations' pages 361, 365, 384, 498, 503, 509, 534; 'Electricity Safety, Quality and Continuity Regulations' pages 261, 262, 305; 'Personal Protective Equipment Regulations' page 20.

Crown copyright: 'Building Regulations' page 338.

Permission to reproduce extracts from BS 7671:2008 is granted by the Institution of Engineering and Technology (IET) and BSI Standards Limited (BSI). No other use of this material is permitted. Pages 266, 332, 342, 397, 498, 499, 508, 533, 576.

BS 7671: 2008 Incorporating Amendment No 1: 2011 can be purchased in hardcopy format only from the IET website http://electrical.theiet.org/ and the BSI online shop: http://shop.bsigroup.com

**From the authors**

***Howard Carey***: I would like to thank Shirley Wakley for her professionalism, patience and guidance in working with my material.

***Paul Cook***: Many thanks to all those individuals on the IET technical regulations committees who gave their scarce and valuable time to help correct many of the errors in my efforts.

***Paul Harris***: I would like to thank my wife, Carol, and sons, George and Alfie, for their support and for putting up with me while writing some of the sections.

***Andrew Hay-Ellis***: Many thanks to my wife, Michelle, for her patience and understanding during the many hours spent writing; also to my sons, Jordan and Brendan, for proofreading sections of text to make sure the words said what I intended.

***Trevor Pickard***: I need to express my sincere thanks to my wife, Sue, whose command of the English language has always been better than mine and without whose help the job of the copyeditor would have been that much harder!

***Peter Tanner***: Many thanks to my wife, Gillian, and daughters, Rebecca and Lucy, for their incredible patience; also to Jim Brooker, my brother-in-law, for showing me the importance of carrying out the correct safe isolation procedure on more than one occasion!

# ABOUT THE AUTHORS

## Howard Carey *MIET, LCGI, CertEd*

My career in electrical installation started when I left school at the age of 16. A five-year apprenticeship with a small electrical engineering company allowed me to gain industrial, commercial, agricultural and domestic knowledge of electrical installation and maintenance. During this period I attended New College Durham on a day release course. I also attended evening classes and attained a City & Guilds 'C cert', a foreman electrician qualification. I have gained many other electrical qualifications, but this was by far my favourite.

After a further five years of experience in the contracting industry, I took a job as an instructional officer in the Civil Service. After six years in this position I became a freelance lecturer, assessor, test engineer and consultant, which I am continuing to enjoy now, around 20 years later.

I have achieved assessor awards and teaching qualifications including City & Guilds and a Certificate in Education. My ultimate achievement was in November 1999 when I became a member of the courses lecturing team with The Institution of Engineering and Technology (IET).

My working experience has involved me teaching in England, Scotland, Wales and Cyprus, on a range of courses including inspection and testing, wiring regulations, health and safety and industrial systems.

My advice to any apprentice or learner is to work hard at college and keep in touch with the real world. Electrical installation work and regulations are ever-changing topics and there is always something new to learn. Remember to dedicate a little time to furthering your qualifications and continuing your professional development.

## Paul Cook *CEng, FIEE*

I have been fortunate to have a varied career, starting as a distribution engineer with an electricity distributor and supplier. I have also worked as an installation contracting engineer and manager with all the hassle that entails. On the break-up (in more ways than one) of the nationalised supply industry, I worked for The Institution of Engineering and Technology (IET) on Wiring Regulations work, including as secretary of the Technical Committee responsible for the publication of the Wiring Regulations. I have written a range of publications for the IET and others, including the IET Commentary on the Wiring Regulations. I hope this guide will ease the entry into a fascinating and rewarding career for very many young people.

## *Eur Ing* **Paul Harris** *BEng (Hons), CEng, FIHEEM, MIEE, MCIBSE*

I am a Chartered Electrical Engineer, having started my career as a JIB Indentured Apprentice in 1978.

Following completion of my apprenticeship I worked in a variety of electrical installation jobs. Throughout my apprenticeship and time as a craftsman, through my enthusiasm to learn, I have always questioned 'why?', 'how?' and 'what will happen if…?' I believe it is important not to separate theory and learning from work life, and to take pride as an individual in work, development and the profession.

I returned to college in 1989, followed by university, and obtained an HNC Electrical Engineering and later BEng (Hons) in Building Services Engineering by part-time study.

In my work life I have held a number of engineer posts including managing direct labour work forces and teams of professional engineers. I have been a part-time lecturer for CGLI and BTEC ONC electrical Installation and have also written technical articles for a variety of institutions.

I currently have my own consultancy where I provide design expertise in specialist areas and installations to a variety of clients and industry professionals. Alongside this I am a Medical Locations expert for Working Group 710, represent the UK at IEC and am a committee member of JPEL64, which is responsible for the production of BS 7671.

My belief held in 1978 remains unchanged – training and development, with its supporting books and documentation, is invaluable to individuals and employers alike and should be valued throughout your career.

## **Andrew Hay-Ellis** *CertEd, QTLS, MIET, LCGI, MinstLM*

When I first left school I joined a firm of chartered accountants (an excuse to play cricket) and after three years I moved to a small mechanical and electrical company primarily to look after the company's day-to-day accounts. As with many small companies, staff members often end up working wherever there is a need and so I soon found myself out on site with the engineers. Looking back to my childhood I always had an enquiring mind and electrical installation work provided the mental and physical challenges that I needed and was no longer finding in accountancy. Some thirty plus years later, having worked as an electrician, a supervisor and a project engineer on both commercial and industrial installations, I can honestly say this was the best decision I could have made.

An insatiable thirst for knowledge and new challenges led me back to college to improve my qualifications and this led to my next career change, moving from electrical contracting into teaching. I have now been teaching for over seventeen years, working both in private training and in further education. While my current roles as Director of Education for Trade Skills 4U and within City & Guilds are more technical, I still very much enjoy the thrill of teaching.

The job of an electrician is becoming ever more technical and the need for quality training is a must for anyone working within the electrical industry. Open your mind to the possibilities and never stop asking why. Set your sights high and strive to achieve and you will never be disappointed with the choices you have made.

## Trevor Pickard *IEng, MIET*

I am an electrical engineering consultant and my interest in all things electrical started when I was quite young. I always had a battery powered model under construction or an electrical motor or some piece of electrical equipment in various stages of being taken apart to see how they operated. Looking back, some of my activities with mains electricity would certainly be considered as unacceptable today!

Upon leaving school in 1966 I commenced work with an electricity distribution company, Midlands Electricity Board (MEB) and after serving a student apprenticeship I held a series of engineering positions. I have never tired of my involvement with electrical engineering and was very fortunate to have had a varied and interesting career in the engineering department of Midlands Electricity and embraced its various changes through privatisation and subsequent acquisition. I held posts in Design, Safety, Production Engineering, as Production Manager of a large urban-based operational division, as General Manager of the Repair and Restoration department, and as General Manager of the Primary Network department (33kV–132kV).

My interest in electrical engineering has extended beyond the '9–5 job' and I have had the opportunity to become involved in the writing of standards in the domestic, European and international arena with BSI, CENELEC and IEC and have for many years lectured for the Institution of Engineering and Technology.

Electricity is with us in almost every aspect of modern life and for those who are just starting their career in this field I would say keep an open mind, be safety conscious in how you carry out your work and who knows where your studies will take you.

## Peter Tanner *MIET, LCGI*

Series Editor

I started in the industry while still at school, chasing walls for my brother-in-law for a bit of pocket-money. This taught me quickly that if I took a career in the industry I needed to progress as fast as I could.

Jobs in the industry were few and far between when I left school so after a spell in the armed forces, I gained a place as a sponsored trainee on the CITB training scheme. I attended block release at Guildford Technical College where the CITB would find me work experience with 'local' employers. My first and only work experience placement was with a computer installation company located over twenty miles away so I had to cycle there every morning but I was desperate to learn and enjoyed my work.

Computer installations were very different in those days. Computers filled large rooms and needed massive armoured supply cables so the range of work I experienced was vast from data cabling, to all types of containment systems and low voltage systems.

In the second year of my apprenticeship I found employment with a company where most of my work centred around the London area. The work was varied, from lift systems in well-known high-rise buildings to lightning protection on the side of even higher ones!

On completion of my apprenticeship I worked for a short time as an intruder alarm installer mainly in domestic dwellings, a role where client relationships and handling information is very important.

Following this I began work with a company where I was involved in shop-fitting and restaurant and pub refurbishments. It wasn't long before I was managing jobs and gaining further qualifications through professional development. I was later seconded to the Property Services Agency designing major installations within some of the most well-known buildings in the UK.

A career-changing accident took me into teaching where I truly found the rewards the industry has to offer. Seeing young trainees maturing into qualified electricians is a worthwhile experience. On many occasions I see many of my old trainees when they attend further training and update courses. Seeing their successes makes it all worthwhile.

I have worked with City & Guilds for over twenty years and represent them on a variety of industry committees such as JPEL64, which is responsible for the production of BS 7671. I am passionate about using my vast experience in the industry to maintain the high standards the industry expects.

# HOW TO USE THIS TEXTBOOK

Welcome to your City & Guilds Level 3 Diploma in Electrical Installations (Buildings and Structures) textbook. It is designed to guide you through your Level 3 qualification and be a useful reference for you throughout your career.

Each chapter covers a unit from the 2365 Level 3 qualification. Each chapter covers everything you will need to understand in order to complete your written or online tests and prepare for your practical assessments. In this book the order of the units, outcomes and criteria sometimes differs from the numbered sequence to enable better presentation of the information.

Throughout this textbook you will see the following features:

**KEY POINT**

An ASHP creates noise and therefore there are restrictions on the number that can be installed on a site.

**KEY POINT** These are particularly useful points that may assist you in revision for your tests or to help you remember something important.

**Efficiency**

The ratio between power in and power out, measured in the same unit.

**DEFINITIONS** Words in bold in the text are explained in the margin to aid your understanding. They also appear in the glossary at the back of the book.

**ACTIVITY**

Why is it important to consider the uplift forces on a roof-mounted system, as well as the weight?

**ACTIVITY** These provide questions or suggested activities to help you learn and practise.

**ASSESSMENT GUIDANCE**

You should be able to draw the basic layout of a micro-hydro system.

**ASSESSMENT GUIDANCE** These highlight useful points that are helpful for your learning and assessment.

**SmartScreen Unit 201**

Presentation 1 and Handout 1

**SMARTSCREEN** These provide references to City & Guilds online learner and tutor resources, which you can access on SmartScreen.co.uk.

**Assessment criteria**

**4.2** Identify typical disadvantages associated with solar thermal systems

**ASSESSMENT CRITERIA** These highlight the assessment criteria coverage through each unit, so you can easily link your learning to what you need to know or do for each Learning outcome.

You will also see the following abbreviation in the running heads:

LO – learning outcome

**LO1** **Principles of a.c. theory**

Where tables and forms in this book have been used directly from other publications such as the IET this has been noted, and the style reflects the original (with kind permission). Always make sure you use the latest information and forms.

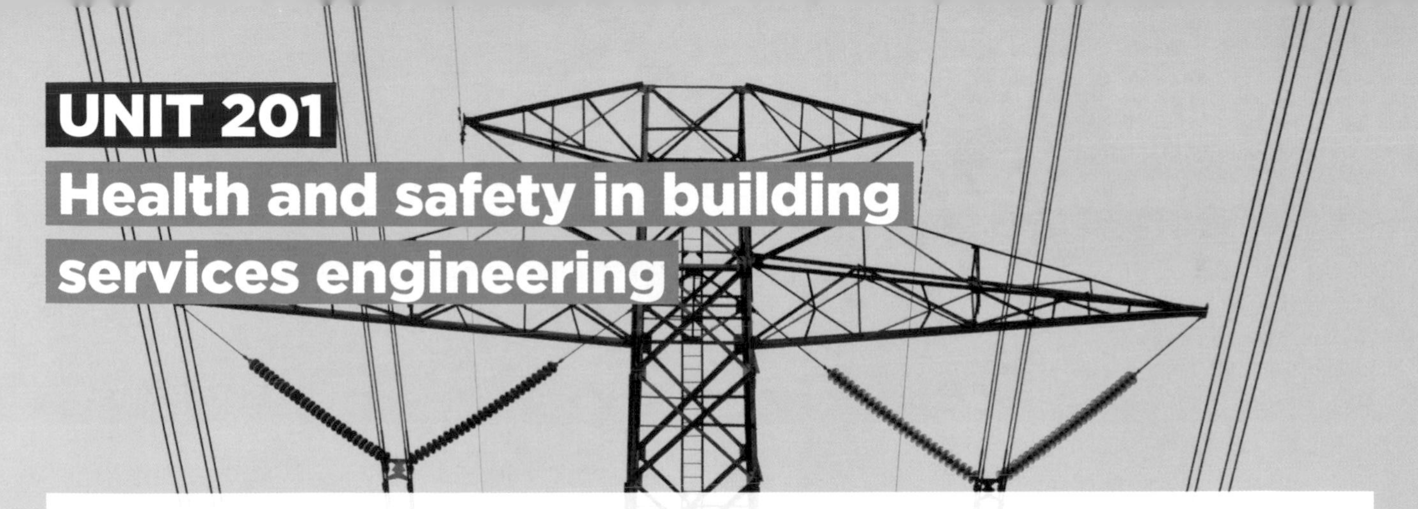

# UNIT 201
# Health and safety in building services engineering

Every year accidents at work result in the deaths of hundreds of people, with several hundred thousand more being injured in the workplace. In 2010/2011, Health and Safety Executive (HSE) statistics recorded 27 million working days being lost due to work-related illness and workplace injury.

Occupational health and safety affects all individuals in the workplace and all aspects of work in the complete range of working environments – hospitals, factories, schools, universities, commercial undertakings, manufacturing plants and offices. As well as the human cost in terms of pain, grief and suffering, accidents in the workplace also have a financial cost, such as lost income, insurance and production disturbance. The HSE put this figure at £13.4 billion for the year 2010/2011. It is therefore important to identify, assess and control the activities that may cause harm in the workplace.

## LEARNING OUTCOMES

There are seven learning outcomes to this unit. The learner will:

1. know health and safety legislation
2. know how to handle hazardous situations
3. know electrical safety requirements when working in the building services industry
4. know the safety requirements for working with gases and heat-producing equipment
5. know the safety requirements for using access equipment in the building services industry
6. know the safety requirements for working safely in excavations and confined spaces in the building services industry
7. be able to apply safe working practice.

This unit will be assessed by:

- workshop-based practical assessments
- online multiple-choice assessment.

# OUTCOME 1

## Know health and safety legislation

The basic concept of health and safety legislation is to provide the legal framework for the protection of people from illness and physical injury that may occur in the workplace.

Assessment criteria

1.1 State the aims of health and safety legislation

## HEALTH AND SAFETY LEGISLATION

There are two sub-divisions of the law that apply to health and safety; civil law and criminal law.

**Civil law** – deals with disputes between individuals, between organisations, or between individuals and organisations, in which compensation can be awarded to the victim. The civil court is concerned with **liability** and the extent of that liability rather than guilt or non-guilt.

**Liability**

A debt or other legal obligation in order to compensate for harm.

**Criminal law** – is the body of rules that regulates social behaviour and prohibits threatening, harming or other actions that may endanger the health, safety and moral welfare of people. The rules are laid down by the government and are enacted by Acts of Parliament as **statutes**. The Health and Safety at Work etc Act 1974 (HSW Act) is an example of criminal law. It is enforced by the Health and Safety Executive (HSE) or Local Authority environmental health officers.

**Statute**

A major written law passed by Parliament.

**KEY POINT**

'So far as is reasonably practicable' involves weighing a risk against the trouble, time and money needed to control it.

There are two sources of law that may be used in the above sub-divisions: common law and statute law.

**Common law** – is the body of law based on custom and decisions made by judges in courts. In health and safety, the legal definitions of terms such as 'negligence', 'duty of care', and 'so as far as is reasonably practicable' are based on legal judgments and are part of common law.

SmartScreen Unit 201
Presentation 1 and Handout 1

**Statute law** – is the name given to law that has been laid down by Parliament as Acts of Parliament.

In terms of health and safety, criminal law is based only on statute law, but civil law may be based on either common law or statute law.

In summary, criminal law seeks to protect everyone in society and civil law seeks to recompense individuals to make amends for loss or harm they have suffered (ie provide compensation).

## The main legal requirements for health and safety at work

The HSW Act is the basis of all British health and safety law. It provides a comprehensive and integrated piece of legislation that sets out the general duties that employers have towards employees, contractors and members of the public, and that employees have towards themselves and each other. These duties are qualified in the HSW Act by the principle of 'so far as is reasonably practicable'.

What the law expects is what good management and common sense would lead employers to do anyway: that is, to look at what the risks are and take sensible measures to tackle those risks. The person(s) who is responsible for the risk and best placed to control that risk is usually designated as the **duty holder**.

The HSW Act lays down the general legal framework for health and safety in the workplace with specific duties being contained in regulations, also called statutory instruments (SIs), which are also examples of laws approved by Parliament.

**KEY POINT**

The duty holder must be competent by formal training and experience and have sufficient knowledge to avoid danger. The appropriate level of competence will differ for different areas of work.

**Duty holder**

The person in control of the danger.

# INDIVIDUALS' RESPONSIBILITIES UNDER HEALTH AND SAFETY LEGISLATION

The HSW Act, which is an **enabling Act**, is based on the principle that those who create risks to employees or others in the course of carrying out work activities are responsible for controlling those risks. The HSW Act places specific responsibilities on:

- employers
- the self-employed
- employees
- designers
- manufacturers and suppliers
- importers.

This section will deal with the responsibilities of employers, the self-employed and employees.

**Assessment criteria**

1.2 Identify the responsibilities of individuals under health and safety legislation

**Enabling Act**

An enabling Act allows the Secretary of State to make further laws (regulations) without the need to pass another Act of Parliament.

**The HSW Act makes provision for securing the health, safety and welfare of persons at work**

## Responsibilities of employers and the self-employed

Under the main provisions of the HSW Act, employers and the self-employed have legal responsibilities in respect of the health and safety of their employees and other people (eg visitors and contractors) who may be affected by their undertaking and exposed to risks as a result. The employers' general duties are contained in Section 2 of the Act.

They are to ensure, 'so far as is reasonably practicable', the health, safety and welfare at work of all their employees, with particular regard to:

- the provision of safe plant and systems of work
- the safe use, handling, storage and transport of articles and substances
- the provision of any required information, instruction, training or supervision
- a safe place of work including safe access and exit
- a safe working environment with adequate welfare facilities.

**ASSESSMENT GUIDANCE**

The exam is multiple choice, but you will need to write for the practical assessment.

These duties apply to virtually everything in the workplace, which therefore includes electrical systems and installations, plant and equipment. An employer does not have to take measures to avoid or reduce the risk if that is technically impossible or if the time, trouble or cost of the measures would be grossly disproportionate to the risk.

## Responsibilities of employees

Employees are required to take reasonable care for the health and safety of themselves and others. To achieve this aim, they have two main duties placed upon them:

- to take reasonable care for the health and safety of themselves and others who may be affected by their acts or omissions at work
- to cooperate with their employer and others to enable them to fulfil their legal obligations.

**KEY POINT**

Omitting/failing to do something required by law or where there is a duty to someone else can be just as bad in law as doing an illegal act or doing something negligently.

In addition there is a duty not to misuse or interfere with safety provisions.

Most of the duties in the HSW Act and the general duties included in the Management of Health and Safety at Work Regulations 1999 (the Management Regulations) are expressed as goals or targets that are to be met 'so far as is reasonably practicable' or through exercising 'adequate control' or taking 'appropriate' (or 'reasonable') steps. This involves making judgments as to whether existing control measures are sufficient and, if not, what else should be done to eliminate or reduce the risk. This risk assessment will be produced using approved codes of practice (ACoP) and published standards, as well as HSE or industry guidance on good practice where available.

**ACTIVITY**

Think of any jobs you have had in the past, such as part-time work in the holidays. What do you think were your responsibilities with regards to health and safety?

## STATUTORY AND NON-STATUTORY HEALTH AND SAFETY MATERIALS

**Assessment criteria**

1.3 Identify statutory and non-statutory health and safety materials

When the HSW Act came into force there were already some 30 statutes and 500 sets of regulations in place. The aim of the Health and Safety Commission (HSC) and the Health and Safety Executive (HSE) was to progressively replace the regulations with a system of regulation that expresses general duties, principles and goals, with any supporting detail set out in ACoPs and guidance.

### Regulations

Statutory instruments (SIs) are laws approved by Parliament. The regulations governing health and safety are usually made under the HSW Act, following proposals from the HSC/HSE. This applies to regulations based on European Commission (EC) Directives as well as those produced in Great Britain.

**ASSESSMENT GUIDANCE**

Make sure you know the difference between statutory (legal) and non-statutory documents.

The HSW Act, and general duties in the Management Regulations, set goals and leave employers the freedom to decide how to control the risks they identify. Guidance and ACoPs give advice.

However, some risks are so great or the proper control measures so costly that it would not be appropriate to leave employers to decide what to do about them. Regulations identify these risks and set out the specific action that must be taken. Often these requirements are absolute – they require something to be done without qualification. The employer has no choice but to undertake whatever action is required to prevent injury, regardless of cost or effort.

Using personal protective equipment (PPE) while working

Some regulations apply across all workplaces. Such regulations include the Manual Handling Operations Regulations 1992, which apply wherever things are moved by hand or bodily force, and the Health and Safety (Display Screen Equipment) Regulations 1992, which apply wherever visual display units (VDUs) are used. Other regulations apply to hazards unique to specific industries, such as mining or the nuclear industry.

The following regulations apply across the full range of workplaces.

- **Control of Noise at Work Regulations 2005:** require employers to take action to protect employees from hearing damage.
- **Control of Substances Hazardous to Health (COSHH) Regulations 2002 (as amended):** require employers to assess the risks from hazardous substances and take appropriate precautions.
- **Electricity at Work Regulations 1989:** require people in control of electrical systems to ensure they are safe to use and maintained in a safe condition.
- **Health and Safety (Display Screen Equipment) Regulations 1992:** set out requirements for work with VDUs.

## ASSESSMENT GUIDANCE

As you can see, there are lots of different regulations that apply to the workplace. You should have a working knowledge of them.

## ACTIVITY

What personal protective equipment (PPE) could be worn to protect against damage to the eyes?

- **Health and Safety (First-Aid) Regulations 1981:** require employers to provide adequate and appropriate equipment, facilities and personnel to ensure their employees receive immediate attention if they are injured or taken ill at work. These regulations apply to all workplaces, including those with fewer than five employees, and to the self-employed.
- **Health and Safety Information for Employees Regulations 1989:** require employers to display a poster telling employees what they need to know about health and safety.
- **Management of Health and Safety at Work Regulations 1999 (as amended):** require employers to carry out risk assessments, make arrangements to implement necessary measures, appoint competent people and arrange for appropriate information and training.
- **Manual Handling Operations Regulations 1992:** cover the moving of objects by hand or bodily force.
- **Personal Protective Equipment at Work Regulations 1992 (as amended):** require employers to provide appropriate protective clothing and equipment for their employees.
- **Provision and Use of Work Equipment Regulations 1998:** require that equipment provided for use at work, including machinery, is suitable and safe.
- **Reporting of Injuries, Diseases and Dangerous Occurrences Regulations 1995 (RIDDOR) (as amended):** require employers to notify the HSE of certain occupational injuries, diseases and dangerous events.
- **Workplace (Health, Safety and Welfare) Regulations 1992:** cover a wide range of basic health, safety and welfare issues such as ventilation, heating, lighting, workstations, seating and welfare facilities.

The following specific regulations cover particular areas, such as asbestos and lead:

- **Chemicals (Hazard Information and Packaging for Supply) Regulations 2002:** require suppliers to classify, label and package dangerous chemicals and provide safety data sheets for them.
- **Construction (Design and Management) Regulations 2007:** cover safe systems of work on construction sites.
- **Control of Asbestos Regulations 2012:** affect anyone who owns, occupies, manages or otherwise has responsibilities for the maintenance and repair of buildings that may contain asbestos.
- **Control of Lead at Work Regulations 2002:** imposes duties on employers to carry out risk assessments, prevent or control exposure to lead and monitor the exposure of employees.
- **Control of Major Accident Hazards Regulations 1999 (as amended):** require those who manufacture, store or transport

dangerous chemicals or explosives in certain quantities to notify the relevant authority.

- **Dangerous Substances and Explosive Atmospheres Regulations 2002:** require employers and the self-employed to carry out a risk assessment of work activities involving dangerous substances.
- **Gas Safety (Installation and Use) Regulations 1998:** cover safe installation, maintenance and use of gas systems and appliances in domestic and commercial premises.
- **Work at Height Regulations 2005:** apply to all work at height where there is a risk of a fall liable to cause personal injury.

## Approved codes of practice (ACoP)

ACoPs offer practical examples of good practice. They were made under Section 16 of the HSW Act and have a special status. They give advice on how to comply with the law by, for example, providing a guide to what is reasonably practicable. For example, if regulations use words such as 'suitable' and 'sufficient', an ACoP can illustrate what is required in particular circumstances. If an employer is prosecuted for a breach of health and safety law, and it is proved that they have not followed the provisions of the relevant ACoP, a court can find them at fault unless they can show that they have complied with the law in some other way.

## Guidance and non-statutory regulations

The HSE publishes guidance on a range of subjects. Guidance can be specific to the health and safety problems of an industry or to a particular process used in a number of industries.

The main purposes of guidance are:

- to interpret and help people to understand what the law says
- to help people comply with the law
- to give technical advice.

Following guidance is not compulsory and employers are free to take other action, but if they do follow the guidance, they will normally be doing enough to comply with the law.

One very good example of guidance and non-statutory regulation is BS 7671 Requirements for Electrical Installations (the IET Wiring Regulations 17th edition). If electrotechnical work is undertaken in accordance with BS 7671, it is likely to meet the requirements of the Electricity at Work Regulations 1989, which deal with work with electrical equipment and systems.

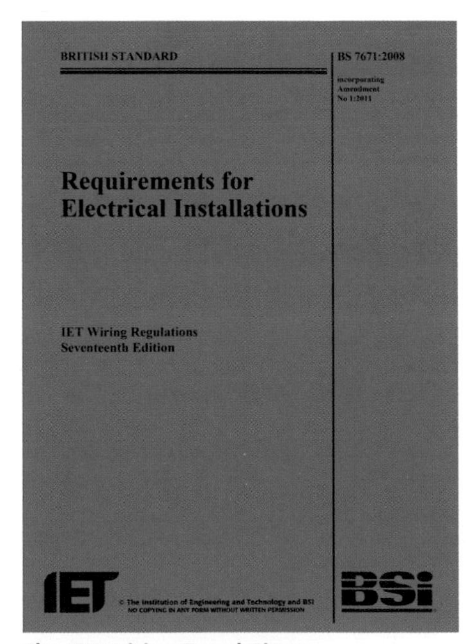

The IET Wiring Regulations

**ASSESSMENT GUIDANCE**

The IET Wiring Regulations 17th edition are non-statutory, but if you carry out work in accordance with the regulations you will meet the requirements of the EAWR 1989.

BS 7671 is the national standard in the UK for low voltage electrical installations. The document is largely based on documents produced by the European Committee for Electrotechnical Standardisation (CENELEC). The regulations deal with the design, selection, erection, inspection and testing of electrical installations operating at a voltage up to 1000 V a.c.

## European law

In recent years much of Great Britain's health and safety law has originated in Europe. Proposals from the EC may be agreed by member states, which are then responsible for making them part of their domestic law.

Modern health and safety law in this country, including much of that from Europe, is based on the principle of risk assessment as required by the Management of Health and Safety at Work Regulations 1999.

**Assessment criteria**

1.4 Identify the different roles of the Health and Safety Executive in enforcing health and safety legislation

## ROLES OF THE HSE IN ENFORCING HEALTH AND SAFETY LEGISLATION

The Health and Safety Executive (HSE) and the Health and Safety Commission (HSC) were, until 2008, the two government agencies responsible for health and safety in Great Britain. The non-executive HSC was there to ensure that relevant legislation was appropriate and understood by conducting and sponsoring research, providing training, providing an information and advisory service, and submitting proposals for new or revised regulations (statutory instruments) and approved codes of practice (ACoPs).

The HSE was the operating arm of the HSC. It advised and assisted the HSC in its functions and had specific responsibility, shared with local authorities, for enforcing health and safety law. The HSC and HSE merged on 1 April 2008 and are now known simply as HSE.

**ACTIVITY**

Find out who the duty holder is at your place of work. It might be more than one person.

Today, the HSE's aim is to prevent death, injury and ill health in Great Britain's workplaces and it has a number of ways of achieving this. Enforcing authorities may offer the duty holder information and advice, both face to face and in writing, or they may warn a duty holder that, in their opinion, the duty holder is failing to comply with the law. (For more on the term 'duty holder' see page 3.)

In carrying out the HSE's enforcement role, inspectors appointed under the HSW Act can:

- enter premises at any reasonable time, accompanied by a police officer if necessary
- examine, investigate and require the premises to be left undisturbed

- take samples, photographs and, if necessary, dismantle and remove equipment or substances
- review relevant documents or information such as risk assessments, accident books, or similar
- seize, destroy or make harmless any substance or article
- issue enforcement notices and start prosecutions.

An inspector may serve one of three types of notice:

- a prohibition notice tells the duty holder to stop an activity immediately
- an improvement notice sets out action needed to remedy a situation and gives the duty holder a date by which they must complete the action
- a Crown notice is issued under the same circumstances that would justify a prohibition or improvement notice, but is only served on duty holders in Crown organisations such as government departments, the Forestry Commission or the Prison Service.

# OUTCOME 2

# Know how to handle hazardous situations

**Risk**

The chance (large or small) of harm actually being done when things go wrong (eg risk of electric shock from faulty equipment).

**Compliance**

The act of carrying out a command or requirement.

**Hazard**

Anything with the potential to cause harm (eg chemicals, working at height, a fault on electrical equipment).

**SmartScreen Unit 201**

Presentation 2 and Handout 2

**KEY POINT**

You must be able to identify hazards as part of a risk assessment. It should then be possible to eliminate, reduce, isolate or control the risk by the application of suitable control measures. The use of PPE should be a last resort.

**Assessment criteria**

2.1 Identify common hazardous situations found on site

The control of **risks** is essential to the provision and maintenance of a safe and healthy workplace and ensures **compliance** with the relevant legal requirements.

Risk assessment starts with the need for **hazard** identification. Risk assessment is usually evaluated in terms of:

- the likelihood of something happening (ie whether a hazard is going to occur)
- the severity of outcome (ie how serious the resulting injury will be).

To control these hazardous situations, duty holders need to:

- find out what the hazards are
- decide how to prevent harm to health
- provide control measures to reduce harm to health
- make sure the control measures are used
- keep all control measures in good working order
- provide information, instruction and training for employees and others
- provide monitoring and health surveillance in appropriate cases
- plan for emergencies.

## COMMON HAZARDOUS SITUATIONS ON SITE

### Housekeeping

This is one of the most important single items influencing safety within the workplace. Poor housekeeping not only causes an increase in the risk of fire, slips, trips and falls, but may also expose members of the public to risks created during building services engineers' work activities. The following are some examples of good housekeeping.

- Stairways, passages and gangways should be kept free from materials, electrical power leads and obstructions of every kind.

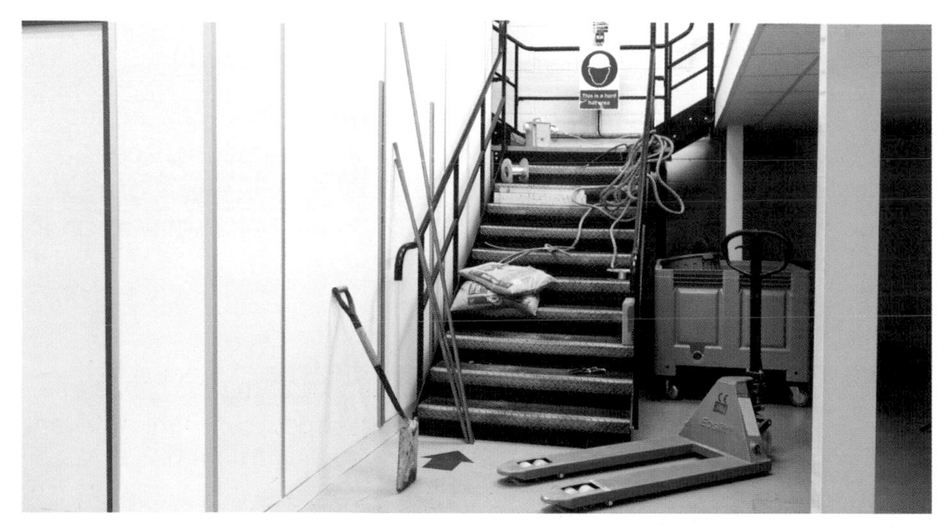

A badly managed site will inevitably lead to accidents

- Materials and equipment should be stored tidily so as not to cause obstruction and should be kept away from the edges of buildings, roofs, ladder access, stairways, floor openings and rising shafts.
- Tools should not be left where they may cause tripping or other hazards. Tools not in use should be placed in a tool belt or tool bag and should be collected and stored in an appropriate container at the end of each working day.
- Working areas should be kept clean and tidy. Scrap and rubbish must be removed regularly and placed in proper containers or disposal areas.
- Rooms and site accommodation should be kept clean. Soiled clothes, scraps of food, etc should not be allowed to accumulate, especially around hot pipes or electric heaters.
- The spillage of oil or other substances must be contained and cleaned up immediately.
- All flammable liquids, liquified petroleum gas (LPG) and gas cylinders must be stored properly.

Gas bottles should be stored properly

Tools should be kept neatly in a tool belt or bag

## ASSESSMENT GUIDANCE

When carrying out the practical assessment, always keep the work area clean and tidy. Dangerous working will mean you could fail.

## Slips, trips and falls

These are the most common hazards to people as they walk around the workplace. They make up over a third of all major workplace injuries. Over 10 000 workers suffered serious injury because of a slip or trip in 2011. Listed below are the main factors that can play a part in contributing to a slip- or trip-free environment.

**ASSESSMENT GUIDANCE**

Floorboards must be lifted correctly and nails removed from them. Be careful not to cut any cables or pipes laid below them. Use a proper floorboard saw.

**ACTIVITY**

Use the internet to find suppliers of cable covers suitable for protecting people from trip hazards.

### Flooring

- The workplace floor must be suitable for the type of work activity that will be taking place on it.
- Floors must be cleaned correctly to ensure they do not become slippery, or be of a non-slip type that it keeps its slip-resistance properties.
- Flooring must be fitted correctly to ensure that there are no trip hazards and any non-slip coatings must be correctly applied.
- Floors must be maintained in good order to ensure that there are no trip hazards such as holes, uneven surfaces, curled-up carpet edges, or raised telephone or electrical sockets.
- Ramps, raised platforms and other changes of level should be avoided. If they cannot be avoided, they must be highlighted.

### Stairs

Stairs should have:

- high-visibility, non-slip, square nosings on the step edges
- a suitable handrail
- steps of equal height
- steps of equal width.

### Contamination

**Contamination**

The introduction of a harmful substance to an area.

Most floors only become slippery once they become contaminated. If **contamination** can be prevented, the slip risk can be reduced or eliminated.

Contamination of a floor can be classed as anything that ends up on a floor, including rainwater, oil, grease, cardboard, product wrapping, dust, etc. It can be a by-product of a poorly controlled work process or be due to bad weather conditions.

### Cleaning

Cleaning is important in every workplace; nowhere is exempt. It is not just a subject for the cleaning team. Everyone's aim should be to keep their workspace clear and deal with contamination such as spillages as soon as they occur.

The process of cleaning can itself create slip and trip hazards, especially for those entering the area being cleaned, including those undertaking the cleaning. Smooth floors left damp by a mop are likely to be extremely slippery. Trailing wires from a vacuum cleaner or polishing machine can present an additional trip hazard.

People often slip on floors that have been left wet after cleaning. Access to smooth wet floors should be restricted by using barriers, locking doors or cleaning in sections. Signs and warning cones only warn of a hazard, they do not prevent people from entering the area. If the water on the floor is not visible, signs and cones are usually ignored.

## Human factors

How people act and behave in their work environments can affect the risk of slips and trips. For example:

- having a positive attitude toward health and safety, for example, dealing with a spillage instead of waiting for someone else to deal with it, can reduce the risk of slip and trip accidents
- wearing the correct footwear can also make a difference
- lack of concentration and distractions, such as being in a hurry, carrying large objects or using a mobile phone, all increase the risk of an accident.

### ACTIVITY

It is very easy to trip or slip, which can result in an injury. What action should you take if you find that a plasterer has left material in a passageway?

## Environmental factors

Environmental issues can affect slips and trips. The following points give an indication of these issues.

- Too much light on a shiny floor can cause glare and stop people from seeing hazards on floors and stairs.
- Too little light will prevent people from seeing hazards on floors and stairs.
- Unfamiliar and loud noises may be distracting.
- Rainwater on smooth surfaces inside or outside a building may create a slip hazard.

## Footwear

Footwear can play an important part in preventing slips and trips. The following points highlight some relevant areas.

- Footwear can perform differently in different situations, for example, footwear that performs well in wet conditions might not be suitable where there are food spillages.
- A good tread pattern on footwear is essential on fluid-contaminated surfaces.
- Sole tread patterns should not be allowed to become clogged with any waste or debris on the floor, as this makes them unsuitable for their purpose.
- Sole material type and hardness are key factors influencing safety.

- The choice of footwear should take into account factors such as comfort and durability, as well as obvious safety features such as protective toecaps and steel mid-soles.

**Assessment criteria**

2.2 Describe safe systems at work

## SAFE SYSTEMS AT WORK

A safe system of work is a work method that results from a systematic examination of the working process to identify the hazards and to specify work methods designed either to completely remove the hazards or control and minimise the relevant risks. Section 2 of the Health and Safety at Work etc Act 1974 (HSW Act) requires employers to provide safe plant and systems of work. Many of the regulations made under the HSW Act have more specific requirements for the provision of safe systems of work.

**Competent person**

Recognised term for someone with the necessary skills, knowledge and experience to manage health and safety in the workplace.

Safe systems of work should be developed by a **competent person**, that is, a person with sufficient training and experience or knowledge to assist with key aspects of safety management and compliance. Staff who are actively involved in the work process also have a valuable role to play in the development of the system. They can help to ensure that it is of practical benefit and that it will be applied diligently.

All safe systems of work need to be monitored regularly to ensure that they are fully observed and effective, and updated as necessary.

Safe systems of work are normally formal and documented but may be given as a verbal instruction. Examples of documented safe systems of work would be for asbestos removal, air-conditioning maintenance, working on live electrical equipment and portable appliance testing.

### Method statements

**ASSESSMENT GUIDANCE**

Practise producing a method statement for a small electrical installation of your choice.

A method statement is a form of safe system of work and describes in a logical sequence exactly how a job is to be carried out, to be safe and without risk to health. It includes all the risks identified in the risk assessment and the measures to control those risks. The statement need be no longer than necessary to achieve these objectives effectively.

### Permit-to-work procedures

**Hot work**

Work that involves actual or potential sources of ignition and done in an area where there is a risk of fire or explosion (eg welding, flame cutting, grinding).

A permit-to-work (PTW) procedure is a specialised written safe system of work that ensures that potentially dangerous work is done safely. Examples of such work include: work in confined spaces, **hot work**, work with asbestos-based materials, work on pipelines with hazardous contents, or work on high voltage electrical systems (above 1000 V) or complex lower voltage electrical systems.

A PTW procedure also serves as a means of communication between site/installation management, plant supervisors and operators and those who carry out the hazardous work. Essential features of PTW systems are:

- clear identification of who may authorise particular jobs (and any limits to their authority) and who is responsible for specifying the necessary precautions
- a PTW should only be issued by a technically competent person, who is familiar with the system and equipment, and is authorised in writing by the employer to issue such documents
- training and instruction in the issue, use and closure of permits
- monitoring and **auditing** to ensure that the system works as intended
- clear identification of the types of work considered hazardous
- clear and standardised identification of tasks, risk assessments, permitted task duration and any additional activity or control measure that occurs at the same time.

The effective operation of a PTW system requires involvement and cooperation from a number of persons. The procedure for issuing a PTW should be written and adhered to.

**Audit**

To conduct a systematic review to make sure standards and management systems are being followed.

**SmartScreen Unit 201**
Worksheet 2

## SAFETY SIGNS

**Assessment criteria**

2.3 Identify the categories of safety signs

The Health and Safety (Safety Signs and Signals) Regulations 1996 require employers to ensure that safety signs are provided (or are in place) and maintained in circumstances where risks to health and safety have not been avoided by other means, for example, safe systems of work. The range of safety signs are shown on the following pages.

**SmartScreen Unit 201**
Presentation 4 and Handout 4

### Prohibition signs

These signs indicate an activity that must not be done. They are circular white signs with a red border and red cross bar.

| | | |
|---|---|---|
| 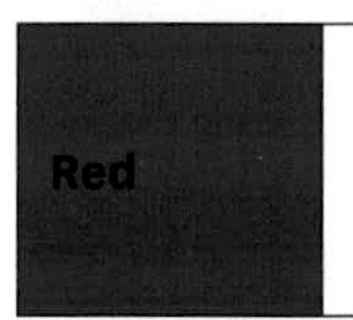 Red | Prohibition sign<br>Danger alarm<br>Fire-fighting equipment | Dangerous behaviour<br>Stop, shutdown, emergency cut-out devices<br>Evacuate identification and location |

Some prohibition signs

**ASSESSMENT GUIDANCE**

Find a suitable website and print or copy a range of safety signs. You may need them in your assessments.

## Warning signs

These provide safety information and/or give warning of a hazard or danger. They are triangular yellow signs with a black border and symbol.

| | | |
|---|---|---|
| **Yellow or amber** | Warning sign | Be careful, take precautions<br>Examine |

Some warning signs

A typical construction site safety sign

**ACTIVITY**

List four other warning and mandatory signs not shown here. You can find more signs on the internet or look around the site where you are based.

## Mandatory signs

These signs give instructions that must be obeyed. They are circular blue signs with a white symbol.

| | | |
|---|---|---|
| **Blue** | Mandatory | Specific behaviour or action<br>Wear personal protective equipment |

Some mandatory signs

## Advisory or safe condition signs

These give information about safety provision. They are square or rectangular signs with a white symbol.

| Green | Emergency escape, first aid sign<br>No danger | Doors, exits, routes, equipment, facilities<br>Return to normal |
|---|---|---|

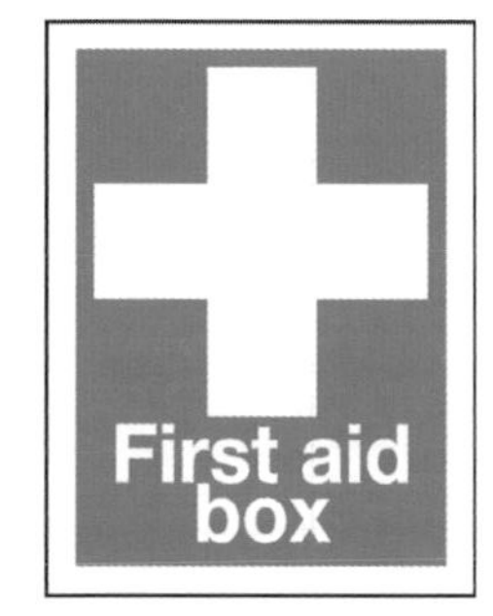

Some safe condition signs

## SYMBOLS FOR HAZARDOUS SUBSTANCES

Assessment criteria

2.4 Identify symbols for hazardous substances

Hazardous substances are given a classification according to the severity and type of hazard they may present to people in the workplace. However, all over the world there are different laws on how to identify the hazardous properties of chemicals. The United Nations has therefore created the Globally Harmonised System of Classification and Labelling of Chemicals (GHS). The aim of the GHS is to have, worldwide, the same:

- criteria for classifying chemicals according to their health, environmental and physical hazards
- hazard communication requirements for labelling and safety data sheets.

The GHS is not a formal treaty, but is a non-legally binding international agreement. In Great Britain the existing legislation is the Chemicals (Hazard Information and Packaging for Supply) Regulations 2009 (CHIPS), but this will gradually be replaced by the European Classification, Labelling and Packaging of Substances and Mixtures Regulations between now and 2015.

**KEY POINT**

You do not need to know all of the detail in the GHS, but it is important that you are aware of it.

SmartScreen Unit 201

Worksheet 4

**Toxic**

Poisonous.

Assessment criteria

2.5 List common hazardous substances used in the building services industry

**Hazardous substance**

Something that can cause ill health to people.

SmartScreen Unit 201

Presentation 8 and Handout 8

## ACTIVITY

Think of some common electrical components that could contain substances hazardous to health.

Examples of the GHS labelling system are shown below:

The hazard signs shown indicate substances that are (top row, left to right) flammable, explosive, oxidising, gas under pressure, **toxic** and (bottom row, left to right) long-term health hazards (causes of cancer), corrosive, require caution (irritants), dangerous to the environment

## COMMON HAZARDOUS SUBSTANCES

**Hazardous substances** at work may include the substances used directly in the work process, such as glue, paints, thinners, solvents and cleaning materials, and those produced by different work activities, such as welding fumes. Health hazards are always present during building services activities due to the nature of the activities and may include other hazards, such as vibration, dust (possibly including asbestos), cement and solvents.

The Control of Substances Hazardous to Health (COSHH) Regulations 2002 provide a framework that helps employers assess risk and monitor effective controls. A COSHH assessment is essential before work starts and should be updated as new substances are introduced.

Hazardous substances include:

- any substance that gives off fumes that may cause headaches or respiratory irritation
- acids that cause skin burns or respiratory irritation (eg battery acid, metal-cleaning materials)
- solvents, for example, for PVC tubes and fittings, that cause skin and respiratory irritation
- man-made fibres that cause eye or skin irritation (eg thermal insulation, optical fibres)
- cement and wood dust that may cause eye irritation and respiratory irritation
- fumes and gases that cause respiratory irritation (eg soldering, brazing and welding fumes, or overheating/burning PVC)
- asbestos.

The COSHH Regulations require employers to assess risk and ensure the prevention or adequate control of exposure to hazardous substances by measures other than the provision of personal protective equipment (PPE) 'so far as is reasonable practicable'.

## PRECAUTIONS TO BE TAKEN WITH HAZARDOUS SUBSTANCES

There is an acknowledged 'hierarchy of control' list of measures designed to control risks. These are considered in order of importance, effectiveness or priority. The measures are listed as follows:

- eliminate the risk by designing out or changing the process
- reduce the risk by substituting less hazardous substances
- isolate the risk using enclosures, barriers or by moving workers away
- control the risk by the introduction of guarding or local exhaust systems
- management control such as safe systems of work, training, etc
- PPE such as eye protection or respiratory equipment.

**Assessment criteria**

**2.6** List precautions to be taken when working with hazardous substances

### ASSESSMENT GUIDANCE

Make sure you can recognise a hazard. Be able to explain what you can do to reduce the risk.

### ACTIVITY

How would you safely dispose of discharge lamps containing mercury?

## ASBESTOS ENCOUNTERED IN THE WORKPLACE

Asbestos is the single greatest cause of work-related deaths in the UK. It is a naturally occurring substance, which is obtained from the ground as a rock-like ore, normally through open-pit mining.

Asbestos was used extensively as a building material in the UK from the 1950s through to the mid-1980s. It was used for a variety of purposes and was ideal for fireproofing and insulation. Any building built before the year 2000 (houses, factories, offices, schools, hospitals, etc) may contain asbestos.

Asbestos materials in good condition are safe unless the asbestos fibres become airborne, which happens when materials are damaged due to demolition or remedial works on or in the vicinity of asbestos ceiling tiles, asbestos cement roofs and wall sheets, sprayed asbestos coatings on steel structures and lagging. In older buildings the presence of asbestos in and around boilers, hot water pipes and structural fire protection must always be anticipated when undertaking electrical work. It is difficult to identify asbestos by colour alone and laboratory tests are normally required for positive identification.

**Assessment criteria**

**2.7** Identify the types of asbestos that may be encountered in the workplace

**SmartScreen Unit 201**

Presentation 9 and Handout 9

Assessment criteria

2.8 Identify the actions to be taken if the presence of asbestos is suspected

## WHAT TO DO IF THE PRESENCE OF ASBESTOS IS SUSPECTED

If asbestos is discovered during a work activity, work should be stopped and the employer or duty holder informed immediately.

The Control of Asbestos Regulations 2012 affect anyone who owns, occupies, manages or otherwise has responsibilities for the maintenance and repair of buildings that may contain asbestos. The regulations cover the need for a risk assessment, the need for method statements for the removal and disposal of asbestos, air monitoring and the control measures required. These control measures include personal protective equipment and training.

### ACTIVITY

In the past, asbestos was often used in fuseboards and as part of the fuse carrier. What action should you take if you discover asbestos during a rewire of an electrical installation?

Assessment criteria

2.9 Describe the implications of being exposed to asbestos

## IMPLICATIONS OF BEING EXPOSED TO ASBESTOS

Asbestos commonly comes in the form of chrysotile (white asbestos), amosite (brown asbestos) and crocidolite (blue asbestos). Chrysotile is the common form of asbestos and accounts for 90% to 95% of all asbestos in circulation. When **abraded** or drilled, asbestos produces a fine dust with fibres small enough to be taken into the lungs. Asbestos fibres are very sharp and can lead to mesothelioma (cancer of the lining of the lung), lung cancer, asbestosis (scarring of the lung), diffuse pleural thickening (thickening and hardening of the lung wall) and death.

**Abrade**

To scrape or wear away.

SmartScreen Unit 201
Worksheet 8

Assessment criteria

2.10 State the application of different types of personal protective equipment

## PERSONAL PROTECTIVE EQUIPMENT AND ITS USE

Virtually all **personal protective equipment (PPE)** is covered by the Personal Protective Equipment at Work Regulations 1992 (PPE Regulations). The exception is respiratory equipment, which is covered by specific regulations relating to specific substances (lead, gases, substances hazardous to health, etc).

**Personal protective equipment (PPE)**

All equipment, including clothing for weather protection, worn or held by a person at work, which protects that person from risks to health and safety.

SmartScreen Unit 201
Presentation 5 and Handout 5

PPE is defined in the PPE Regulations as 'all equipment (including clothing affording protection against the weather) which is intended to be worn or held by a person at work and which protects them against one or more risks to his health or safety'. Such equipment includes safety helmets, gloves, eye protection, high-visibility clothing, safety footwear and safety harnesses. Employers are responsible for providing, replacing and paying for PPE.

Hearing protection and respiratory protective equipment provided for most work situations are not covered by these regulations because other regulations apply to them. However, these items need to be compatible with any other PPE provided.

PPE should be used only when all other measures are inadequate to control exposure. It protects only the wearer while it is being worn and, if it fails, PPE offers no protection at all. The provision of PPE is only one part of the protection package; training, selection of the correct equipment in all work situations, good supervision, monitoring and supervision of its use, all play a part in the success of PPE as a control measure.

## PPE for different types of protection

### Protection for the eyes

**Hazards** – chemical or metal splash, dust, projectiles, gas and vapour, radiation.

**PPE options** – safety spectacles, goggles, face shields, visors.

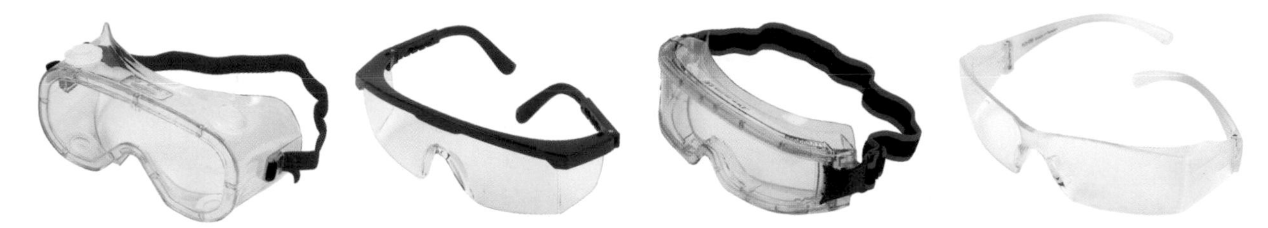

Different types of eye protection

### Protection for the head

**Hazards** – impact from falling or flying objects, risk of head bumping or hair getting caught.

**PPE options** – a range of helmets and bump caps.

A safety helmet and bump cap

### Protection for breathing

**Hazards –** dust, vapour, gas, oxygen-deficient atmospheres.

**PPE options** – disposable filtering face-piece or respirator, half or full-face respirators, air-fed helmets, breathing apparatus.

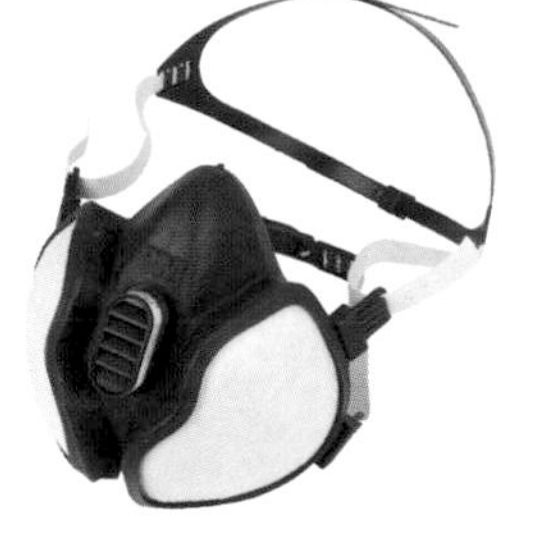

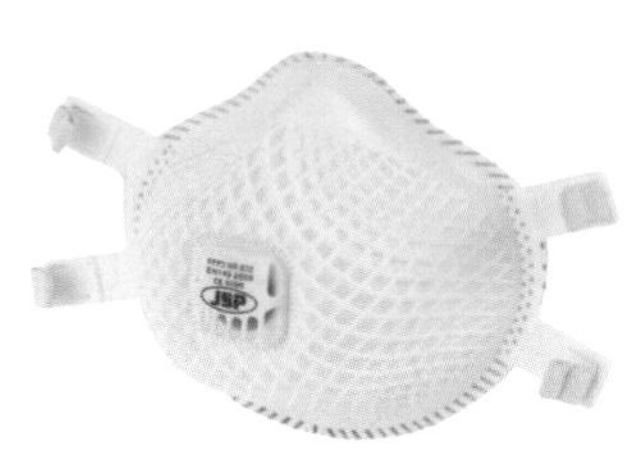

Examples of protective breathing equipment

**ASSESSMENT GUIDANCE**

Keep all your PPE in good condition and wear it correctly. Hard hats worn backwards provide no protection to the face.

## Protection for the body

**Hazards** – temperature extremes, adverse weather, chemical or metal splash, spray from pressure leaks or spray guns, impact or penetration, contaminated dust, excessive wear or entanglement of own clothing.

**PPE options** – conventional or disposable overalls, boiler suits, specialist protective clothing such as chain-mail aprons, flame-retardant or high-visibility clothing.

## Protection for hands and arms

**ACTIVITY**

At what stages might gloves be required during the electrical installation process?

**Hazards** – abrasion, temperature extremes, cuts and punctures, impact, chemicals, electric shock, skin infection, disease or contamination.

**PPE options** – gloves, gauntlets, mitts, wrist cuffs, armlets.

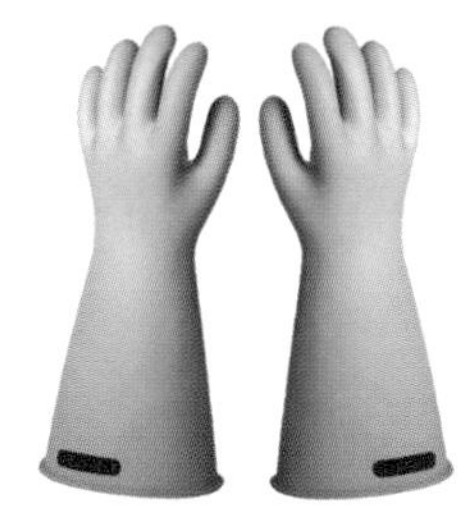

Riggers' gloves – for heavy duty/manual handling work, PVC gloves – for acids and oil, IEC 60903 gloves – made from insulating material, for live work

## Protection for feet and legs

**Hazards** – spillages underfoot, electrostatic build-up, slipping, cuts and punctures, falling objects, metal and chemical splash, abrasion.

**ASSESSMENT GUIDANCE**

Safety footwear is likely to protect against falling objects and against nails penetrating from below. Nails should be removed from wood that has been stripped out, but sometimes one will be missed. A nail in the foot is a very painful experience as well as carrying the risk of tetanus.

**PPE options** – safety boots and shoes with protective toecaps and penetration-resistant mid-sole, gaiters, leggings, spats.

It is important that employees know why PPE is needed and are trained to use it correctly, as otherwise it is unlikely to protect as required. The following points should be considered.

- Does it fit correctly?
- How does the wearer feel?

- Is it comfortable?
- Do all items of PPE work well together?
- Does PPE interfere with the job being done?
- Does PPE introduce another health risk, for example, of overheating or getting caught up in machinery?
- If PPE needs maintenance or cleaning, how is it done?

SmartScreen Unit 201
Worksheet 5

When employees find PPE comfortable they are far more likely to wear it.

## PROCEDURES FOR MANUALLY HANDLING HEAVY AND BULKY ITEMS

**Assessment criteria**

**2.11** Identify the procedures for manually handling heavy and bulky items

**Manual handling**

The movement of items by lifting, lowering, carrying, pushing or pulling by human effort alone.

**Manual handling** is one of the most common causes of injury at work and causes over a third of all workplace injuries. Manual handling injuries can occur almost anywhere in the workplace. Heavy manual labour, awkward postures and previous or existing injury can increase the risk. Work-related manual handling injuries can have serious implications for both the employer and the person who has been injured.

The introduction of the Manual Handling Operations Regulations 1992 saw a change from reliance on safe lifting techniques to an analysis, using risk assessment, of the need for manual handling. The regulations established a clear hierarchy of manual handling measures:

- avoid manual handling operations so as far as is reasonably practicable by re-engineering the task to avoid moving the load or by mechanising the operation
- if manual handling cannot be avoided, a risk assessment should be made
- reduce the risk of injury so as far as is reasonably practicable either by the use of mechanical aids or by making improvements to the task (eg using two persons), the load and the environment.

Even if mechanical handling methods are used to handle and transport equipment or materials, there may still be hazards. These hazards may still be present in the four elements that make up mechanical handling:

- Handling equipment – mechanical handling equipment must be suitable for the task, well maintained and inspected on a regular basis.
- The load – the load needs to be prepared in such a way as to minimise accidents, taking into account such things as security of the load, flammable materials and stability of the load.

SmartScreen Unit 201
Worksheet 3a

- The workplace – if possible, the workplace should be designed to keep the workforce and the load apart.
- The human element – employees who use the equipment must be properly trained.

Assessment criteria

2.12 Identify the actions that should be taken when an accident or emergency is discovered

## WHAT TO DO IN AN ACCIDENT OR EMERGENCY

An accident is defined by the HSE as, 'any unplanned event that results in injury or ill health of people, or damage or loss to property, plant, materials or the environment, or a loss of a **business opportunity**'.

**Business opportunity**

In this context, the opportunity to make profit from the work or contract.

### Emergencies

SmartScreen Unit 201
Presentation 6 and Handouts 6a and 6b

Emergency procedures are there to limit the damage to people and property caused by an incident. Although the most likely emergency to be dealt with is fire (see pages 29 and 41), there are many more emergency situations that need to be considered, including the following.

### Electrical fire or explosion

Fires involving electricity are often caused by lack of care in the maintenance and use of electrical equipment and installations. The use of electrical equipment should be avoided in potentially flammable atmospheres as far as is possible. However, if the use of electrical equipment in these areas cannot be avoided, then equipment purchased in accordance with the Equipment and Protective Systems Intended for Use in Potentially Explosive Atmospheres Regulations 1996 must be used.

**ASSESSMENT GUIDANCE**

When carrying out a practical assessment, always wear your PPE.

### Escape of toxic fumes or gases

Some gases are poisonous and can be dangerous to life at very low concentrations. Some toxic gases have strong smells such as the distinctive 'rotten eggs' smell of hydrogen sulphide ($H_2S$). The measurements most often used for the concentration of toxic gases are parts per million (ppm) and parts per billion (ppb). More people die from toxic gas exposure than from explosions caused by the ignition of flammable gas. With toxic substances, the main concern is the effect on workers of exposure to even very low concentrations. These could be inhaled, ingested (swallowed) or absorbed through the skin. Since adverse effects can often result from cumulative, long-term exposure, it is important not only to measure the concentration of gas, but also the total time of exposure.

**KEY POINT**

Toxic substances can stop one or more organs in the human body working. Mercury, lead and carbon monoxide, for example, are all toxic.

## Gas explosion

A gas explosion is an explosion resulting from a gas leak in the presence of an ignition source. The main explosive gases are natural gas, methane, propane and butane because they are widely used for heating purposes in temporary and permanent situations. However, many other gases, such as hydrogen, are **combustible** and have caused explosions in the past. The source of ignition can be anything from a naked flame to the electrical energy in a piece of equipment.

**Combustible**

Able to catch fire and burn easily.

Industrial gas explosions can be prevented with the use of intrinsic safety barriers to prevent ignition. The principle behind intrinsic safety is to ensure that the electrical and thermal energy from any electrical equipment in a hazardous area is kept low enough to prevent the ignition of flammable gas. Items such as electric motors would not be permitted in a hazardous area.

SmartScreen Unit 201
Worksheet 6

## PROCEDURES FOR HANDLING INJURIES SUSTAINED ON SITE

Assessment criteria

2.13 State procedures for handling injuries sustained on site

The type of accident that can occur in the workplace is dependent on the work activity being undertaken but can range from a cut finger to a **fatality**, or from a vehicle collision to the collapse of a structure. The person in control of the premises needs to be prepared to deal with all types of accidents to ensure that the injured person can be treated quickly and effectively and that all the legal obligations are met.

**Fatality**

Death.

Having a well-established procedure that everyone on site is aware of and understands will enable the person in control of premises to cope calmly and effectively when dealing with an accident. Good management following an accident will ensure that appropriate records are made, the accident is reported correctly and any lessons to be learned from the accident are understood and communicated to the workforce.

### ACTIVITY

How should you deal with a cut finger sustained while threading a piece of conduit?

The procedures to be followed in the event of any accident or incident should be clear and specific to the project or site and should detail the following as a minimum:

- name of the appointed person(s) who will take control when someone is injured or falls ill
- name of the person(s) who will administer first aid
- location of the first-aid boxes and name of the person(s) responsible for ensuring they are fully stocked
- course of action that must be followed by the appointed person who takes control in the event of an accident
- guidance on action to take after the accident
- how information should be recorded and by whom (F2508 RIDDOR Form, which can be found on the HSE website www.hse.gov.uk/forms/incident/).

Assessment criteria

2.14 State the procedures for recording accidents and near misses at work

## PROCEDURES FOR RECORDING ACCIDENTS AND NEAR MISSES AT WORK

The HSE definition of an accident has been given in the section 'What to do in an accident or emergency' on page 24.

**Near miss**

Any incident that could, but does not, result in an accident.

A **near miss** is an unplanned event that does not result in injury, illness or damage, but had the potential to do so. Normally only a fortunate break in the chain of events prevents an injury, fatality or damage taking place. So a near miss could be defined as any incident that could have resulted in an accident. The keeping of information on near misses is very important in helping to prevent accidents occurring. Research has shown that damage and near miss accidents occur much more frequently than injury accidents and therefore give an indication of hazards.

Employers are required to keep a record of all accidents in the Statutory Accident Book, or an equivalent

The Social Security Act 1975 specifically requires employers to keep information on accidents. This should be the Statutory Accident Book for all Employee Accidents (Form B1510, found on the HSE website www.hse.gov.uk/forms/incident/) or equivalent. Each entry should be made on a separate page and the completed page securely stored to protect personal data (under the Data Protection Act 2003). An entry may be made by the employee or by anyone acting on their behalf. This information should be kept for a period of not less than three years.

### Reporting the incident

The reporting of certain types of injury and incidents to the enforcing authority (the HSE or the local authority) is a legal requirement under the Reporting of Injuries, Diseases and Dangerous Occurrences Regulations 1995 (RIDDOR). Failure to comply with these regulations is a criminal offence.

**ACTIVITY**

Would a small cut on a finger, which did not require time off work, be reportable under RIDDOR?

RIDDOR states that deaths, major injuries (specified in the regulations) and injuries resulting in absence from work for over three days, and dangerous occurrences (specified in the regulations) must be reported. It is the responsibility of employers or the person in control of premises to report these types of incidents. Reportable major injuries include those resulting from an electric shock or electrical burn leading to unconsciousness, resuscitation or admittance to hospital for more than 24 hours. Electrical burns include burns caused by arcing or arcing products. A dangerous occurrence is a near miss that could have led to serious injury or loss of life.

**KEY POINT**

The types of injury that are reported are under review during 2013 and may be changed by an Act of Parliament. Consult the HSE website for more up-to-date information.

A death, reportable major injury or a dangerous occurrence must be reported immediately. (From October 2013, 'major injury' is likely to change to 'specified injury'.) Injuries resulting in absence from work for over seven days must be reported within 15 days.

The police and HSE have the right to investigate fatal accidents at work. Therefore all fatal accidents must also be notified to the police. The police will often notify the HSE, but it is always a sensible precaution to ensure that the HSE has been notified.

## Investigating accidents

There is nearly always something to be learned following an accident and ideally the causes of all accidents should be established regardless of whether or not injury or damage resulted. The level and nature of an investigation should reflect the significance of the event being investigated. The results of the accident investigation may lead to a review, possible amendment to the risk assessment and appropriate action to prevent similar accidents from occurring.

## Keeping records

There are numerous records to keep following even a minor accident. Easily accessible records should be maintained for all accidents that have occurred. In addition to the legal requirements, accident information can help an organisation identify key risk areas within the business. The accident book must be kept for three years following the last entry.

- The F2508 for reportable incidents should be kept for a period of not less than three years from the date the accident occurred.
- Appointed trade union representatives and representatives of employee safety should be provided with a copy of the F2508 if it relates to the workplace or the group of employees represented.
- It is advisable to keep copies of any accident investigation reports for the same period as above (three years).

**KEY POINT**

Try to keep notes of what happened during any incident. People are very bad at remembering what they actually saw.

OUTCOME 3

# Know electrical safety requirements when working in the building services industry

Assessment criteria

3.1 Identify the common electrical dangers to be aware of on site

SmartScreen Unit 201

Presentation 7 and Handout 7

## COMMON ELECTRICAL DANGERS ON SITE

Modern living is shaped by electricity. It is a safe, clean and immensely powerful source of energy and is in use in every factory, office, workshop and home in the country. However this energy source also has the potential to be very hazardous, with a possibility of death, if it is not handled with care. The Electricity at Work Regulations 1989 were made under the Health and Safety at Work etc Act 1974 (HSW Act) and came into force on 1 April 1990.

The 1989 Regulations are goal-setting in that they specify the objectives concerning the design, specification and construction of, and work activities on, electrical systems, in order to prevent injury caused by electricity. They do not specify the means for achieving these objectives. The 1989 Regulations are supported by a Memorandum of Guidance, HSR25 (2nd edition 2007).

Electrical injuries can be caused by a wide range of voltages, and are dependent upon individual circumstances, but the risk of injury is generally greater with higher voltages. Alternating current (a.c.) and direct current (d.c.) electrical supplies can cause a range of injuries including:

- electric shock
- electrical burns
- loss of muscle control
- fires arising from electrical causes
- arcing and explosion.

### Electric shock

Electric shocks may arise either by direct contact with a live part or indirectly by contact with an exposed conductive part (eg a metal equipment case) that has become live as a result of a fault condition. Faults can arise from a variety of sources:

- broken equipment case exposing internal bare live connections
- cracked equipment case causing 'tracking' from internal live parts to the external surface

- damaged supply cord insulation, exposing bare live conductors
- broken plug, exposing bare live connections.

The magnitude (size) and duration of the shock current are the two most significant factors determining the severity of an electric shock. The magnitude of the shock current will depend on the contact voltage and impedance (electrical resistance) of the shock path. A possible shock path always exists through ground contact (eg hand-to-feet). In this case the shock path impedance is the body impedance plus any external impedance. A more dangerous situation is a hand-to-hand shock path when one hand is in contact with an exposed conductive part (eg an earthed metal equipment case), while the other simultaneously touches a live part. In this case the current will be limited only by the body impedance.

As the voltage increases, so the body impedance decreases, which increases the shock current. When the voltage decreases, the body impedance increases, which reduces the shock current. This has important implications concerning the voltage levels that are used in work situations and highlights the advantage of working with reduced low voltage (110 V) systems or battery-operated hand tools.

At 230 V, the average person has a body impedance of approximately 1 300 Ω. At mains voltage and frequency (230 V–50 Hz), currents as low as 50 milliamps (0.05 A) can prove fatal, particularly if flowing through the body for a few seconds.

**ACTIVITY**

Battery-powered equipment is much safer and more convenient than mains equipment, but generally lacks the constant power available with 110 V equipment. Are there any circumstances when 230 V equipment can be used on site?

## Electrical burn

Burns may arise due to:

- the passage of shock current through the body, particularly if at high voltage
- exposure to high-frequency radiation (eg from radio transmission antennas).

**ASSESSMENT GUIDANCE**

Static electricity can also cause electric shock, though it is more of a nuisance than a danger. How is static electricity produced?

## Loss of muscle control

People who experience an electric shock often get painful muscle spasms that can be strong enough to break bones or dislocate joints. This loss of muscle control often means the person cannot 'let go' or escape the electric shock. The person may fall if they are working at height or be thrown into nearby machinery and structures.

## Fire

Electricity is believed to be a factor in the cause of over 30 000 fires in domestic and commercial premises in Britain each year. One of the principal causes of such fires is wiring with defects such as insulation failure, the overloading of conductors, lack of electrical protection, poor connections.

## Arcing

Arcing frequently occurs due to short-circuit flashover accidentally caused while working on live equipment (either intentionally or unintentionally). Arcing generates UV radiation, causing severe sunburn. Molten metal particles are also likely to be deposited on exposed skin surfaces.

## Explosion

There are two main electrical causes of explosion: short circuit due to an equipment fault, and ignition of flammable vapours or liquids caused by sparks or high surface temperatures.

## Controlling current flow

It is necessary to include devices in circuits to control current flow, that is, to switch the current on or off by making or breaking the circuit. This may be required:

- for functional purposes (to switch equipment on or off)
- for use in an emergency (switching in the event of an accident)
- so that equipment can be switched off to prevent its use and allow maintenance work to be done safely on the mechanical parts
- to isolate a circuit, installation or piece of equipment to prevent the risk of shock where exposure to electrical parts and connections is likely for maintenance purposes.

The preparation of electrical equipment for maintenance purposes requires effective disconnection from all live supplies and the means for securing that disconnection (by locking off).

There is an important distinction between switching and isolation. Switching is cutting off the supply. Isolation is the secure disconnection and separation from all sources of electrical energy.

**ACTIVITY**

Why are light switches unsuitable as devices for isolation?

A variety of control devices are available for switching, isolation or a combination of these functions, some incorporating protective devices. Before starting work on a piece of isolated equipment, checks should be made to ensure that the circuit is dead, using an approved testing device.

In the case of portable equipment connected via a supply cord and plug, removal of the plug from the socket provides a ready means of isolation.

# ELECTRICAL SUPPLY FOR TOOLS AND EQUIPMENT

**Assessment criteria**

**3.2** List different sources of electrical supply for tools and equipment

Portable electric tools can provide valuable assistance with much of the physical effort required in electrotechnical activities. These tools can use different sources of electrical supply (mains or battery) and different means of maintaining safety in relation to that electrical supply.

Basic equipment constructions, all aimed at preventing the risk of electric shock, are specified in BS 2754 1976: Construction of Electrical Equipment for Protection against Electric Shock, as detailed below.

## Class I

The basic insulation may be an air gap and/or some form of insulating material. External conductive parts (eg the metal case) must be earthed by providing the supply through a three-core supply lead incorporating a protective conductor. The most important aspect of any periodic inspection/testing of Class I equipment is to check the integrity of this protective conductor. There is no recognised symbol for Class I equipment, though some appliances may show the symbol on the right.

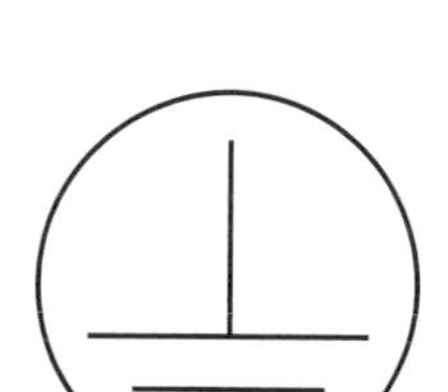

This symbol may appear on Class I items

## Class II

Class II equipment has either no external conductive parts apart from fixing screws (insulation-encased equipment) or there is adequate insulation between any external conductive parts and the internal live parts to prevent the former becoming live as a result of an internal fault (metal-encased equipment). Periodic inspection or testing needs to focus on the integrity of the insulation. Class II equipment is identified by the symbol shown on the right.

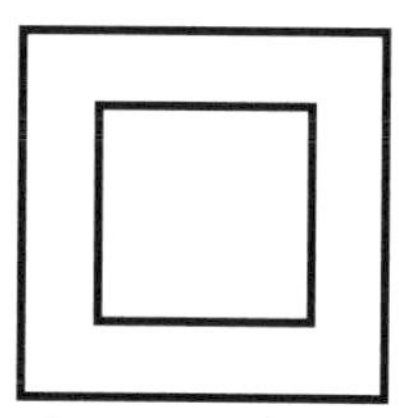

Class II symbol

## Class III

This method of protection is not designed to prevent shock but to reduce its severity and therefore make the shock more survivable. The supply (no greater than 50 V a.c.) must be provided from a separated extra-low voltage (SELV) source such as a safety isolating transformer conforming to BS EN 61558, or a battery. Class III equipment is identified by the diamond symbol and the safety isolating symbol by two interlinked circles, as shown on the right.

Class III symbol

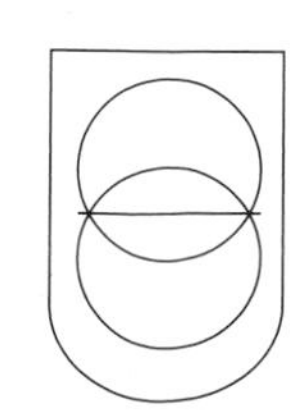

Class III supply symbol

Another way of reducing the risk of electric shock is by using a reduced low voltage system. This is *not* a Class III system but is a safer arrangement than using mains-operated (230 V) equipment because of the lower potential shock voltage. Supply is provided via a mains-powered (230 V) step-down transformer with the centre point of the secondary winding connected to earth.

**Assessment criteria**

**3.3** Describe reasons for using reduced low voltage electrical supplies for tools and equipment on site

**KEY POINT**

The concept of resistance is associated with direct currents, whereas impedance is associated with alternating currents.

Electrical resistance resists the flow of electric current when a voltage is applied. Different materials have different levels of resistance (resistivity). Resistance is measured in ohms (Ω).

Alternating current oscillates as a sine wave, so inductance and capacitance have to be considered as well as resistance. Together, these components are termed impedance. When alternating current goes through an impedance, a voltage drop is produced 0–90° out of phase with the current. Impedance is also measured in ohms but is given the symbol *Z*. More detail regarding this is provided in Unit 202.

Make sure you note the difference between mΩ (milliohms) and MΩ (megohms).

1 mΩ = 0.001 Ω but 1MΩ = 1 000 000 Ω.

## REDUCED LOW VOLTAGE ELECTRICAL SUPPLIES FOR TOOLS AND EQUIPMENT

The impedance of the human body is dependent on the touch voltage. Although 50 V can be dangerous in certain circumstances, if the system voltage can be reduced to around this level, then the magnitude (size) of any current flow through the body will be significantly reduced.

The most common low voltage system of this type in use in the UK uses a 230/110 V double-wound step-down transformer with the secondary winding centre-tapped and connected to earth (CTE). While the supply voltage for equipment supplied from such a transformer is 110 V, the maximum voltage to earth is 55 V (63.5 V for a 110 V three-phase system). This is covered in BS 7671 (Regulation 411.8 Reduced low voltage systems). Overcurrent protection for the 110 V supply may be provided by fuses at the transformer's 110 V output terminals or by a thermal trip to detect excessive temperature rise in the secondary winding. The latter method is generally employed for portable units.

Such reduced voltage supplies may be provided:

- through fixed installations (workshops, plant rooms, lift rooms or other areas where portable electrical equipment is in frequent use)
- through small portable transformers designed to supply individual portable tools. See BS 4363:1998 and BS 7375:1996, which give a specification for the distribution of electricity on building and construction sites, based on the use of 230/110 V CTE transformers.

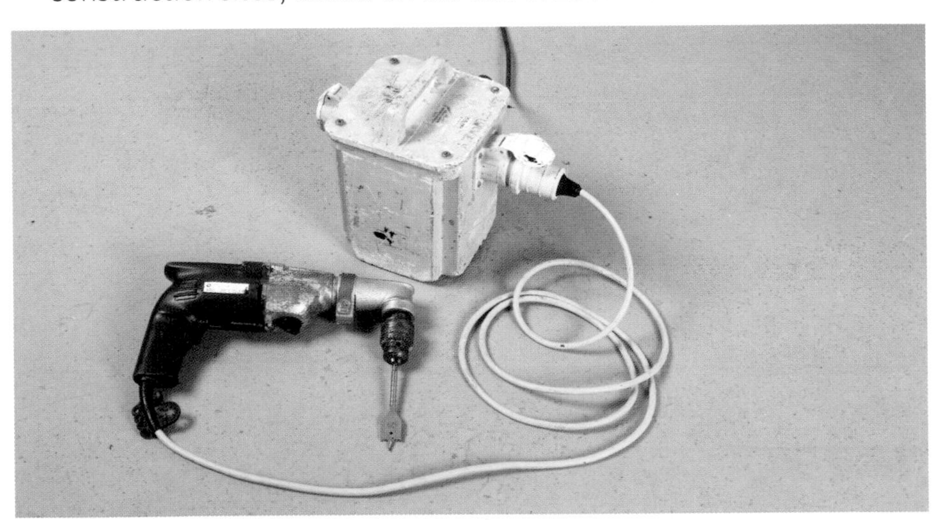

**A portable CTE transformer and drill**

One additional advantage of using a reduced low voltage system is that this safeguard applies to all parts of the system on the load side of the transformer, including the flexible leads, as well as the tools, hand lamps, etc.

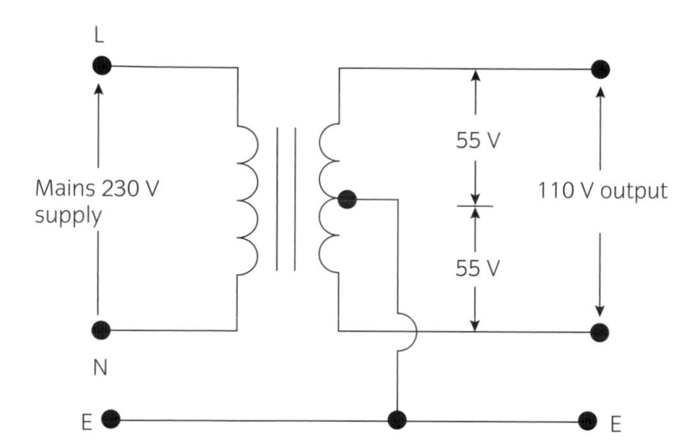


**Schematic diagram of a 110/55 V centre-tapped transformer in a reduced low voltage system**

Alternatively, cordless or battery-powered tools offer a convenient way of providing a powered hand tool without the inconvenience of using a mains supply and without the hazard of trailing power leads.

## HOW TO CONDUCT A VISUAL INSPECTION OF PORTABLE ELECTRICAL EQUIPMENT

**Assessment criteria**

3.4 Identify how to conduct a visual inspection of portable electrical equipment for safe condition before use

**ACTIVITY**

The flex used to supply portable equipment is liable to mechanical damage. Use wholesalers' catalogues or websites to select suitable flexible cables.

To maintain the safety and integrity of tools and equipment, regular in-service inspection and testing should be undertaken to confirm that the equipment remains in a safe condition. Portable appliance testing (PAT) is the term used to describe the examination of electrical appliances and equipment to ensure they are safe to use. There are three categories of in-service inspection or testing for portable tools and equipment:

- user checks (pre-use inspections) – users play an important role by checking equipment before use for signs of damage or obvious defects liable to affect safety in use
- periodic formal inspection or checks
- periodic combined inspection and testing.

The user checks are a visual check only, and should deal with the following inspection requirements separately for the equipment, supply lead and plug.

### Equipment

- Equipment should be manufactured to relevant standards (BS or BS EN).
- The casing should have no visible damage and be free from dents and cracks.
- The switches should operate correctly.
- The supply lead should be secure and correctly connected.
- It should be suitable for the task.

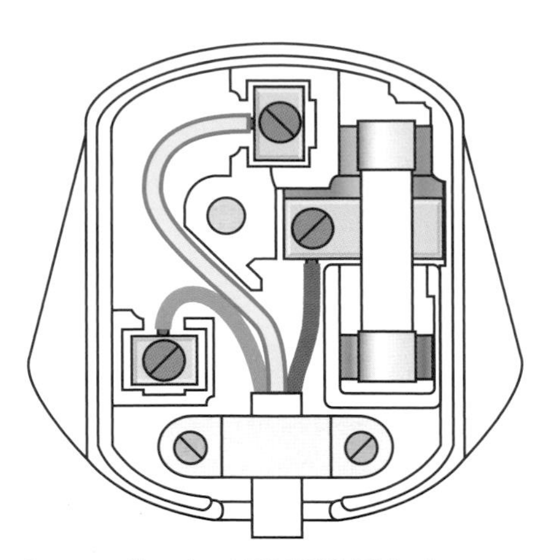

A correctly wired BS 1363 13 A plug

**ASSESSMENT GUIDANCE**

When you make a connection, make sure all the strands are under the terminal. This can done by a visual inspection.

**ACTIVITY**

What are the colours used to identify plugs and sockets for the following voltages?

- 110 V single-phase
- 230 V single-phase
- 400 V three-phase

### Supply leads

- Supply leads should be manufactured to relevant standards (BS or BS EN).
- They should be suitable for the environment.
- They should be free from cuts or fraying.
- There should be no visible exposed conductor insulation (damaged sheath) or exposed live conductors.
- There should be no signs of damage to the cord sheath.
- There should be no joints evident.

### Supply lead plugs

- Supply lead plugs should be manufactured to relevant standards (BS or BS EN).
- The insulation should have no damage.
- The cord/cable connections should be correct and secure.

The user checks must be backed up by periodic inspection and, where appropriate, testing. At that time, a thorough examination is undertaken by a nominated person, competent for that purpose. Schedules giving details of inspection and maintenance periods, together with records of the inspection, should form part of the procedure.

**Assessment criteria**

**3.5** State actions to take when portable electrical equipment fails visual inspection

## WHAT TO DO WHEN PORTABLE ELECTRICAL EQUIPMENT FAILS VISUAL INSPECTION

Most electrical safety defects can be found by visual examination, but some types of defect can only be found by testing. However, it is essential to understand that visual examination is an essential part of the process because some types of electrical safety defect cannot be detected by testing alone.

A relatively brief user check, based upon simple training and the use of a brief checklist, is a very useful part of any electrical maintenance regime. If the user checks detailed on page 33 and above are carried out, 95% of all faults will be identified and the appropriate action taken. No record is needed if there are no defects found. However, if equipment is found to be unsafe, it must be removed from service immediately. It must be labelled to show that it must not be used and the fault must be reported to a responsible person.

**Assessment criteria**

**3.6** Outline the Safe Isolation Procedure

## THE SAFE ISOLATION PROCEDURE

Safe operating procedures for the isolation of plant and machinery during both electrical and mechanical maintenance must be prepared and followed. Wherever possible, electrical isolators should be fitted with a means by which the isolating mechanism can be locked in the open/off position. If this is not possible, an agreed procedure must be followed for the removal and storage of fuse links.

Regulation 12 of the Electricity at Work Regulations 1989 requires that, where necessary to prevent danger, suitable means (including, where appropriate, methods of identifying circuits) must be available for:

- cutting off the supply of electrical energy to any electrical equipment
- the isolation of any electrical equipment.

The aim of Regulation 12 is to ensure that work can be undertaken on an electrical system without danger, in compliance with Regulation 13 (work when equipment has been made dead).

**Cutting off the supply** – Depending on the equipment and the circumstances, this may be no more than normal functional switching (on/off) or emergency switching by means of a stop button or a trip switch.

**Isolation** – This means the disconnection and separation of the electrical equipment from every source of electrical energy in such a way that this disconnection and separation is secure.

**From every source of electrical energy** – Many accidents occur due to a failure to isolate all sources of supply to or within equipment (eg control and auxiliary supplies, uninterruptable power supply (UPS) systems or parallel circuit arrangements giving rise to back feeds).

**Secure** – Security can best be achieved by locking off with a safety lock (ie a lock with a unique key). The posting of a warning notice also serves to alert others to the isolation.

The steps below should be followed.

1. Select an approved voltage indicator to GS38 and confirm operation on a known source such as a proving unit.
2. Locate the correct source of supply and isolator for the section needing isolation.
3. Confirm that the device used for isolation is suitable and may be secured effectively.
4. Power down circuit loads if the isolator is not suitable for on-load switching.
5. Disconnect using the located isolator (from step 2).
6. Secure in the off position. If it is a lockable device*, keep the key on your person. Post warning signs.
7. Using the voltage indicator, confirm isolation by checking all combinations.
8. Prove the voltage indicator on a known source such as a proving unit.

**KEY POINT**

GS38 is a document published by the HSE giving guidance on the safety of test instruments and leads.

**ACTIVITY**

What type of device is it preferable to use when more than one person is working on an electrical circuit?

* If the device is a fuse or removable handle instead of a lockable device, keep this securely under supervision while work is undertaken.

Clear identification and labelling of circuits, switchgear and protective devices will minimise the risk of incorrect isolation.

**Assessment criteria**

3.7 State the procedures for dealing with electric shocks

## HOW TO DEAL WITH ELECTRIC SHOCKS

If all of the correct requirements, precautions and training of staff are taken, it is unlikely that an electrical accident will occur. However, procedures should be in place to deal with electric shock injury in the event of an accident. The recommended procedure for dealing with a person who has received a low voltage shock is as follows.

- Raise the alarm (colleagues and a trained first-aider).
- Make sure the area is safe by switching off the electricity supply.
- Request colleagues to call an ambulance (999 or 112).
- If it is not possible to switch off the power supply, move the person away from the source of electricity by using a non-conductive item.
- Check if the person is responsive, whether their airway is clear and whether they are breathing.
- If the person is unconscious but breathing, move them into the recovery position.
- If they are unconscious but not breathing, start to give cardiopulmonary resuscitation (CPR):
  - CPR is undertaken by interlocking the hands and giving 30 chest compressions in the centre of the chest, between the two pectoral muscles, at a rate of about 100 pulses per minute.
  - Tilt the casualty's head back gently, by placing one hand on the forehead and the other under the chin, to open the airway and give two mouth-to-mouth breaths.
  - Repeat the cycle of 30 compressions to two breaths until either help arrives or the patient recovers.
- Any minor burns should be treated by placing a sterile dressing over the burn and securing with a bandage.

**SmartScreen Unit 201**

Worksheet 7

### ACTIVITY

Find out the latest advice on giving CPR from the British Heart Foundation website at www.bhf.org.uk or the St John Ambulance website at www.sja.org.uk.

# OUTCOME 4

# Know the safety requirements for working with gases and heat-producing equipment

## DIFFERENT TYPES OF GASES USED ON SITE

Industrial gas makes up a group of gases that is used for a large range of applications in the building services sector. The gases may be organic or inorganic and are produced either by extracting them from air or making them synthetically from chemicals. They can take various forms, such as compressed, liquid or solid.

The most common industrial gases are stored under high pressure in metal containers (gas cylinders) and are used as a single gas or as a mixture (eg oxy-acetylene for welding). The table below lists the properties of the most common gases in use.

**Assessment criteria**

4.1 Identify different types of gases used on site

**SmartScreen Unit 201**

Presentation 10 and Handout 10

| Gas | Colour | Chemical formula | Smell | Respiratory hazard | Flammability | Weight vs air |
|---|---|---|---|---|---|---|
| Acetylene | **Maroon** | $C_2H_2$ | Pungent with hint of garlic | Asphyxiant | Highly flammable | Lighter |
| Argon | **Dark green shoulder** | $Ar_2$ | None | Asphyxiant | Not flammable | Heavier |
| Butane | **Blue or yellow** | $C_4H_{10}$ | Odourised | Asphyxiant | Highly flammable | Heavier |
| Carbon dioxide | **Grey Shoulder** | $CO_2$ | None | Asphyxiant | Not flammable | Lighter |
| Chlorine | **Yellow shoulder** | $Cl_2$ | Irritating, pungent odour | Toxic by inhalation | Can react violently with other substances | Heavier |
| Hydrogen | **Red shoulder** | $H_2$ | None | None | Highly flammable | Lighter |
| Helium | **Brown shoulder** | He | None | Non-toxic | Not flammable | Lighter |
| Nitrogen | **Black shoulder** | $N_2$ | None | Asphyxiant | Not flammable | Equal |
| Oxygen | **White shoulder** | $O_2$ | None | None | Not flammable but supports combustion | Equal |
| Propane | **Red shoulder** | $C_3H_8$ | Odourised | Asphyxiant | Highly flammable | Heavier |

The design and manufacture of pressure equipment and assemblies must be in compliance with the Pressure Equipment Regulations 1999.

Gas cylinders are a convenient way to transport and store these gases and they may be used in many different applications:

- chemical processes
- soldering, welding and flame-cutting (oxy-acetylene)
- breathing (eg diving, emergency rescue)
- medical and laboratory uses
- dispensing drinks and beverages
- fuel for vehicles such as fork-lift trucks (propane)
- extinguishing fires (carbon dioxide)
- heating and cooking (butane, propane).

The main causes of accidents involving gas cylinders are:

- inadequate training and supervision
- poor installation
- poor examination and maintenance
- faulty equipment and/or design (eg badly fitted valves and regulators)
- poor handling
- poor storage
- inadequately ventilated working conditions
- incorrect filling procedures.

**ACTIVITY**

Soldering is rarely used now in electrical installation work. Small and large joints are usually crimped either by hand crimping or by hydraulic crimps for large joints. Why do you think soldering is no longer used routinely in installation work?

**ACTIVITY**

Buildings have been burnt down as a result of contractors leaving a gas flame unattended. What should you do in the event of the fire alarm being sounded while you are using gas-flamed equipment?

**Assessment criteria**

**4.2** Describe how bottled gases and equipment should be safely transported and stored

## HOW TO SAFELY TRANSPORT AND STORE BOTTLED GASES AND EQUIPMENT

The most likely hazards that will be encountered due to the use of gas cylinders in the building services environment are in: handling, storage and transport.

### Handling and use

- Gas cylinders should be used in a vertical position, unless specifically designed to be used otherwise.
- Cylinder colours are only a guide. Labels should be used as the primary means of identifying cylinder content.
- Valves, regulators and pipework must be suitable for the type of gas and pressure being used.
- Cylinders must not be used for any purpose other than the transport and storage of gas.

- Cylinders should not be dropped, rolled or dragged.
- Cylinder valves should be closed and dust caps should be replaced when a gas cylinder is not in use.
- Valves should be protected by a valve cap or collar.

### Transport

- Stow gas cylinders securely when in transit.
- Gas cylinders should be contained within the body of the vehicle.
- Gas cylinders should be clearly marked to show their contents (including their United Nations unique number).
- The transport of gas cylinders is subject to the Carriage of Dangerous Goods and Use of Transportable Pressure Equipment Regulations 2009.

### Storage

- Gas cylinders should not be stored for excessive periods of time.
- Stocks should be rotated to ensure that the first in is the first used.
- Gas cylinders should be stored in a dry, safe place on a flat surface in the open air or in an adequately ventilated building or part of a building specifically reserved for this purpose.
- Cylinders containing flammable gas should not be stored in part of a building used for other purposes.
- Gas cylinders should be protected from external heat sources that might stop them working properly.
- Gas cylinders should be stored away from sources of ignition and other flammable materials.
- Do not allow gas cylinders to be stored so that they stand or lie in water.
- Valves should be kept shut on empty cylinders to prevent contaminants getting in.
- Store gas cylinders securely when they are not in use. They should be properly fixed to prevent them falling over, unless designed to be freestanding.
- Store cylinders where they are not vulnerable to hazards caused by impact (eg from vehicles such as fork-lift trucks).

## HOW TO CONDUCT A VISUAL INSPECTION ON HEAT-PRODUCING EQUIPMENT

**Assessment criteria**

4.3 Describe how to conduct a visual inspection on heat-producing equipment for safe condition

An external visual inspection of the surfaces of gas cylinders and any attachments (valves, connecting hoses, flashback arresters and regulators) should identify the majority of in-service faults. The inspection is looking for signs of impact damage, scrapes, fire damage, corrosion, erosion and wear. These signs should be noted and any cylinder with defects that could cause a failure should be removed from service and returned to the supplier.

**Assessment criteria**

**4.4** Describe how combustion takes place

## HOW COMBUSTION TAKES PLACE

Most fires are preventable and, by adopting the right behaviours and procedures, prevention can easily be achieved.

**The fire triangle shows the three elements of a fire. If one element is not present, or removed, the fire will not ignite or will stop burning**

A fire needs three elements to start: a source of ignition (heat), a source of fuel (something that burns) and oxygen. If any one of these elements is removed, a fire will not ignite or will cease to burn.

- Sources of ignition include heaters, lighting, naked flames, electrical equipment, smokers' materials (cigarettes, matches) and anything else that can get very hot or cause sparks.
- Sources of fuel include wood, paper, plastic, rubber or foam, loose packaging materials, waste rubbish, combustible liquids and furniture.
- Sources of oxygen include the air surrounding us.

A fire safety risk assessment using the same approach as used in the health and safety risk assessment should be used to determine the risks. Based on the findings of the assessment, employers must ensure that adequate and appropriate fire safety measures are in place to minimise the risk of injury or loss of life in the event of a fire. These findings must be kept up to date.

**Assessment criteria**

**4.5** State the dangers of working with heat-producing equipment

## DANGERS OF WORKING WITH HEAT-PRODUCING EQUIPMENT

The use of heat-producing equipment (**hot work**) is a common occurrence in building services work activities. Hot work is work that might generate sufficient heat, sparks or flame to cause a fire. It includes welding, flame cutting, soldering, brazing, grinding and other equipment that incorporates a flame, such as boilers for bitumastic materials.

**Hot work**

Work that involves actual or potential sources of ignition and undertaken in an area where there is a risk of fire or explosion (eg welding, flame cutting, grinding).

The flames, sparks and heat produced during the hot work are ignition sources that can cause fires and explosions in many different situations. For example:

- sparks produced during hot work can land on combustible materials and cause fires and explosions
- hot work performed on tanks and vessels with residual flammable substances and vapours can cause the tanks to explode.

To help prevent fire in the workplace, the risk assessment should identify what could cause a fire to start – the sources of ignition (heat or sparks), the substances that burn – and the people who may be at risk.

Once the risks have been identified, appropriate action can be taken to control them. Actions should be based on whether the risks can be avoided altogether or, if this is not possible, how they can be reduced. The checklist below will help with an appropriate action plan.

- Keep sources of ignition and flammable substances apart.
- Avoid accidental fires (eg make sure heaters cannot be knocked over).
- Ensure good housekeeping at all times, for example, avoid build-up of rubbish that could burn.
- Consider how to detect fires and how to warn people quickly if a fire starts, for example, install smoke alarms and fire alarms or bells.
- Have the correct fire-fighting equipment available.
- Keep fire exits and escape routes clearly marked and unobstructed at all times.
- Ensure workers receive appropriate training on procedures and fire drills.
- Review and update the risk assessment on a regular basis.

The Regulatory Reform (Fire Safety) Order 2005 covers general fire safety in England and Wales.

## PROCEDURES ON DISCOVERY OF FIRES ON SITE

**Assessment criteria**

**4.6** State the procedures to follow on discovery of fires on site

How people react in the event of fire depends on how well they have been prepared and trained for a fire emergency. It is therefore imperative that all employees (and visitors and contractors) are familiar with the company procedure to follow in the event of an emergency. A basic procedure is:

- on discovery of a fire, raise the alarm immediately
- if staff are trained and it is considered safe to do so, attempt to fight the fire using the equipment provided
- if fire fighting fails, evacuate (leave the area) immediately
- ensure that no one is left in the fire area and close doors on exit in order to prevent the spread of fire
- go straight to the designated assembly point. These points are specially chosen as they are in locations of safety and where the emergency services are not likely to be obstructed on arrival.

**KEY POINT**

Remember never to ignore a fire alarm, even if it has gone off many times before that day. It may be the last one you ignore.

**Assessment criteria**

**4.7** Identify different classifications of fires
**4.8** Identify types of fire extinguisher for different classifications of fires

## DIFFERENT CLASSIFICATIONS OF FIRES

All fires are grouped into classes, according to the type of materials that are burning.

The grid below is a guide to the different types of fire and the type of extinguisher that should be used.

| Fire classification | Water | Foam | Powder | Carbon dioxide |
|---|---|---|---|---|
| **Class A** – Combustible materials such as paper, wood, cardboard and most plastics | ✓ | ✓ | ✓ | |
| **Class B** – Flammable or combustible liquids such as petrol, kerosene, paraffin, grease and oil | | ✓ | ✓ | ✓ |
| **Class C** – Flammable gases, such as propane, butane and methane | | | ✓ | ✓ |
| **Class D** – Combustible metals, such as magnesium, titanium, potassium and sodium | | | Specialist dry powder | |
| **Class F** – Cooking oils and fats | | Specialist wet chemical | | |

**SmartScreen Unit 201**
Worksheet 9

### ACTIVITY

Why should you not use water extinguishers on oil fires?

Fires involving equipment such as electrical circuits or electronic equipment are often referred to as Class E fires, although the category does not officially exist under the BS EN 3 rating system. This is because electrical equipment is often the cause of the fire, rather than the actual type of fire. Most modern fire extinguishers specify on the label whether they should be used on electrical equipment. Normally carbon dioxide or dry powder are suitable agents for putting out a fire involving electricity.

**OUTCOME 5**

# Know the safety requirements for using access equipment in the building services industry

Working at height remains one of the biggest causes of fatalities and major injuries within the construction industry with almost 50% of fatalities resulting from falls from ladders, stepladders and through fragile roofs. Work at height means work in any place, including at or below ground level (eg in underground workings), where a person could fall a distance liable to cause injury.

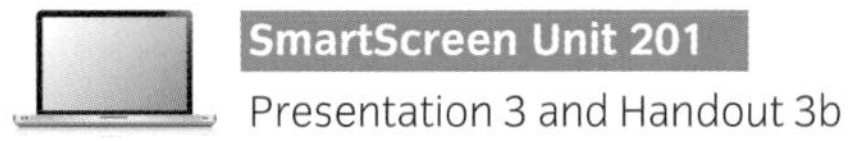


The Work at Height Regulations 2005 require duty holders to ensure that:

- all work at height is properly planned and organised
- those involved in work at height are competent
- the risks from work at height are assessed and appropriate work equipment is selected and used
- the risks of working on or near fragile surfaces are properly managed
- the equipment used for work at height is properly inspected and maintained.

**ASSESSMENT GUIDANCE**

Remember that standing on the floor of a six-storey building is not really working at height, but standing on a stepladder on the ground floor is.

## ACCESS EQUIPMENT FOR DIFFERENT TYPES OF WORK

**Assessment criteria**

5.1 Identify different types of access equipment

5.2 Select suitable equipment for carrying out work at heights based on the work being carried out

Many different types of access equipment are used in the building services industry, such as mobile elevated work platforms (MEWPs), ladders, mobile tower scaffolds, tube and fitting scaffolding, personal suspension equipment (harnesses).

### MEWPs

There is a wide range of MEWPs (vertical scissor lift, self-propelled boom, vehicle-mounted boom and trailer-mounted boom) and if any of these are to be used it is important to consider:

- the height from the ground
- whether the MEWP is appropriate for the job
- the ground conditions
- training of operators

- overhead hazards such as trees, steelwork or overhead cables
- the use of a restraint or fall arrest system
- closeness to passing traffic.

**ACTIVITY**

Never use any access equipment which you are unfamiliar with or which is damaged. Think of three items that could cause danger when using stepladders.

**ASSESSMENT GUIDANCE**

A wooden ladder is found to have a split in it. What action should you take?

A vertical scissor lift

The correct angle for a ladder is 1:4

## Ladders

It is recommended that ladders are only used for low-risk, short-duration work (between 15 and 30 minutes depending upon the task). Common causes of falls from ladders are:

- overreaching – maintain three points of contact with the ladder
- slipping from the ladder – keep the rungs clean, wear non-slip footwear, maintain three points of contact with the ladder, make sure the rungs are horizontal
- the ladder slips from its position – position the ladder on a firm level surface, secure the ladder top and bottom, check the ladder daily
- the ladder breaks – position the ladder properly using the 1:4 rule (four units up for every one unit out), do not exceed the maximum weight limit of the ladder, only carry light materials or tools.

## Stepladders

Many of the common causes of falls from ladders can also be applied to stepladders. If stepladders are used, ensure that:

- they are suitable, in good condition and inspected before use
- they are sited on stable ground
- they are face onto the work activity
- knees are never above the top tread of the stepladder
- the stepladder is open to the maximum
- wooden stepladders (or ladders) are not painted as this may hide defects.

Correct use of stepladders

## Scaffolding

Some work activities, such as painting, roof work, window replacement or brickwork, are almost certainly more conveniently undertaken from a fixed external scaffold. Tube and fitting scaffolding must only be erected by competent people who have attended recognised training courses. Any alterations to the scaffolding must also be carried out by a competent person. Regular inspections of the scaffold must be made and recorded.

# SAFETY CHECKS AND SAFE ERECTION METHODS FOR ACCESS EQUIPMENT

**Assessment criteria**

**5.3** Describe the safety checks to be carried out on access equipment
**5.4** Describe safe erection methods for access equipment

The Work at Height Regulations 2005 require that all scaffolding and equipment that is used for working at height, where a person could fall 2 metres or more, is inspected on a regular basis, using both formal and pre-use inspections to ensure that it is fit for use.

A marking, coding or tagging system is a good method of indicating when the next formal inspection is due. However, regular pre-use checks must take place as well as formal inspections.

The following safety checks must be carried out.

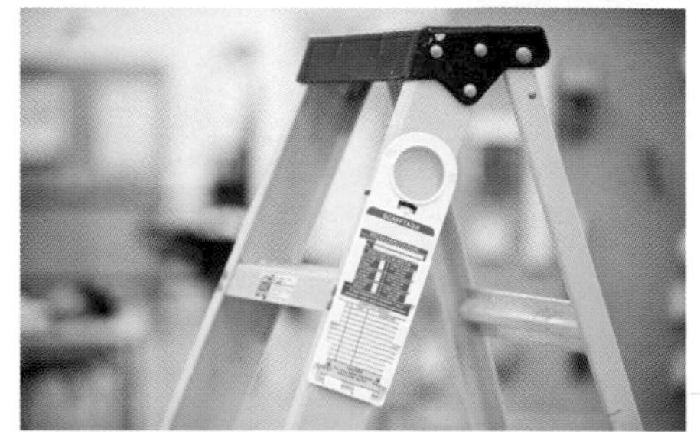

Tags to record that a ladder inspection has taken place

## MEWPs

These must only be operated by trained and competent persons, who must also be competent to carry out the following pre-use checks.

- The ground conditions must be suitable for the MEWP, with no risk of the MEWP becoming unstable or overturning.
- Check for overhead hazards such as trees, steelwork and overhead cables.
- Guard rails and toe boards must be in place.
- Outriggers should be fully extended and locked in position.
- The tyres must be properly inflated and the wheels immobilised.
- Check the controls to make sure they work as expected.

**ACTIVITY**

What is the qualification that allows a person to use mobile elevating work platforms (MEWPs)?

- Check the fluid and/or battery charge levels.
- Check that the descent alarm and horn are working.
- Check that the emergency or ground controls are working properly.

## Tube and fitting scaffolds

This equipment must only be erected by competent people who have attended recognised training courses. Any alterations to the scaffolding must also be carried out by a competent person. However, users should undertake the following fundamental checks to prevent accidents.

- Toe boards, guard rails and intermediate rails must be in place to prevent people and materials from falling.
- The scaffold must be on a stable surface and the uprights must be fitted with base plates and sole plates.
- Safe access and exit (ladders) must be in place and secured.

## Ladders and step ladders

Users must check that:

- ladders and stepladders are of the right classification (trade/ industrial)
- the styles, rungs or steps are in good condition
- the feet are not missing, loose, damaged or worn
- rivets are in place and secure
- the locking bars are not bent or buckled.

## Prefabricated mobile scaffold towers

Mobile scaffold towers are a convenient means of undertaking repetitive tasks in the building services industry. The erection and dismantling must only be undertaken by a competent person. The following pre-use checks will ensure safe use.

- The maximum height-to-base ratios must not be exceeded.
- Diagonal bracing and stabilisers must not be damaged or bent.
- The brace claws must work properly.
- Internal access ladders must be in place.
- Wheels must be locked when work is in progress.
- The working platform must be boarded, with guard rails and toe boards fitted.
- The towers must be tied in windy conditions.
- The platform trap door must be operating correctly.
- Rivets must be checked visually to ensure they are in place and not damaged.

**SmartScreen Unit 201**
Worksheet 3b

OUTCOME 6

# Know the safety requirements for working safely in excavations and confined spaces in the building services industry

## WHERE IT MAY BE NECESSARY TO WORK IN EXCAVATIONS

Assessment criteria

6.1 Identify the situations in which it may be necessary to work in excavations

Every year people are killed or seriously injured by collapses and falling materials while working in excavations. These excavations may be required in the course of ground source heating projects or the installation of drains and soakaways, septic tanks, electrical distribution networks and retaining structures.

SmartScreen Unit 201

Presentation 11 and Handout 11

## HOW TO PREPARE EXCAVATIONS FOR SAFE WORKING

Assessment criteria

6.2 Describe how excavations should be prepared for safe working

The hazards associated with excavations are:

- excavations collapsing and burying or injuring people working in them
- material falling from the sides into any excavation
- people or plant falling into excavations
- contact with underground services (electricity, high pressure water and gas)
- undermining other structures
- exhaust fumes from petrol or diesel-engined equipment such as compressors or generators.

Planning is the key to the safety of any excavation. Before work commences a decision will be required on what temporary support will be needed and what precautions need to be taken. The equipment and precautions needed (trench sheets, props, baulks, etc) should be available on site before work starts.

The sides of the excavation must be prevented from collapsing either by battering at an angle of between 5° and 45° or by shoring the sides up with timber, sheeting or a proprietary support system. In **granular soils**, the angle of slope (**batter**) should be less than the natural angle of **repose** of the material being excavated. In wet ground, a considerably flatter slope will be required. Loose materials may fall from spoil heaps into the excavation, therefore edge protection should include toe boards or other means to protect against falling materials. Head protection should be worn in excavations.

### ASSESSMENT GUIDANCE

Gas can accumulate in underground workings. Always beware of this and test the atmosphere in the underground workings first.

**Granular soils**

Gravel, sand or silt (coarse grained soil) with little or no clay content. Although some moist granular soils exhibit apparent cohesion (grains sticking together forming a solid), they have no cohesive strength. Granular soil cannot be moulded when moist and crumbles easily when dry.

**Batter or slope**

The angle in relation to the horizontal surface, of the trench walls of an excavation, to prevent the walls collapsing.

**Repose**

The angle to the horizontal at which the material in the cut face is stable and does not fall away. Different materials have different angles of repose, for example, 90° for solid rock and 30° for sand.

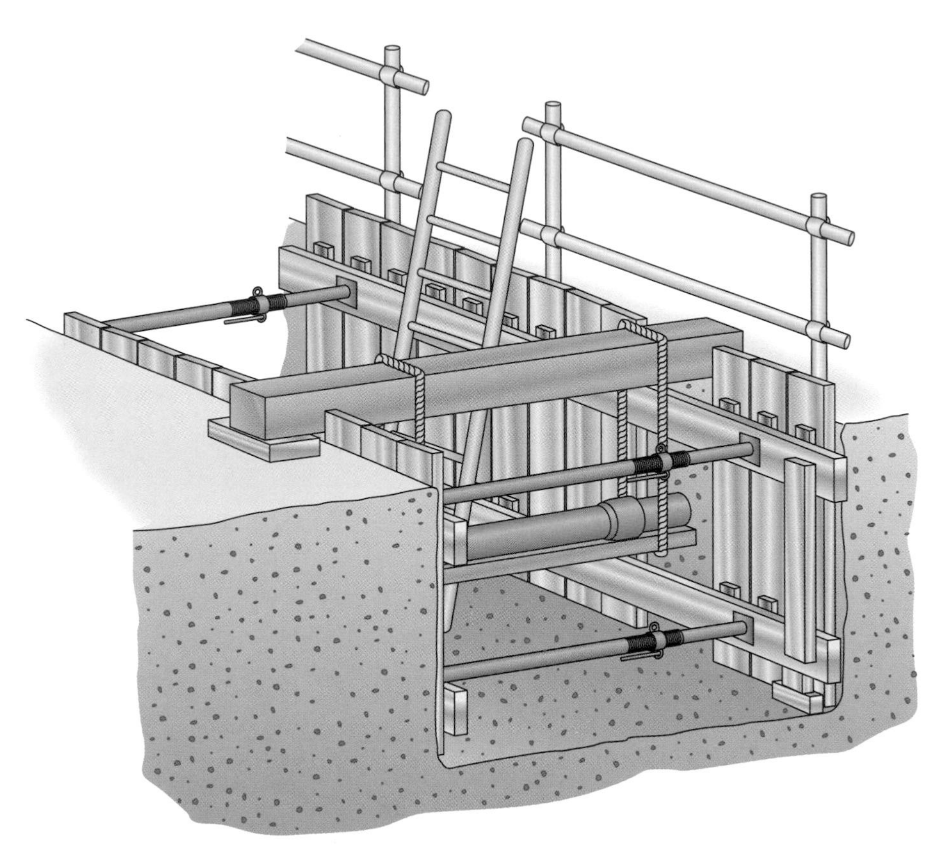

A timbered excavation with ladder access, barriers, wooden boards and supported services

Assessment criteria

**6.3** State precautions to be taken to make excavations safe

## PRECAUTIONS TO MAKE EXCAVATIONS SAFE

To prevent people from falling into an excavation, the edges should be protected in one of the following ways:

- with guard rails and toe boards inserted into the ground immediately next to the supported excavation side
- by a fabricated guard rail assembly that connects to the sides of the trench box
- by using the trench support system itself (eg using trench box extensions or trench sheets longer than the trench depth).

**KEY POINT**

Never take a chance with an excavation. It only takes a second for an unsafe working area to collapse and suffocate you.

A competent person who fully understands the dangers and necessary precautions should inspect the excavation at the start of each shift and after any event that may have affected their strength or stability, or after a fall of rock or earth.

A record of the inspections will be required and any faults that are found should be corrected immediately.

## AREAS WHERE WORKING IN CONFINED SPACE MAY BE A CONSIDERATION

**Assessment criteria**

6.4 Identify areas where working in confined space may be a consideration

A confined space is a place which is substantially enclosed (though not always entirely) and where serious injury can occur from hazardous substances or conditions within the space or nearby (eg lack of oxygen). Examples include chambers, tanks, vessels, furnaces, boilers or cisterns, inspection chambers, pits, roof spaces and under suspended timber floors. Entry into a confined space requires the correct equipment, including PPE and a harness. Activities below ground require a descent control and rescue system.

## SAFETY CONSIDERATIONS WHEN WORKING IN CONFINED SPACES

**Assessment criteria**

6.5 State safety considerations when working in confined spaces

**SmartScreen Unit 201**

Worksheet 10

The two main hazards associated with confined spaces are the presence of toxic or other dangerous substances and the absence of adequate oxygen.

The Confined Space Regulations 1997 detail the specific controls that are necessary when people enter confined spaces. These can be summarised as:

- avoid entry to confined spaces (eg by doing the work from the outside)
- if entry to a confined space is unavoidable, follow a safe system of work which should include rigorous preparation, isolation, air testing and other precautions, and the use of a confined space entry permit
- put in place adequate emergency arrangements before the work starts.

Adequate ventilation, sufficient lighting, the provision of the correct PPE, emergency evacuation procedures and medical conditions must all be catered for. Lone working should not be allowed.

Section 706 of BS 7671 gives the requirements for supplies to portable electrical equipment used in confined spaces that are restrictive and/ or conducting.

# OUTCOME 7

# Be able to apply safe working practice

This outcome is about you being able to demonstrate your understanding of health and safety by practical activity. The contents of this section have been covered in previous outcomes but are outlined again here to act as a reminder of the key points. Safe working practices are systems of work that are carried out in a safe manner for any given task.

Assessment criteria

7.1 Perform manual handling techniques

**Manual handling**

The movement of items by lifting, lowering, carrying, pushing or pulling by human effort alone.

## MANUAL HANDLING TECHNIQUES

**Manual handling** is one of the most common causes of injury at work and causes over a third of all workplace injuries. Manual handling injuries can occur almost anywhere in the workplace and heavy manual labour, awkward postures and previous or existing injury can increase the risk. Work-related manual handling injuries can have serious implications for both the employer and the person who has been injured.

### ACTIVITY

What should you check before moving an object from one place to another by manual handling?

The introduction of the Manual Handling Operations Regulations 1992 saw a change from reliance on safe lifting techniques to an analysis, using risk assessment, of the need for manual handling. The regulations established a clear hierarchy of manual handling measures:

- avoid manual handling operations so as far as is reasonably practicable by re-engineering the task to avoid moving the load or by mechanising the operation
- if manual handling cannot be avoided, a risk assessment should be made
- reduce the risk of injury so as far as is reasonably practicable either by the use of mechanical aids or by making improvements to the task (eg using two persons), the load and the environment.

Assessment criteria

7.2 Manually handle loads using mechanical lifting aids

## HOW TO MANUALLY HANDLE LOADS USING MECHANICAL LIFTING AIDS

Even if mechanical handling methods are used to handle and transport equipment or materials, hazards may still be present in the four elements that make up the mechanical handling: the handling equipment, the load, the workplace and the human element.

- Mechanical handling equipment must be suitable for the task, well maintained and inspected on a regular basis.
- The load needs to be prepared in such a way as to minimise accidents, taking into account such things as security of the load, flammable materials and stability of the load.
- If possible, the workplace should be designed to keep the workforce and the load apart.
- Employees who are to use the equipment must be properly trained.

**ASSESSMENT GUIDANCE**

When you are manually handling a load, always make sure the route you will be taking is clear and all obstacles are removed. Never try to lift more than you can handle. Use safe handling techniques.

## THE SAFE METHOD OF ASSEMBLY OF ACCESS EQUIPMENT

**Assessment criteria**

7.3 Demonstrate the safe method of assembly of access equipment

The different types of access equipment likely to be used in electrotechnical work are:

- ladders
- fixed scaffolds
- mobile tower scaffolds
- mobile elevated work platforms (MEWPs).

The main cause of accidents involving ladders is ladder movement while in use. This normally occurs when a ladder has not been secured to a fixed point. The ladder needs to be stable in use and this can be achieved by making sure the inclination is as near as possible to a 1 in 4 ratio of distance from the wall to the distance up the wall. The foot of the ladder should be tied to a rigid support and not used in high winds or heavy rain.

**The correct angle for a ladder is one unit out for every four units up**

Fixed scaffolds are often more effective for long duration work than a ladder. Tube and fitting scaffolding must only be erected by competent people who have attended recognised training courses. Any alterations to the scaffolding must also be carried out by a competent person. Regular inspections of the scaffold must be made and recorded.

As scaffolding must only be erected by competent people, it will be difficult for non-competent persons to check the security of the scaffolding. However certain areas can be checked prior to use, as follows.

- Are there adequate toe boards, guard rails and intermediate rails to prevent people or materials from falling?
- Is the scaffold on a stable surface with base plates and timber sole plates?
- Is there safe access and exit?
- Is the working platform fully boarded?
- Are the lower level uprights prominently marked?
- Is the scaffold braced or secured to the building?

Tube and fitting scaffold

Mobile scaffold towers are a convenient means of undertaking repetitive tasks in the building services industry. The following must be adhered to when mobile towers are used:

- erection and dismantling must be done by competent persons only
- the maximum height-to-base ratios must not be exceeded
- diagonal bracing and stabilisers must always be used
- internal access ladders must always be used
- wheels must be locked when work is in progress
- the working platform must be boarded and guard rails and toe boards fitted
- towers must be tied in windy conditions
- the manufacturer's instructions must be followed at all times.

The following factors must be considered when MEWPs are used:

- whether the MEWP is appropriate for the job
- whether the ground conditions affect stability
- operation should only be by trained and competent persons
- overhead hazards such as trees, steelwork or overhead cables
- restraint or fall arrest systems should be used
- closeness to passing traffic
- the height from the ground
- tyres should be properly inflated and the wheels immobilsed.

## HOW TO USE ACCESS EQUIPMENT SAFELY

Assessment criteria

7.4 Use access equipment safely

Working at height remains one of the biggest causes of fatalities and major injuries within the construction industry with almost 50% of fatalities resulting from falls from ladders, stepladders and through fragile roofs. Work at height means work in any place, including at or below ground level (eg in underground workings), where a person could fall a distance liable to cause injury.

The Work at Height Regulations 2005 require duty holders to ensure that:

- all work at height is properly planned and organised
- those involved in work at height are competent
- the risks from work at height are assessed and appropriate work equipment is selected and used
- the risks of working on or near fragile surfaces are properly managed
- the equipment used for work at height is properly inspected and maintained.

Before deciding if access equipment is necessary to work at height, it will be necessary to carry out a risk assessment. This will consist of:

- a careful examination of the work at height task to identify hazards
- consideration of whether the hazards pose a risk that could cause harm to people.

If it is absolutely necessary to work at height, the following points should be considered.

- Use an existing safe place of work to access work at height. If there is already a safe means of access such as a permanent stair and guard-railed platform, it should be used.
- Provide or use work equipment, such as scaffolding, mobile access towers or mobile elevated work platforms (MEWPs) with guardrails around the working platform, to prevent falls.
- Minimise distance and consequences of a fall by using, for example, a properly set up stepladder or ladder within its limitations for low-level, short-duration work only.

Having decided that access equipment is necessary, the condition of all the equipment required will also need to be determined. To achieve this it may be necessary to find out how much equipment is in circulation. This may require an initial audit, giving each piece of equipment a thorough examination to establish if it is safe to use and fit for purpose. A database can then be constructed recording the type, location and condition of each piece of access equipment to enable the equipment to be tracked and managed proactively.

For those employers who have no in-depth knowledge of working at height, the HSE has produced a Work at Height Access and Information Toolkit (WAIT). It gives practical advice on the factors to consider when selecting access equipment and how to work at height safely, as well as information on some of the different types of access equipment available.

Job-specific training will ensure that employees undertake their job in a safe manner. This must include training for the competent person in erecting and dismantling access equipment and also for the user of that equipment.

**Always observe requirements of site safety signs**

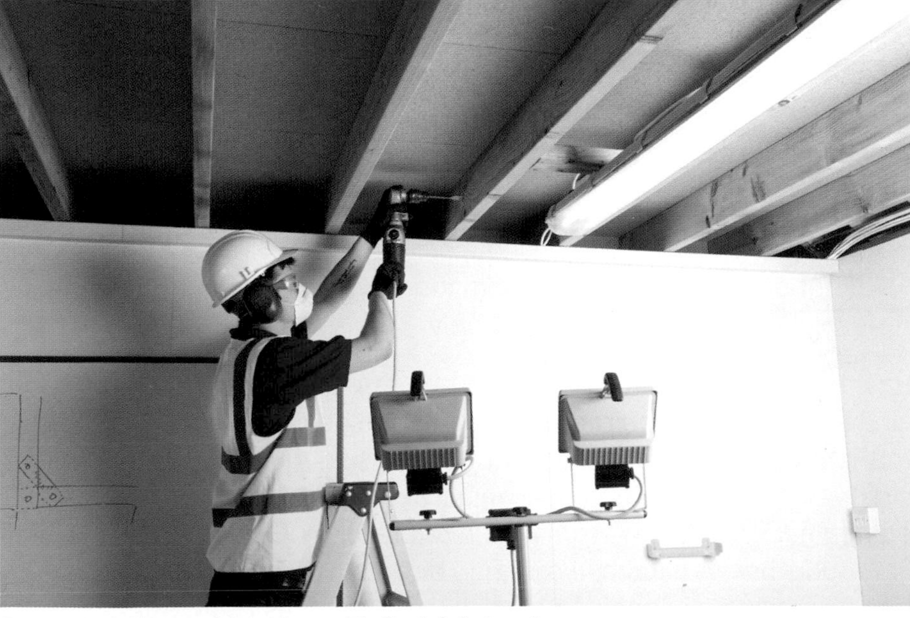

**Correct PPE does not interfere with the job being done**

# ASSESSMENT CHECKLIST

## WHAT YOU NOW KNOW/CAN DO

| Learning outcome | Assessment criteria | Page number |
|---|---|---|
| 1 Know health and safety legislation | *The learner can:* | |
| | 1 State the aims of health and safety legislation | 2 |
| | 2 Identify the responsibilities of individuals under health and safety legislation | 3 |
| | 3 Identify statutory and non-statutory health and safety materials | 5 |
| | 4 Identify the different roles of the Health and Safety Executive in enforcing health and safety legislation. | 8 |
| 2 Know how to handle hazardous situations | *The learner can:* | |
| | 1 Identify common hazardous situations found on site | 10 |
| | 2 Describe safe systems at work | 14 |
| | 3 Identify the categories of safety signs | 15 |
| | 4 Identify symbols for hazardous substances | 17 |
| | 5 List common hazardous substances used in the building services industry | 18 |
| | 6 List precautions to be taken when working with hazardous substances | 19 |
| | 7 Identify the types of asbestos that may be encountered in the workplace | 19 |
| | 8 Identify the actions to be taken if the presence of asbestos is suspected | 20 |
| | 9 Describe the implications of being exposed to asbestos | 20 |
| | 10 State the application of different types of personal protective equipment | 20 |
| | 11 Identify the procedures for manually handling heavy and bulky items | 23 |
| | 12 Identify the actions that should be taken when an accident or emergency is discovered | 24 |
| | 13 State procedures for handling injuries sustained on site | 25 |
| | 14 State the procedures for recording accidents and near misses at work. | 26 |

| Learning outcome | Assessment criteria | Page number |
|---|---|---|
| 3 Know electrical safety requirements when working in the building services industry | *The learner can:* | |
| | 1 Identify the common electrical dangers to be aware of on site | 28 |
| | 2 List different sources of electrical supply for tools and equipment | 31 |
| | 3 Describe reasons for using reduced low voltage electrical supplies for tools and equipment on site | 32 |
| | 4 Identify how to conduct a visual inspection of portable electrical equipment for safe condition before use | 33 |
| | 5 State actions to take when portable electrical equipment fails visual inspection | 34 |
| | 6 Outline the Safe Isolation Procedure | 34 |
| | 7 State the procedures for dealing with electric shocks. | 36 |
| 4 Know the safety requirements for working with gases and heat-producing equipment | *The learner can:* | |
| | 1 Identify different types of gases used on site | 37 |
| | 2 Describe how bottled gases and equipment should be safely transported and stored | 38 |
| | 3 Describe how to conduct a visual inspection on heat-producing equipment for safe condition | 39 |
| | 4 Describe how combustion takes place | 40 |
| | 5 State the dangers of working with heat-producing equipment | 40 |
| | 6 State the procedures to follow on discovery of fires on site | 41 |
| | 7 Identify different classifications of fires | 42 |
| | 8 Identify types of fire extinguisher for different classifications of fires. | 42 |

| Learning outcome | Assessment criteria | Page number |
|---|---|---|
| 5 Know the safety requirements for using access equipment in the building services industry | *The learner can:* | |
| | 1 Identify different types of access equipment | 43 |
| | 2 Select suitable equipment for carrying out work at heights, based on the work being carried out | 43 |
| | 3 Describe the safety checks to be carried out on access equipment | 45 |
| | 4 Describe safe erection methods for access equipment. | 45 |
| 6 Know the safety requirements for working safely in excavations and confined spaces in the building services industry | *The learner can:* | |
| | 1 Identify the situations in which it may be necessary to work in excavations | 47 |
| | 2 Describe how excavations should be prepared for safe working | 47 |
| | 3 State precautions to be taken to make excavations safe | 48 |
| | 4 Identify areas where working in confined space may be a consideration | 49 |
| | 5 State safety considerations when working in confined spaces. | 49 |
| 7 Be able to apply safe working practice | *The learner can:* | |
| | 1 Perform manual handling techniques | 50 |
| | 2 Manually handle loads using mechanical lifting aids | 50 |
| | 3 Demonstrate the safe method of assembly of access equipment | 51 |
| | 4 Use access equipment safely. | 53 |

# ASSESSMENT GUIDANCE

This unit is common to the Level 2 and Level 3 qualifications. You do not need to be reassessed on this unit if you have passed it at Level 2.

The assessment of this unit is in two parts.

## Practical assessment (201)

| Health & safety (Unit 201) task | Description | Assessment criteria | Recommended assignment time |
|---|---|---|---|
| H&S 1 | Identify hazards | AC 2.1 | 1 hour, 30 minutes |
| H&S 2 | Manual handling | AC 7.1 | 45 minutes |
| H&S 3 | Manual handling using lifting aids | AC 7.2 | 45 minutes |
| H&S 4 | Ladders | AC 7.3, 7.4 | 45 minutes |
| H&S 5 | Stand steps | AC 7.3, 7.4 | 45 minutes |
| H&S 6 | Scaffold | AC 7.3, 7.4 | 1 hour, 30 minutes |
| H&S 7 | Safety signs | AC 2.3 | 1 hour |

- The above table shows the practical assessments in which you are required to demonstrate the ability to carry out the tasks listed.
- Failure to work safely will cause the assessment to be terminated.
- Make sure you understand all the verbal instructions given to you by your assessor before you begin.
- If you are not sure of anything, ask.
- Make sure that all personal protective equipment given to you fits properly before you begin.
- You will be given feedback at the end of the assessment.

## Online multiple-choice assessment (501)

- This is a closed book online e-volve multiple-choice assessment.
- Attempt all questions.
- Do not leave until you are confident that you have completed all questions.
- Keep an eye on the time as it moves quickly when you are concentrating.
- Make sure you read each question fully before answering.

- Ensure you know how the e-volve system works. Ask for a demonstration if you are not sure.
- Do not take any paperwork into the exam with you.
- If you need paper to work anything out, ask the invigilator to provide some.
- Make sure your mobile phone is switched *off* (not on silent) during the exam. You may be asked to give it to the invigilator.

## Before the assessment

- You will find sample questions on SmartScreen and some questions in the section below to test your knowledge of the learning outcomes.
- Make sure you go over these questions in your own time.
- Spend time on revision in the run-up to the assessments.

## OUTCOME KNOWLEDGE CHECK

1 Under the Health and Safety at Work etc Act 1974 employers are responsible for:
   a) transport to and from work
   b) payment of trade union fees
   c) subsidised canteen and rest facilities
   d) a safe working environment with adequate welfare facilities.

2 A risk assessment should be completed at the start of a job and:
   a) filed away until the work is finished
   b) continuously reviewed as work progresses
   c) posted to the Health and Safety Executive
   d) fixed to the office notice board.

3 Which of the following is a non-statutory document?
   a) Personal Protective Equipment at Work Regulations 1992.
   b) Provision and Use of Work Equipment Regulations 1998.
   c) Manual Handling Operations Regulations 1992.
   d) BS 7671: IET Wiring Regulations.

4 Work at Height Regulations 2005: apply to all work at height:
  a) inside buildings only
  b) on roofs when fixing PV systems only
  c) where there is a risk of a fall liable to cause personal injury
  d) when there is no one around to hold the ladder.

5 Compliance with the IET Wiring Regulations will meet the requirements of:
  a) Health and Safety at Work etc Act 1974
  b) Electricity at Work Regulations 1989
  c) Control of Substances Hazardous to Health (COSHH) Regulations 2002 (as amended)
  d) Health and Safety (Display Screen Equipment) Regulations 1992.

6 A prohibition notice requires the duty holder to:
  a) stop an activity immediately
  b) stop an activity at the end of the day
  c) carry on but put up notices
  d) contact the Health and Safety Executive for advice.

7 The word which collectively means keeping a clean and tidy working environment is:
  a) maintenance
  b) tidysafe
  c) housekeeping
  d) sweepclear.

8 Scrap and rubbish should be:
  a) left on the stairs for collection later
  b) pushed into a corner out of the way
  c) placed in an appropriate container
  d) collected in a bin and burnt at night.

9 When working on **any** construction site, one essential item of PPE required would be:
  a) overalls
  b) earplugs
  c) respirator
  d) safety footwear.

10 A sign consists of a red circle with a red diagonal line. This type of sign is:

a) a warning
b) a prohibition
c) mandatory
d) advisory.

11 A sign consists of white symbols on a green background. This type of sign is:

a) a warning
b) a prohibition
c) mandatory
d) advisory.

12 Upon discovering asbestos on site, the action to take is:

a) continue working
b) cover with a damp cloth
c) stop work and inform duty holder
d) remove the asbestos at the end of work.

13 The safety wear shown is used for protecting:

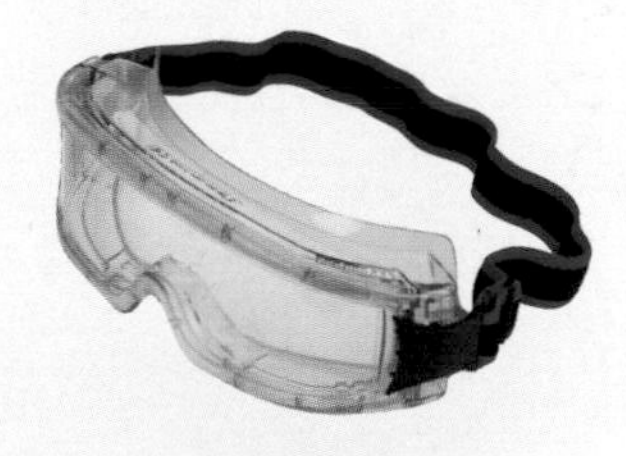

a) ears
b) nose
c) face
d) eyes.

14 This type of protective equipment is known as a:

a) baseball cap
b) hard hat
c) bump cap
d) soft hat.

15 The minimum shock current that may be fatal is considered to be:

a) 1 mA
b) 30 mA
c) 50 mA
d) 6 A.

16 Class I power tools must have a protective conductor connection to the:

a) plastic parts of the casing
b) metallic parts of the casing
c) neutral conductor connection
d) line conductor connection.

17 What standard must approved voltage indicators comply with?

a) Guidance Note 3.
b) GS58.
c) BS38.
d) GS38.

18 Upon discovering an unconscious colleague who has received an electric shock and is still touching an electric cable. The first action to take is:

a) remove the person from the supply
b) apply CPR
c) shout for help
d) dial 999 or 112.

19 The three parts of a fire triangle are:

a) fuel, oxygen, heat
b) fuel, nitrogen, carbon
c) oxygen, helium, carbon
d) oxygen, carbon, fuel.

20 A Class I 'C' fire (flammable gases) can be fought using:

a) powder, carbon dioxide
b) water, carbon dioxide
c) water, foam
d) powder, foam.

# UNIT 301
# Understand the fundamental principles and requirements of environmental technology systems

The UK government is committed to carbon dioxide ($CO_2$) reduction in order to address the issue of climate change. House building is an area that has been identified as a sector that can make a significant contribution to this goal of carbon dioxide reduction. It is estimated that 43% of $CO_2$ emissions in the UK are attributable to domestic properties. The government has introduced the Code for Sustainable Homes, which is the national standard for the sustainable design and construction of new homes. It aims to reduce $CO_2$ emissions and promote standards of sustainable design higher than the current minimum standards set out by the building regulations. The target is that all new homes will be carbon neutral by 2016. Because of this, any operative working within the construction industry is likely to come into contact with one or more forms of environmental technology systems. A knowledge of the working principles, the advantages and disadvantages, the requirements of the location and an overview of the planning and regulatory requirements for each environmental technology is therefore a prerequisite to working in the construction industry.

## LEARNING OUTCOMES

There are four learning outcomes to this unit. The learner will:

1. know the fundamental working principles of micro-renewable energy and water conservation technologies
2. know the fundamental requirements of building location/building features for the potential to install micro-renewable energy and water conservation systems
3. know the fundamental regulatory requirements relating to micro-renewable energy and water conservation technologies
4. know the typical advantages and disadvantages associated with micro- renewable energy and water conservation technologies.

This unit will be assessed by:

- an online multiple-choice test 2365-301.

# How this unit is organised

**SmartScreen Unit 201**

Additional resources to support this unit are available on SmartScreen.

This unit of the book is divided up according to technology, rather than consecutively by outcome.

This has been done to enable the learner to appreciate fully the principles of each technology and how these impact on the installation requirements and the regulatory requirements. The regulatory requirements (Outcome 3) are covered first, followed by a discussion of the advantages and disadvantages of each technology.

The environmental technologies that are discussed in this chapter are described below.

**Heat-producing**

- Solar thermal
- Ground source heat pump
- Air source heat pump
- Biomass

**Electricity-producing**

- Solar photovoltaic
- Micro-wind
- Micro-hydro

**Co-generation**

- Micro-combined heat and power (heat-led)

**Water conservation**

- Rainwater harvesting
- Greywater re-use

**Environmental technology systems covered**

OUTCOME 3

# Know the fundamental regulatory requirements relating to micro-renewable energy and water conservation technologies

Learning outcome 3 deals with planning requirements and building regulations for each technology. This section explains the terminology used, provides an insight into the workings of both planning and building regulations and explains the differences across the UK.

## PLANNING AND PERMITTED DEVELOPMENT

In general, under the Town and Country Planning Act 1990, before any building work that increases the size of a building is carried out, a planning application must be submitted to the local authority. A certain amount of building work is, however, allowed without the need for a planning application. This is known as *permitted development*. Permitted development usual comes with criteria that must be met. When building an extension, for example, it may be possible to do so under permitted development, if the extension is under a certain size, is a certain distance away from the boundary of the property and is not at the front of the property. If the extension does not meet these criteria, then a full application must be made.

The permitted development is intended to ease the burden on local authorities and to smooth the process for the builder or installer. Permitted development exists for renewable technologies and these are outlined within each technology section.

**Assessment criteria**

**3.1** Confirm what would typically be classified as 'permitted development' under Town and Country Planning Regulations for each technology

### ACTIVITY

Find out who your local authority and water company are. See if you can find any reference to environmental controls on their websites.

## BUILDING REGULATIONS

The Climate Change and Sustainable Energy Act 2006 brought micro-generation under the requirements of the building regulations.

Even if a planning application is not required, because the installation meets the criteria for permitted development, there is still a requirement to comply with the relevant building regulations.

Local Authority Building Control (LABC) is the body responsible for checking that building regulations have been met. The person carrying out the work is responsible for ensuring that approval is obtained.

Building regulations are statutory instruments that seek to ensure that the policies and requirements of the relevant legislation are complied with.

**Assessment criteria**

**3.2** Confirm which sections of the current building regulations/building standards apply for each technology

The building regulations themselves are rather brief and are currently divided into 14 sections, each of which is accompanied by an approved document. The approved documents are non-statutory and give guidelines on how to comply with the statutory requirements.

The 14 parts of the Building Regulations in England and Wales are listed below.

| Part | Title |
|---|---|
| A | Structure |
| B | Fire safety |
| C | Site preparation and resistance to contaminants and moisture |
| D | Toxic substances |
| E | Resistance to the passage of sound |
| F | Ventilation |
| G | Sanitation, hot-water safety and water efficiency |
| H | Drainage and waste disposal |
| J | Combustion appliances and fuel-storage systems |
| K | Protection from falling, collision and impact |
| L | Conservation of fuel and power |
| M | Access to and use of building |
| N | Glazing – safety in relation to impact, opening and cleaning |
| P | Electrical safety – dwellings |

There is a 15th Approved Document, which relates to Regulation 7 of the Building Regulations, entitled Approved Document 7 Materials and Workmanship.

Compliance with building regulations is required when installing renewable technologies but not all will be applicable and different technologies will have to comply with different building regulations. Building regulations applicable to each technology are indicated in each section.

**ASSESSMENT GUIDANCE**

The relationship between the Building Regulations and the approved documents is similar to that between the Electricity at Work Regulations and BS 7671.

# DIFFERENCES IN BUILDING REGULATIONS ACROSS THE UK

It should be noted, that due to devolution of government in the UK, each country's government takes responsibility for building regulations, so there are differences between the individual countries.

## England and Wales

England and Wales currently follow the same legal structure when it comes to building regulations.

| | |
|---|---|
| Primary legislation: | Building Act 1984 |
| Secondary legislation: | Building Regulations 2010 |
| Guidance: | 15 Approved Codes of Practice |

Although Wales follows the same model as England, the Welsh Government is now responsible for the majority of functions under the Building Act, including making Building Regulations in Wales. The functions that have remained with the UK Government are as set out in The Welsh Ministers (Transfer of Functions) (No. 2) Order 2009 (S.I. 2009/3019).

## Scotland

| | |
|---|---|
| Primary legislation: | Building (Scotland) Act 1984 |
| Secondary legislation: | Building (Scotland) Regulations 2004 – Amended 2009 |
| Non-statutory guidance: | 2 Technical Guide Books – Dwellings, Non-dwellings |

## Northern Ireland

| | |
|---|---|
| Primary legislation: | Building Regulations (Northern Ireland) 1979 Order (Amended 2009) |
| Secondary legislation: | The Building Regulations (Northern Ireland) 2012 |
| Non-statutory guidance: | 15 Technical Booklets |

The technical booklets have similar content to the approved documents used in England and Wales but the order of the documents is different. A comparison is included overleaf, along with any differences in title.

| England & Wales | | Northern Ireland | |
|---|---|---|---|
| A | Structure | D | |
| B | Fire safety | E | |
| C | Site preparation and resistance to contaminants and moisture | C | |
| D | Toxic substances | No comparable document | |
| E | Resistance to the passage of sound | G | |
| F | Ventilation | K | |
| G | Sanitation, hot-water safety and water efficiency | P | Sanitary appliances, unvented hot-water storage systems and reducing the risk of scalding |
| H | Drainage and waste disposal | J<br>N | Solid waste in buildings<br>Drainage |
| J | Combustion appliances and fuel storage systems | L | |
| K | Protection from falling, collision and impact | H | Stairs ramps guarding and protection from impact |
| L | Conservation of fuel and power | F1<br>F2 | Dwellings<br>Non-dwellings |
| M | Access to and use of building | R | |
| N | Glazing – safety in relation to impact, opening and cleaning | V | |
| P | Electrical safety – dwellings | No comparable document | |
| 7 | Materials and workmanship | B | |

Within this unit, reference to parts within the building regulations is applicable to the part designations used in the Building Regulations of England and Wales.

# OUTCOMES 1 TO 4

# Heat-producing micro-renewable energy technologies

## SOLAR THERMAL (HOT-WATER) SYSTEMS

A solar thermal hot-water system uses solar radiation to heat water, directly or indirectly.

### Working principles

The key components of a solar thermal hot water system are:

1 solar collector
2 differential temperature controller
3 circulating pump
4 hot-water storage cylinder
5 auxiliary heat source.

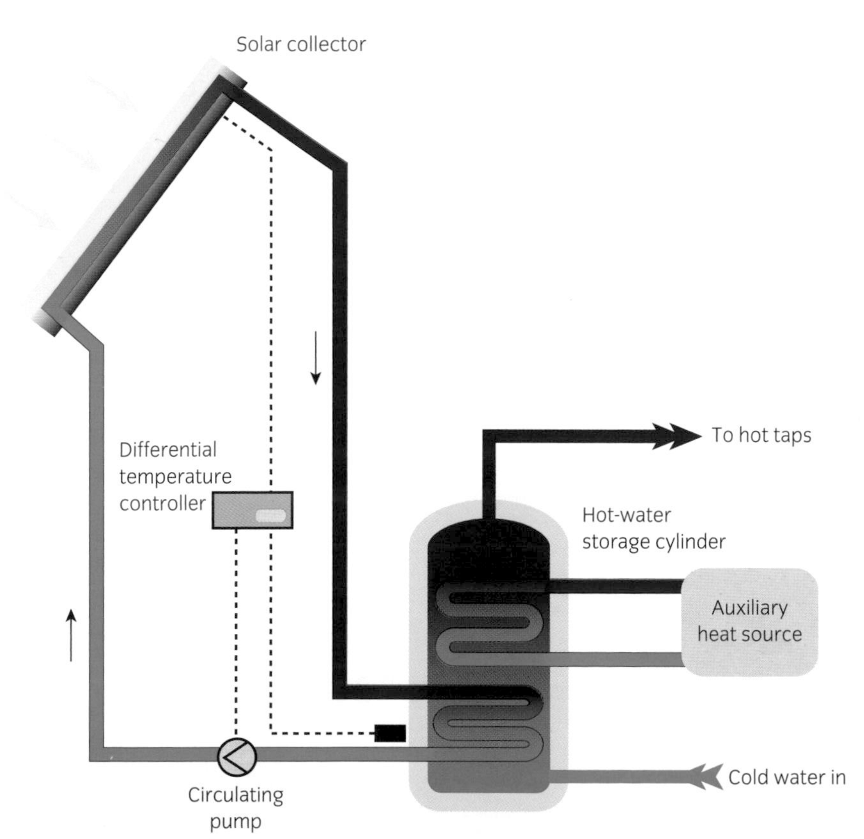


**Solar thermal system components**

**Assessment criteria**

1.1 Identify the fundamental working principles for solar thermal (hot water) systems

**ASSESSMENT GUIDANCE**

A lot of older systems will have a gravity hot-water system, whereby the primary water circulates through the heating coil due to hot water being lighter than cold water. The system shown here uses a pump to circulate the primary water much more quickly.

**ACTIVITY**

Why is a circulating pump required with a solar thermal system?

## 1 Solar thermal collector

A solar thermal collector is designed to collect heat by absorbing heat radiation from the Sun. The heat energy from the Sun heats the heat-transfer fluid contained in the system.

Two types of solar collector are used.

**Flat-plate collectors** are less efficient but cheaper than evacuated tube collectors.

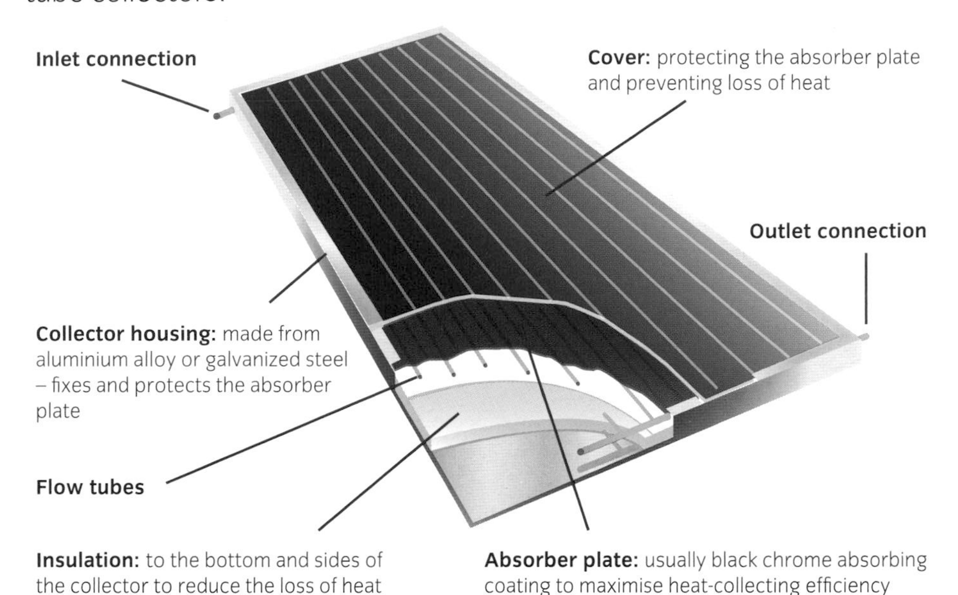


**Cutaway diagram of a flat-plate collector**

With this type of collector, the heat-transfer fluid circulates through the collectors and is directly heated by the Sun. The collectors need to be well insulated to avoid heat loss.

**Evacuated-tube collectors** are more efficient but more expensive than flat-plate collectors.

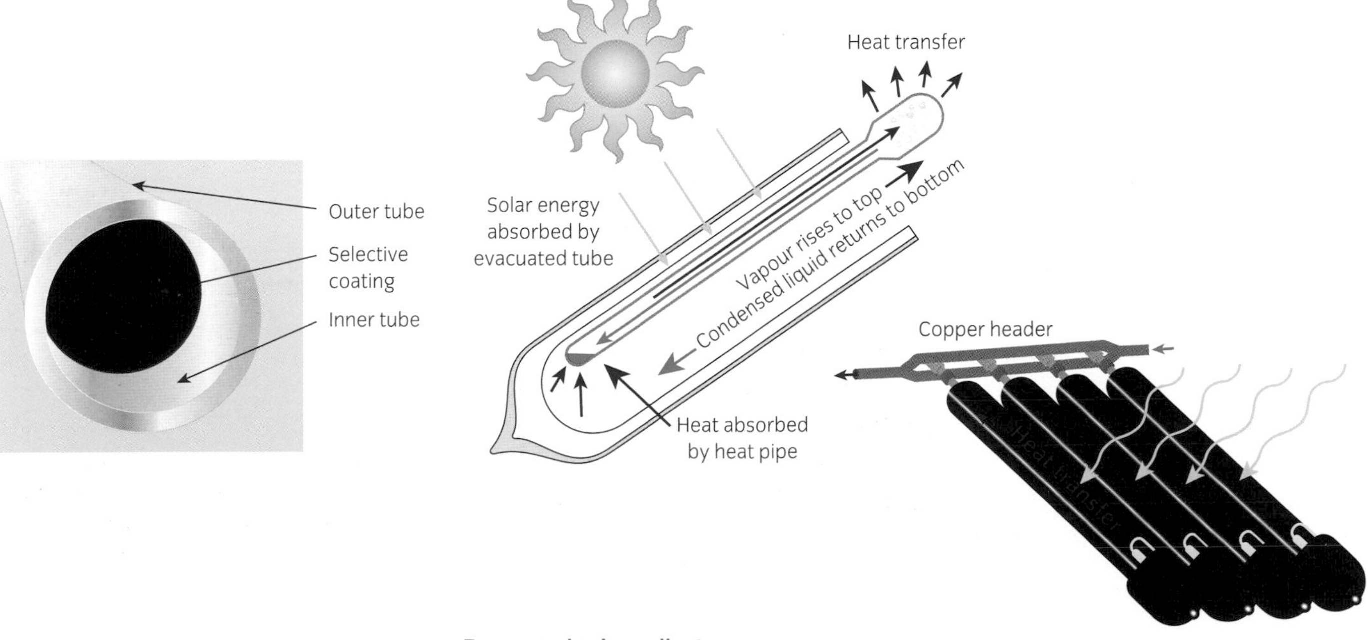


**Evacuated-tube collector**

An evacuated-tube collector consists of a specially coated, pressure-resistant, double-walled glass tube. The air is evacuated from the tube to aid the transfer of heat from the Sun to a heat pipe housed within the glass tube. The heat pipe contains a temperature-sensitive medium, such as methanol, that, when heated, vaporises. The warmed gas rises within the tube. A solar collector will contain a number of evacuated tubes in contact with a copper header tube that is part of the solar heating circuit. The heat tube is in contact with the header tube. The heat from the methanol vapour in the heat tubes is transferred by conduction to the heat-transfer fluid flowing through the solar heating circuit. This process cools the methanol vapour, which condenses and runs back down to the bottom of the heat tubes, ready for the process to start again. The collector must be mounted at a suitable angle to allow the vapour to rise and the condensed liquid to flow back down the heat pipes.

## 2 Differential temperature controller

The differential temperature controller (DTC) has sensors connected to the solar collector (high level) and the hot-water storage system (low level). It monitors the temperatures at the two points. The DTC turns the circulating pump on when there is enough solar energy available and there is a demand for water to be heated. Once the stored water reaches the required temperature, the DTC shuts off the circulating pump.

## 3 Circulating pump

The circulating pump is controlled by the DTC and circulates the system's heat-transfer fluid around the solar hot-water circuit. The circuit is a closed loop between the solar collector and the hot-water storage tank. The heat-transfer fluid is normally water-based but, depending on the system type, usually also contains glycol so that at night, or in periods of low temperatures, it does not freeze in the collectors.

## 4 Hot-water storage cylinder

The hot-water storage cylinder enables the transfer of heat from the solar collector circuit to the stored water. Several different types of cylinder or cylinder arrangement are possible.

### Twin-coil cylinder

With this type of cylinder the lower coil is the solar heating circuit and the upper coil is the auxiliary heating circuit. Cold water enters at the base of the cylinder and is heated by the solar heating coil. If the solar heating circuit cannot meet the required demand, then the boiler will provide heat through the upper coil. Hot water is drawn off, by the taps, from the top of the cylinder.

**ASSESSMENT GUIDANCE**

When the hot tap is opened the cold-water pressure at the bottom of the tank forces the hot water out at the top.

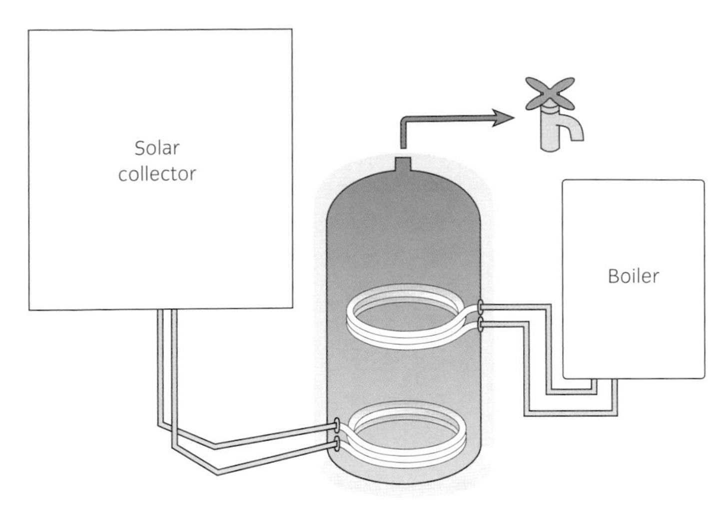


Twin-coil cylinder

**Alternatives:** One alternative arrangement is to use one cylinder as a solar preheat cylinder, the output of which feeds a hot-water cylinder. The auxiliary heating circuit is connected to the second cylinder.

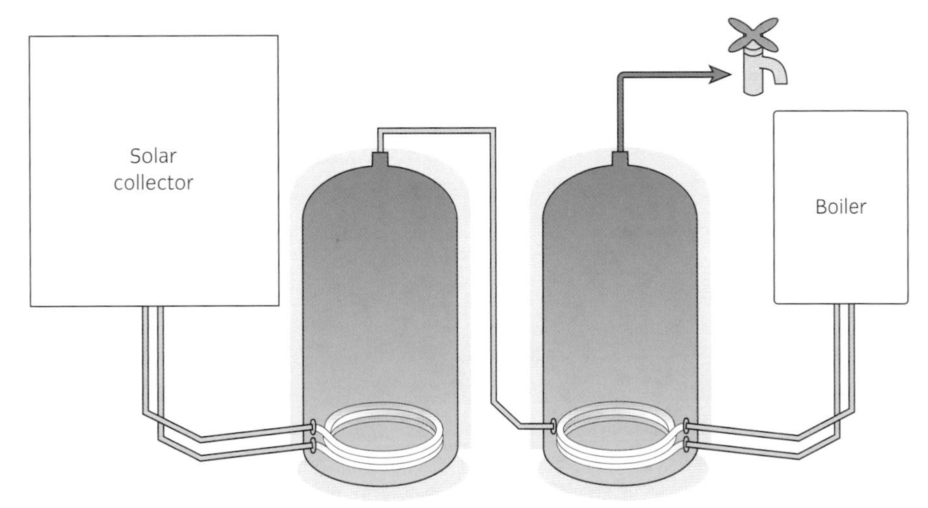


Using two separate cylinders

The two arrangements that have been described are indirect systems, with the solar heating circuit forming a closed loop.

**Direct system:** An alternative to the indirect system is the direct system, in which the domestic hot water that is stored in the cylinder is directly circulated through the solar collector and is the same water that is drawn off at the taps. Due to this fact, antifreeze (glycol) cannot be used in the system, so it is important to use freeze-tolerant collectors.

## 5 Auxiliary heat source

In the UK there will be times when there is insufficient solar energy available to provide adequate hot water. On these occasions an auxiliary heat source will be required. Where the premises have space-heating systems installed, the auxiliary heat source is usually this boiler. Where no suitable boiler exists, the auxiliary heat source will be an electric immersion heater.

## Location and building requirements

Assessment criteria

2.1 Clarify the fundamental requirements for the potential to install a solar water heating system to exist

When deciding whether or not a solar thermal hot-water system is suitable for particular premises, the following factors should be considered.

### The orientation of the solar collectors

The optimum direction for the solar collectors to face is due south. However, as the Sun rises in the east and sets in the west, any location with a roof facing east, south or west is suitable for mounting a solar thermal system, although the efficiency of the system will be reduced for any system not facing due south.

**ACTIVITY**

Which orientation is best for solar thermal systems in the UK?

### The tilt of the solar collectors

During the year, the maximum elevation or height of the Sun, relative to the horizon, changes. It is lowest in December and highest in June. Ideally, solar collectors should always be perpendicular to the path of the Sun's rays. As it is generally not practical to change the tilt angle of a solar collector, a compromise angle has to be used. In the UK, the angle is 35°; however, the collectors will work, but less efficiently, from vertical through to horizontal.

### Shading of the solar collectors

Any structure, tree, chimney, aerial or other object that stands between the collector and the Sun will block the Sun's energy. The Sun shines for a limited time and any reduction in the amount of heat energy reaching the collector will reduce its ability to provide hot water to meet the demand.

| Shading | % of sky blocked by obstacles | Reduction in output |
|---|---|---|
| Heavy | > 80% | 50% |
| Significant | > 60–80% | 35% |
| Modest | > 20–60% | 20% |
| None or very little | ⩽ 20% | No reduction |

### The suitability of the structure for mounting the solar collector

The structure has to be assessed as to its suitability for the chosen mounting system. Consideration needs to be given to the strength and condition of the structure and the suitability of fixings. The effect of wind must also be taken into account. The force exerted by the wind on the collectors, an upward force known as 'wind uplift', affects both the solar collector fixings and the fixings holding the roof members to the building structure.

**ACTIVITY**

Why is it important to consider the uplift forces on a roof-mounted system, as well as the weight?

How much hot water is needed

In the case of roof-mounted systems on flats and other shared properties, the ownership of the structure on which the proposed system is to be installed must be considered.

The space needed to mount the collectors is dependent on demand for hot water. The number of people occupying premises determines the demand for hot water and, therefore, the number of collectors required and the space need to mount them.

## Compatibility with the existing hot-water system

Solar thermal systems provide stored hot water rather than instantaneous hot water.

- Premises using under/over-sink water heaters and electric showers will not be suitable for the installation of a solar thermal hot-water system.
- Premises using a combination boiler to provide hot water will not be suitable for the installation of a solar thermal hot-water system unless substantial changes are made to the system.

**Assessment criteria**

**3.1** Confirm what would be typically classified as 'permitted development' under town and country planning regulations in relation to the deployment of a solar thermal system

## Planning permission

Permitted development applies where a solar thermal system is installed:

- on a dwelling house or block of flats
- on a building within the grounds of a dwelling house or block of flats
- as a stand-alone system in the grounds of a dwelling house or block of flats.

However there are criteria to be met in each case.

For building mounted systems:

- the solar thermal system cannot protrude more than 200 mm from the wall or the roof slope
- the solar thermal system cannot protrude past the highest point of the roof (the ridgeline), excluding the chimney.

The criteria that must be met for stand-alone systems are that:

- only one stand-alone system is allowed in the grounds
- the array cannot exceed 4 m in height
- the array cannot be installed within 5 m of the boundary of the grounds
- the array cannot exceed 9 $m^2$ in area
- no dimension of the array can exceed 3 m in length.

For both stand-alone and building mounted systems:

- the system cannot be installed in the grounds or on a building within the grounds of a listed building or a scheduled monument
- if the dwelling is in a conservation area or a World Heritage Site, then the array cannot be closer to a highway than the house or block of flats.

In every other case, planning permission will be required.

## Compliance with building regulations

**Assessment criteria**

**3.2** Confirm which sections of the current building regulations/building standards apply in relation to the deployment of a solar thermal system

The following building regulations will apply to solar thermal hot-water systems.

| Part | Title | Relevance |
|---|---|---|
| A | Structure | Where solar collectors and other components can put extra load on the structure, particularly the roof structure, not only the additional downwards load but also the uplift caused by the wind must be considered. |
| B | Fire safety | Where holes for pipes are made, this may reduce the fire resistance of the building fabric. |
| C | Resistance to contaminants and moisture | Where holes for pipes and fixings for collectors are made, this may reduce the moisture resistance of the building and allow ingress of water. |
| G | Sanitation, hot-water safety and water efficiency | Hot-water safety and water efficiency |
| L | Conservation of fuel and power | Energy efficiency of the system and the building as a whole |
| P | Electrical safety | The installation of electrical controls and components |

## Other regulatory requirements to consider

- BS 7671 Requirements for Electrical Installations
- Approved document Part G3: Unvented hot-water storage systems
- Water Regulations (WRAS)

Assessment criteria

4.1 Identify typical advantages associated with solar thermal systems

## Advantages

- It reduces $CO_2$ emissions.
- It reduces energy costs.
- It is low maintenance.
- It improves the energy rating of the building.

Assessment criteria

4.2 Identify typical disadvantages associated with solar thermal systems

## Disadvantages

- It may not be compatible with the existing hot-water system.
- It may not meet demand for hot water in the winter.
- There are high initial installation costs.
- It requires a linked auxiliary heat source.

Assessment criteria

1.1 Identify the fundamental working principles for heat pumps

# HEAT PUMPS

## Working principles

**A heat pump moves heat from one location to another, just as a water pump moves water from one location to another.**

A water pump moves water from a lower level to a higher level, through the application of energy. Pumping the handle draws water up from a lower level to a higher level through the application of kinetic energy.

As the name suggests, a heat pump moves heat energy from one location to another by the application of energy. In most cases, the applied energy is electrical energy.

Heat energy from the Sun exists in the air that surrounds us and in the ground beneath our feet. At *absolute zero* or 0 K (kelvin), there is no heat in a system. This temperature is equivalent to –273 °C so, even with outside temperatures of –10 °C, there is a vast amount of free heat energy available.

–273 –10 0 20

**Heat energy exists down to absolute zero (0 K ≈ –273 °C)**

Using a relatively small amount of energy, that stored heat energy in the air or in the ground can be extracted and put to use in heating our living accommodation.

Heat pumps extract heat from outside and transfer it inside, in much the same way that a refrigerator extracts heat from the inside of the refrigerator and releases it at the back of the refrigerator via the heat-exchange fins.

A basic rule of heat transfer is that heat moves from warmer spaces to colder spaces.

### ASSESSMENT GUIDANCE

Look at your refrigerator or freezer at home. The inside is cold because the heat energy has been removed from it. The tubes at the back are hot.

A heat pump contains a refrigerant. The external air or ground is the medium or heat source that gives up its heat energy. When the refrigerant is passed through this heat source the refrigerant is cooler than its surroundings and so absorbs heat. The compressor on the heat pump then compresses the refrigerant, causing the gas to heat up. When the refrigerant is passed to the interior, the refrigerant is now hotter than its surroundings and gives up its heat to the cooler surroundings. The refrigerant is then allowed to expand, where it once again turns into a liquid. As the refrigerant expands it cools and the cycle starts all over again.

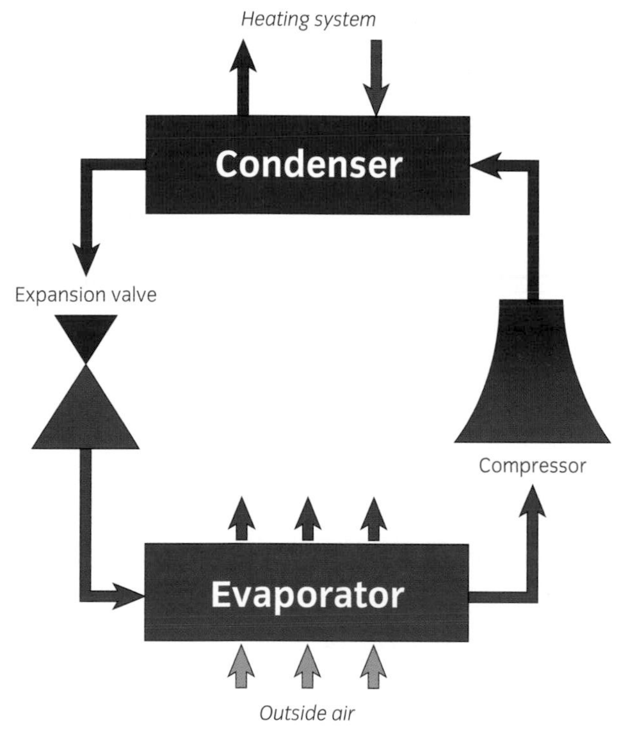


**The refrigeration process**

The only energy needed to drive the system is what is required by the compressor. The greater the difference in temperature between the refrigerant and the heat-source medium from where heat is being extracted, the greater the efficiency of the heat pump. If the heat-source medium is very cold then the refrigerant will need to be colder, to be able to absorb heat, so the harder the compressor must work and the more energy is needed to accomplish this.

Two main types of heat pump are in common use are:

- ground source heat pumps (GSHP)
- air source heat pumps (ASHP).

Heat pumps extract heat energy from the air or the ground, but the energy extracted is replaced by the action of the Sun.

It is not uncommon for heat pumps to have efficiencies in the order of 300%; for an electrical input of 3 kW, a heat output of 9 kW is achievable. If we compare this to other heat appliances we can see where the savings are made.

Electricity input 1 kW

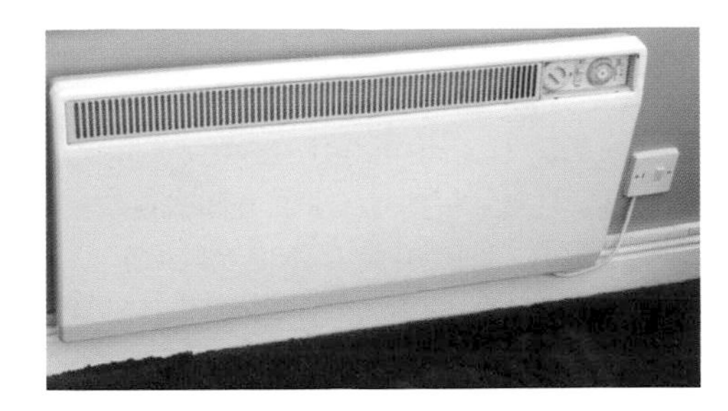

Heat output 1 kW

100% efficiency

The efficiency of an electric panel heater

Gas input 1 kW

Heat output 0.95 kW

95% efficiency

The efficiency of an A-rated condensing gas boiler

Electricity input 1 kW

+2 kW free heat extracted from air

Heat output 3 kW

equates to a 300% efficiency

The efficiency of an air source heat pump (centre image courtesy of Bosch Thermotechnology Limited)

## ACTIVITY

Visit the website of a gas boiler supplier and compare the efficiency of a conventional boiler to that of a condensing gas boiler.

The efficiency of a heat pump is measured in terms of the *coefficient of performance* (COP), which is the ratio between the heat delivered and the power input of the compressor.

$$\text{COP} = \frac{\text{heat delivered}}{\text{compressor power}}$$

The higher the COP value, the greater the efficiency. Higher COP values are achieved in mild weather than in cold weather because, in cold weather, the compressor has to work harder to extract heat.

## Storing excess heat produced

Heat pumps are not able to provide instant heat and so therefore work best when run continuously. Stop–start operations will shorten the lifespan of a heat pump. A buffer tank, simply a large water-storage vessel, is incorporated into the circuit so that, when heat is not required within the premises, the heat pump can 'dump' heat to it and thus keep running. When there is a need for heat, this can be drawn from the buffer tank. A buffer tank can be used with both ground source and air source heat pumps.

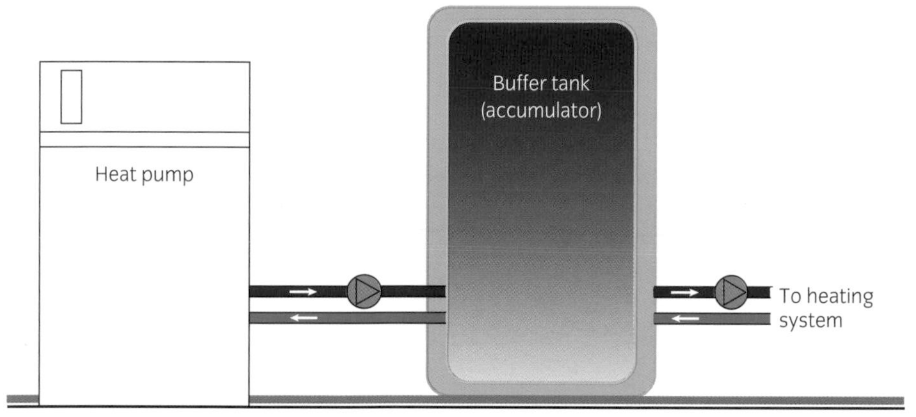


**Storing heat in a buffer tank**

# GROUND SOURCE HEAT PUMPS

**Assessment criteria**

1.1 Identify the fundamental working principles for ground source heat pumps

## Working principles

A ground source heat pump (GSHP) extracts low-temperature free heat from the ground, upgrades it to a higher temperature and releases it, where required, for space heating and water heating.

**ASSESSMENT GUIDANCE**

Although the ground or air may 'feel' cold, there is still energy available to be collected.

The key components of a GSHP are:

- heat-collection loops and a pump
- heat pump
- heating system.

The collection of heat from the ground is accomplished by means of pipes containing a mixture of water and antifreeze, which are buried in the ground. This type of system is known as a 'closed-loop' system. Three methods of burying the pipes are used, with each having advantages and disadvantages.

## Horizontal loops

Piping is installed in horizontal trenches that are generally 1.5–2 m deep. Horizontal loops require more piping than vertical loops – around 200 m of piping for the average house.

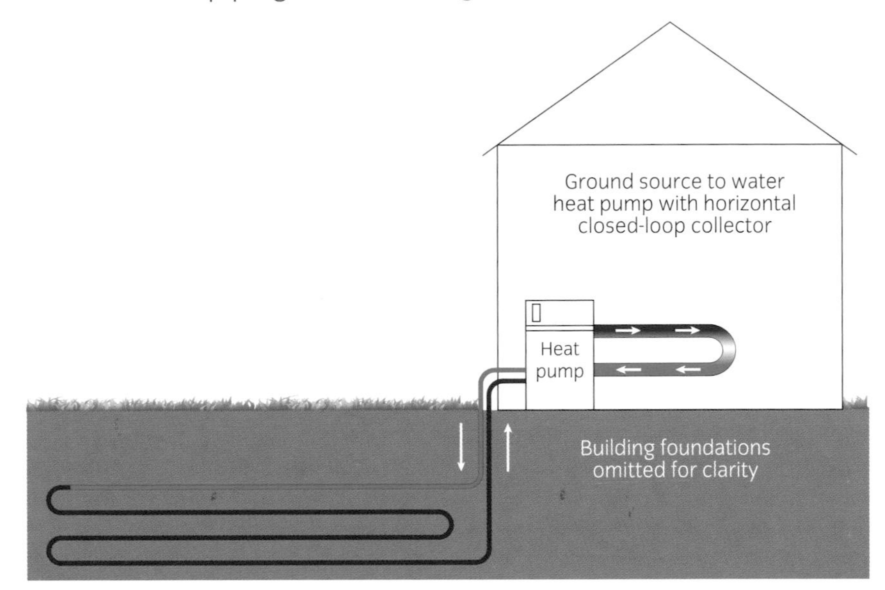


**Horizontal ground loops**

## Vertical loops

Most commercial installations use vertical loops. Holes are bored to a depth of 15–60 m, depending on soil conditions, and spaced approximately 5 m apart. Pipe is then inserted into these bore holes. The advantage of this system is that less land is needed.

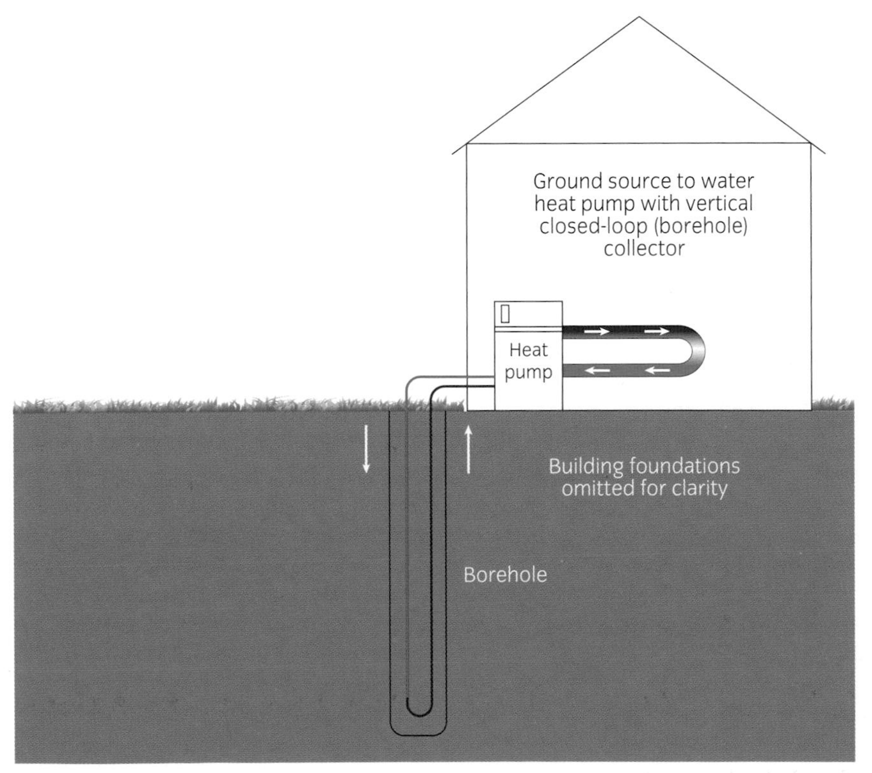


**Vertical ground loops**

### ACTIVITY

What ground structure/material would make vertical loops impracticable or very expensive?

## Slinkies

Slinky coils are flattened, overlapping coils that are spread out and buried, either vertically or horizontally. They are able to concentrate the area of heat transfer into a small area of land. This reduces the length of trench needing to be excavated and therefore the amount of land required. Slinkies installed in a 10 m long trench will yield around 1 kW of heating load.

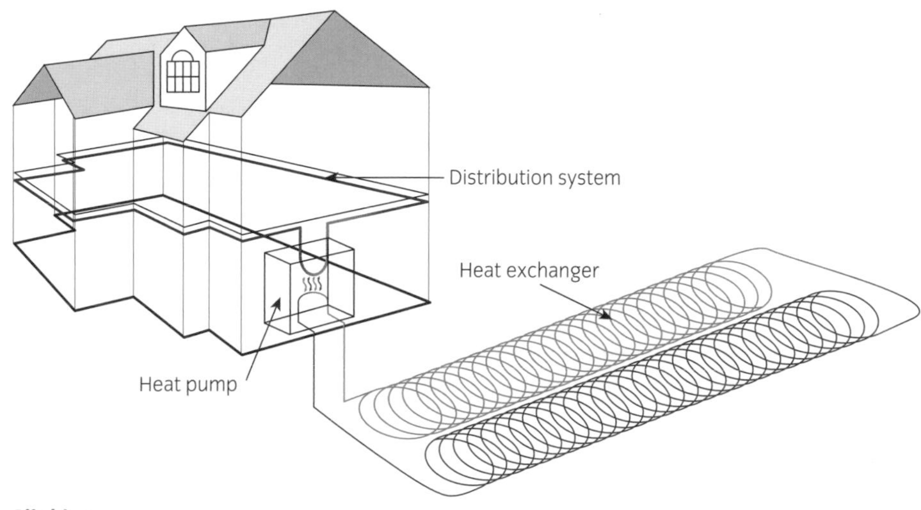


Slinkies

Slinkies being installed in the ground

**ASSESSMENT GUIDANCE**

Looking at this photograph, it is obvious that unless you want your garden to be given the 'Time Team' treatment it is best to have the ground source heat pump pipes installed during the original building stage.

The water–antifreeze mix is circulated around these ground pipes by means of a pump. The low-grade heat from the ground is passed over a heat exchanger, which transfers the heat from the ground to the refrigerant gas. The refrigerant gas is compressed and passed across a second heat exchanger, where the heat is transferred to a pumped heating loop that feeds either radiators or under-floor heating.

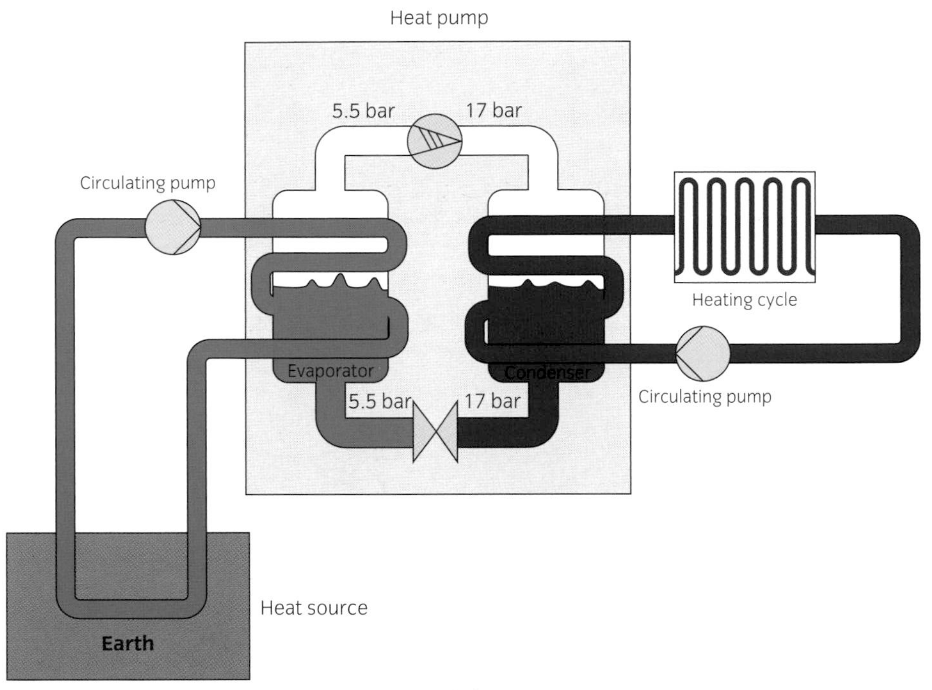


**Ground source heat pump operating principle**

Final heat output from the GSHP is at a lower temperature than would be obtained from a gas boiler. GSHP heat output is at 40 °C, compared to a gas boiler at 60–80 °C. For this reason, under-floor heating, which requires temperatures of 30–35 °C, is the most suitable form of heating arrangement to use with a GSHP. Low-temperature or oversized radiators could also be used. A GSHP system in itself is unable to heat hot water directly to a suitable temperature. Hot water needs to be stored at a temperature of 60 °C. An ancillary heating device will be required to reach the required temperatures.

## ACTIVITY

What is another name for a buffer tank?

A GSHP is unable to provide instant heat and, for maximum efficiency, should run all the time. In some cases it is beneficial to fit a buffer tank to the output so that any excess heat is stored, ready to be used when required.

By reversing the refrigeration process, a GSHP can also be used to provide cooling in summer.

## Location and building requirements

For a GSHP system to work effectively, and due to the fact that the output temperature is low, the building has to be well insulated.

A suitable amount of land has to be available for trenches or, alternatively, land that is suitable for bore holes. In either case, access for machinery will be required.

Assessment criteria

**2.3** Clarify the fundamental requirements for the potential to install a ground source heat pump system to exist

## Planning permission

The installation of a ground source heat pump is usually considered to be permitted development and will not require a planning application to be made.

If the building is a listed building or in a conservation area, the local area planning authority will be need to be consulted.

Assessment criteria

**3.1** Confirm what would be typically classified as 'permitted development' under town and country planning regulations in relation to the deployment of ground source heat pumps

## Compliance with building regulations

The following building regulations will apply.

| Part | Title | Relevance |
|---|---|---|
| A | Structural safety | Where heat pumps and other components put additional load on the building structure or where openings are formed to pass from outside in inside |
| B | Fire safety | Where holes for pipes may reduce the fire-resistant integrity of the building structure |
| C | Resistance to contaminants and moisture | Where holes for pipes may reduce the moisture-resistant integrity of the building structure |
| E | Resistance to sound | Where holes for pipes may reduce the soundproof integrity of the building structure |
| G | Sanitation, hot-water safety and water efficiency | Hot-water safety and water efficiency<br><br>Unvented hot-water system |
| L | Conservation of fuel and power | Energy efficiency of the system and the building as a whole |
| P | Electrical safety | The installation of electrical controls and components as well as the supply to the heat pump |

Assessment criteria

**3.2** Confirm which sections of the current building regulations/building standards apply in relation to the deployment of ground source heat pumps

**ASSESSMENT GUIDANCE**

It is not good practice to install a heat pump via a plug and socket. Any load over 3 kW (13 A) will need its own circuit wired back to the consumer unit by a competent electrician.

## Other regulatory requirements to consider

- BS 7671 Requirements for Electrical Installations
- F (fluorinated) gas requirements if working on refrigeration pipework

**Assessment criteria**

4.1 Identify typical advantages associated with ground source heat pumps

## Advantages

- There is high efficiency.
- There is a reduction in energy bills – cheaper to run than electric, gas or oil boilers.
- There is a reduction in $CO_2$ emissions.
- There are no $CO_2$ emissions on site.
- They are safe, as no combustion takes place and no emission of potentially dangerous gases.
- They are low maintenance, compared with combustion devices.
- They have a long lifespan.
- There is no requirement for fuel storage, so less installation space is required.
- They can be used to provide cooling in summer.
- They are more efficient than air source heat pumps.

**Assessment criteria**

4.2 Identify typical disadvantages associated with ground source heat pumps

## Disadvantages

- The initial costs are high.
- They require large ground area or boreholes.
- The design and installation are complex tasks.
- They are unlikely to work efficiently with an existing heating system.
- They use refrigerants, which could be harmful to the environment.
- They are more expensive to install than air source heat pumps.

**Assessment criteria**

1.1 Identify the fundamental working principles for air source heat pumps

# AIR SOURCE HEAT PUMPS

## Working principles

An air source heat pump (ASHP) extracts free heat from low-temperature air and releases it where required, for space heating and water heating.

The key components of an ASHP are:

- a heat pump containing a heat exchanger, a compressor and an expansion valve
- a heating system.

Air source heat pump

An ASHP works in a similar way to a refrigerator, but the cooled area becomes the outside world and the area where the heat is released is the inside of a building. The steps of the ASHP process are as follows.

- The pipes of the pump system contain refrigerant that can be a liquid or a gas, depending on the stage of the cycle. The refrigerant, as a gas, flows through a heat exchanger (evaporator), where low-temperature air from outside is drawn across the heat exchanger by means of the unit's internal fan. The heat from the air warms the refrigerant. Any liquid refrigerant boils to gas.
- The warmed refrigerant vapour then flows to a compressor, where it is compressed, causing its temperature to rise further.
- Following this pressurisation stage, the refrigerant gas passes through another heat exchanger (condenser), where it loses heat to the heating-system water, because it is hotter than the system water. At this stage, some of the refrigerant has condensed to a liquid. The heating system carries heat away to heat the building.
- The cooled refrigerant passes through an expansion valve, where its pressure drops suddenly and its temperature falls. The refrigerant flows once more to the evaporator heat exchanger, continuing the cycle.

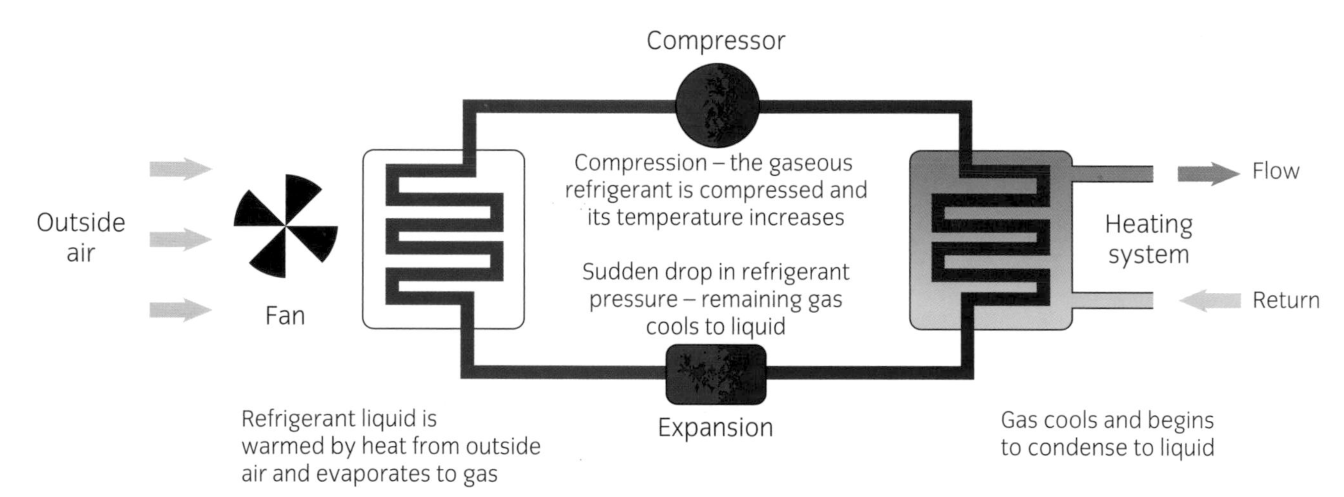


**Air source heat pump operating principle**

The two types of ASHP in common use are:

- air-to-water – the type described above, which can be used to provide both space heating and water heating
- air-to-air – this type is not suitable for providing water heating.

The output temperature of an ASHP will be lower than that of a gas-fired boiler. Ideally, the ASHP should be used in conjunction with an under-floor heating system. Alternatively, it could be used with low-temperature radiators.

**Assessment criteria**

2.4 Clarify the fundamental requirements for the potential to install an air source heat pump system to exist

**ASSESSMENT GUIDANCE**

Older houses in the UK are generally not well insulated. Those with solid walls obviously cannot have cavity wall insulation, but insulation can still be fitted. Loft insulation is the most effective and there are government schemes and grants available for the supply and installation of the insulation material.

**ACTIVITY**

Name four different methods used to insulate a dwelling.

**Assessment criteria**

3.1 Confirm what would be typically classified as 'permitted development' under town and country planning regulations in relation to the deployment of air source heat pumps

## Location and building requirements

When deciding whether or not an ASHP is suitable for the premises, the following should be considered.

- The premises must be well insulated.
- There must be space to fit the unit on the ground outside the building, or mounted on a wall. There will also need to be clear space around the unit to allow an adequate airflow.
- The ideal heating system to couple to an ASHP is either under-floor heating or warm-air heating.
- An ASHP will pay for itself in a shorter period of time if it replaces an electric, coal or oil heating system than if it is replacing a gas-fired boiler.

Air sourced heat pumps are an ideal solution for new-build properties, where high levels of insulation and under-floor heating are to be installed.

## Planning permission

Permitted development applies where an air source heat pump is installed:

- on a dwelling house or block of flats
- on a building within the grounds of a dwelling house or block of flats
- in the grounds of a dwelling house or block of flats.

There are, however, criteria to be met, mainly due to noise generation by the ASHP.

- The air source heat pump must comply with the MCS Planning Standards or equivalent.
- Only one ASHP may be installed on the building or within the grounds of the building.
- A wind turbine must not be installed on the building or within the grounds of the building.
- The volume of the outdoor unit's compressor must not exceed $0.6\,m^3$.
- It cannot be installed within 1 m of the boundary.
- It cannot be installed on a pitched roof.
- If it is installed on a flat roof, it must not be within 1 m of the roof edge.
- If installed on a wall that fronts a highway, it cannot be mounted above the level of the ground storey.
- It cannot be installed on a site designated as a monument.

- It cannot be installed on a building that is a listed building, or in its grounds.
- It cannot be installed on a roof or a wall that fronts a highway, or within a conservation area or World Heritage Site.
- If the dwelling is in a conservation area or a World Heritage Site, then the ASHP cannot be closer to a highway than the house or block of flats.

**KEY POINT**

An ASHP creates noise and therefore there are restrictions on the number that can be installed on a site.

## Compliance with building regulations

**Assessment criteria**

**3.2** Confirm which sections of the current building regulations/building standards apply in relation to the deployment of air source heat pumps

The following building regulations will apply.

| Part | Title | Relevance |
|---|---|---|
| A | Structural safety | Where heat pumps and other components put additional load on the building structure, for instance, where the heat pump is installed on the roof or on a wall |
| B | Fire safety | Where holes for pipes may reduce the fire-resistant integrity of the building structure |
| C | Resistance from contaminants and moisture | Where holes for pipes may reduce the moisture-resistant integrity of the building structure |
| E | Resistance to sound | Where holes for pipes may reduce the soundproof integrity of the building structure |
| G | Sanitation, hot-water safety and water efficiency | Hot-water safety and water efficiency |
| L | Conservation of fuel and power | Energy efficiency of the system and the building as a whole |
| P | Electrical safety | The installation of electrical controls and components as well as the supply to the heat pump |

## Other regulatory requirements to consider

- BS 7671 Requirements for Electrical Installations.
- F (fluorinated) Gas Regulations if working on refrigeration pipework.

**Assessment criteria**

4.1 Identify typical advantages associated with air source heat pumps

## Advantages

- There is high efficiency.
- There is a reduction in energy bills – they are cheaper to run than electric, gas or oil boilers.
- There is a reduction in $CO_2$ emissions.
- There are no $CO_2$ emissions on site.
- They are safe, as no combustion takes place and there is no emission of potentially dangerous gases.
- They are low maintenance, compared with combustion devices.
- There is no requirement for fuel storage, so less installation space is required.
- They can be used to provide cooling in summer.
- They are cheaper and easier to install than GSHPs.

**ACTIVITY**

Why is it important to reduce $CO_2$ emissions?

**Assessment criteria**

4.2 Identify typical disadvantages associated with air source heat pumps

## Disadvantages

- They are unlikely to work efficiently with an existing heating system.
- They are not as efficient as GSHPs.
- The initial cost is high.
- They are less efficient in winter.
- There is noise from fans.
- They have to incorporate a defrost cycle to stop the heat exchanger freezing in winter.

**Assessment criteria**

1.1 Identify the fundamental working principles for biomass

# BIOMASS

What is biomass? Biomass is biological material from living or recently living organisms. Biomass fuels are usually derived from plant-based material but could be derived from animal material.

The major difference between biomass and fossil fuels, both of which are derived from the same source, is time. Fossil fuels, such as gas, oil and coal, have taken millions of years to form. Demand for these fuels is outstripping supply and replenishment. Biomass is derived from recently living organisms. As long as these organisms are replaced by replanting, and demand does not outstrip replacement time, the whole process is sustainable. Biomass is therefore rightly regarded as a renewable energy technology.

Both fossil fuels and biomass fuels are burnt to produce heat and both produce carbon dioxide. This is a greenhouse gas that has been linked to global warming. During their lives, plants and trees absorb carbon dioxide from the atmosphere, to enable growth to take place. When these plants are burnt, the carbon dioxide is released once again into the atmosphere.

So how does biomass have a carbon advantage over fossil fuels? The answer again is time. Fossil fuels absorbed carbon dioxide from the atmosphere millions of years ago and have trapped that carbon dioxide ever since. When fossil fuels are burnt, they release the carbon dioxide from all those millions of years ago and so add to the present-day atmospheric carbon dioxide level.

Biomass absorbs carbon dioxide when it grows, reducing current atmospheric carbon dioxide levels. When biomass is turned into fuel and burnt, it releases the carbon dioxide back into the atmosphere. The net result is that there is no overall increase in the amount of carbon dioxide in the atmosphere.

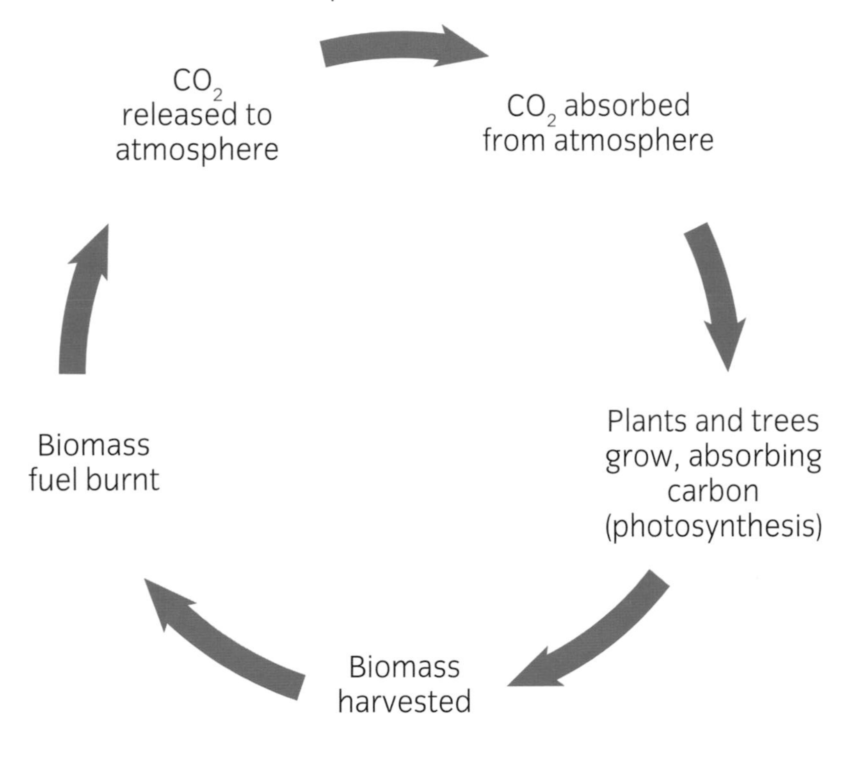


**The carbon cycle**

A disadvantage of biomass is that the material is less dense than fossil fuels so, to achieve the same heat output, a greater quantity of biomass than fossil fuel is required. However, with careful management, the use of biomass is sustainable, whereas the use of fossil fuels is not.

**ASSESSMENT GUIDANCE**

Biomass is in some ways a further development from the wood-burning back boiler, with more controls.

The classes of biomass raw material that can be turned into biomass fuels are:

- wood
- crops such as elephant grass, reed canary grass and oil-seed rape
- agricultural by-products such as straw, grain husks, forest product waste, animal waste such as chicken litter and slurry
- food waste – it is estimate that some 35% of food purchased ends up as waste
- industrial waste.

## Woody biomass

For domestic use, wood-related products are the primary biomass fuels.

For wood to work as a sustainable material, the trees used need to be relatively fast growing, so short-rotation coppice woodlands containing willow, hazel and poplar are used on a 3–5 year rotation. Because of this, large logs are not available, neither can slow-growing timbers that would have a higher calorific value be used. Woody biomass as a fuel is generally supplied as:

- small logs
- wood chips – mechanically shredded trees, branches, etc
- wood pellets – formed from sawdust or shavings that are compressed to form pellets.

The *calorific value* (energy given off by burning) of woody biomass is generally low. The greener (wetter) the wood is, the lower the calorific value will be.

### Woody biomass boilers

A biomass boiler can be as simple as a log-burner providing heat to a single room or may be a boiler heating a whole house.

Woody biomass boilers can be automated so that a constant supply of fuel is available. Wood pellets are transferred to a combustion chamber by means of an auger drive or, if the fuel storage is remote from the boiler, by a suction system. The combustion process is monitored via thermostats in the flue gases and adjustments are made to the fan speed, which controls air intake, and to the fuel-feed system, to control the feed of pellets. All of this is controlled by a microprocessor.

The hot flue gases are passed across a heat exchanger, where the heat is transferred to the water in the central-heating system. From this point the heated water is circulated around a standard central-heating system.

In automated biomass boilers, heat exchangers are self-cleaning and the amount of ash produced is relatively small. As a result, the boilers require little maintenance. The waste gases are taken away from the boiler by the flue and are then dispersed via the flue terminal.

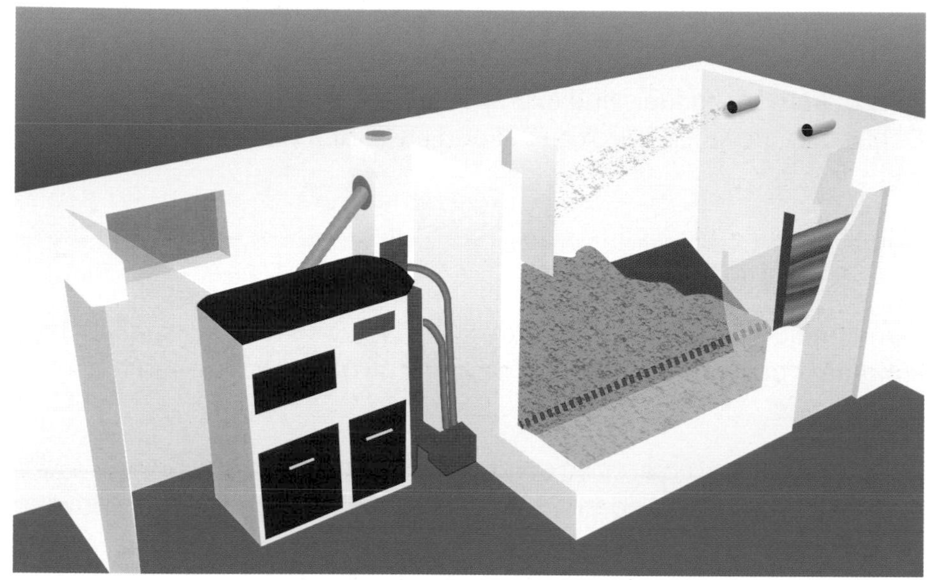

Biomass boiler with suction feed system

## Location and building requirements

When considering the installation of a biomass boiler, the following considerations should be taken into account.

- Space will be required for storage of biomass fuel.
- Easy access will be required for delivery of biomass fuel.
- A biomass boiler may not be permitted in a designated smokeless zone.

**Assessment criteria**

2.5 Clarify the fundamental requirements for the potential to install a biomass system to exist

**ACTIVITY**

An existing building is to be converted to be heated by biomass. The fuel is to be delivered by a heavy goods vehicle via country lanes. Identify two problems that may have to be considered.

### Smoke-control areas and exempt appliances

In the past, when it was common to burn coal as a source of domestic heat or for the commercial generation of heat and power, many cities suffered from very poor air quality and *smogs*, which are a mixture of winter fog and smoke, were common. These smogs contained high levels of sulphur dioxide and smoke particles, both of which are harmful to humans. In December 1952, a period of windless conditions prevailed, resulting in a smog in London that lasted for five days. Apart from very poor visibility, it was estimated at the time that this smog resulted in some 4000 premature deaths and another 100 000 people suffered smog-related illnesses. Recent reports have found that these figures were seriously underestimated and as many as 12 000 may have died. Not surprisingly, there was a massive public outcry, which lead to the government introducing the Clean Air Act 1956 and local authorities declaring areas as 'smoke-control areas'.

The Clean Air Act of 1956 was replaced in 1993. Under the Clean Air Act 1993 it is an offence to sell or burn an *unauthorised fuel* in a smoke control area unless it is burned in what is known as an 'exempt appliance'. These appliances are able to burn fuels that would normally be 'smoky', without emitting smoke to the atmosphere. Each appliance is designed to burn a specific fuel.

**KEY POINT**

Before fitting a biomass boiler in a smoke control area it is necessary to confirm that it is classified as an exempt appliance by Defra.

Lists of authorised fuels and exempt appliances can be found on the Department for the Environment Food and Rural Affairs (Defra) website.

**Assessment criteria**

**3.1** Confirm what would be typically classified as 'permitted development' under town and country planning regulations in relation to the deployment of biomass technology

## Planning permission

Planning permission will not normally be required for the installation of a biomass boiler in a domestic dwelling if all of the work is internal to the building.

If the installation requires an external flue to be installed, it will normally be classed as permitted development as long as the following criterion is met.

- The flue is to the rear or side elevation and does not extend more than 1 m above the highest part of the roof.

**KEY POINT**

Before deciding on any particular system, contact the local authority to see if there are any special requirements.

### Listed building or buildings in a designated area

Check with the local planning authority for both internal work and external flues.

### Buildings in a conservation area or in a World Heritage site

Flues should not be fitted on the principle or side elevation if they would be visible from a highway.

If the project includes the construction of buildings for storage of the biofuels, or to house the boiler, then the same planning requirements as for extensions and garden outbuildings will apply.

## Compliance with building regulations

Assessment criteria

3.2 Confirm which sections of the current building regulations/building standards apply in relation to the deployment of biomass technology

The following building regulations will apply.

| Part | Title | Relevance |
|---|---|---|
| A | Structural safety | Where the biomass appliance and other components put load on the structure |
| B | Fire safety | Where holes for pipes, etc. may reduce the fire-resisting integrity of the building structure |
| C | Resistance to contaminants and moisture | Where holes for pipes, etc may reduce the moisture-resisting integrity of the building structure |
| E | Resistance to sound | Where holes for pipes, etc may reduce the soundproof integrity of the building structure |
| G | Sanitation, hot-water safety, and water efficiency | Hot-water safety and water efficiency |
| J | Heat-producing appliances | Biomass boilers produce heat and therefore must be installed correctly |
| L | Conservation of fuel and power | Energy efficiency of the installed system and the building |
| P | Electrical safety | Safe installation of the electrical supplies and any controls |

## Advantages

Assessment criteria

4.1 Identify typical advantages associated with biomass technology

- It is carbon neutral.
- It is a sustainable fuel source.
- When it is burnt, the waste gases are low in nitrous oxide, with no sulphur dioxide – both are greenhouse gases.

**Assessment criteria**

4.2 Identify typical disadvantages associated with biomass technology

## Disadvantages

- Transportation costs are high – wood pellets or chips will need to be delivered in bulk to make delivery costs viable.
- Storage space is needed for the fuel. As woody biomass has a low calorific value, a large quantity of fuel will be required. Consideration must be given to whether or not adequate storage space is available.
- Control – when a solid fuel is burnt, it is not possible to have instant control of heat, as would be the case with a gas boiler. The fuel source cannot be instantly removed to stop combustion.
- It requires a suitable flue system.

OUTCOMES 1 TO 4

# Electricity-producing micro-renewable energy technologies

The electricity-producing micro-renewable energy technologies that will be discussed in this section are:

- solar photovoltaic
- micro-wind
- micro-hydro.

The major advantages of these technologies are that they do not use any of the planet's dwindling fossil fuel resources. They also do not produce any carbon dioxide ($CO_2$) when running.

With each of the electricity producing micro-renewable energy technologies, two types of connection exist:

1 on-grid or grid tied – where the system is connected in parallel with the grid supplied electricity
2 off-grid – where the system is not connected to the grid but supplies electricity directly to current-using equipment or is used to charge batteries and then supplies electrical equipment via an inverter.

The batteries required for off-grid systems need to be deep-cycle type batteries, which are expensive to purchase. The other downside of using batteries to store electricity is that the batteries' life span may be as short as five years, after which the battery bank will require replacing.

With on-grid systems, any excess electricity generated is exported back to the grid. At times when the generation output is not sufficient to meet the demand, electricity is imported from the grid.

While the following sections will be focused primarily on on-grid or grid-tied systems, which are the most common type in use, an overview of the components required for off-grid systems is included to provide a complete explanation of the technology.

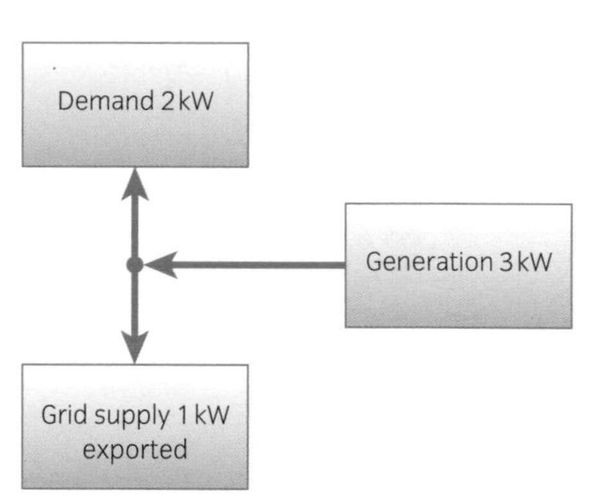

**Generation exceeds demand**

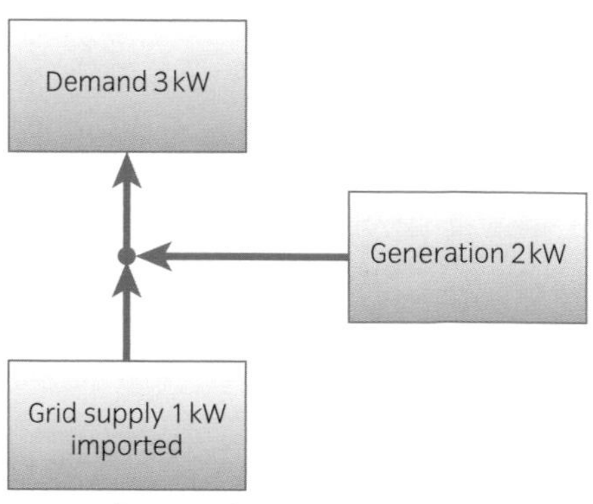

**Demand exceeds generation**

Assessment criteria

1.2 Identify the fundamental working principles for solar photovoltaic technology

# SOLAR PHOTOVOLTAIC (PV)

Solar photovoltaic (PV) is the conversion of light into electricity. Light is electromagnetic energy and, in the case of visible light, is electromagnetic energy that is visible to the human eye. The electromagnetic energy released by the Sun consists of a wide spectrum, most of which is not visible to the human eye and cannot be converted into electricity by PV modules.

## ASSESSMENT GUIDANCE

Photovoltaic is probably the most common system being installed. It is fairly simple to install and requires minimum disruption when connecting the system to the meter position.

## Working principles

The basic element of photovoltaic energy production is the PV cell, which is made from semiconductor material. A semiconductor is a material with resistivity that sits between that of an insulator and a conductor. Whilst various semiconductor materials can be used in the making of PV cells, the most common material is silicon. Adding a small quantity of a different element (an impurity) to the silicon, a process known as 'doping', produces n-type or p-type semiconductor material. Whether it is n-type (negative) or p-type (positive) semiconductor material is dependent on the element used to dope the silicon. Placing an n-type and a p-type semiconductor material together creates a p-n junction. This forms the basis of all semiconductors used in electronics.

When *photons*, which are particles of energy from the Sun, hit the surface of the PV cell they are absorbed by the p-type material. The additional energy provided by these photons allows electrons to overcome the bonds holding them and move within the semiconductor material, thus creating a potential difference or – in other words – generating a voltage.

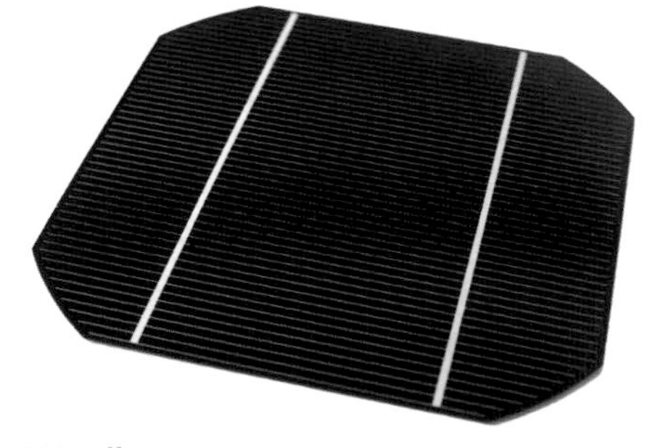

PV cell

Photovoltaic cells have an output voltage of 0.5 V, so a number of these are linked together to form modules with resulting higher voltage and power outputs. Modules are connected together in series to increase voltage. These are known as 'strings'. All the modules together are known as an 'array'. An array therefore can comprise a single string or multiple strings. The connection arrangements are determined by the size of the system and the choice of inverter. It should be noted that PV arrays can attain d.c. voltages of many hundreds of volts.

## ACTIVITY

What is one danger associated with photovoltaic systems?

There are many arrangements for PV systems but they can be divided into two categories:

- off-grid systems, where the PV modules are used to charge batteries
- on-grid systems, where the PV modules are connected to the grid supply via an inverter.

The key components of an off-grid PV system are:

- PV modules
- a PV module mounting system
- d.c. cabling
- a charge controller
- a deep-discharge battery bank
- an inverter.

Other components, such as isolators, will also be required.

**Off-grid system components**

Off-grid systems are ideal where no mains supply exists and there is a relatively small demand for power. Deep-discharge batteries are expensive and will need replacing within 5–10 years, depending on use.

## On-grid systems where the PV modules are connected to the grid supply via an inverter

The key components of an on-grid PV system are:

- PV modules
- a PV module mounting system
- d.c. cabling
- an inverter
- a.c. cabling
- metering
- a connection to the grid.

Other components such as isolators will also be required.

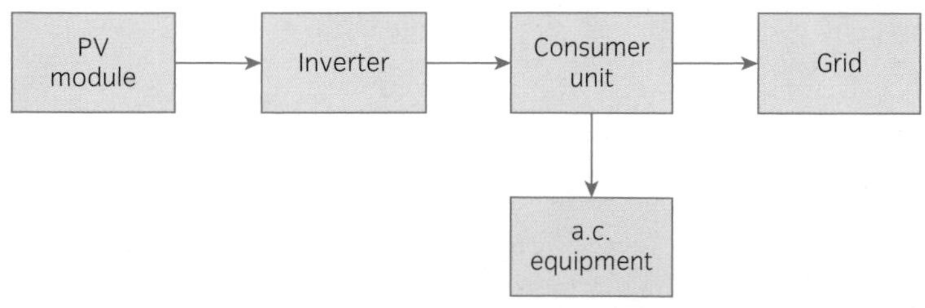


**On-grid system components**

## PV modules

A range of different types of module, of various efficiencies, is available. The performance of a PV module is expressed as an *efficiency percentage*: the higher the percentage the greater the efficiency.

- Monocrystalline modules range in efficiency, from 15% to 20%.
- Polycrystalline modules range in efficiency, from 13% to 16%, but are cheaper to purchase than monocrystalline modules.
- Amorphous film ranges in efficiency, from 5% to 7%. Amorphous film is low efficiency but is flexible, so it can be formed into curves and is ideal for surfaces that are not flat.

Whilst efficiencies may appear low, the maximum theoretical efficiency that can be obtained with a single junction silicon cell is only 34%.

## PV module mounting system

Photovoltaic modules can be fitted as on-roof systems, in-roof systems or ground-mount systems.

- On-roof systems are the common method employed for retrofit systems. Various different mounting systems exist for securing the modules to the roof structure. Most consist of aluminium rails, which are fixed to the roof structure by means of roof hooks. Mounting systems also exist for fitting PV modules to flat roofs. Checks will need to be made to ensure that the existing roof structure can withstand the additional weight and also the uplift forces that will be exerted on the PV array by the wind.
- In in-roof systems the modules replace the roof tiles. The modules used are specially designed to interlock, to ensure that the roof structure is watertight. The modules are fixed directly to the roof structure. Several different systems are on the market, from single-tile size to large panels that replace a whole section of roof tiles. In-roof systems cost more than on-roof systems but are more aesthetically pleasing. In-roof systems are generally only suitable for new-build projects or where the roof is to be retiled.

**On-roof mounting system**

### ASSESSMENT GUIDANCE

It is essential that the integrity of the roof is maintained and that no leaks occur due to the installation. Only approved contractors with the relevant insurance should be employed.

### ACTIVITY

With regards to a PV installation, who would normally be responsible for

a) mounting the roof brackets and panels

b) testing and connecting the electrical system?

**In-roof mounting system**

- Ground-mount systems and pole-mount systems are available for free-standing PV arrays.

Ground-mount system

Tracking systems are the ultimate in PV mounting systems. They are computer-controlled motorised mounting systems that change both **azimuth** and tilt to track the Sun as it passes across the sky.

**Azimuth**

Ideally the modules should face due south, but any direction between east and west will give acceptable outputs.

Azimuth refers to the angle that the panel direction diverges from facing due south

## Inverter

The inverter's primary function is to convert the d.c. input to a 230 V a.c. 50 Hz output, and synchronise it with the mains supply frequency. The inverter also ensures that, in the event of mains supply failure, the PV system does not create a danger by continuing to feed power onto the grid. The inverter must be matched to the PV array with regard to power and d.c. input voltage, to avoid damage to the inverter and to ensure that it works efficiently. Both d.c. and a.c. isolators will be fitted to the inverter, to allow it to be isolated for maintenance purposes.

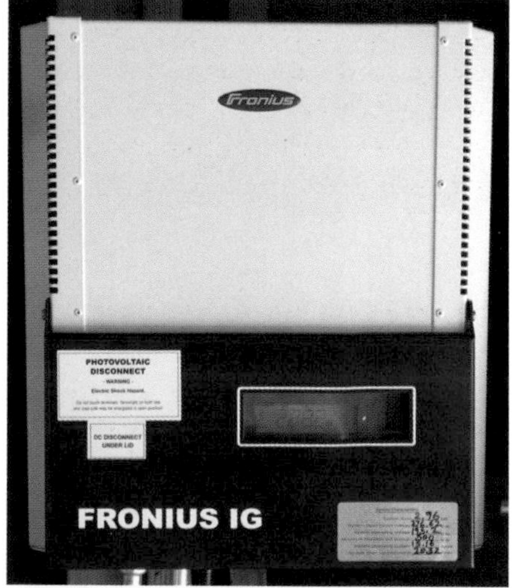


PV inverter

**ACTIVITY**

What is the purpose of an inverter?

## Metering

A generation meter is installed on the system to record the number of units generated, so that the feed in tariff can be claimed.

## Connection to the grid

Connection to the grid within domestic premises is made via a spare way in the consumer unit and a 16 A overcurrent protective device. An isolator is fitted at the intake position to provide emergency switching, so that the PV system can easily be isolated from the grid.

**Assessment criteria**

**2.2** Clarify the fundamental requirements for the potential to install a solar photovoltaic system to exist

## Location and building requirements

When deciding on the suitability of a location or building for the installation of PV, the following considerations should be taken into account.

### Adequate roof space available

The roof space available determines the maximum size of PV array that can be installed. In the UK, all calculations are based on 1000 Wp (watts peak) of the Sun's radiation on 1 $m^2$ so, if the array uses modules with a 15% efficiency, each 1 kWp of array will require approximately 7 $m^2$ of roof space. The greater the efficiency of the modules, the less roof space that is required.

**Ideal orientation is south**

### The orientation (azimuth) of the PV array

The optimum direction for the solar collectors to face is due south; however, as the Sun rises in the east and sets in the west, any location with a roof facing east, south or west is suitable for mounting a PV array, but the efficiency of the system will be reduced for any system not facing due south.

### The tilt of the PV array

Throughout the year, the height of the Sun relative to the horizon changes from its lowest in December through to its highest in June. As it is generally not practical to vary the tilt angle throughout the year, the optimum tilt for the PV array in the UK is between 30° and 40°; however, the modules will work outside the optimum tilt range and will even work if vertical or horizontal, but they will be less efficient.

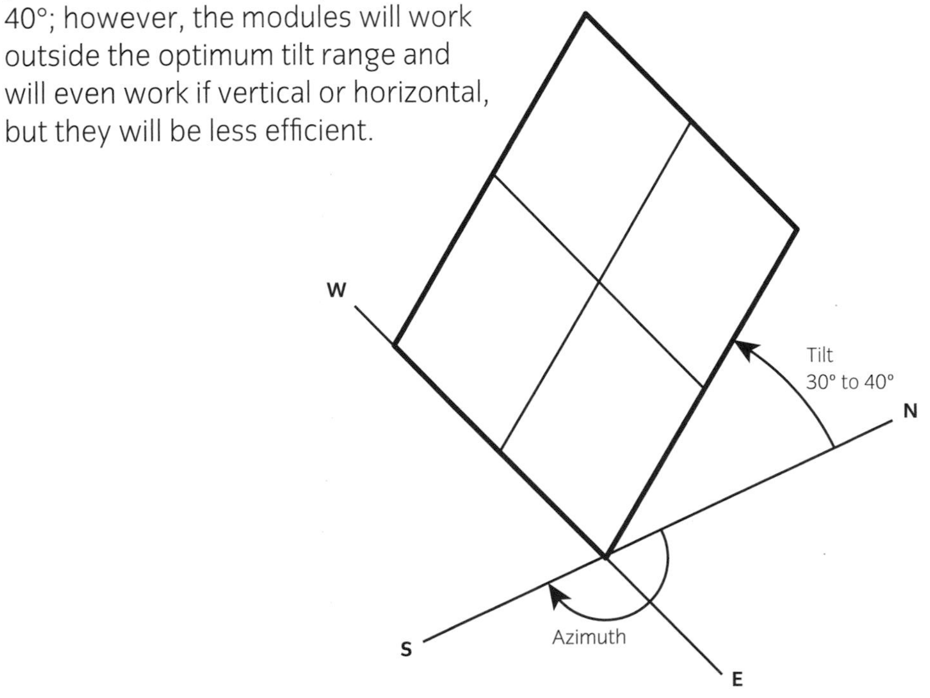


PV array facing south at fixed tilt

**Ideal tilt is 30–40°**

## Shading of the PV array

Any structure, tree, chimney, aerial or other object that stands between the PV array (collector) and the Sun will prevent some of the Sun's energy from reaching the collector. The Sun shines for a limited time and any reduction in the amount of sunlight landing on the collector will reduces its ability to produce electricity.

Shading of PV system

## Location within the UK

The location within the UK will determine how much sunshine will fall, annually, on the PV array and, in turn, this will determine the amount of electricity that can be generated. For example, a location in Brighton will generate more electricity than one in Newcastle, purely because Brighton receives more sunshine.

## The suitability of the structure for mounting the solar collector

The structure has to be assessed for its suitability for fixing the chosen mounting system. Consideration needs to be given to the strength of the structure, the suitability of fixings and the condition of the structure. Consideration also needs to be given to the effect known as 'wind uplift', an upward force exerted by the wind on the module and mounting system. The strength of the PV array fixings and the fixings holding the roof members to the building structure must be great enough to allow for wind uplift.

In the case of roof-mounted systems on flats and other shared properties, consideration must also be given to the ownership of the structure on which the proposed system is to be installed.

## A suitable place to mount the inverter

The inverter is usually mounted either in the loft space or at the mains position.

## Connection to the grid

A spare way within the consumer unit will need to be available for connection of the PV system. If one is not available then the consumer unit may need to be changed.

**Assessment criteria**

**3.1** Confirm what would be typically classified as 'permitted development' under town and country planning regulations in relation to solar photovoltaic systems

## Planning permission

Permitted development applies where a PV system is installed:

- on a dwelling house or block of flats
- on a building within the grounds of a dwelling house or block of flats
- as a stand-alone system in the grounds of a dwelling house or block of flats.

However, there are criteria to be met in each case.

For building-mounted systems:

- the PV system must not protrude more than 200 mm from the wall or the roof slope
- the PV system must not protrude past the highest point of the roof (the ridgeline), excluding the chimney.

For stand-alone systems the following criteria must be met.

- Only one stand-alone system is allowed in the grounds.
- The array must not exceed 4 m in height.
- The array must not be installed within 5 m of the boundary of the grounds.
- The array must not exceed 9 $m^2$ in area.
- No dimension of the array may exceed 3 m in length.

For both stand-alone and building-mounted systems the following criteria must be met.

- The system must not be installed in the grounds or on a building within the grounds of a listed building or a scheduled monument.
- If the dwelling is in a conservation area or a World Heritage Site, then the array must not be closer to a highway than the house or block of flats.
- In every other case, planning permission will be required.

## Compliance with building regulations

The following building regulations will apply.

| Part | Title | Relevance |
|---|---|---|
| A | Structural safety | The PV modules will impose both downward force and wind uplift stresses on the roof structure. |
| B | Fire safety | The passage of cables through the building fabric could reduce the fire-resisting integrity of the structure. |
| C | Resistance to contaminants and moisture | The fixing brackets for on-roof systems and the passage of cables through the building fabric could reduce the moisture-resisting integrity of the structure. |
| E | Resistance to sound | The passage of cables through the building fabric could reduce the sound-resisting properties of the structure. |
| L | Conservation of fuel and power | The efficiency of the system and the building overall |
| P | Electrical safety | The installation of the components and wiring system |

**Assessment criteria**

**3.2** Confirm which sections of the current building regulations/building standards apply in relation to the deployment of solar photovoltaic systems

### ACTIVITY

What documentation should be completed by the electrical installer after testing the new PV installation?

### ASSESSMENT GUIDANCE

All systems will require some penetration of the building fabric, be it the roof or walls, depending on the building type and construction. You should be able to describe the methods of making good for all building fabrics.

## Other regulatory requirements

- BS 7671 Wiring Regulations will apply to the PV installation.
- G83 requirements will apply to on-grid systems up to 3.68 kW per phase; above this size the requirements of G59 will need to be complied with.
- Micro Generation Certification Scheme requirements will apply.

## Advantages

**Assessment criteria**

**4.1** Identify typical advantages associated with solar photovoltaic systems

- They can be fitted to most buildings.
- There is a feed-in tariff available for electricity generated, regardless of whether it is used on site or exported to the grid.
- Excess electricity can be sold back to the distribution network operator (DNO).
- There is a reduction in electricity imported.
- It uses zero carbon technology.
- It improves energy performance certificate ratings.
- There is a reasonable payback period on the initial investment.

Assessment criteria

4.2 Identify typical disadvantages associated with solar photovoltaic systems

## Disadvantages

- Initial cost is high.
- The system size is dependent on available, suitable roof area.
- It requires a relatively large array to offset installation costs.
- It gives variable output that is dependent on the amount of sunshine available. Lowest output is at times of greatest requirement, such as at night and in the winter. Savings need to be considered over the whole year.
- There is an aesthetic impact (on the appearance of the building).

# MICRO-WIND

Wind turbines harness energy from the wind and turn it into electricity. The UK is an ideal location for the installation of wind turbines, as about 40% of Europe's wind energy passes over the UK. A micro-wind turbine installed on a suitable site could easily generate more power than would be consumed on site.

Assessment criteria

1.2 Identify the fundamental working principles for micro-wind technology

## Working principles

The wind passing the rotor blades of a turbine causes it to turn. The hub is connected by a low-speed shaft to a gearbox. The gearbox output is connected to a high-speed shaft that drives a generator which, in turn, produces electricity. Turbines are available as either horizontal-axis wind turbines (HAWT) or vertical-axis wind turbines (VAWT).

A HAWT has a tailfin to turn the turbine so that it is facing in the correct direction to make the most of the available wind. The gearbox and generator will also be mounted in the horizontal plane.

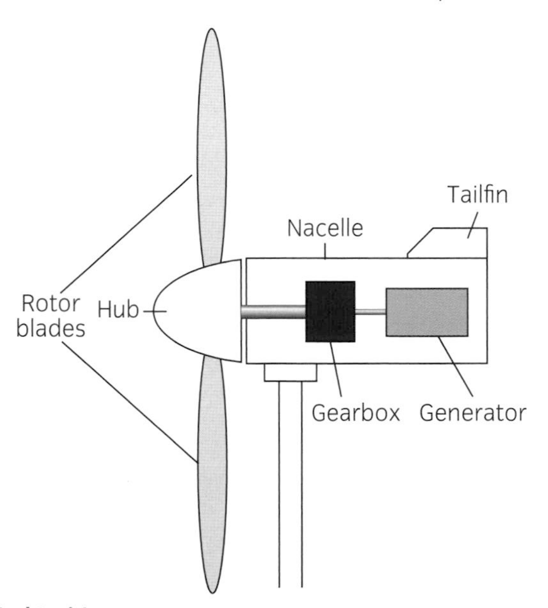


**Horizontal-axis wind turbine**

Vertical-axis wind turbines, of which there are many different designs, will work with wind blowing from any direction and therefore do not require a tailfin. A VAWT also has a gearbox and generator.

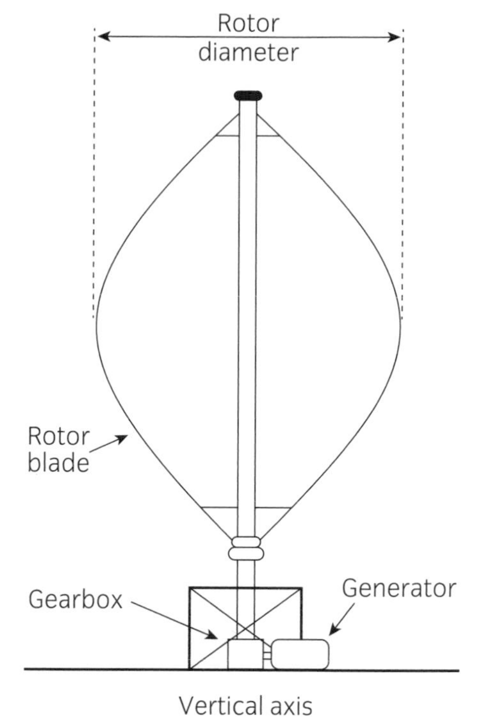

**Vertical-axis wind turbine**

The two types of micro-wind turbines suitable for domestic installation are:

- pole-mounted, freestanding wind turbines
- building-mounted wind turbines, which are generally smaller than pole-mounted turbines.

Micro-wind generation systems fall into two basic categories:

- on-grid (grid tied), which is connected in parallel with the grid supply via an inverter
- off-grid, which charge batteries to store electricity for later use.

The output from a micro-wind turbine is *wild* alternating current (a.c.). 'Wild' refers to the fact that the output varies in both voltage and frequency. The output is connected to a system controller, which rectifies the output to d.c.

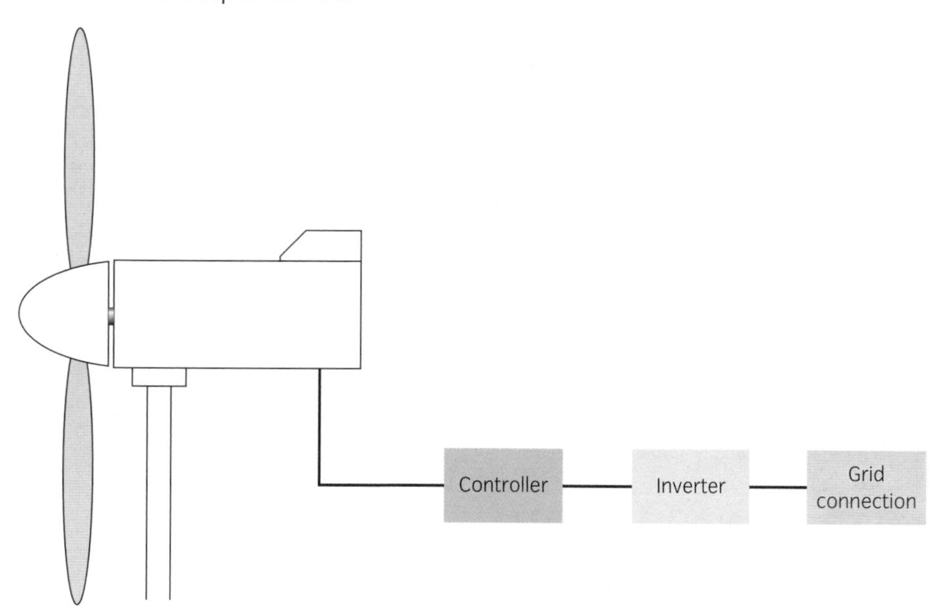

**Block diagram of an on-grid micro-wind system**

In the case of an on-grid system the d.c. output from the system controller is connected to an inverter which converts d.c. to a.c. at 230 V 50 Hz, for connection to the grid supply via a generation meter and the consumer unit.

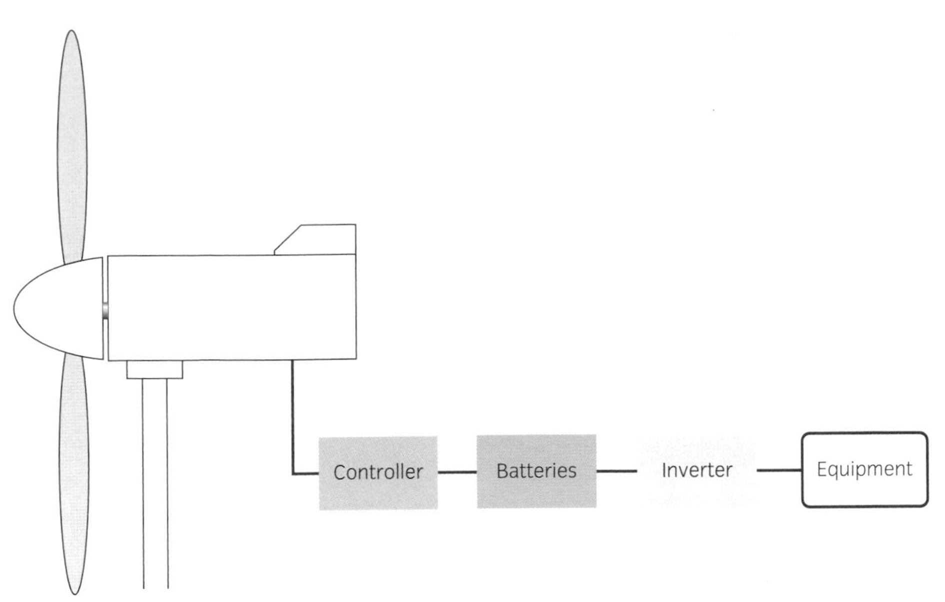


Block diagram of an off-grid micro-wind system

With off-grid systems the output from the controller is used to charge batteries so that the output is stored for when it is needed. The output from the batteries then feeds an inverter so that 230 V a.c. equipment can be connected.

## ACTIVITY

The first wind turbine is believed to have been installed in Scotland in July 1887 by James Blythe. Find out what he used it for.

**Assessment criteria**

**2.6** Clarify the fundamental requirements for the potential to install a micro-wind system to exist

## Location and building requirements

When considering the installation of a micro-wind turbine, it is important to consider the location or building requirements, including:

- the average wind speed on the site
- any obstructions and turbulence
- the height at which the turbine can be mounted
- turbine noise, vibration, flicker.

### Wind speed

Wind is not constant, so the average wind speed on a site, measured in metres per second (m/s), is a prime consideration when deciding on a location's suitability for the installation of a micro-wind turbine.

Wind speed needs to be a minimum of 5 m/s for a wind turbine to generate electricity. Manufacturers of wind turbines provide power curves for their turbines, which show the output of the turbine at different wind speeds. Most micro-wind turbines will achieve their maximum output when the wind speed is around 10 m/s.

## ASSESSMENT GUIDANCE

There is plenty of wind in the UK for wind generators. Surrounding buildings, other structures and trees can cut down the effectiveness of micro-wind systems.

## Obstructions and turbulence

For a wind turbine to work efficiently, a smooth flow of air needs to pass across the turbine blades.

The ideal site for a wind turbine would be at the top of a gentle slope. As the wind passes up the slope it gains speed, resulting in a higher output from the turbine.

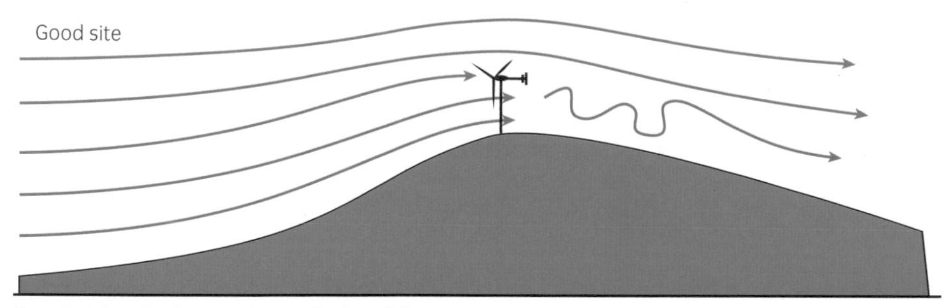


**A suitable site for a micro-wind turbine**

The diagram below illustrates the effect on the wind when a wind turbine is poorly sited. The wind passing over the turbine blades is disturbed and thus the efficiency is reduced.

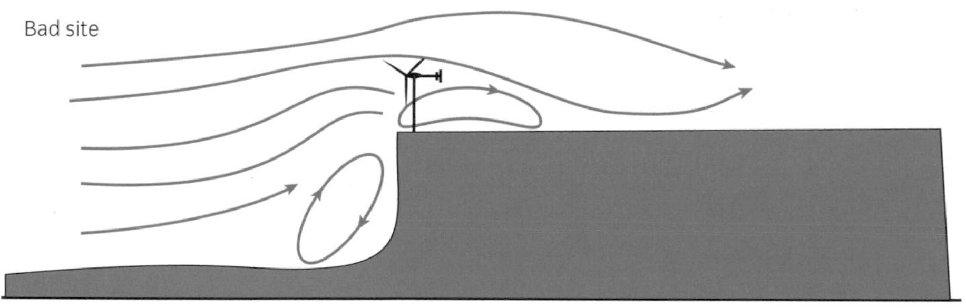


**An unsuitable site for a micro-wind turbine**

Any obstacles, such as trees or tall buildings, will affect the wind passing over the turbine blades.

Where an obstacle is upwind of the wind turbine, in the direction of the prevailing wind, the wind turbine should be sited at a minimum distance of 10 × the height of the obstacle away from the obstacle. In the case of an obstacle that is 10 m in height, this would mean that the wind turbine should be sited a minimum of 10 × 10 m away, which is 100 m from the obstacle.

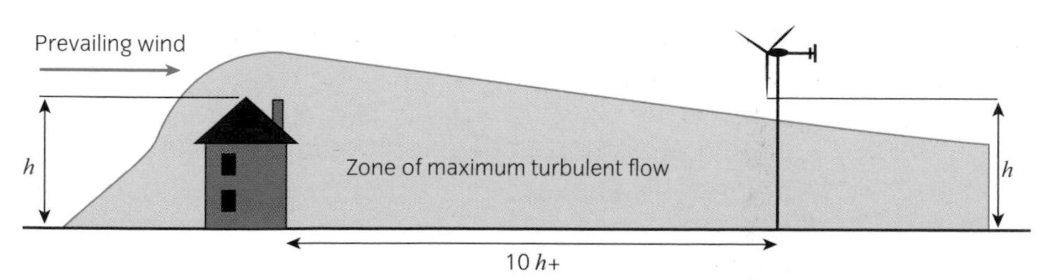


**Placement of micro-wind turbine to avoid obstacles**

## Turbine mounting height

Generally, the higher a wind turbine is mounted, the better. The minimum recommended height is 6–7 m but, ideally, it should be mounted at a height of 9–12 m. As a wind turbine has moving parts, consideration needs to be given to access for maintenance. Where an obstacle lies upwind of the turbine, the bottom edge of the blade should be above the height of the obstacle.

## Turbine noise

Consideration needs to be given to buildings sited close to the wind turbine as the turbine will generate noise in use.

## Turbine vibration

Consideration needs to be given to vibration when the wind turbine is building mounted. It may be necessary to consult a structural engineer.

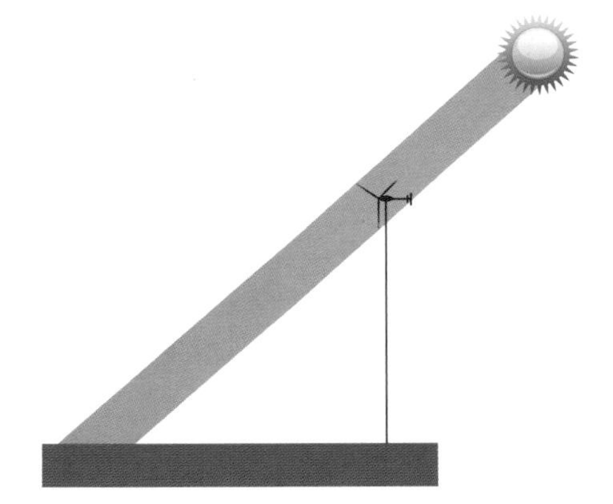

The area affected by shadow flicker

## Shadow flicker

Shadow flicker is the result of the rotating blades of a turbine passing between a viewer and the Sun. It is important to ensure that shadow flicker does not unduly affect a building sited in the shadow-flicker zone of the wind turbine.

The distance of the shadow-flicker zone from the turbine will be at its greatest when the Sun is at its lowest in the sky.

**Assessment criteria**

**3.1** Confirm what would be typically classified as 'permitted development' under town and country planning regulations in relation to micro-wind systems

## Planning permission

Whilst permitted development exists for the installation of wind turbines, it is severely restricted so, in the majority of installations, a planning application will be required. The permitted development criteria are detailed below.

Permitted development applies where a wind turbine is installed:

- on a detached dwelling house
- on a detached building within the grounds of a dwelling house or block of flats
- as a stand-alone system in the grounds of a dwelling house or block of flats.

It is important to note that permitted development for building-mounted wind turbines only applies to detached premises. It does not apply to semi-detached houses or flats.

Even with detached buildings or stand-alone turbines there are criteria to be met.

- The wind turbine must comply with the MCS planning standards, or equivalent.
- Only one wind turbine may be installed on the building or within the grounds of the building.
- An air source heat pump may not be installed on the building or within the grounds of the building.
- The highest part of the wind turbine (normally the blades) must not protrude more than 3 m above the ridge line of the building or be more than 15 m in height.
- The lowest part of the blades of the wind turbine must be a minimum of 5 m from ground level.
- The wind turbine must be a minimum of 5 m from the boundary of the premises.
- The wind turbine cannot be installed on or within:
  - land that is safeguarded land (usually designated for military or aeronautical reasons)
  - a site that is designated as a scheduled monument
  - a listed building
  - the grounds of a listed building
  - land within a national park
  - an area of outstanding natural beauty
  - the Broads (wetlands and inland waterways in Norfolk and Suffolk).
- The wind turbine cannot be installed on the roof or wall of a building that fronts a highway, if that building is within a conservation area.

The following conditions also apply.

- The blades must be made of non-reflective material.
- The wind turbine should be sited so as to minimise its effect on the external appearance of the building.

**ACTIVITY**

It is possible that wind turbines may have an adverse effect on wildlife. Identify which group could be affected.

Assessment criteria

**3.2** Confirm which sections of the current building regulations/building standards apply in relation to the deployment of micro-wind systems

## Compliance with building regulations

The following building regulations will apply.

| Part | Title | Relevance |
|---|---|---|
| A | Structural safety | A wind turbine mounted on a building will exert additional structural load, as well as forces, due to its operation. |
| B | Fire safety | Cable entries and fixings may reduce the fire-resisting integrity of the building structure. |
| C | Resistance to contaminants and moisture | Cable entries and fixings may reduce the moisture-resisting integrity of the building fabric. |
| E | Resistance to sound | Cable entries may reduce the sound-resisting integrity of the building fabric. |
| L | Conservation of fuel and power | The efficiency of the system and the building |
| P | Electrical safety | Installation of wiring and components |

**KEY POINT**

Rats, mice and insects can enter a building through surprisingly small holes. A wasps' nest can cause considerable distress to people.

## Other regulatory requirements

- For on-grid systems, the requirements of the Distribution Network Operator (DNO) will apply.
- Wiring Regulations BS 7671 will apply to the installation of micro-wind turbines.

Assessment criteria

**4.1** Identify typical advantages associated with micro-wind systems

## Advantages

- They can be very effective on a suitable site as the UK has 40% of Europe's wind resources.
- There are no carbon dioxide emissions.
- They produce most energy in winter, when consumer demand is at its maximum.
- A feed-in tariff is available.
- This can be a very effective technology where mains electricity does not exist.

**ACTIVITY**

What is meant by a 'feed-in tariff'?

## Disadvantages

Assessment criteria

4.2 Identify typical disadvantages associated with micro-wind systems

- Initial costs are high.
- The requirements of the site are onerous.
- Planning can be onerous.
- Performance is variable and is dependent on wind availability.
- Micro-wind turbines cause noise, vibration and flicker.

# MICRO-HYDRO-ELECTRIC

All rivers flow downhill. This movement of water from a higher level to a lower level is a source of free kinetic energy that hydro-electric generation harnesses. Water passing across or through a turbine can be used to turn a generator and thus produce electricity. Given the right location, micro-hydro-electric is the most constant and reliable source of all the micro-generation technologies and is the most likely of the technologies to meet all of the energy needs of the consumer.

As with the other micro-generation technologies, there are two possible system arrangements for micro-hydro schemes: on-grid and off-grid systems.

## Working principles

Assessment criteria

1.2 Identify the fundamental working principles for micro-hydro technology

Whilst it is possible to place generators directly into the water stream, it is more likely that the water will be diverted from the main stream or river, through the turbine, and back into the stream or river at a lower level. Apart from the work involved with the turbines and generators, there is also a large amount of civil engineering and construction work to be carried out to route the water to where it is needed.

The main components of the water course construction are:

- intake – the point where a portion of the river's water is diverted from the main stream
- the canal that connects the intake to the forebay
- the forebay, which holds a reservoir of water that ensures that the penstock is pressurised at all times and allows surges in demand to be catered for
- the penstock, which is pipework taking water from the forebay to the turbines
- the powerhouse, which is the building housing the turbine and the generator
- the tailrace, which is the outlet that takes the water exiting the turbines and returns it to the main stream of the river.

See the diagram on the next page.

### ACTIVITY

What types of fish are most likely to be adversely affected by the installation of turbines in rivers?

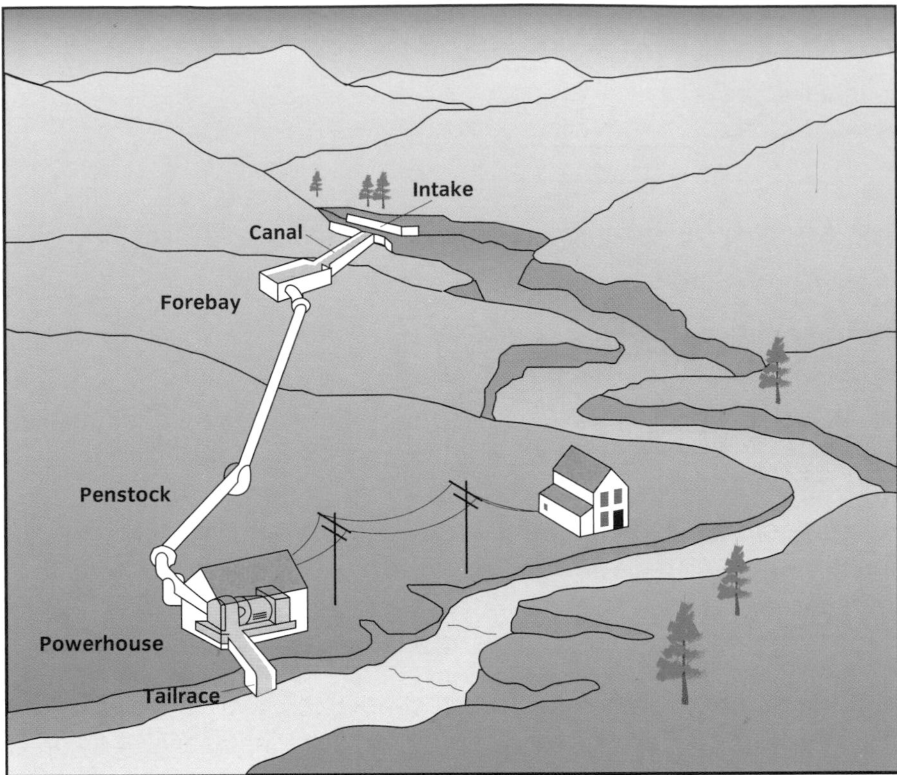


The component parts of a micro-hydro system

The idea of water turbines sounds great but of course few people live near a water course that is suitable for this application. PV and solar thermal systems are more adaptable for different locations.

To ascertain the suitability of the water source for hydro-electric generation, it is necessary to consider the head and the flow of the water source.

## Head

The head is the vertical height difference between the proposed inlet position and the proposed outlet. This measurement is known as 'gross head'.

Head height is generally classified as:

- low head – below 10 m
- medium head – 10–50 m
- high head – above 50 m.

There is no absolute definition for each classification. The Environment Agency, for example, classifies low head as below 4 m. Some manufacturers specify high head as above 300 m.

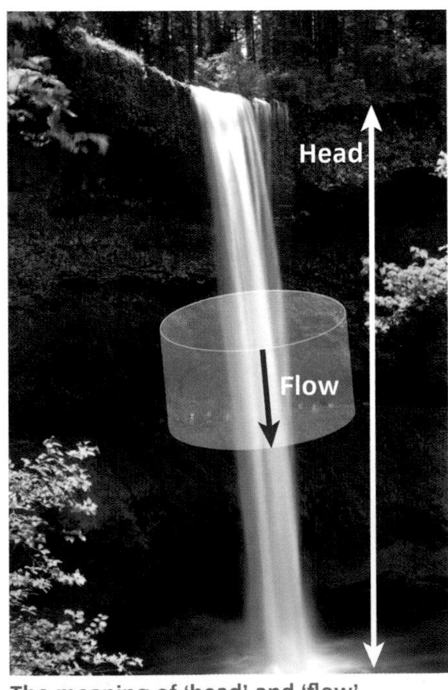


The meaning of 'head' and 'flow'

## Net head

This is used in calculations of potential power generation and takes into account losses due to friction, as the water passes through the penstock.

## Flow

This is the amount of water flowing through the water course and is measured in cubic metres per second ($m^3/s$).

## Turbines

There are many different types of turbine but they fall into two primary design groups, each of which is better suited to a particular type of water supply.

### Impulse turbine

In an impulse turbine, the turbine wheel or *runner* operates in air, with water jets driving the runner. The water from the penstock is focused on the blades by means of a nozzle. The velocity of the water is increased but the water pressure remains the same so there is no requirement to enclose the runner in a pressure casing. Impulse turbines are used with high-head water sources.

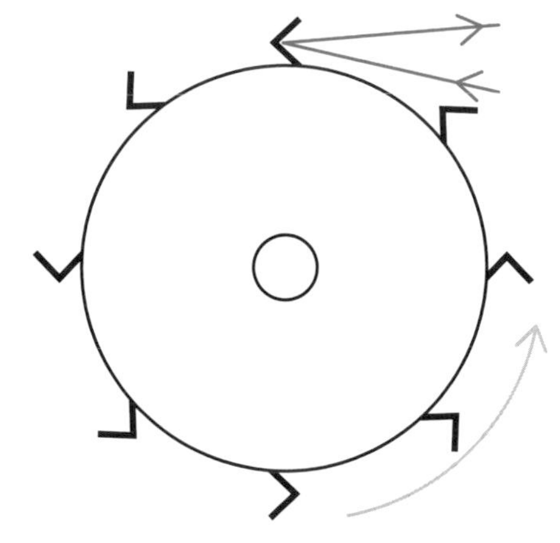

Impulse turbine

Examples of impulse turbines are described below.

#### Pelton

This consists of wheel with bucket-type vanes set around the rim. The water jet hits the vane and turns the runner. The water gives up most of its energy and falls into a discharge channel below. A multi-jet Pelton turbine is also available. This type of turbine is used with water sources with medium or high heads of water.

#### Turgo

This is similar to the Pelton but the water jet is designed to hit the runner at an angle and from one side of the turbine. The water enters at one side of the runner and exits at the other, allowing the Turgo turbine to be smaller than the Pelton for the same power output. This type of turbine is used with water sources with medium or high heads of water.

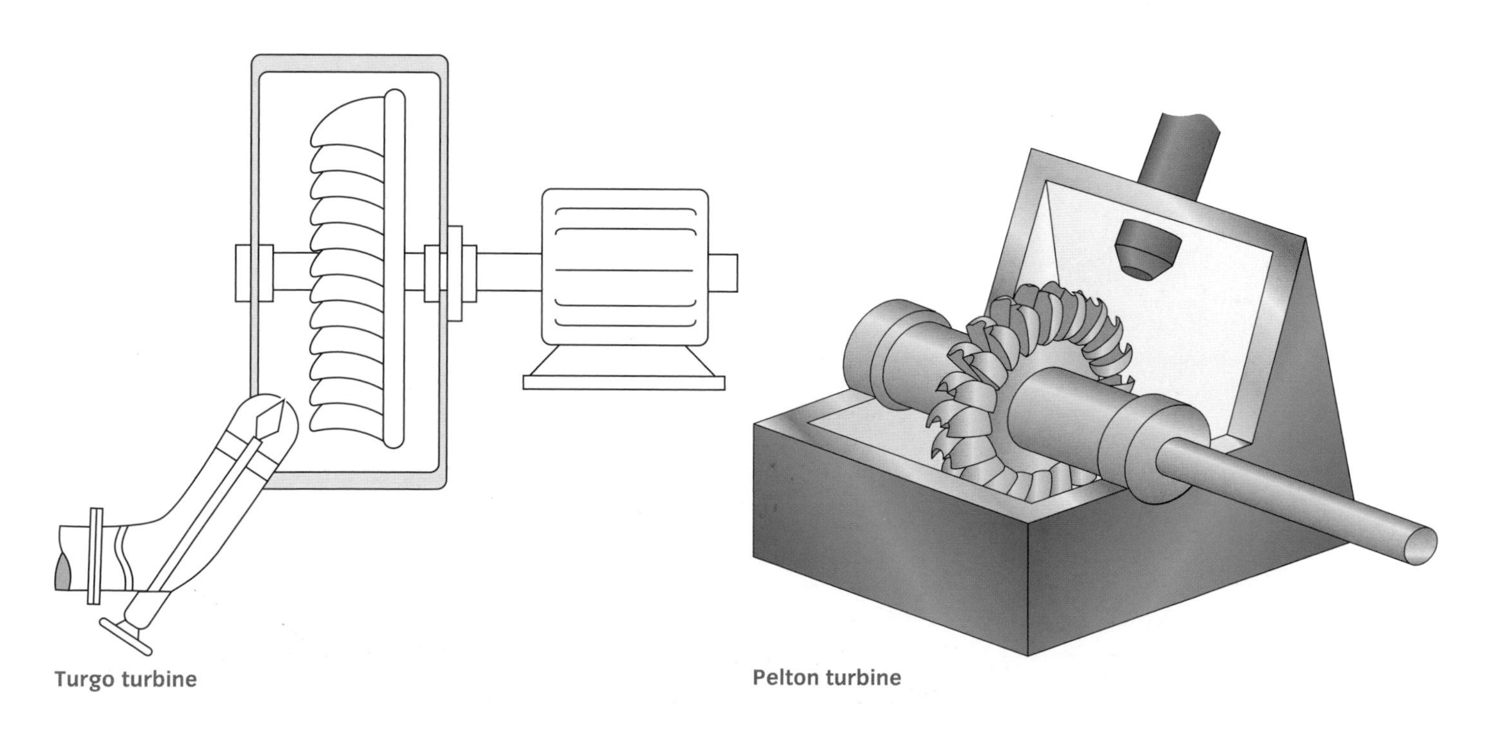

Turgo turbine

Pelton turbine

### Cross-flow or Banki

With this type of turbine the runner consists of two end-plates with slats, set at an angle, joining the two discs, much like a water wheel. Water passes through the slats, turning the runner and then exiting from below. This type of turbine is used with water sources with low or medium heads of water.

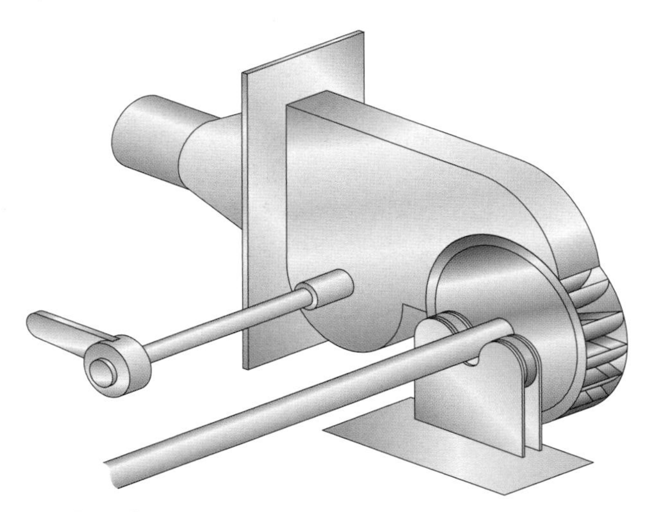

Cross-flow or Banki turbine

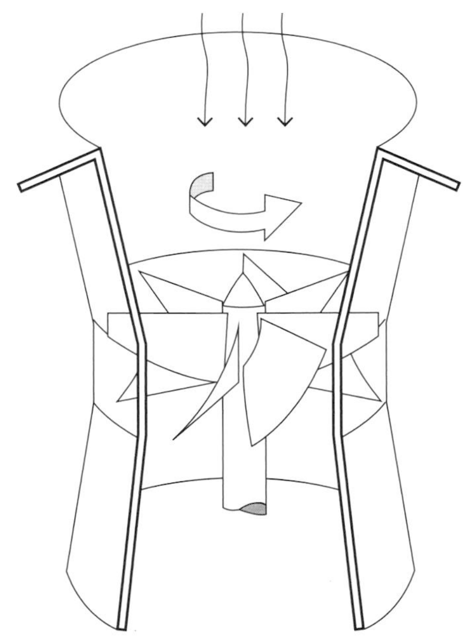

Reaction turbine

## Reaction turbine

In the reaction turbine, the runners are fully immersed in water and are enclosed in a pressure casing. Water passes through the turbine, causing the runner blades to turn or react.

Examples of reaction turbines are described below.

### Francis wheel

Water enters the turbine housing and passes through the runner, causing it to turn. This type of turbine is used with water sources with low heads of water.

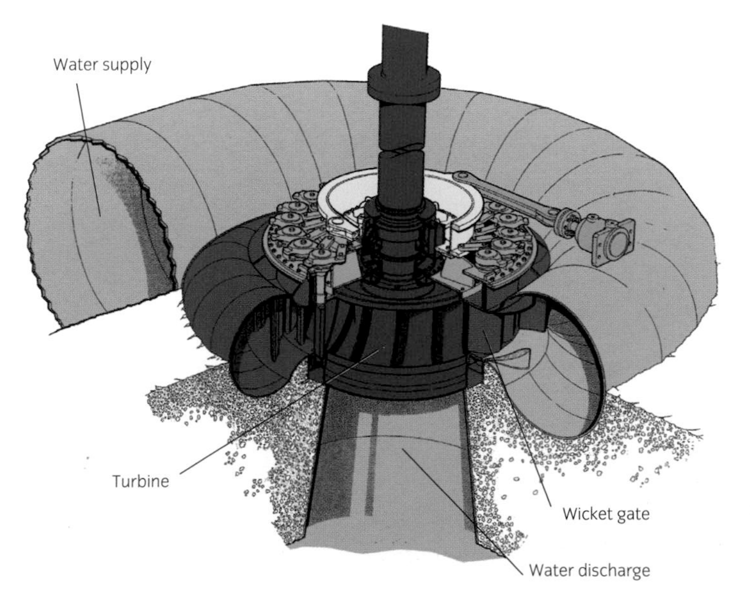


Francis wheel turbine

### Kaplan (propeller)

This works like a boat propeller in reverse. Water passing the angled blades turns the runner. This type of turbine is used with water sources with low heads of water.

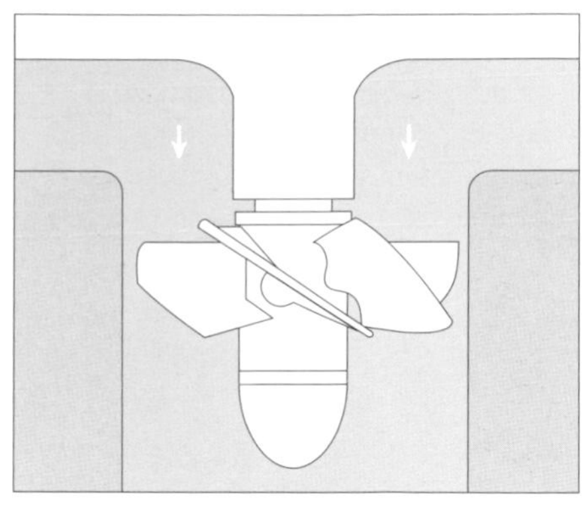

Kaplan or propeller turbine

### Reverse Archimedes' screw

The Archimedes' screw consists of a helical screw thread, which was originally designed so that turning the screw – usually by hand – would draw water up the thread to a higher level. In the case of hydro-electric turbines, water flows down the screw, hence *reverse*, turning the screw, which is connected to the generator. This type of turbine is particularly suited to low-head operations but its major feature is that, due to its design, it is 'fish-friendly' and fish are able to pass through it, so it may be the only option if a hydro-electric generator is to be fitted on a river that is environmentally sensitive.

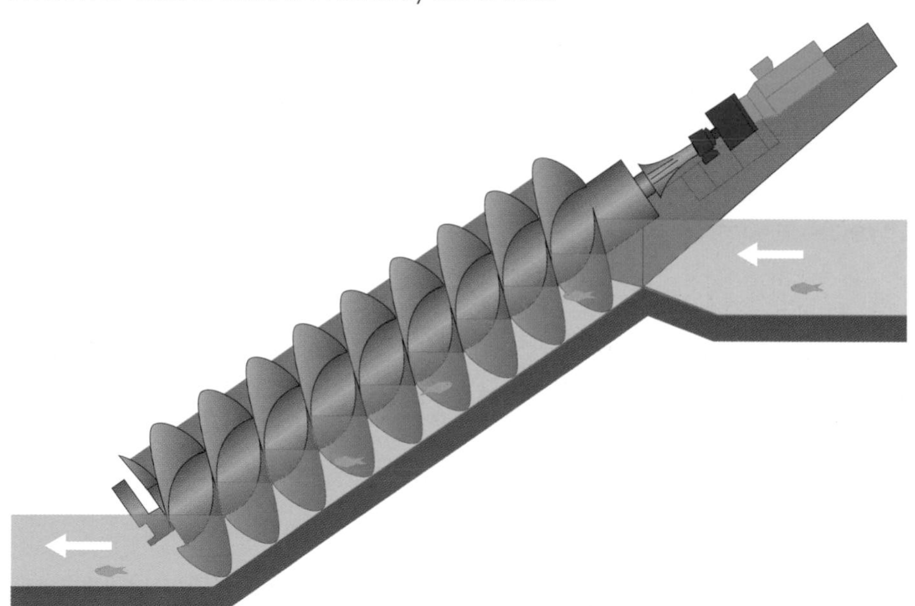

Reverse Archimedes' screw

## Location building requirements

When considering the installation of a micro-hydro turbine, the following location or building requirements should be taken into account.

- The location will require a suitable water source with:
    - a minimum head of 1.5 m
    - a minimum flow rate of 100 litres/second.

The water source should not be subject to seasonal variation that will take the water supply outside of the above parameters.

- The location has to be suitable to allow construction of:
    - the water inlet
    - the turbine/generator building
    - the water outlet or tailrace.

**Assessment criteria**

**2.7** Clarify the fundamental requirements for the potential to install a micro-hydro system to exist

**ASSESSMENT GUIDANCE**

You should be able to draw the basic layout of a micro-hydro system.

**Assessment criteria**

3.1 Confirm what would be typically classified as 'permitted development' under town and country planning regulations in relation to micro-hydro systems

## Planning permission

Planning permission will be required.

A micro-hydro scheme will have an impact on:

- the landscape and visual amenity
- nature conservation
- the water regime.

The planning application will need to be accompanied by an environmental statement detailing any environmental impact and what measures will be taken to minimise these. The environmental statement typically covers:

- flora and fauna
- noise levels
- traffic
- land use
- archaeology
- recreation
- landscape
- air and water quality.

**Assessment criteria**

3.2 Confirm which sections of the current building regulations/building standards apply in relation to the deployment of micro-hdyro systems

**ACTIVITY**

What are the requirements of BS 7671 regarding repairing holes made in walls for the passage of cables?

## Compliance with building regulations

The following building regulations will apply.

| Part | Title | Relevance |
|---|---|---|
| A | Structural safety | If any part of the system is housed in or connected to the building, then structural considerations will need to be taken into account. |
| B | Fire safety | Where cables pass through the building fabric they may reduce the fire-resisting properties of the building fabric. |
| C | Resistance to contaminants and moisture | Where cables pass through the building fabric they may reduce the moisture-resisting properties of the building fabric. |
| E | Resistance to sound | Where cables pass through the building fabric they may reduce the sound-resisting properties of the building fabric. |
| P | Electrical safety | Installation of components and cables |

## Other regulatory requirements

- BS 7671 Wiring Regulations
- G83 requirements for grid-tied systems
- Micro Generation Certification Scheme requirements
- Environment Agency requirements

In England and Wales, all waterways of any size are controlled by the Environment Agency. To remove water from these waterways, even though it may be returned – as in the case of a hydro-electric system – will usually require permission and a licence.

There are three types of licence that may apply to a hydro-electric system.

- An *abstraction licence* will be required if water is diverted away from the main water course. The major concern will be the impact that the project has on fish migration, as the majority of turbines are not fish-friendly. This requirement may affect the choice of turbine (*see* Reverse Archimedes' screw, page 115). It may mean that fish screens are required over water inlets or, where the turbine is in the main channel of water, a fish pass around the turbine may need to be constructed.
- An *impoundment licence* – an impoundment is any construction that changes the flow of water, so if changes or additions are made to sluices, weirs, etc that control the flow within the main stream of water, an impoundment licence will be required.
- A *land drainage licence* will be required for any changes made to the main channel of water.

An Environment Site Audit (ESA) will be required as part of the initial assessment process. The ESA covers:

- water resources
- conservation
- chemical and physical water quality
- biological water quality
- fisheries
- managing flood risk
- navigation of the waterway.

**ASSESSMENT GUIDANCE**

Micro-hydro systems are no good on rivers which suffer seasonal droughts or are subject to winter flooding or freezing, unless preventative measures are taken to protect the generator and control gear.

Assessment criteria

4.1 Identify typical advantages associated with micro-hydro systems

## Advantages

- There are no on-site carbon emissions.
- Large amounts of electricity are output, usually more than required for a single dwelling. The surplus can be sold.
- A feed-in tariff is available.
- There is a reasonable payback period.
- It is an excellent system where no mains electricity exists.
- It is not dependent on weather conditions or building orientation.

Assessment criteria

4.2 Identify typical disadvantages associated with micro-hydro systems

## Disadvantages

- It requires a high head or fast flow of water on the property.
- It requires planning permission, which can be onerous.
- Environment Agency permission is required for water extraction.
- It may require strengthening of the grid for grid-tied systems.
- Initial costs are high.

## CO-GENERATION TECHNOLOGIES - MICRO-COMBINED HEAT AND POWER (HEAT-LED)

In micro-combined heat and power (mCHP) technology, a fuel source is used to satisfy the demand for heat but, at the same time, generates electricity that can either be used or sold back to the supplier.

Currently, mCHP units used in domestic dwellings are powered by means of natural gas or liquid propane gas (LPG), but could be fuelled by using biomass, liquid propane gas (LPG) or other fuels.

The diagram below represents, from left to right, an old, inefficient gas boiler, a modern condensing boiler and an mCHP unit.

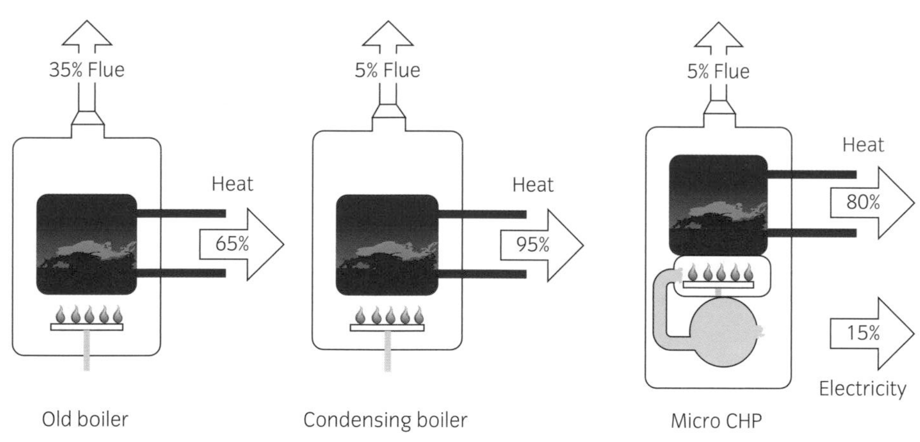


**The efficiency of different boilers**

With the old, inefficient boiler, 65% of the input energy is used to provide heating for the premises, 35% is lost up the flue. With the condensing boiler, this lost heat is re-used so that the output to the heating is 95%. The mCHP unit will achieve the same efficiencies as the modern condensing boiler but 80% of the input is used to provide heat and 15% is used to power a generator to provide power.

There are obvious savings to be made in replacing an old, inefficient boiler with a mCHP unit, but could the same savings not be made by fitting a condensing boiler? At first sight this may appear to be feasible; however, on a unit-by-unit comparison, gas is cheaper than electricity, so any electricity generated by using gas means a proportionally greater financial saving over using electricity.

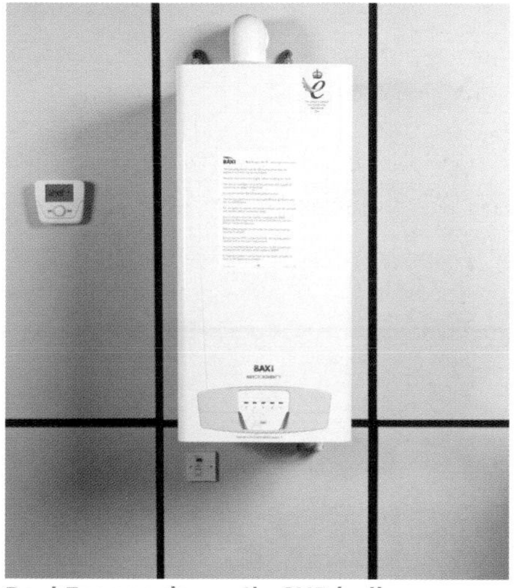

**Baxi Ecogen domestic CHP boiler**

## ASSESSMENT GUIDANCE

The feed-in tariff for mCHP is slightly less than the PV rate and considerably less than wind and hydro tariffs at 2013 rates. For latest rates visit the Energy Saving Trust: www.energysavingtrust.org.uk/.

In addition to this saving, locally generated power reduces transmission losses and consequently creates less carbon dioxide ($CO_2$) than if the electricity were generated at a power station some distance away.

This type of generation, using a mCHP unit, is known as 'heat-led', as the primary function of the unit is to provide space heating, while the generation of electricity is secondary. The more heat that is produced, the more electricity is generated. The unit only generates electricity when there is a demand for heating. Most domestic mCHP units will generate between 1 kW and 1.5 kW of electricity. Micro-combined heat and power is a carbon-reduction technology rather than a carbon-free technology.

**Assessment criteria**

1.3 Identify the fundamental working principles of the following co-generation technologies: micro-combined heat and power (heat-led)

## Working principles

Combined heat and power (CHP) units have been available for a number of years but it is only recently that domestic versions have become available. Domestic versions are usually gas-fired and use a Stirling engine to produce electricity, though other fuel sources, and types of generator combinations, are available.

## ACTIVITY

Find out an alternative name for a Stirling engine.

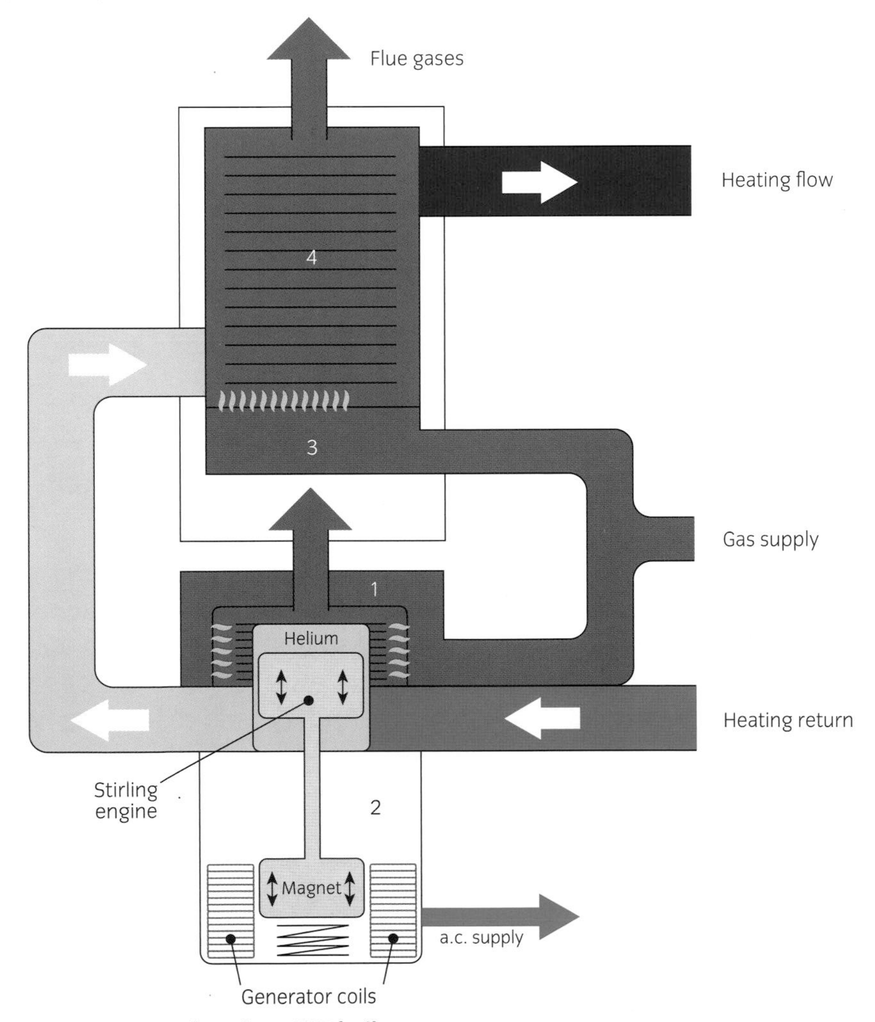


**Component parts of a micro-CHP boiler**

The key components of the mCHP unit are:

1 the engine burner
2 the Stirling engine generator
3 the supplementary burner
4 the heat exchanger.

When there is a call for heat, the engine burner fires and starts the Stirling engine generator. The engine burner produces about 25% of the full heat output of the unit. The burner preheats the heating-system return water before passing it to the main heat exchanger. The hot flue gases from the engine burner are passed across the heat exchanger to heat the heating-system water further. If there is greater demand than is being supplied by the engine burner, then the supplementary burner operates to meet this demand.

## How the Stirling engine generator works

The first Stirling engine was invented by Robert Stirling in 1816 and is very different from the internal combustion engine. The Stirling engine uses the expansion and contraction of internal gases, due to changes in temperature, to drive a piston. The gases within the engine do not leave the engine and no explosive combustion takes place, so the Stirling engine is very quiet in use.

In the case of the Stirling engine used in an mCHP unit, the gas contained within the engine is helium. When the engine burner fires, the helium expands, forcing the piston downwards. The return water from the heating system passing across the engine cools the gas, causing it to contract. A spring arrangement within the engine returns the piston to the top of the cylinder and the process starts all over again.

The piston is used to drive a magnet up and down between coils of wire, generating an electromotive force (emf) in the coils.

## Connection of the mCHP unit to the supply

The preferred connection method between the mCHP unit and the supply is via a dedicated circuit, directly from the consumer unit. This method will allow for easy isolation of the generator from the incoming supply.

Assessment criteria

2.8 Clarify the fundamental requirements for the potential to install a micro-combined heat and power (heat-led) system to exist

## Location and building requirements

For mCHP to be viable, the following criteria must be met.

- The building should have a high demand for space heating. The larger the property the greater the carbon savings.
- A building that is well insulated will not usually be suitable, as a well-insulated building is unlikely to have a high demand for space heating.

If an mCHP unit is fitted to a building that is either too small or is well insulated, this will mean that the demand for heat will be small and the mCHP unit will cycle on and off, resulting in inefficient operation.

**KEY POINT**

Micro-CHP boilers are only suitable where there is a high demand for heat.

Assessment criteria

3.1 Confirm what would be typically classified as 'permitted development' under town and country planning regulations in relation to micro-combined heat and power (heat-led) systems

## Planning permission

Planning permission will not normally be required for the installation of an mCHP unit in a domestic dwelling if all of the work is internal to the building. If the installation requires an external flue to be installed, this will normally be classed as permitted development as long as the following criterion is met.

- Flues to the rear or side elevation do not extend more than 1 m above the highest part of the roof.

### Listed building or buildings in a designated area

Check with the local planning authority regarding both internal work and external flues.

### Buildings in a conservation area or in a World Heritage Site

Flues should not be fitted on the principle or side elevation if they would be visible from a highway.

If the project includes the construction of buildings for fuel storage, or to house the mCHP unit then the same planning requirements for extensions and garden outbuildings will apply.

**ASSESSMENT GUIDANCE**

It is necessary to consider not only the purchase costs but also the running and maintenance costs of any system.

## Compliance with building regulations

The following building regulations will apply.

| Part | Title | Relevance |
|---|---|---|
| A | Structural safety | Where the components increase load on the structure or where holes reduce the structural integrity of the building |
| B | Fire safety | Where installation of the system decreases the fire-integrity of the structure, for example, where pipes or cables pass through fire compartments |
| C | Resistance to contaminants and moisture | Holes for pipes or cables could reduce the moisture-resisting integrity of the building. |
| E | Resistance to sound | Holes for pipes or cables could reduce the sound-resisting qualities of the building. |
| G | Sanitation, hot-water safety, and water efficiency | Hot-water safety and water efficiency |
| J | Heat-producing appliances | mCHP units are heat-producing systems. |
| L | Conservation of fuel and power | Energy efficiency of the system and the building |
| P | Electrical safety | Electrical installation of controls and supply |

Assessment criteria

**3.2** Confirm which sections of the current building regulations/building standards apply in relation to the deployment of micro-combined heat and power (heat-led) systems

## Other regulatory requirements

- Gas regulations will apply to the installation of the mCHP unit. The gas installation work will need to be carried out by an operative registered on the Gas Safe register.
- Water regulations (WRAS) will apply to the water systems.
- BS 7617 Wiring Regulations will apply to the installation of control wiring and the wiring associated with the connection of the mCHP electrical generation output.
- G83 requirements will apply to the connection of the generator, although mCHP units do have a number of exemptions.
- Micro Generation Certification Scheme requirements.

Assessment criteria

4.1 Identify typical advantages associated with micro-combined heat and power (heat-led) systems

## Advantages

- The ability to generate electricity is not dependent on building direction or weather conditions.
- The system generates electricity whilst there is a need for heat.
- A feed-in tariff is available but is limited to generator outputs of less than 2 kW and is only applicable to the first 30 000 units.
- Saves carbon over centrally generated electricity.
- Reduces the building's carbon footprint.

Assessment criteria

4.2 Identify typical disadvantages associated with micro-combined heat and power (heat-led) systems

## Disadvantages

- The initial cost of an mCHP is high, when compared to an efficient gas boiler.
- It is not suitable for properties with low demand for heat – small or very well insulated properties.
- There is limited capacity for generation of electricity.

**OUTCOMES 1 TO 4**

# Water conservation technologies

Many people regard the climate of the United Kingdom as wet. It is a common perception that the UK has a lot of rain and, in some locations, this is true, especially towards the west, where average annual rainfall is in excess of 1000 mm, but along the east coast the average is less than half of this.

The population of the UK is expanding and the demands on the water supply systems are ever increasing. Hose-pipe bans in many parts of the country are a regular feature of the summer months. In the UK, unlike many of our European neighbours, the water supplied is suitable for consumption straight from the tap, but we use it not only for drinking, but for bathing, washing clothes, watering gardens and washing cars.

Even in the UK, clean, fresh water is a limited resource. With growing demand, the pressure on this vital resource is increasing. Water conservation is one way of ensuring that demand does not outstrip supply and that shortages are avoided.

The two methods of water conservation covered in this unit are:

- rainwater harvesting
- re-use of greywater.

Water conservation is one way of reducing water bills. Whether calculated as measured (metered) or unmeasured, water bills, both contain two charges:

- charges for fresh water supplied
- charges for sewage or waste water taken away.

The amount of water taken away, which includes surface water (rainwater), is assumed to be 95% of the water supplied.

By conserving water, the waste-water charge, as well as the charge for fresh water supplied, can be reduced. Besides producing a reduction in household bills, water conservation will also help to relieve the pressure on a vital resource.

**ACTIVITY**

Water conservation has been used for hundreds of years, especially by gardeners. How was this achieved?

## ASSESSMENT GUIDANCE

Given the demands for water and the problems water extraction from rivers may cause, it is essential that better use is made of our available water resources. The historic use of wholesome water for flushing toilets has to change.

## ACTIVITY

Water is a scarce resource which we cannot live without. Most of the water on Earth is seawater which is not fit for human consumption. Find out where your water supply comes from.

The Code for Sustainable Homes sets a target for reducing average drinking water consumption from 150 litres per person per day to an optimum 80 litres. The adopted target is currently 103 litres. Part G of the Building Regulations sets the level at 125 litres. Whichever target is used, the conclusion to be drawn is that a reduction in consumption is vital.

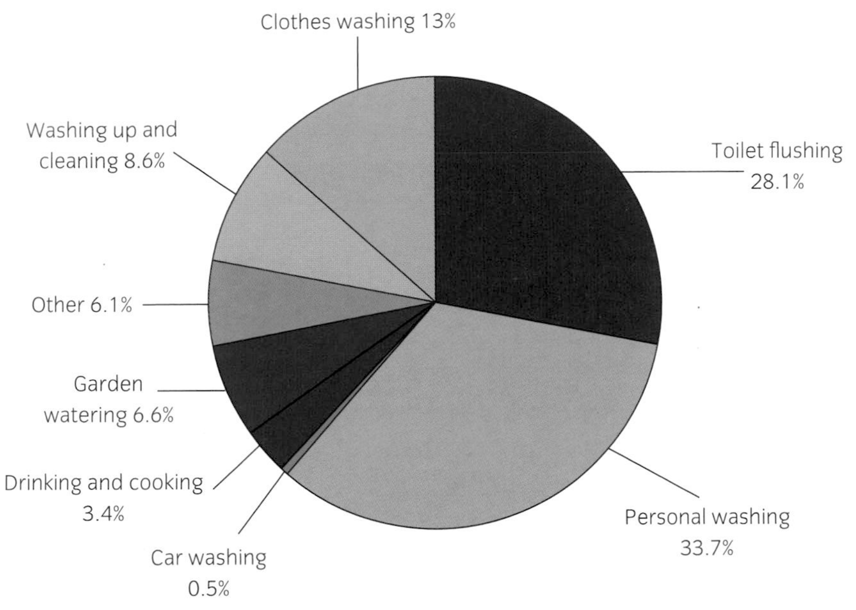


How water is used

Within the average home the amount of water used for drinking or food preparation is estimated at only 3.4% of total consumption, but there are obvious opportunities elsewhere for water savings.

The technologies to be covered are not concerned with direct carbon reduction or with financial savings, but are concerned with the reduction in consumption of a valuable resource.

| Terminology used | Meaning |
|---|---|
| Wholesome water | Water that is palatable and suitable for human consumption. The water that is obtained from the utility company supply and is also known as 'wholesome', 'potable' and 'white water' |
| Rainwater | Water captured from rainwater gutters and downpipes |
| Greywater | Waste water from wash basins, showers, baths, sinks and washing machines |
| Black water | Sewage |

# RAINWATER HARVESTING

Rainwater harvesting refers to the process of capturing and storing rainwater from the surface it falls on, rather than letting it run off into drains or allowing it to evaporate. Re-use of rainwater can result in sizeable reductions in wholesome water usage and thus monetary savings as well as carbon reductions.

If harvested rainwater is filtered, stored correctly and used regularly, so that it does not remain in the storage tanks for an excessive period of time, it can be used for:

- flushing toilets
- car washing
- garden watering
- supplying a washing machine.

Harvested rainwater cannot be used for:

- drinking water
- washing of dishes
- food preparation or washing of food
- personal hygiene, ie washing, bathing or showering.

Rainwater is classified as 'fluid category 5' risk, which is the highest risk category.

## Working principles

**Assessment criteria**

**1.4** Identify the fundamental working principles for rainwater harvesting

The process steps in re-using rainwater are:

- collection
- filtration
- storage
- re-use.

### Collection or capture of rainwater

Rainwater can be captured from roofs or hard standings. In the case of roofs the water is captured by means of gutters and flows to the water-harvesting tank via the property's rainwater downpipes. The amount of water that can be collected will be governed by:

- the size of the capture area
- the annual rainfall in the area.

Not all of the water that falls on the surface can be captured. During periods of very heavy rainfall, water may overflow gutters or merely bounce off the roof surface and avoid the guttering system completely.

Water collected from roofs covered in asbestos, copper, lead or bitumen may not be suitable for re-use and may pose a health risk, or result in discolouration of the water or odour problems. Water collected from hard standings such as driveways may be contaminated with oil or faecal matter.

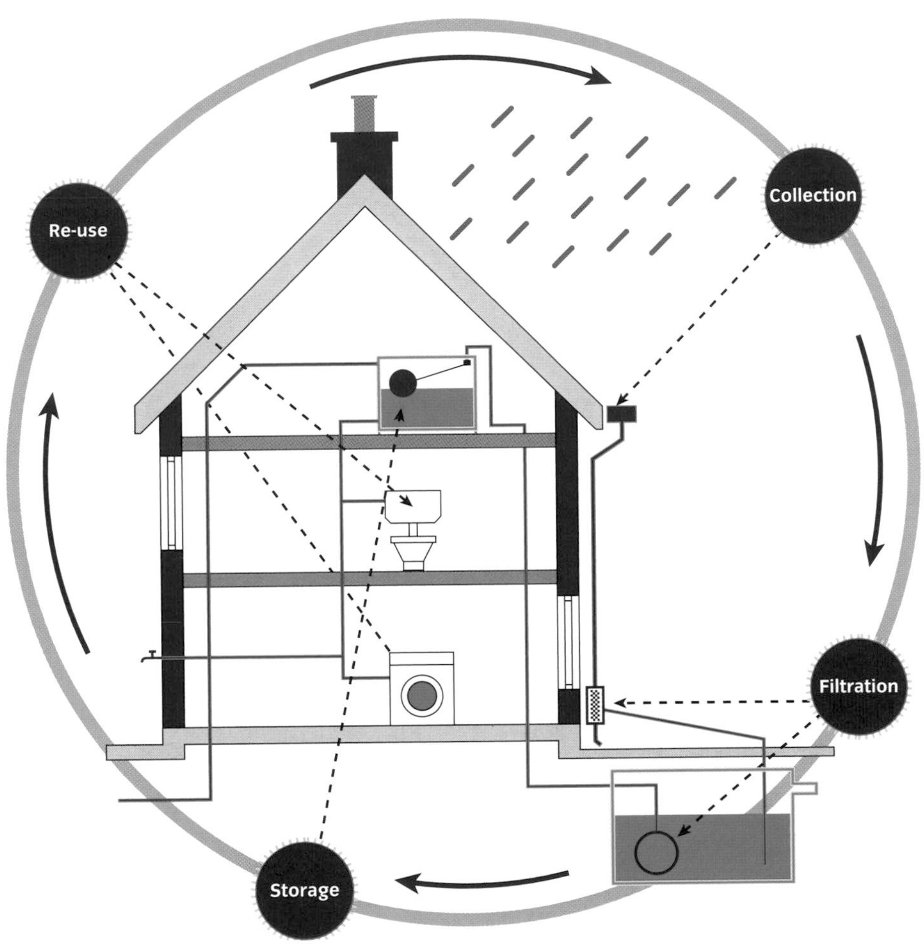


Rainwater-harvesting cycle

## Filtration

As the rainwater passes from the rainwater-capture system to the storage tanks, it passes through an in-line filter to remove debris such as leaves. These are flushed out. The efficiency of this filter will determine how much of the captured water ends up in the storage tank. Manufactures usually quote figures in excess of 90% efficiency.

## Storage of rainwater

Rainwater storage tanks can be either above-ground or below-ground types and can vary in size from a small tank next to a house, to a buried tank that is able to hold many thousands of litres of water. Below-ground tanks will require excavation works, whilst above-ground tanks will need a suitably sized space to be available. Whichever type of tank is used, it will need to be protected against freezing, heating from direct sunlight and contamination.

**KEY POINT**

Most people with a garden will have a water butt to collect rainwater. Special adaptors are available that can be cut into the drainpipe to divert the water into the water butt.

The size of the tank will be determined by the rainwater available and the annual demand. It is common practice to size a tank to be 5% of annual rainwater supply or the anticipated annual demand.

A submersible pump is used to transport water from the storage tank to the point of demand.

The tank will incorporate an overflow pipe connected to the drainage system of the property, for times when the harvested rainwater exceeds the capacity of the storage tank.

## Re-use of stored rainwater

With rainwater harvesting, two system options are available for the re-use of the collected and stored water: indirect and direct distribution.

### Indirect distribution system

With indirect distribution systems, water is pumped from the storage tank to a supplementary storage tank or header tank located within the premises. This, in turn, feeds the water outlets via pipework separated from the wholesome water supply pipes.

**ACTIVITY**

Rainwater is not always as clean as it looks. As it falls through the air it picks up pollutants. Would it be wise to drink rainwater directly after it had fallen?

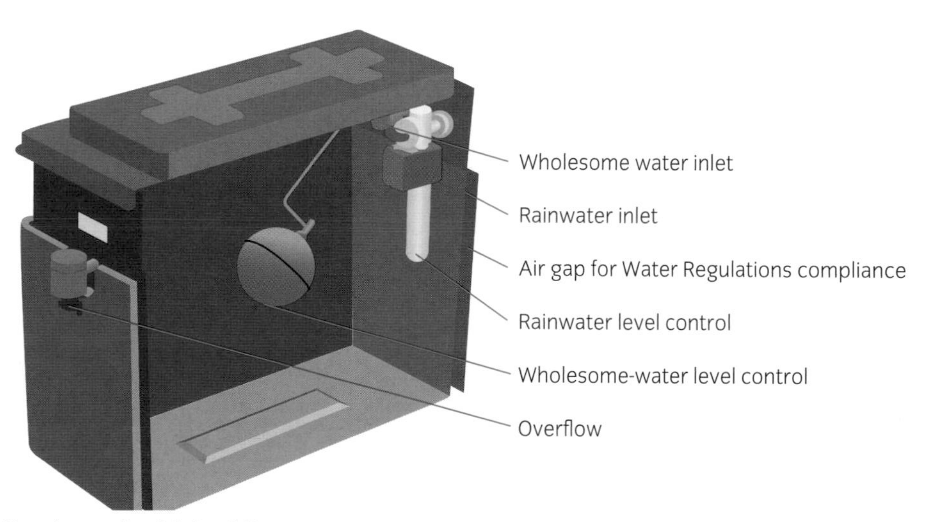


**Header tank with backflow protection**

The header tank will incorporate a backflow prevention air gap to meet the Water Regulation requirements. The arrangement of water control and overflow pipes will ensure that the air gap is maintained. The rainwater level control connects to a control unit that operates the submersible pump, so that water is drawn from the main storage tank when required. At times when there is not enough rainwater to meet the demand, fresh water is introduced into the system via the wholesome water inlet, which is controlled by means of the wholesome-water level control.

### Direct distribution system

In direct distribution systems, the control unit pumps rainwater directly to the outlets on demand. At times of low rainwater availability, the control unit will provide water from the wholesome supply to the outlets. The backflow prevention methods to meet the requirements of the Water Regulations will be incorporated into the control unit. This type of system uses more energy than the indirect distribution system.

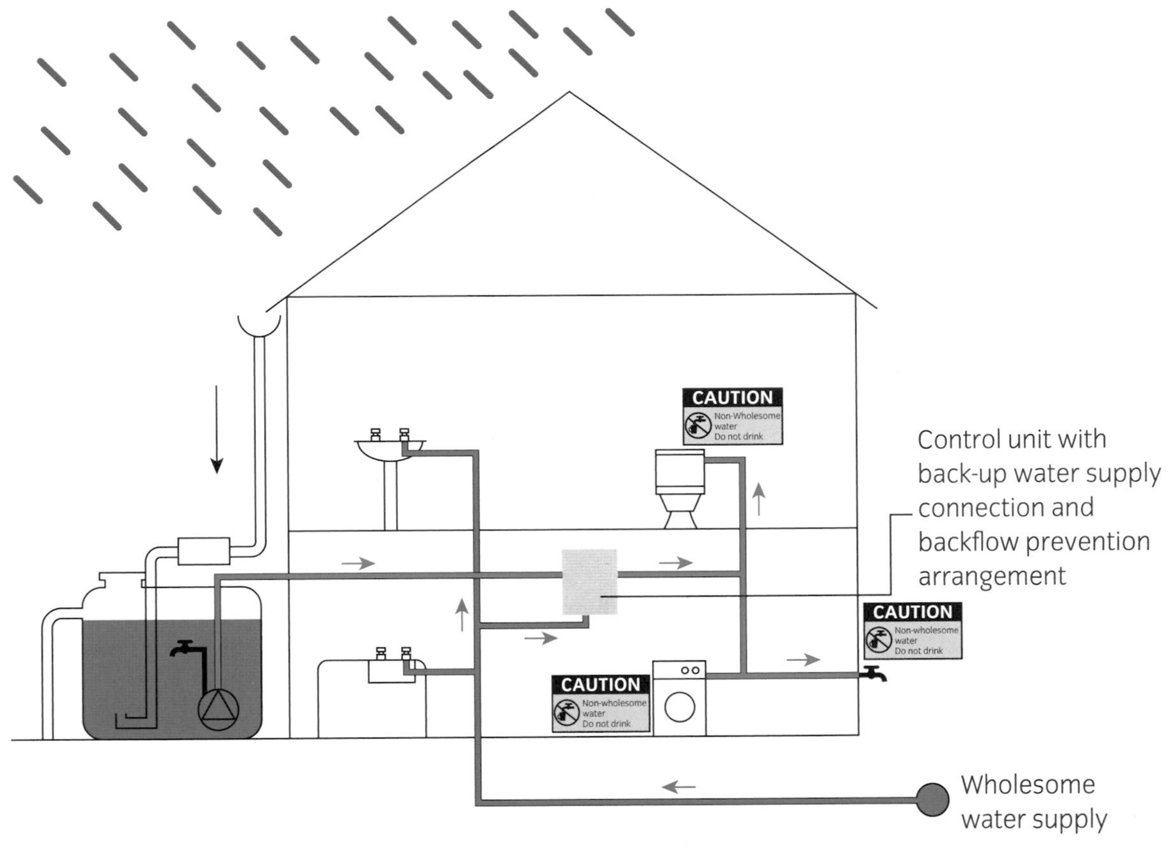


Rainwater-harvesting pipework

**Assessment criteria**

**2.9** Clarify the fundamental requirements for the potential to install a rainwater harvesting system to exist

## Location and building requirements

When deciding on the suitability of a location for the installation of a water-harvesting system, the following points will need to be taken into account.

- Is there suitable supply of rainwater to meet the demand? This is determined by finding the rainfall available and the level of water use by the occupants.
- A suitable supply of wholesome water will be required to provide back-up at times of drought.
- For above-ground storage tanks, the chosen location must avoid the risk of freezing, or the warming effects of sunlight, which may encourage algal growth.
- For below-ground tanks, consideration will need to be given to access for excavation equipment.

## Planning permission

In principle, planning permission is not normally required for the installation of a rainwater-harvesting system if it does not alter the outside appearance of the property. It is, however, always worth enquiring, especially if the system is installed above ground, the building is in a designated area or the building is listed.

**Assessment criteria**

3.1 Confirm what would be typically classified as 'permitted development' under town and country planning regulations in relation to the deployment of rainwater harvesting systems

## Compliance with building regulations

The following building regulations will apply.

| Part | Title | Relevance |
|---|---|---|
| A | Structural safety | Where the components affect the loadings placed on the structure of the building or excavations are in close proximity to the building |
| B | Fire safety | Where holes for pipework reduce the fire-resisting integrity of the structure |
| C | Resistance from contaminants and moisture | Where holes for pipework reduce the moisture-resisting integrity of the structure, for example, pipework passing through vapour barriers |
| E | Resistance to sound | Where holes for pipework reduce the sound-resisting integrity of the structure |
| G | Sanitation, hot-water safety, and water efficiency | Water efficiency |
| H | Drainage and waste disposal | Where gutters and rainwater pipes are connected to the system. |
| P | Electrical safety | Installation of supply and control wiring for the system |

**Assessment criteria**

3.2 Confirm which sections of the current building regulations/building standards apply in relation to the deployment of rainwater harvesting systems

## Other regulatory requirements

- The Water Supply (Water Fittings) Regulations 1999 apply to rainwater-harvesting systems. The key area of concern will be the avoidance of cross-contamination between rainwater and wholesome water. This is known as 'backflow prevention' and, as rainwater is classified as a category 5 risk, the usual method of providing backflow prevention is with a type AA air-gap between the wholesome water and the rainwater.

Pipework labels

Any pipework used to supply outlets with the rainwater will need to be labelled to distinguish it from wholesome water. Outlets will also need to be labelled to indicate that the water supplied is not suitable for drinking.

## Other regulatory requirements

- Wiring Regulations BS 7671 will apply to the installation of supplies and control systems for the rainwater-harvesting system.
- BS8515:2009 Rainwater Harvesting Systems – Code of Practice

**Assessment criteria**

4.1 Identify typical advantages associated with each of rainwater harvesting systems

## Advantages

- There is a reduction in use of wholesome water.
- Water bills are reduced if the supply is metered.
- Water does not require any further treatment before use.
- The system is less complicated than greywater re-use systems.

**Assessment criteria**

4.2 Identify typical disadvantages associated with each of rainwater harvesting systems

## Disadvantages

- The quantity of available water is limited by roof area. It may not meet the demand in dry periods.
- Initial costs are high.
- A water meter should be fitted.

**ASSESSMENT GUIDANCE**

Combined systems divert the surface water (rainwater) into the main drainage system. This can lead to flooding of premises in times of heavy rain.

# GREYWATER RE-USE

Greywater is the waste water from baths, showers, hand basins, kitchen sinks and washing machines. It gets its name from its cloudy, grey appearance. Capturing and re-using the water for permitted uses reduces the consumption of wholesome (drinking) water.

**KEY POINT**

Whilst kitchen sink waste is classified as greywater, it is not usually collected and recycled. This is because the FOGS (fats, oils and greases) contained within it will emulsify as the water cools and will not be kind to the filtration systems.

Greywater collected from washbasins, showers and baths will often be contaminated with human intestinal bacteria and viruses, as well as organic material such as skin particles and hair. As well as these contaminants, it will also contain soap, detergents and cosmetic products, which are ideal nutrients for bacteria growth. Add to this the relatively high temperature of the greywater and the ideal conditions exist to encourage the growth of bacteria.

For these reasons, untreated greywater cannot be stored for more than a few hours. The less-polluted water from washbasins, showers and baths is usually used in greywater re-use systems. This is known as 'bathroom greywater'. Where a greater supply of greywater is required, washing-machine waste water is collected.

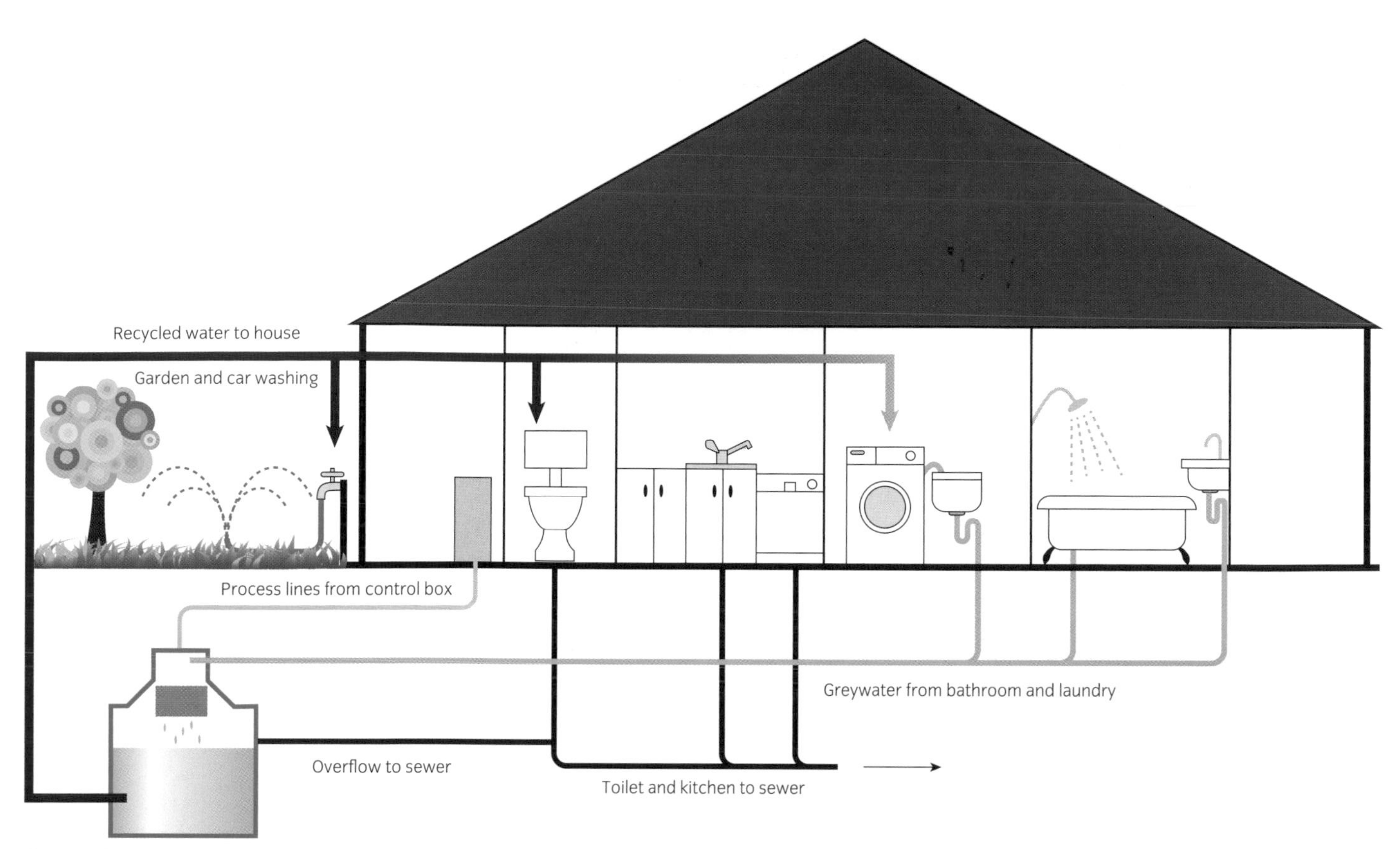


**Greywater re-use system**

Greywater is classified as fluid category 5 risk (the highest) under the Water Supply (Water Fittings) Regulations 1999. Greywater can pose a serious health risk, due to its potential pathogen content. Untreated greywater deteriorates rapidly when stored, so all systems that store greywater will need to incorporate an appropriate level of treatment.

If greywater is filtered and stored correctly then it can be used for:

- flushing toilets
- car washing
- garden watering
- washing clothes (after additional processing).

Greywater cannot be used for:

- drinking water
- washing of dishes
- food preparation or washing of food
- personal hygiene – washing, bathing or showering.

## ACTIVITY

Currently, greywater is normally discharged directly into the drainage system of the house and then to the main drainage system. Think of the changes that would have to be made to direct all the greywater to a central collection point.

Assessment criteria

**1.4** Identify the fundamental working principles for greywater re-use

## Working principles

Several types of greywater re-use systems exist but, apart from the direct re-use system, they all have similar common features:

- a tank for storage of the treated water
- a pump
- a distribution system for moving the water from storage to where it is to used
- some form of treatment.

### Direct re-use

Greywater is collected from appliances and directly re-used, without treatment or storage, and can be used for such things as watering the garden. Even so, the greywater is not considered suitable for watering fruit or vegetable crops.

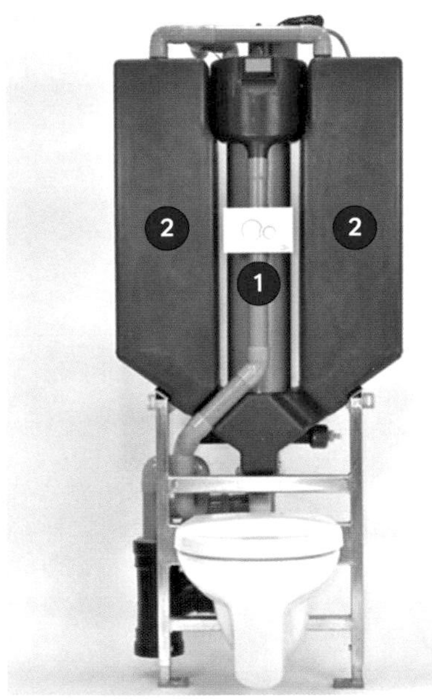


**Ecoplay greywater system**

1 **Cleaning tank**
2 **Storage tank**

### Short retention system

Greywater from baths and showers is collected in a cleaning tank, where it is treated, by means such as surface skimming, to remove debris such as soap, hair and foam. Heavier particles are allowed to settle to the bottom, where they are flushed away as waste. The remaining water is then transferred to a storage tank, ready for use. The storage tanks are usually relatively small, at around 100 litres, which is enough for 18–20 toilet flushes. If the water is not used within a short time, generally 24 hours, the stored greywater is purged, the system is cleaned and a small amount of flush water is introduced to allow toilet flushing. This avoids the greywater deteriorating, and beginning to smell, at times when the premises are unoccupied for a lengthy period of time. This type of system can result in water savings of 30%.

This type of system would be ideal for installation in a new-build project but would be more difficult to retrofit. It is usually fitted in the same room as the source of greywater.

### Physical and chemical system

This system uses a filter to remove debris from the collected water (physical cleaning). After the greywater has been filtered, chemical disinfectants such as chlorine or bromine are added, to inhibit bacterial growth during storage.

### Biomechanical system

This type of system is the most advanced of the greywater re-use systems. It uses both biological and physical methods to treat the collected greywater. An example of such a system is the German system 'AquaCycle® 900', which comprises an indoor unit about the size of a large refrigerator.

**ACTIVITY**

Go to Water Works UK www.wwuk.co.uk or alternative websites and see the range of systems available.

Greywater enters the system and passes through the filtering unit (1), where particles such as hair and textiles debris are filtered out. The filtering unit is electronically controlled to provide automatic flushing of the filter.

Water enters the main recycling chamber (2), where organic matter is decomposed by bio-cultures. The water remains in this chamber for 3 hours before being pumped to the secondary recycling chamber (3), for further biological treatment. Biological sediment settles to the bottom of each chamber (4), where it is sucked out and transferred to a drain.

After a further period of 3 hours, the water passes through a UV filter (5), to the final storage chamber (7), where it is ready for use. When there is demand for the treated water this is pumped (8) to the point of demand. At times when treated water availability is low, fresh water (6) can be introduced to the system.

Water from this unit can be used for washing clothes as well as the other uses previously stated.

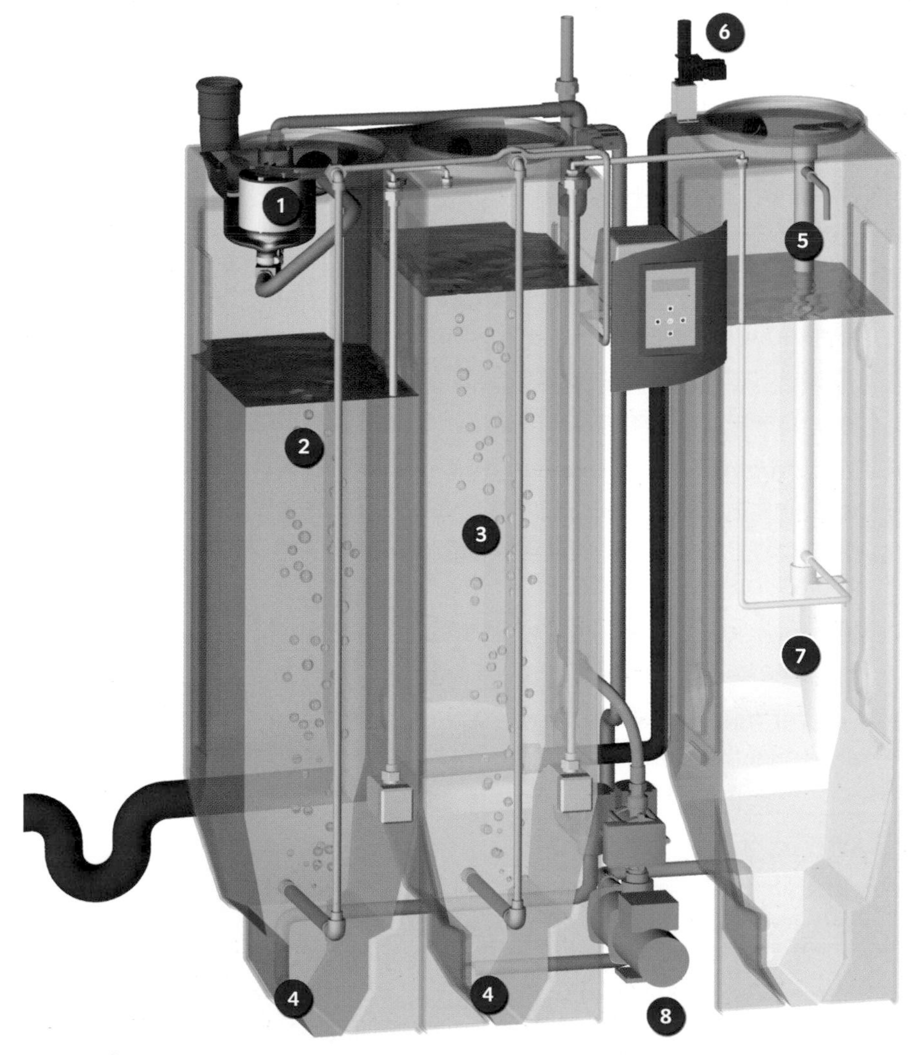


AquaCycle® 900 system

## Biological system

This type of system uses some of the principles employed by sewage treatment works. In this case, bacterial growth is encouraged rather than inhibited, by the introduction of oxygen to the waste water. Oxygen can be introduced by means of pumps pushing air through the storage tanks. Bacteria then 'digest' the organic matter contained within the greywater.

A more 'natural' method of oxygenating the water is by the use of reed beds. In nature, reeds, which thrive in waterlogged conditions, transfer oxygen to their roots. The greywater is allowed to infiltrate reed beds. The added oxygen and naturally occurring bacteria will remove any organic matter contained in the waste water. The disadvantages of using reed beds are the land area required for the reed beds and the expertise required to maintain them.

GROW system

Alternative Water Solutions produces a system based on these principles, but on a smaller scale, called the Green Roof Recycling System (GROW). This uses a system of tiered gravel-filled troughs planted with native plants, the roots of which can perform the same function of filtering as a reed bed would.

**Assessment criteria**

**2.9** Clarify the fundamental requirements for the potential to install a greywater re-use system to exist

## Location and building requirements

When considering the installation of a greywater re-use system, the following location or building requirements should be taken into account.

- There needs to be a suitable supply of greywater to meet the demand. Premises with a low volume of greywater are not suitable.
- Suitability of the location and availability of space to store enough greywater to meet the demand of the premises must be assessed.
- Storage tanks need to be located away from heat, including direct sunshine, to avoid the growth of algae. They need to be located so they are not subject to freezing in cold weather.There needs to be a wholesome water supply.
- Where greywater tanks are retrofitted, access for excavation equipment will need to be considered.
- A water meter will need to be fitted on the water supply to maximise the benefits.

**Assessment criteria**

**3.1** Confirm what would be typically classified as 'permitted development' under town and country planning regulations in relation to the deployment of greywater re-use systems

## Planning permission

In principle, planning permission is not normally required for the installation of a greywater re-use system, if the system does not alter the outside appearance of the property. It is, however, always worth enquiring, especially if the system is installed above ground, the building is in a designated area or the building is listed.

If a building is required to house the greywater storage system, then a planning application will need to be submitted.

Assessment criteria

3.2 Confirm which sections of the current building regulations/building standards apply in relation to the deployment of greywater re-use systems

## Compliance with building regulations

The following building regulations will apply.

| Part | Title | Relevance |
|---|---|---|
| A | Structural safety | Where the components affect the loadings placed on the structure of the building or excavations are in close proximity to the building |
| B | Fire safety | Where holes for pipework reduce the fire-resisting integrity of the structure |
| C | Resistance to contaminants and moisture | Where holes for pipework reduce the moisture-resisting integrity of the structure, for example, pipework passing through vapour barriers |
| E | Resistance to sound | Where holes for pipework reduce the sound-resisting integrity of the structure |
| G | Sanitation, hot-water safety, and water efficiency | Water efficiency |
| H | Drainage and waste disposal | Where waste pipes are connected to the system |
| P | Electrical safety | Installation of supply and control wiring for the system |

## Other regulatory requirements

- The Water Supply (Water Fittings) Regulations 1999 apply to greywater recycling installations. The key area of concern will be the avoidance of cross-contamination between greywater and wholesome water. This is known as 'backflow prevention' and, as greywater is classified as a category 5 risk, the usual method of providing backflow prevention is with an air gap between the wholesome water and the greywater.
- Any pipework used to supply outlets with the treated greywater will need to be labelled to distinguish it from pipework for wholesome water. Outlets will need to be clearly identified by means such as labelling. Outlets will also need to be labelled to indicate that the water supplied is not suitable for drinking.

Greywater warning label

## Other regulatory requirements

- The local water authority must be notified when a greywater re-use system is to be installed.
- Wiring Regulations BS 7671 will apply to the installation of supplies and control systems for the greywater re-use system.

Assessment criteria

4.1 Identify typical advantages associated with each of greywater re-use systems

## Advantages

- There will be a reduction in water bills if the supply is metered.
- It reduces demands on the wholesome water supply.
- A wide range of system options exists.
- It has the potential to provide more re-usable water than a rainwater harvesting system.

Assessment criteria

4.2 Identify typical disadvantages associated with each of greywater re-use systems

**KEY POINT**

Traditionally, water charges were based on the rateable value of the house. Since this system was abolished, most companies kept similar charges but many are now fitting water meters.

## Disadvantages

- There are long payback periods.
- It can be difficult to integrate into an existing system.
- Only certain types of appliance or outlet can be connected. This causes additional plumbing work.
- Cross-contamination can be a problem.
- A water meter will need to be fitted to make maximum gains.
- The need for filtering and pumping may actually increase rather than decrease the carbon footprint.

# ASSESSMENT CHECKLIST

## WHAT YOU NOW KNOW/CAN DO

| Learning outcome | Assessment criteria | Page number |
|---|---|---|
| 1 Know the fundamental working principles of micro-renewable energy and water conservation technologies | *The learner can:* | |
| | 1 Identify the fundamental working principles for each of the following heat-producing micro-renewable energy technologies: | |
| | ■ solar thermal (hot water) | 69 |
| | ■ ground source heat pump | 76, 79 |
| | ■ air source heat pump | 76, 84 |
| | ■ biomass | 88 |
| | 2 Identify the fundamental working principles for each of the following electricity producing micro-renewable energy technologies: | |
| | ■ solar photovoltaic | 73 |
| | ■ micro-wind | 104 |
| | ■ micro-hydro | 111 |
| | 3 Identify the fundamental working principles of the following co-generation technologies: | |
| | ■ micro-combined heat and power (heat-led) | 120 |
| | 4 Identify the fundamental working principles for each of the following water conservation technologies: | |
| | ■ rainwater harvesting | 127 |
| | ■ greywater re-use. | 134 |

| Learning outcome | Assessment criteria | Page number |
|---|---|---|
| 2 Know the fundamental requirements of building location/building features for the potential to install micro-renewable energy and water conservation systems to exist | *The learner can:* | |
| | 1 Clarify the fundamental requirements for the potential to install a solar water heating system to exist | 73 |
| | 2 Clarify the fundamental requirements for the potential to install a solar photovoltaic system to exist | 100 |
| | 3 Clarify the fundamental requirements for the potential to install a ground source heat pump system to exist | 83 |
| | 4 Clarify the fundamental requirements for the potential to install an air source heat pump system to exist | 86 |
| | 5 Clarify the fundamental requirements for the potential to install a biomass system to exist | 91 |
| | 6 Clarify the fundamental requirements for the potential to install a micro-wind system to exist | 106 |
| | 7 Clarify the fundamental requirements for the potential to install a micro-hydro system to exist | 115 |
| | 8 Clarify the fundamental requirements for the potential to install a micro-combined heat and power (heat-led) system to exist | 122 |
| | 9 Clarify the fundamental requirements for the potential to install a rainwater harvesting/ greywater re-use system to exist. | 130 |

| Learning outcome | Assessment criteria | Page number |
|---|---|---|
| 3 Know the fundamental regulatory requirements relating to micro-renewable energy and water conservation technologies | *The learner can:* | |
| | 1 Confirm what would be typically classified as 'permitted development' under town and country planning regulations in relation to the deployment of the following technologies: | 65 |
| | ■ solar thermal (hot water) | 74 |
| | ■ solar photovoltaic | 102 |
| | ■ ground source heat pump | 83 |
| | ■ air source heat pump | 86 |
| | ■ micro-wind | 108 |
| | ■ biomass | 92 |
| | ■ micro-hydro | 116 |
| | ■ micro-combined heat and power (heat-led) | 122 |
| | ■ rainwater harvesting | 131 |
| | ■ greywater re-use | 136 |
| | 2 Confirm which sections of the current building regulations/building standards apply in relation to the deployment of the following technologies: | 65 |
| | ■ solar thermal (hot water) | 75 |
| | ■ solar photovoltaic | 103 |
| | ■ ground source heat pump | 83 |
| | ■ air source heat pump | 87 |
| | ■ micro-wind | 110 |
| | ■ biomass | 93 |
| | ■ micro-hydro | 116 |
| | ■ micro-combined heat and power (heat-led) | 123 |
| | ■ rainwater harvesting | 131 |
| | ■ greywater re-use. | 137 |

| Learning outcome | Assessment criteria | Page number |
|---|---|---|
| 4 Know the typical advantages and disadvantages associated with micro-renewable energy and water conservation technologies | *The learner can:* | |
| | 1 Identify typical advantages associated with each of the following technologies: | |
| | ■ solar thermal (hot water) | 76 |
| | ■ solar photovoltaic | 103 |
| | ■ ground source heat pump | 84 |
| | ■ air source heat pump | 88 |
| | ■ micro-wind | 110 |
| | ■ biomass | 93 |
| | ■ micro-hydro | 118 |
| | ■ micro-combined heat and power (heat-led) | 124 |
| | ■ rainwater harvesting | 132 |
| | ■ greywater re-use | 138 |
| | 2 Identify typical disadvantages associated with each of the following technologies: | |
| | ■ solar thermal (hot water) | 76 |
| | ■ solar photovoltaic | 104 |
| | ■ ground source heat pump | 84 |
| | ■ air source heat pump | 88 |
| | ■ micro-wind | 111 |
| | ■ biomass | 94 |
| | ■ micro-hydro | 118 |
| | ■ micro-combined heat and power (heat-led) | 124 |
| | ■ rainwater harvesting | 132 |
| | ■ greywater re-use. | 138 |

## ASSESSMENT GUIDANCE

- The assessment for this unit is by a multiple-choice exam.
- You will find sample questions in this book and on SmartScreen.
- Make sure you go over these questions in your own time.
- Spend time on revision in the run-up to the exam.
- Ensure you know how the e-volve system works. Ask for a demonstration if you are not sure.
- It is better to have time left over at the end rather than have to rush to finish.
- Make sure you read every question carefully.
- If you need paper ask the invigilator to provide some.
- Make sure you have a scientific (non-programmable) calculator.
- Do not take any paperwork into the exam with you.
- Make sure your mobile phone is switched off during the exam. You may be asked to give it to the invigilator.

## OUTCOME KNOWLEDGE CHECK

1 A system which uses a fluid to capture heat from the sun is known as:

**a)** photovoltaic
**b)** solarvoltaic
**c)** solar thermal
**d)** solar thermostatic.

2 A buffer tank can be used for storing:

**a)** cold water
**b)** rainwater
**c)** electricity
**d)** hot water.

3 A system which uses sunlight to generate electricity is known as:

**a)** photovoltaic
**b)** photosynthetics
**c)** solar thermal
**d)** solar chemical.

4 Excess electricity can be sold back to the grid using:
   a) an off-peak tariff
   b) a feed-in tariff
   c) a feed-out tariff
   d) a daytime tariff.

5 One disadvantage of ground source heat pumps using horizontal pipework is:
   a) the need for unshaded ground
   b) the need for large ground area
   c) the need for deep boreholes
   d) that it is less than 100% efficient.

6 The air source heat pump works on the principle of:
   a) cold exchange
   b) electromagnetic induction
   c) the refrigeration cycle
   d) the Otto cycle.

7 Which one of the following fuels is commonly used in biomass systems?
   a) Coal.
   b) Oil.
   c) Shale.
   d) Wood chips.

8 Rainwater harvesting should not be used for:
   a) car washing
   b) drinking water
   c) toilet flushing
   d) garden watering.

9 A micro-hydro system relies upon:
   a) a constant wind speed
   b) reliable water flow
   c) a deep well
   d) sufficient sunlight.

10 A micro-wind system generates d.c. that is changed into a.c. by the use of:

a) a rectifier
b) an inverter
c) a commutator
d) slip rings.

11 Which orientation is NOT suitable for mounting a photovoltaic panel?

a) North.
b) South.
c) East.
d) West.

12 The maximum distance a solar thermal panel may extend above the roof surface is:

a) 50 mm
b) 150 mm
c) 200 mm
d) 250 mm.

13 A ground source heat pump installed on a building with limited grounds will require:

a) deep boreholes
b) long pipe trenches
c) an extra low-voltage supply
d) shallow trenches.

14 An air source heat pump has an electrical input of 2.5 kW and output of 7.5 kW. The coefficient of performance will be:

a) 0.33
b) 3
c) 5
d) 0.

**15** The position of the biomass flue should be:

a) at the front of the building
b) at the side or rear of a building
c) roof-mounted at least 2 m above roof level
d) roof-mounted at least 3 m above roof level.

**16** An 'on-grid' generating system is one where the system:

a) is connected in a 4 × 4 grid
b) is connected to a battery bank only
c) can be moved within an installed grid
d) is connected to the national grid.

**17** Installation of greywater systems would depend upon:

a) total black and greywater available
b) adequate greywater availability
c) sufficient rainwater only
d) direct mixing with wholesome water.

**18** One disadvantage associated with biomass systems is:

a) water produced by burning fuel
b) inability to control heat output
c) need for large storage area
d) large amount of ash produced.

**19** One disadvantage of photovoltaic systems is:

a) they cannot be installed facing South
b) there is little or no output at night
c) they will always require roof strengthening
d) they cannot be exposed to snow or ice.

**20** The minimum recommended height of wind turbine blades from the ground is:

a) 2 m
b) 3 m
c) 4 m
d) 5 m.

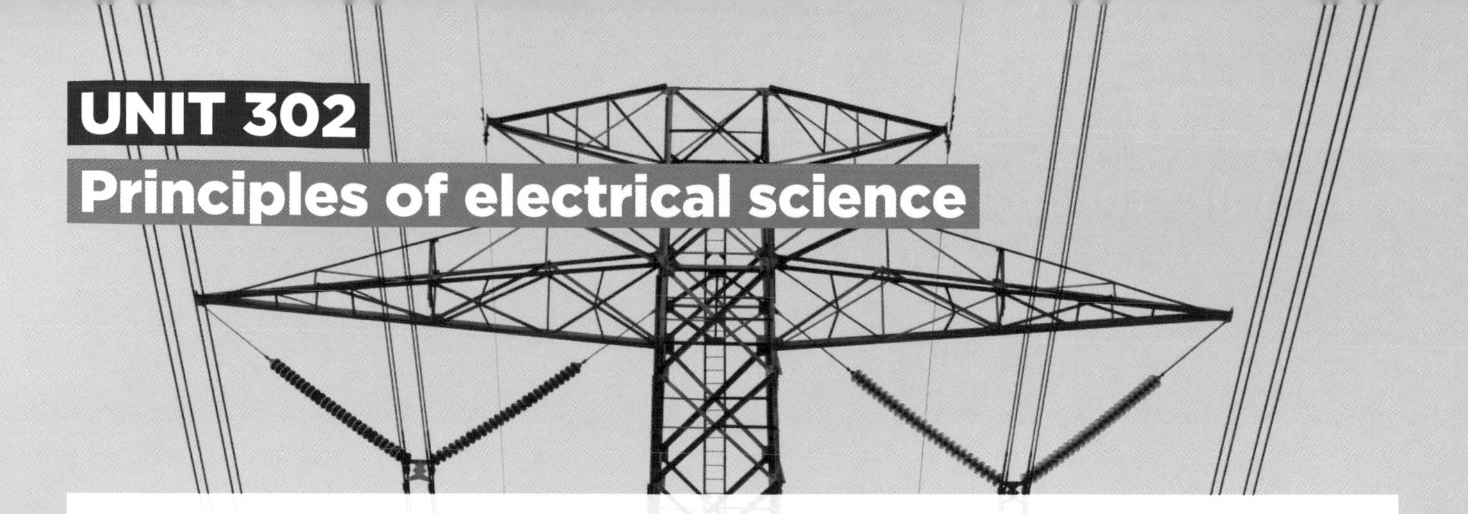

# UNIT 302
# Principles of electrical science

The aim of this unit is to enable you to understand the principles of electrical science related to a.c. theory, machines, devices and systems. This understanding is applied when designing wiring systems for clients and fault diagnosis.

## LEARNING OUTCOMES

There are seven learning outcomes to this unit. The learner will:

1. Understand the principles of a.c. theory
2. Understand the principles of lighting systems
3. Understand electrical quantities in star delta configurations
4. Understand the principles of electrical machines
5. Understand the principles of electrical devices
6. Understand the principles of electrical heating systems
7. Understand the principles of electronic components in electrical systems.

The unit will be assessed by:

- practical assignment
- written short-answer examination.

# OUTCOME 1

# Understand the principles of a.c. theory

**SmartScreen Unit 201**

Additional resources to support this unit are available on SmartScreen.

## INTRODUCTION

Alternating currents (a.c.) are universally used in the supply industry and also in most industrial, commercial and domestic applications. In the past, supplies in the UK were direct current (d.c.) but as d.c. cannot be *transformed*, that is stepped up or stepped down, a.c. was soon adopted.

The pure a.c. waveform follows a sine wave, rising and falling twice per cycle in both the voltage and current components. Understanding the effects that different types of load have on an a.c. waveform is important when considering power factor and any necessary correction required.

**Assessment criteria**

1.1 Explain the effects of components in a.c. circuits

## THE EFFECTS OF COMPONENTS IN A.C. CIRCUITS

Different components have different effects on a.c. circuits; these effects vary with frequency, depending on the components in the circuit. Other factors, such as resistance, have no effect on the waveform except to resist current flow.

### Resistance

When a resistor is used in an a.c. circuit, the voltage drop and the current through the resistor are in phase, as the resistor has no effect on the circuit voltage or current, other than to restrict current flow proportionally to the voltage. The sinusoidal voltages and currents have no phase shift and are described as being 'in phase'. This in-phase or *unity* relationship can be understood using Ohm's law for the voltage and current:

$$V = I \times R$$

As there is no phase shift effect, there is no requirement in a resistive circuit to include the power factor. Resistive loads in a.c. circuits include items such as incandescent lamps, water heater (immersion) elements and electric kettles. Any load that relies on current passing through a material and producing heat is resistive.

**ASSESSMENT GUIDANCE**

Ohm's law states that the current flowing in a circuit is proportional to the applied voltage and inversely proportional to the resistance.

**ASSESSMENT GUIDANCE**

Power factor is the ratio of true power (watts) to volt-amperes:

Pf = W/VA

The relationship between the voltage, current and resistance in an a.c. circuit can be shown as a circuit diagram, phasor diagram and sine wave, together with the appropriate formula.

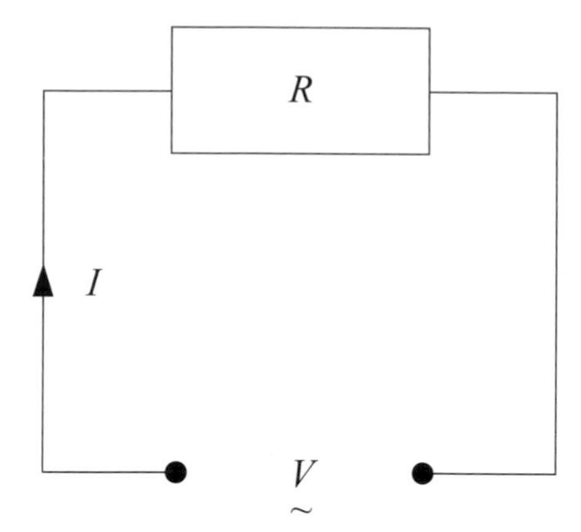


**Relationship between voltage, current and resistance as shown by a circuit diagram**

The circuit diagram shows the resistor connected to an a.c. supply.

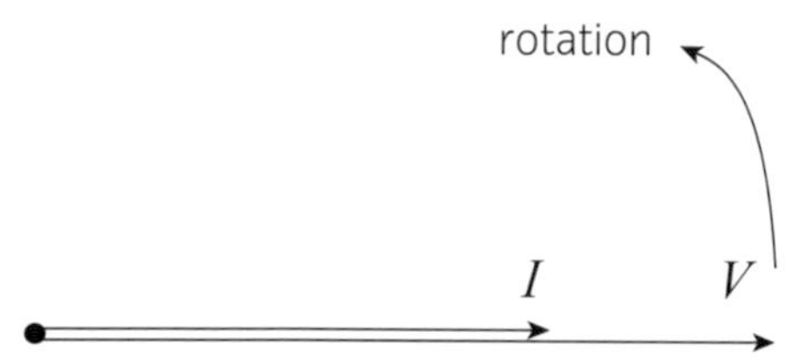


**Relationship between the voltage and current as shown by a phasor diagram**

## ASSESSMENT GUIDANCE

A phasor always rotates anti-clockwise.

The phasor diagram shows the relationship between the voltage and current. As the voltage and current are together, they are in phase or unity.

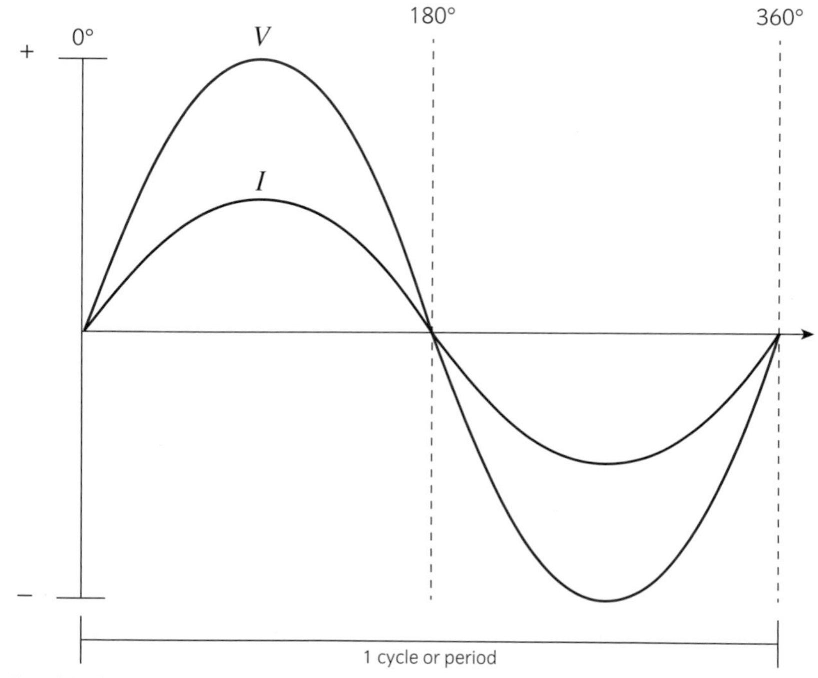


**Relationship between voltage and current as shown by a wave form**

The sine wave shows the voltage and current rising and falling at the same time. In a.c. circuits involving resistance only, Ohm's law applies as:

$$V = I \times R$$

## Inductance

An inductor is a length of wire; sometimes this is a longer wire wound into a coil with a core of iron or air. An inductor is a component, such as a solenoid or field winding in a motor or a ballast unit in a fluorescent luminaire, that induces (produces) magnetism. Inductance is proportional to the inductor's opposition to a.c. current flow.

- The symbol for inductance is $L$.
- The unit of inductance is the henry (H).

As inductance increases and all other factors, such as frequency, remain the same, a.c. current flow reduces. This opposition to a.c. current flow is indicated as an increase in inductive reactance $X_L$.

We will assume that an inductor in a circuit produces pure inductance but this is actually impossible; as an inductor is a coiled wire, it also has resistance.

### ASSESSMENT GUIDANCE

Self-inductance occurs when a current flowing in a coil induces an emf which opposes the force setting it up. In other words, the back emf opposes the supply voltage. This occurs all the time in an a.c. circuit and also in a d.c. circuit when the current changes in value.

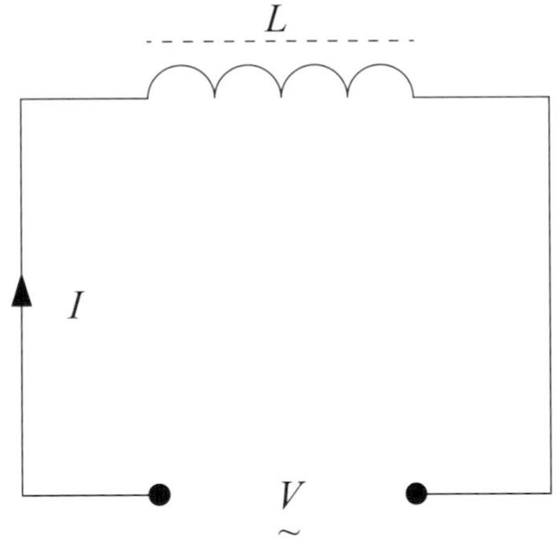


An inductor in a circuit

A pure inductance, as shown in the circuit, will create a phase shift that causes the current to lag behind the voltage by 90°.

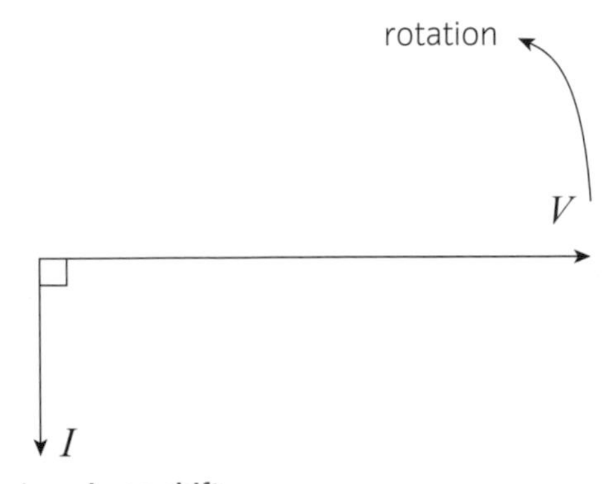


Phasor diagram showing phase shift

The phasor diagram shows the phase shift. When the phasor is rotated in the direction shown, the voltage leads the current or, as we normally say, the current *lags* behind the voltage. If the inductance was pure, this lag would be 90°. We will explore how to create a phasor diagram later in this outcome.

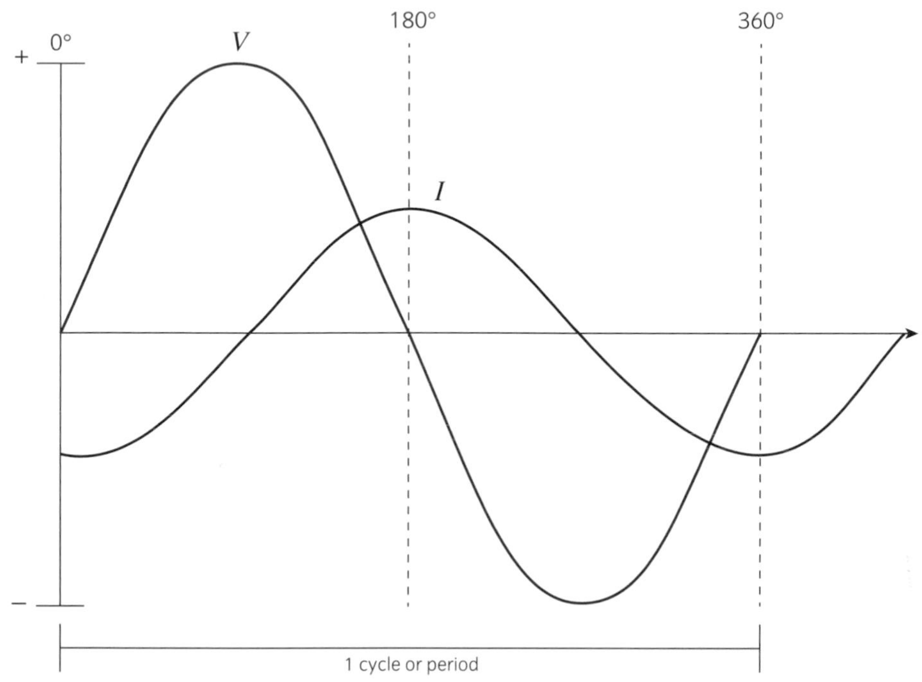


**Sine wave diagram showing phase shift**

## ACTIVITY

A sine wave or sinusoidal wave can be plotted using sine values from 0 to 360°.

Using a calculator find the values from 0 to 360° and plot a sine wave in which the peak voltage is 100 V.

If you look at the lagging current sine wave, you will see that the current does not begin its cycle until the voltage is 90° into its cycle. Where the voltage begins its negative half-cycle, the current is still in a positive cycle – meaning the two values are opposing or reacting to one another. This opposition is known as reactance, $X$, measured in ohms ($\Omega$). As the reactance is linked to an inductor, we call this inductive reactance ($X_L$).

The reactance in the circuit can be determined using the value of inductance ($L$) in henrys (H) and the frequency of the supply ($f$) measured in hertz (Hz). So

$$X_L = 2\pi fL$$

## Example calculation

Determine the reactance when an inductance of 40 mH is connected to a 230 V 50 Hz supply.

As:

$$X_L = 2\pi fL$$

then:

$$X_L = 2\pi \times 50 \times 40 \times 10^{-3} = 12.56\,\Omega$$

If the inductance is pure, the reactance replaces resistance in Ohm's law so:

$$I = \frac{V}{X_L}$$

So using the example above, the value of circuit current would be:

$$I = \frac{230}{12.56} = 18.3\,\text{A}$$

Assuming the resistance to be pure, this current would lag the voltage by 90°. Remember, however that the inductance cannot be pure as the coil or winding is made up of a conductor which has resistance. We will study the effect of resistance on an inductor later in this outcome.

## Capacitance

A capacitor is a device that is used to store and discharge energy. It does not contain a resistance so it does not dissipate energy. A capacitor *can* be pure. Capacitors are used in circuits for many reasons; most commonly, capacitors are found in fluorescent luminaires for power factor correction purposes. Capacitance ($C$) is measured in farads (F) – usually expressed in micro-farads. One micro-farad (μF) is one-millionth ($10^{-6}$) of a farad.

### ASSESSMENT GUIDANCE

The amount of energy stored in a capacitor is very small. In a typical fluorescent capacitor it is about 0.1 J.

A capacitor has the opposite effect to an inductor when connected to an a.c. circuit; it causes the current to lead the voltage by 90°.

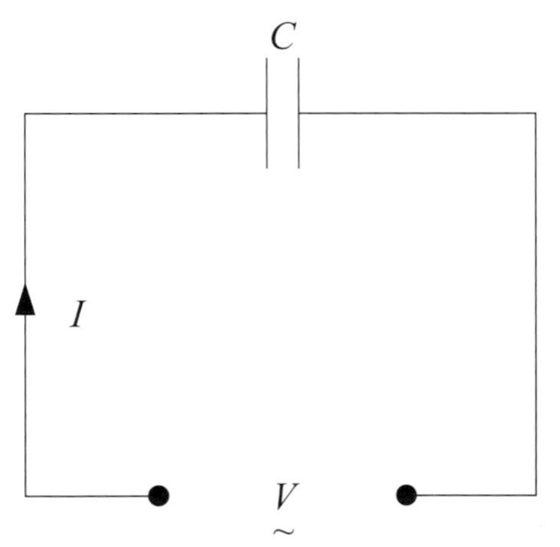


**Circuit diagram with capacitor**

The capacitor produces a reactance. This reactance, known as capacitive reactance ($X_C$), affects current flow.

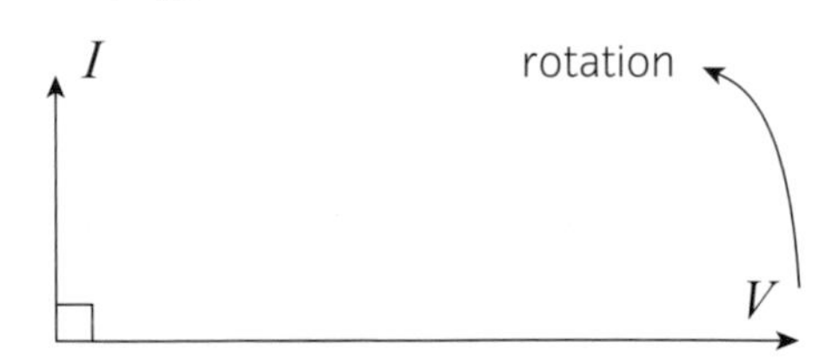


**Phasor diagram showing the effects of a capacitor**

The phasor shows that the current leads the voltage by 90° in the direction of rotation.

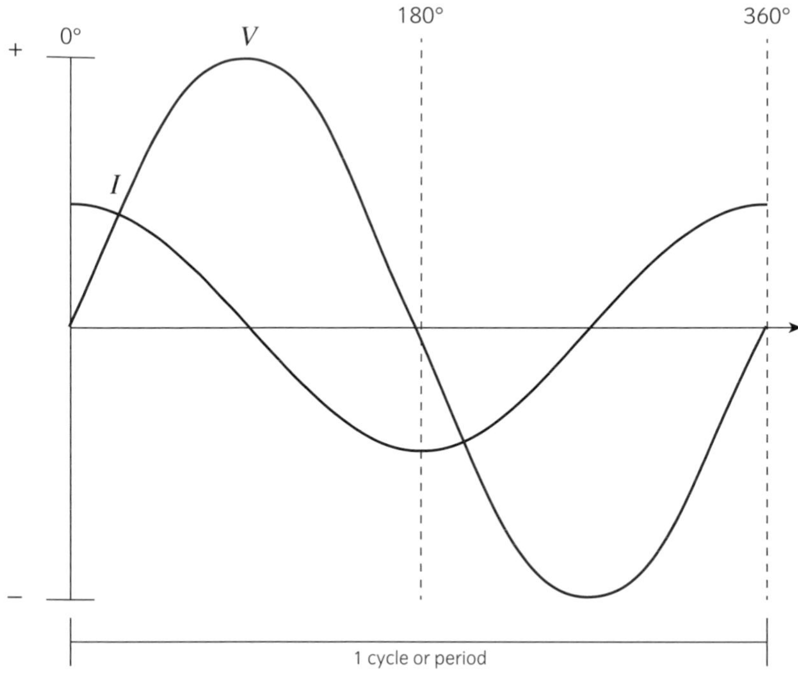


**Sine wave diagram showing phase shift due to a capacitor**

The sine wave diagram shows the current leading the voltage and, once again, the current and voltage are in opposition. At various points in the cycle there will be reactance.

The capacitive reactance of a capacitive circuit can be determined using

$$X_C = \frac{1}{2\pi fC}$$

## Example calculations

Determine the reactance when a 120 μF capacitor is connected to a 230 V 50 Hz supply.

As:

$$X_C = \frac{1}{2\pi fC}$$

then:

$$X_C = \frac{1}{2\pi \times 50 \times 120 \times 10^{-6}} = 26.52\Omega$$

Where the capacitance is pure (that is no other components are connected in that part of the circuit), then $X_C$ replaces resistance in Ohm's law.

Determine, using the example above, the current drawn by the capacitor.

So as:

$$I = \frac{V}{X_C}$$

then:

$$I = \frac{230}{26.52} = 8.6\,\text{A}$$

## Impedance

Impedance is the product of resistance and one or more of the other two components in a circuit. Where resistance and another component are connected in a circuit, the effect reduces the angle by which the current leads or lags. As resistance is in unity and the other components cause the current to lead or lag the voltage, and since more than one current cannot exist, the resulting current will fall in between, depending on the values of all the components.

**ASSESSMENT GUIDANCE**

Most loads are either resistive (heaters and tungsten lamps) or a mixture of resistive and inductive.

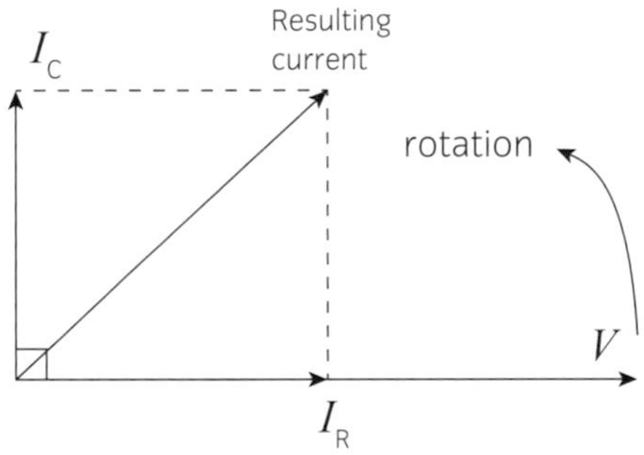

**Phasor diagram showing a leading current due to a capacitor and resistor in the circuit**

**ASSESSMENT GUIDANCE**

A two-core cable has capacitance as it consists of two plates (conductors) and a dielectric (insulation).

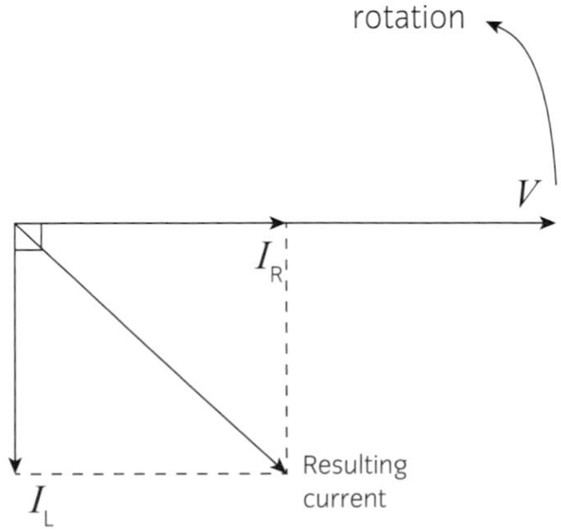

**Phasor diagram showing the resultant lagging current due to the inductor and resistor in the circuit**

This current is determined by using the circuit impedance ($Z$) measured in ohms. Impedance can be determined by:

$$Z = \sqrt{R^2 + X^2}$$

The value of reactance ($X$) depends on the component used so, if an inductance is connected with an impedance, $X_L$ is used and for capacitance $X_C$ is used.

Where a circuit contains both, the resulting reactance is used; this is the smallest value taken from the largest.

## Example calculation

As an example, determine the impedance if a circuit contained a 30 Ω resistance, an inductor having a 12.56 Ω reactance and a capacitor having a 26.52 Ω reactance.

The total reactance is:

$$X = X_C - X_L$$

as the capacitive reactance is the larger value.

So as:

$$Z = \sqrt{R^2 + X^2}$$

then:

$$Z = \sqrt{30^2 + (26.52 - 12.56)^2} = 33.08\,\Omega$$

## Impedance triangle

The relationship between the different components in a circuit can be explored using an impedance triangle, as shown in the diagram.

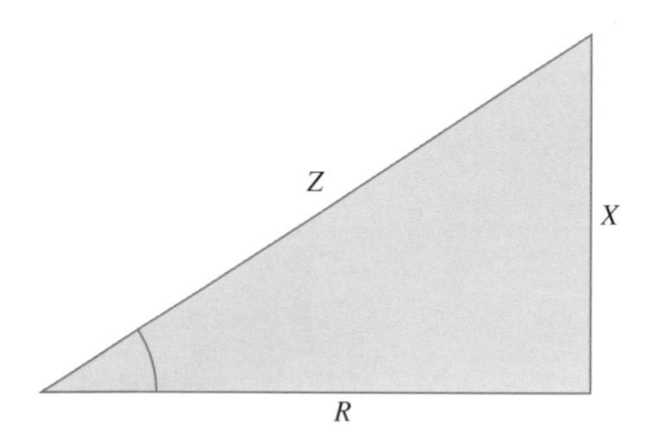


Impedance triangle

Using the triangle and Pythagoras' theorem, we can see that the impedance value alters depending on the values of resistance or impedance. Impedance triangles are drawn to scale. If the value of reactance increases or decreases, this affects the angle between the resistance ($R$) and impedance ($Z$). We will explore this relationship later in this outcome when we look at power factor.

Assessment criteria

1.2 Calculate quantities in a.c. circuits

ACTIVITY

Calculate the current drawn by a 10 µF capacitor when connected to a 230 V 50 Hz a.c. supply.

# CALCULATE QUANTITIES IN A.C. CIRCUITS

Components in a.c. circuits can be arranged in series, parallel or in combination. Each configuration has different effects on circuit properties.

## Series connected circuits (RL series)

When an inductor is connected into a circuit, we consider that the inductor has pure inductance (pages 150–152). This, in reality, is not possible as the conductor that is used to form the inductor's coil has a resistance. These two properties are in series with one another. As with d.c. circuits, if components are connected in series, the current is constant (that is, it is the same through each component), but the voltage changes as it is 'lost' across each component, creating voltage drop or a potential difference.

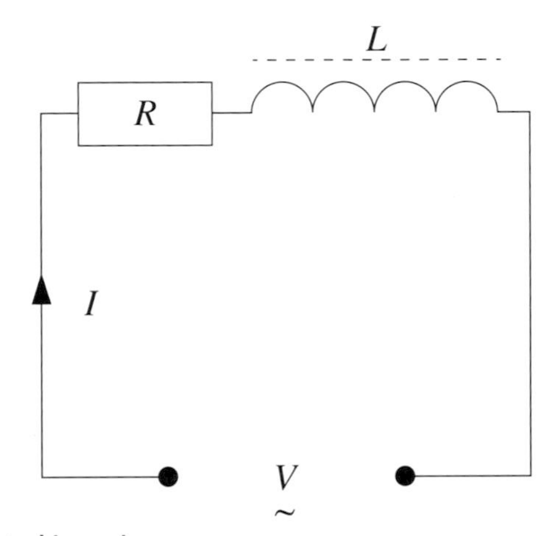


Components connected in series

The circuit shown in the diagram has the following values:

$R = 40\ \Omega$

$L = 38$ mH

Supply = 230 V 50 Hz

### Example calculation

Determine:

a) the inductive reactance ($X_L$)

b) the impedance ($Z$)

c) the total circuit current ($I$)

d) the value of voltage across each component ($R_R$) and ($V_L$).

**a)** As $X_L = 2\pi fL$ then $X_L = 2 \times \pi \times 50 \times 38 \times 10^{-3} = 11.93\,\Omega$

**b)** As $Z = \sqrt{R^2 + X^2}$ then $Z = \sqrt{40^2 + 11.93^2} = 41.74\,\Omega$

**c)** As $I = \frac{V}{Z}$ then $I = \frac{230}{41.74} = 5.51\,\text{A}$

**d)** The value of voltage across each component part is determined from Ohm's law:

$V_R = I \times R$ so $V_R = 5.51 \times 40 = 220.4\,\text{V}$

$V_L = I \times R_L$ so $V_L = 5.51 \times 11.93 = 65.73\,\text{V}$

As you can see, the two voltages do not add up to the supply voltage of 230 V. This is because the supply voltage would be the phasor sum of the two values as each component reacts differently to a.c. current.

We can determine the phasor sum in two ways, by calculation (using Pythagoras' theorem) or by constructing a phasor to represent the two components.

## Determining the phasor sum by calculation

$$V_{supply} = \sqrt{V_R^2 + V_L^2}$$

so

$$V_{supply} = \sqrt{220.4^2 + 65.61^2} = 229.99\,\text{V or } 230\,\text{V}$$

## Constructing a phasor

As we have seen, a phasor diagram is a representation of the component values in an a.c. circuit and how the components lead or lag. Also, it shows the resulting supply characteristics.

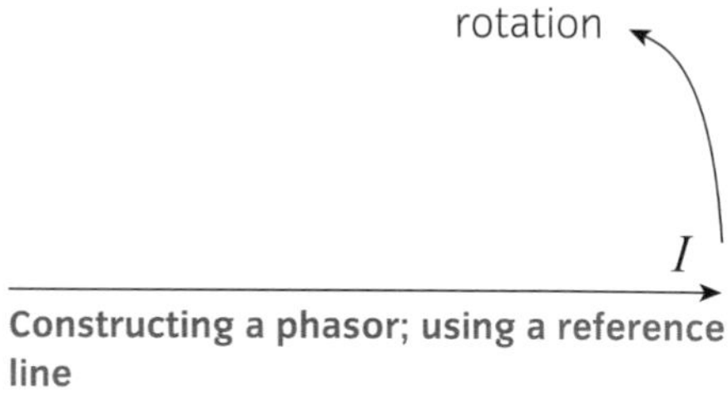

**Constructing a phasor; using a reference line**

To construct a phasor, we must first decide on a reference line. In a series circuit the common component is current as it is the same throughout the circuit; in a parallel circuit, this would be voltage.

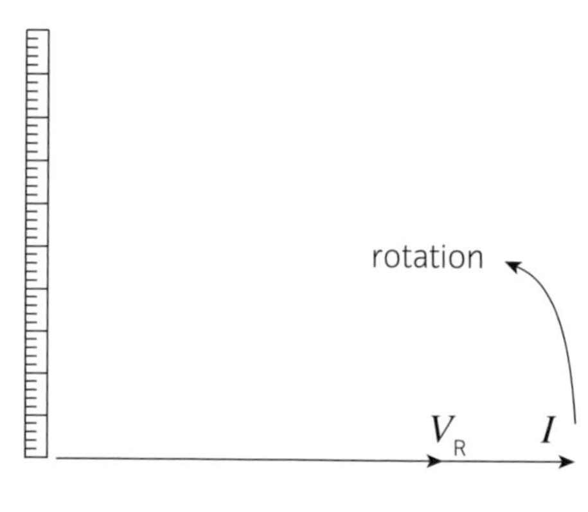

**Drawing a phasor to scale**

A phasor must be drawn to scale. We draw the resistor voltage line which is in unity to the supply (reference line).

Then we add the value of the inductor voltage to the same scale. As the current lags in an inductive circuit, the voltage $V_L$ is drawn upwards. This is to show that the current (reference line) lags the voltage (given the direction of rotation). As we must assume this component to be pure, the line is drawn 90° from the reference line.

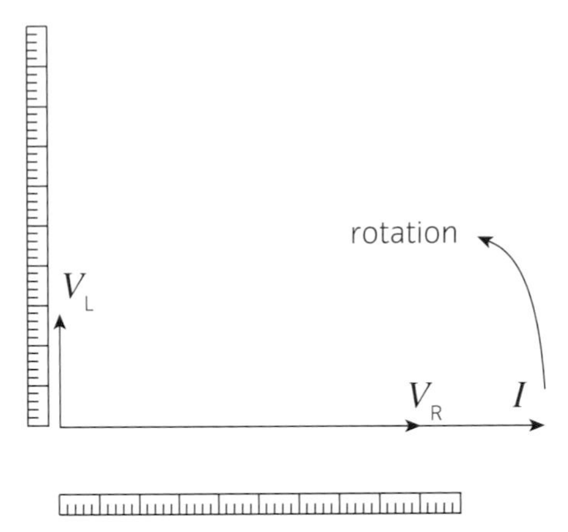


**Drawing the voltage value**

Then we construct a parallelogram from the two voltage values.

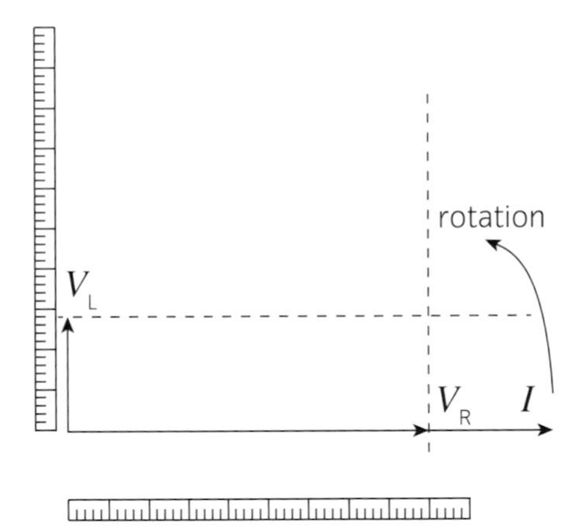


**Constructing a parallelogram**

Finally, the resulting supply voltage is drawn in, connecting the origin to the point where the two sides of the parallelogram intersect. The length of this line, to the scale used, represents the supply voltage $V_S$. We can see from the phasor that the current lags behind the supply voltage by a particular angle less than 90° – the resistance acts against the pure inductance, meaning the current lag is not fully 90°.

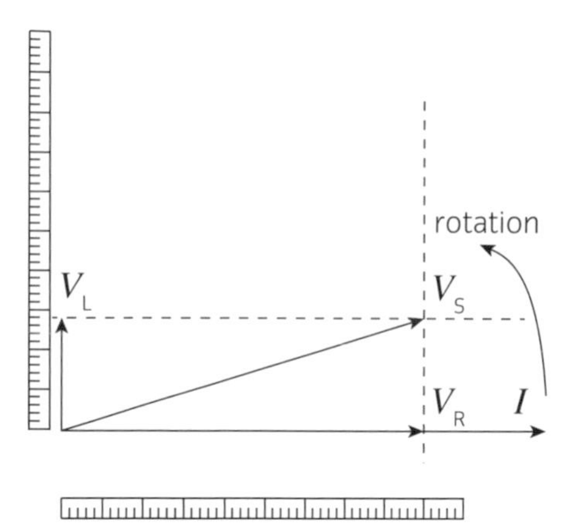

Drawing in the supply voltage

## RLC series circuits

If a circuit contains all three components, as in the circuit shown in the diagram, the components behave differently as the circuit current changes.

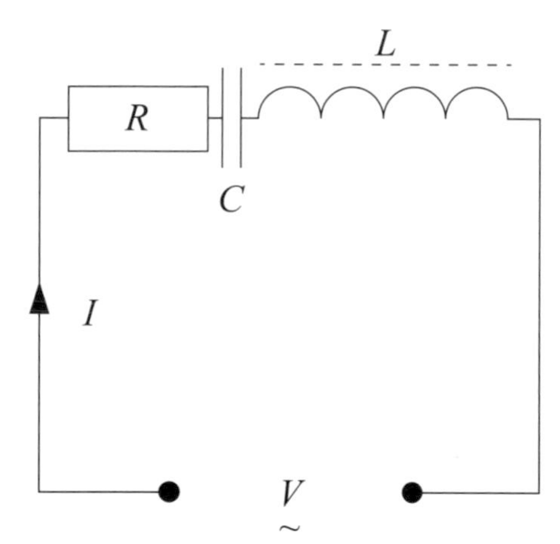

RLC circuit containing all three components

The circuit shown in the diagram has the following values:

$R = 40\ \Omega$

$L = 38$ mH

$C = 120\ \mu$F

Supply = 230 V 50 Hz

## Example calculation

Determine:

**a)** the inductive reactance ($X_L$)

**b)** the capacitive reactance ($X_C$)

**c)** the impedance ($Z$)

**d)** the total circuit current ($I$)

**e)** the value of voltage across each component ($R_R$), ($V_L$) and ($V_C$).

### ASSESSMENT GUIDANCE

In an RLC series circuit where the inductive and capacitive reactance cancel each other out, $R = Z$ and the power factor is 1 (unity). The current is $V/R$. High voltage may appear across the reactive components.

### ACTIVITY

A capacitor and inductor each of 200 Ω reactance are connected in series to a 115 Ω resistor. What is the voltage across the capacitor if the supply voltage is 230 V?

a) As $X_L = 2\pi fL$ then $X_L = 2 \times \pi \times 50 \times 38 \times 10^{-3} = 11.93\,\Omega$

b) As $X_C = \dfrac{1}{2\pi fC}$ then $X_C = \dfrac{1}{2 \times \pi \times 50 \times 120 \times 10^{-6}} = 26.52\,\Omega$

c) As $Z = \sqrt{R^2 + X^2}$ and $X = X_C - X_L$ then

$X = X_C - X_L = 26.52 - 11.93 = 14.59\,\Omega$

So $Z = \sqrt{40^2 + 14.59^2} = 42.57\,\Omega$

d) As $I = \dfrac{V}{Z}$ then $I = \dfrac{230}{42.57} = 5.4\,\text{A}$

e) The value of voltage across each component part is determined from Ohm's law:

$V_R = I \times R$ so $V_R = 5.4 \times 40 = 216\,\text{V}$

$V_L = I \times X_L$ so $V_L = 5.4 \times 11.93 = 64.42\,\text{V}$

$V_C = I \times R_C$ so $V_C = 5.4 \times 26.52 = 143.2\,\text{V}$

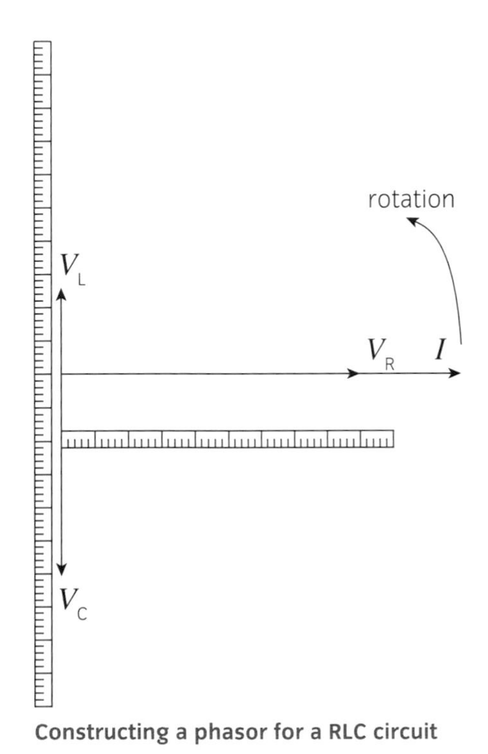

Constructing a phasor for a RLC circuit

Once again, we could prove the circuit supply voltage by calculation or phasor.

## Determining the phasor sum by calculation

$$V_{supply} = \sqrt{V_R^2 + V_X^2} \text{ where } V_X = V_C - V_L$$

Once again, subtract the smallest value from the largest:

$$V_X = 143.2 - 64.42 = 78.78\,\text{V}$$

So

$$V_{supply} = \sqrt{216^2 + 78.78^2} = 229.5 \text{ or } 230\,\text{V}$$

## Constructing a phasor

We construct the phasor as before, but this time we insert the voltage across the capacitor before we construct the parallelogram. Remember, as the current leads the voltage in a capacitor, the voltage lags the reference line (in the downwards direction) by 90°.

Following this, we take the smallest value of voltage in the capacitor or inductor from the largest, just as in the calculation, so we end up with a resulting voltage $V_X$ from which the parallelogram may be formed.

We can see that the current now leads the voltage by a particular angle as the capacitor is the stronger component.

Forming the parallelogram

## Components in parallel

When components are connected in parallel, the voltage becomes the common component and current is split through each component.

The circuit shown in the diagram has the following values:

$R = 40\ \Omega$

$C = 120\ \mu F$

Supply = 230 V 50 Hz

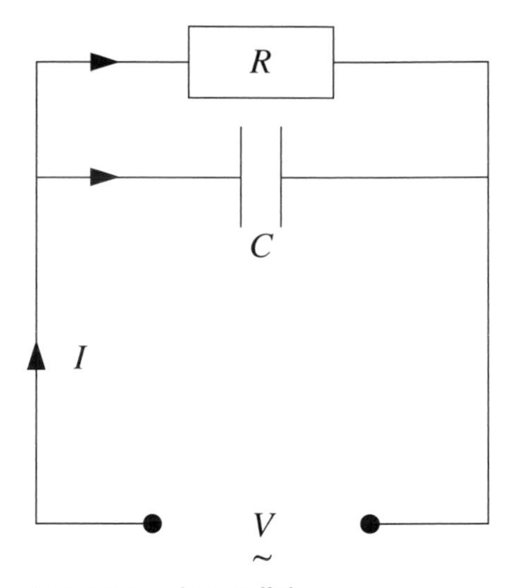


Components in parallel

### Example calculation

Determine:

a) the capacitive reactance $X_C$

b) the value of current through each component

c) the total circuit current.

**a)** As $X_C = \frac{1}{2\pi fC}$ then $X_C = \frac{1}{2 \times \pi \times 50 \times 120 \times 10^{-6}} = 26.52\ \Omega$

**b)** $I_R = \frac{V}{R}$ so $I_R = \frac{230}{40} = 5.75\text{ A}$

**c)** $I_C = \frac{V}{X_L}$ so $I_C = \frac{230}{26.52} = 8.67\text{ A}$

In the same way as we did for voltage, we can prove the circuit supply current by calculation or phasor.

### Proving the circuit supply current by calculation

$$I_{supply} = \sqrt{I_R^2 + I_X^2}$$

So

$$I_{supply} = \sqrt{5.75^2 + 8.67^2} = 10.4\text{ A}$$

### Constructing a phasor

Once again, drawn to a suitable scale, the phasor is used to determine the supply current. Notice that, this time, the voltage is the reference line as it is common to all components. The capacitor is the stronger component so, as a result, the supply current ends up leading.

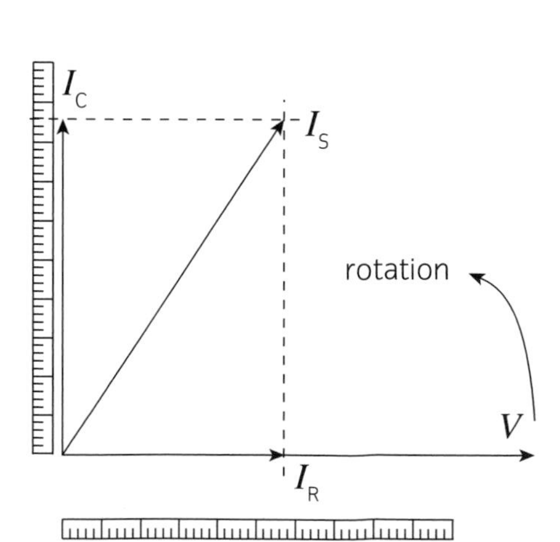


Constructing a phasor for a parallel circuit

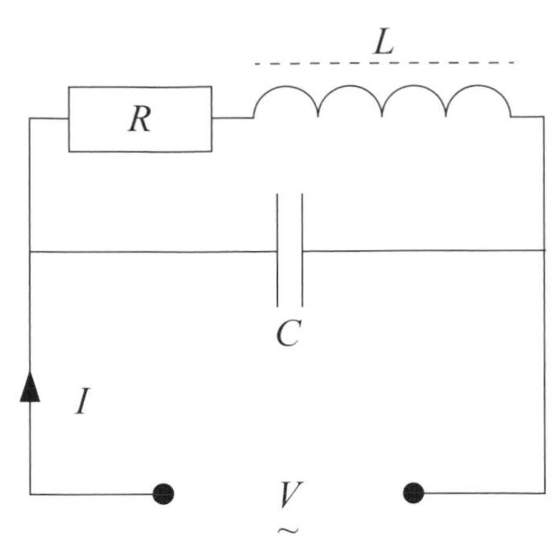

Circuit with components in both series and in parallel

## Circuits with both series and parallel components

In practice, in electrical installations, electricians are more likely to come across circuits where the inductor and resistor are in series, such as in motor winding or in the choke/ballast in a fluorescent luminaire, and the capacitor is in parallel for power factor correction purposes.

In this situation, the current in the series section of the circuit is determined using impedance and the capacitor is determined using capacitive reactance.

The circuit shown in the diagram has the following values:

$R = 12\,\Omega$

$L = 88$ mH

$C = 150\,\mu$F

Supply = 230 V 50 Hz

### ACTIVITY

*R* and *L* in series with *C* in parallel is a typical power factor correction arrangement. It is not usually required to correct to unity power factor. Explain why this is.

### Example calculation

First, we determine the inductive reactance:

$$X_L = 2\pi f L$$

so:

$$X_L = 2\pi \times 50 \times 88 \times 10^{-3} = 27.64\,\Omega$$

and:

$$Z = \sqrt{R^2 + X_L^2}$$

so:

$$Z = \sqrt{12^2 + 27.64^2} = 30.13\,\Omega$$

The current in the inductive/resistive section of the circuit is:

$$I = \frac{V}{Z}$$

so:

$$I = \frac{230}{30.13} = 7.63\,\text{A}$$

The current drawn by the capacitor is based on the capacitive reactance:

$$X_C = \frac{1}{2\pi fC}$$

so:

$$X_C = \frac{1}{2\pi \times 50 \times 150 \times 10^{-6}} = 21.22\,\Omega$$

therefore as:

$$I = \frac{V}{X_C}$$

then:

$$I = \frac{230}{21.22} = 10.83\,\text{A}$$

Showing this as a phasor to determine the total current requires a slightly different approach to before; we need first to construct a phasor using the series circuit values, then draw in the capacitance values. Before we do this, we need to understand how power factor affects the circuit and, therefore, the resulting phase angle.

## POWER FACTOR

The power factor is defined as the cosine of the angle by which the current leads or lags the voltage (cos $\theta$). The power factor does not have a unit of measurement as it is a factor. The value can range from 0.01 to 0.99. A power factor of 1 is unity, the same as having a resistive circuit where the current rises and falls in phase with the voltage.

Power factors are used to express the effect of leading and lagging currents and many machines have a power-factor rating stated on the rating plate. The value is used to determine the current demand of the machine. As circuits with leading or lagging currents introduce reactance, this creates the effect of additional loading and, therefore, additional current demand in a circuit. We need to understand and allow for this additional load when selecting equipment and cables for a circuit or installation.

**Assessment criteria**

**1.4** Calculate power factor

### ACTIVITY

Use your calculator to find the cosine of 1°, 10°, 30°, 45°, 70° and 90°. Can you see that the greater the angle, the smaller the factor; the smaller the angle, the closer the factor is to 1, or unity?

If you reverse the process and choose a factor, you can determine the angle by using the $\cos^{-1}$ feature on your calculator. Try $\cos^{-1}$ 0.75.

### KEY POINT

A power factor cannot go beyond 0.00 (which represents 90°) as any current that lags by more than 90° becomes a leading current.

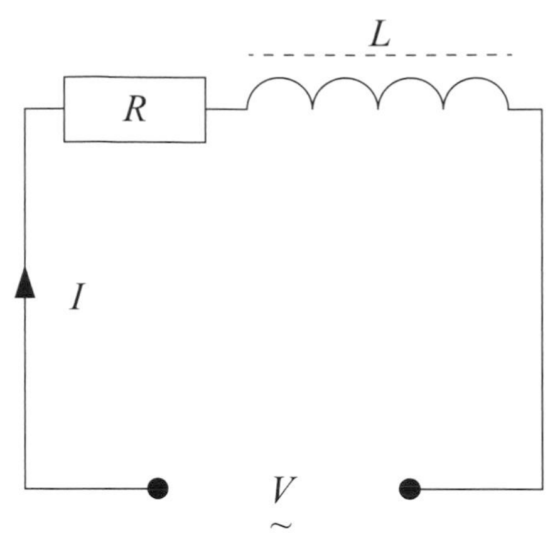

We will calculate power factor for this RL series circuit

## Calculating power factor

To determine the values of the power factor at circuit level, we can apply the following equation:

$$\text{power factor} = \cos\theta = \frac{R}{Z}$$

The circuit shown in the diagram has the following values:

$R = 40\ \Omega$

$L = 38\ \text{mH}$

Supply = 230 V 50 Hz

## Example calculation

Determine:

a) the inductive reactance ($X_L$)

b) the impedance ($Z$)

c) the total circuit current ($I$)

d) the power factor and angle by which the current lags the voltage.

**a)** As $X_L = 2\pi fL$ then $X_L = 2 \times \pi \times 50 \times 38 \times 10^{-3} = 11.93\Omega$

**b)** As $Z = \sqrt{R^2 + X^2}$ then $Z = \sqrt{40^2 + 11.93^2} = 41.74\Omega$

**c)** As $I = \frac{V}{Z}$ then $I = \frac{230}{41.74} = 5.51\ \text{A}$

**d)** As the power factor $= \cos\theta = \frac{R}{Z}$ then $\cos\theta = \frac{40}{41.74} = 0.958\ \text{A}$

So if the power factor is 0.95, the angle by which the current lags the voltage (inductive circuit) is:

$$\cos^{-1} 0.958 = 16.7°$$

## ASSESSMENT GUIDANCE

Power factor correction only takes effect up to the point of connection. In a fluorescent fitting, the supply current is reduced but the lamp current remains the same.

## KEY POINT

$\cos^{-1}$ is evaluated on most calculators by pressing the shift or 2nd key before the cos button. Look at your calculator to see if $\cos^{-1}$ is written above the cos key. What colour is it written in and what colour is the shift key?

## Power factors and impedance triangles

Power factors can also be determined from impedance triangles as the angle formed by the $R$ and $Z$ lines represents the angle by which the current leads or lags the voltage. The cosine of this angle is the power factor.

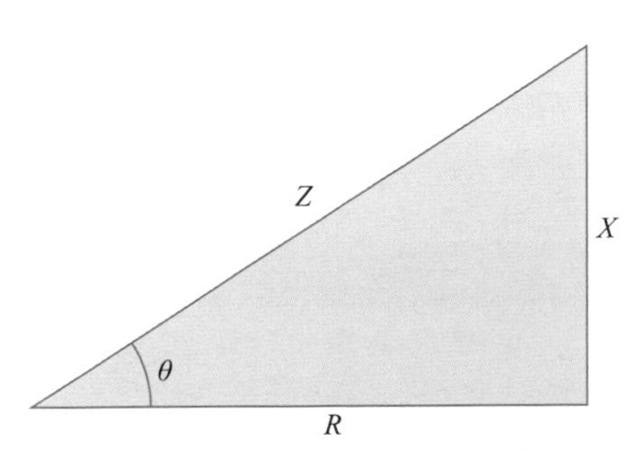

Using an impedance triangle to calculate the power factor

We will look again at power factors once we have looked at values of power quantities in the next section.

In an earlier example (page 162), we examined a circuit with components in series and in parallel. Here is the circuit again.

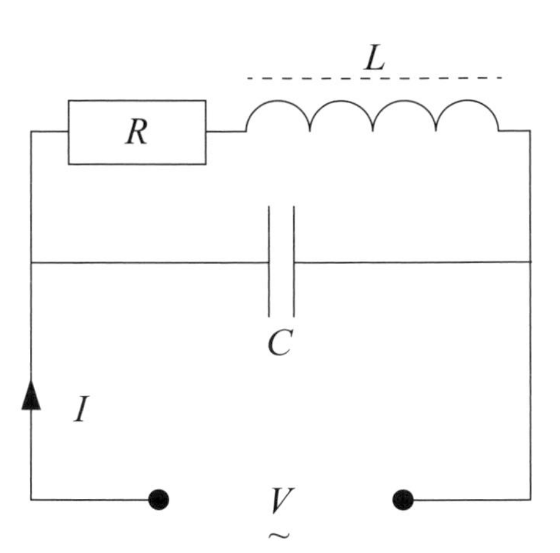


**Circuit with components in series and in parallel**

Recall the information given for this circuit:

$R = 12\,\Omega$
$L = 88$ mH
$C = 150\,\mu$F
Supply = 230 V 50 Hz

As we have previously determined:

$$X_L = 2\pi f L$$

so:

$$X_L = 2\pi \times 50 \times 88 \times 10^{-3} = 27.64\,\Omega$$

and:

$$Z = \sqrt{R^2 + X_L^2}$$

so:

$$Z = \sqrt{12^2 + 27.64^2} = 30.13\,\Omega$$

The current in the inductive/resistive section of the circuit is:

$$I = \frac{V}{Z}$$

so:

$$I = \frac{230}{30.13} = 7.63\text{ A}$$

The current drawn by the capacitor is based on the capacitive reactance:

$$X_C = \frac{1}{2\pi f C}$$

so:

$$X_C = \frac{1}{2\pi \times 50 \times 150 \times 10^{-6}} = 21.22\,\Omega$$

As:

$$I = \frac{V}{X_C}$$

then:

$$I = \frac{230}{21.22} = 10.83\text{ A}$$

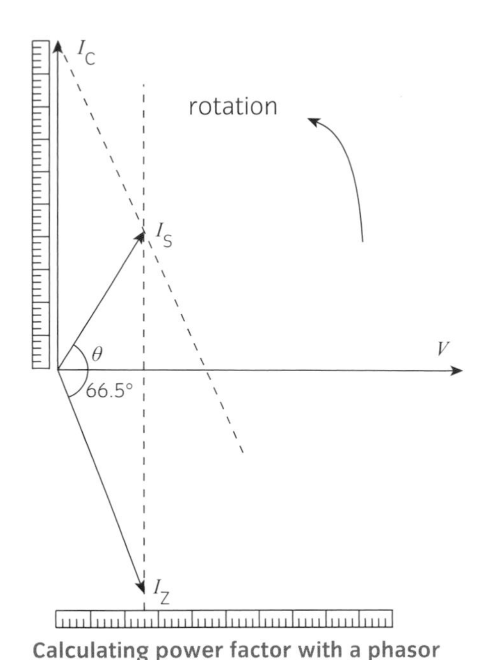

Calculating power factor with a phasor

## ACTIVITY

Construct this phasor to scale and measure the new supply current angle. From this angle, determine the power factor of the circuit.

We can now go on to determine the power factor and angle of the lagging current in the series branch of the circuit in order to construct our phasor.

As the power factor $= \cos\theta = \dfrac{R}{Z}$

then $\cos\theta = \dfrac{12}{30.13} = 0.398$

So the angle is $\cos^{-1} 0.398 = 66.55°$ lagging (inductive circuit)

We can construct the phasor, firstly, by showing that the current drawn by the impedance (series branch) part of the circuit lags the voltage reference line by 66.55° to a scale value of 7.63 A.

We can then add the capacitor in parallel, drawing 10.83 A and leading the voltage by 90°. Forming a parallelogram from these two points gives us the point from which to draw the supply circuit current. This should work out to be 4.8 A to scale. From this, we can also see the angle at which it leads the voltage. The cosine of this angle is the power factor of the circuit.

We will return to power factor once we have explored power quantities.

**Assessment criteria**

**1.3** Explain the relationship of power quantities in a.c. circuits

## THE RELATIONSHIP OF POWER QUANTITIES IN A.C. CIRCUITS

When we explored power at Level 2, we described the relationship between voltage and current as:

$$P = V \times I$$

If a load was purely resistive, this would be true. However, as many a.c. circuits have impedances and capacitors, the power behaves differently as an element of the power dissipated, due to the reactance of the circuit.

**KEY POINT**

Remember that, when an inductor or capacitor causes the current to lead or lag the voltage, at various points in the sine wave the voltage and current are in opposition and, therefore, are reacting (giving reactance).

### Power triangle

In power terms, this reactive part of the circuit (together with the true power relationship) can be explained using a power triangle.

As we can see, a resistive load gives the true power. The reactive (capacitive or inductive) load creates the reactive power element that affects the overall impedance. This is the apparent power.

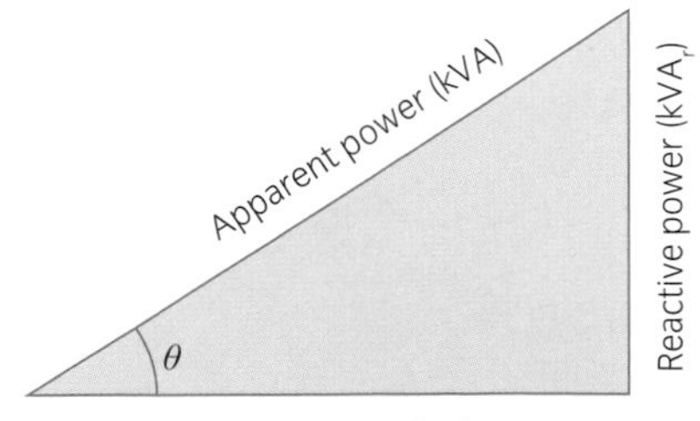

The true power relationship expressed in a power triangle

Therefore, having a reactive component in the load creates an apparent power. This draws more current from the load than if this was a purely resistive load. The apparent load is measured in volt-amperes (VA) or kilovolt-amperes (kVA). We can also see that an angle forms between the true power and apparent power. The cosine of this angle gives the power factor for the circuit or load.

**KEY POINT**

$kVA_r$ stands for reactive kilovolt-amperes.

So in order to determine the true power of a circuit we must apply this equation:

$$\text{true power (watts, W)} = V \times I \times \cos\theta$$

And from this, we could also state:

$$\text{power factor } \cos\theta = \frac{\text{true power}}{\text{apparent power}} \text{ or } \frac{\text{kW}}{\text{kVA}}$$

**ACTIVITY**

Make a list of five items in a domestic installation that would have an inductive component in the load.

Like a phasor, a power triangle could help us to determine appropriate values of capacitance to improve the power factor.

## Example calculation

A 230 V 4 kW motor has a power factor rating of 0.4. Determine a suitably sized capacitor to improve the power factor to 0.85. Supply frequency is 50 Hz.

**ASSESSMENT GUIDANCE**

The power factor of a motor depends on the load. Off-load the power factor is poor but this improves as the load increases. Any capacitor calculation should be based on the expected load (or consult the manufacturer).

Let us first work out some values to see the extent of the problem.

If we ignore reactance and, therefore, power factor, this motor should draw a current of:

$$\frac{4000\,\text{W}}{230\,\text{V}} = 17.4\,\text{A}$$

But in reality it draws a current of:

$$\frac{4000\,\text{W}}{230\,\text{V} \times 0.4} = 43.5\,\text{A}$$

So, you can see that the reactance in the circuit causes the motor to draw 43.5 A instead of 17.4 A. This is a huge difference. By installing a correctly sized capacitor into the circuit, we could improve this, reducing the overall current demand. As the motor is an inductive load causing the current to lag, a capacitor will draw the current back towards unity and, therefore, closer to the value of true power.

To work out the size of capacitor needed, we need to determine the amount of reactive power the capacitor consumes. Remember, this reactive power drawn by the capacitor doesn't increase overall power demand, it simply off-sets the reactance caused by the impedance as it draws the current towards leading and, therefore, reducing reactance.

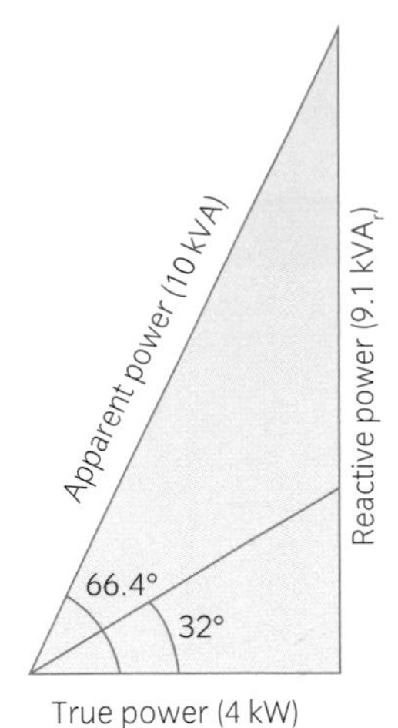

**Power triangle for example calculation**

To determine the amount of reactive power, we need to draw a power triangle to show the relationship before correction.

Using a suitable scale, we draw true power to represent 4 kW. Then we measure an angle from this of $\cos^{-1} 0.4 = 66.4°$.

The line from this angle represents the apparent power. We then draw a line up at 90° from the other end of the true power line. This line represents the apparent power. The two lines meet to form the power triangle, as seen in the diagram.

As we need to improve power factor to 0.85, we need to measure another angle from the true power line at $\cos^{-1} 0.85 = 31.79$ (32°). This line represents the new apparent power following correction, and gives a value by which reactance must be reduced.

By measurement, this line represents 6.6 kVA$_r$ so:

$$\frac{\text{kVA}_r \times 100}{V} = I_C$$

so:

$$\frac{6.6 \times 1000}{230} = 28.70\text{ A}$$

So, we need a capacitor that will draw a current of 28.70 A.

Then we need to carry out some of the previous calculations in reverse.

So:

$$\text{as } I_C = \frac{V}{X_C} \text{ then } X_C = \frac{V}{I_C} \text{ so } \frac{230}{28.70} = 8\,\Omega$$

And then:

$$X_C = \frac{1}{2\pi fC} \text{ so } C = \frac{1}{2\pi f X_C} = \frac{1}{2\pi \times 50 \times 8} = 398 \times 10^{-6}$$

A 398 μF capacitor is needed.

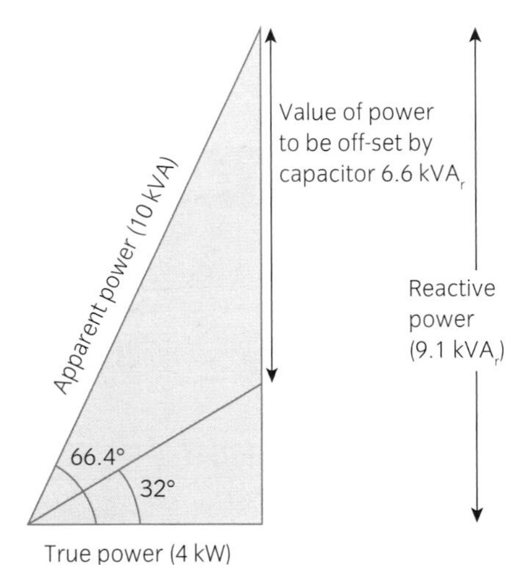

**Power triangle showing correction of power factor**

# POWER FACTOR CORRECTION

**Assessment criteria**

1.5 Explain power factor correction

1.6 Compare methods of power factor correction

There are different types of power factor (PF) correction that the designer can use; however, not all methods are practicable in all circumstances.

In order to correct the power factor on the supply, it is necessary to measure it. In order to do this, metering can be applied, as indicated in the diagram.

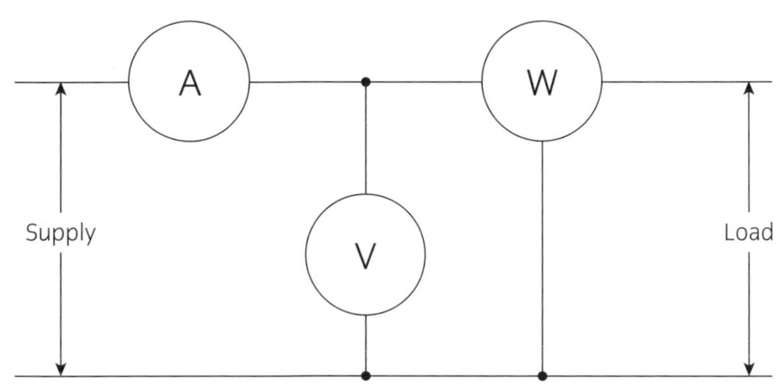


**Connection of instruments to calculate PF**

The wattmeter is used to indicate true power ($P$ in watts, W), whereas the volt and ammeter are used to calculate the apparent power (volt-amperes in VA).

Since:

$$\text{PF} = \frac{\text{true power}}{\text{apparent power}}$$

then the above connected meters will enable the PF to be calculated.

## ACTIVITY

Power factor correction can be carried out by individual capacitors mounted at the load or by bulk correction at the intake. Name one advantage and one disadvantage of each arrangement.

## ASSESSMENT GUIDANCE

A power factor meter can be used to get a direct reading.

## Correction by capacitors

The application of power factor correction capacitor banks is employed in electrical installations to correct the power factor.

This is due to the fact that most circuits are inductive in nature, which creates a lagging power factor. By adding power factor correction capacitors to the circuit, the $kVA_r$ is reduced as the capacitive $kVA_r$ cancels out the inductive $kVA_r$.

In the past, power factor correction was also carried out using synchronous motors to supplement capacitor banks but this practice is now very rare.

Larger automatic switchable unit

Power factor correction in many installations is achieved by a bank or banks of fixed capacitors. Alternatively, where the load changes or more accurate correction is required, power factor correction is achieved through automatically switched capacitors. Automatic switching units use monitoring technology to switch capacitors in and out of the load automatically as the load profile changes. These units are in banks, normally in multiples of 50kVA$_r$, so that the first bank of fixed capacitors deals with the base load. Remember that going too far with capacitors causes a leading power factor, which is again chargeable.

**KEY POINT**

If too much capacitance is added, the result will be an increase in apparent power as a lagging current becomes a leading current.

Therefore, it is ideal to balance the load and switch in capacitors throughout the changing load profile to ensure the PF stays around 0.95. These units are normally installed at or near the intake position in a building in order to deal with the overall power factor correction.

As well as using capacitor banks to correct power factor at source, the power factor may be corrected through the equipment. Installing suitably rated capacitors in parallel with a load improves the power factor. As an example, fluorescent luminaires contain capacitors for this reason. The capacitor is connected between line and neutral in the luminaire. If the capacitor is removed, the luminaire will still operate but it will draw slightly more current due to the power factor.

# OUTCOME 2

## Understand the principles of lighting systems

Lighting is a specialist area in electrical installations work. Many designers and installers rely on specialists to manage the design of lighting, but there are some basic areas of knowledge that an electrician needs to know in order to install and maintain luminaires effectively.

## LAWS OF ILLUMINATION AND ILLUMINATION QUANTITIES

**Assessment criteria**

**2.1** Explain the laws of illumination

**2.2** Calculate illumination quantities

Two laws explain how light behaves when it is emitted from a luminaire onto a surface: the inverse square law and the cosine law. First, look at the terms and units used to explain and quantify lighting.

**Lighting terms and units**

| Term | Symbol | Unit | Description |
|---|---|---|---|
| Luminous intensity | $I$ | candela (cd) | The amount of light emitted per solid angle or in a given direction |
| Luminous flux | $F$ | lumen (lm) | The total amount of light emitted from a source |
| Illuminance | $E$ | lumens per metre$^2$ (lux) | The amount of light falling on a surface |
| **Efficacy** | $K$ | lumens per watt (lm/W) | This is a term is used to measure the **efficiency** of a lamp or luminaire. It compares the amount of light emitted to the electrical power consumed. |
| Maintenance factor<br>Or<br>Light loss factor | Mf<br><br>llf | none | These factors are used to de-rate the light output of a lamp, allowing for dust. The factor used depends on the environment. An average office environment would have a factor of 0.8 whereas a factory where lots of dust accumulates may be 0.4. |
| Coefficient of utilisation or utilisation factor | Uf | none | This factor takes the surfaces in a room, such as walls and ceilings, into account. Emitted light bounces off walls that reflect light well, giving more light. An average factor for a room is 0.6. The lighter the colour of the room, the higher the factor. |
| Space–height ratio | | | This ratio is used to determine how close together luminaires need to be, taking into account their height from a given surface, in order to illuminate a room with an even spread of light from multiple luminaires. |

**Efficiency**

The ratio between power in and power out, measured in the same unit.

**Efficacy**

The ratio of power in and power out, measured in two different units. For example, the ratio of light output in lumens to the electrical power measured in watts.

Some lamps, such as light-emitting diode (LED) lamps, are rated in lumens whereas others are rated in candela. Take care with this as the choice of lamp depends on its application. To illuminate a particular point, such as a kitchen work surface, the candela rating is important as it rates the intensity of light in a particular direction. For illuminating a general area, such as a driveway, the luminous flux (lumens) is a better indicator as it measures the total light output in all directions.

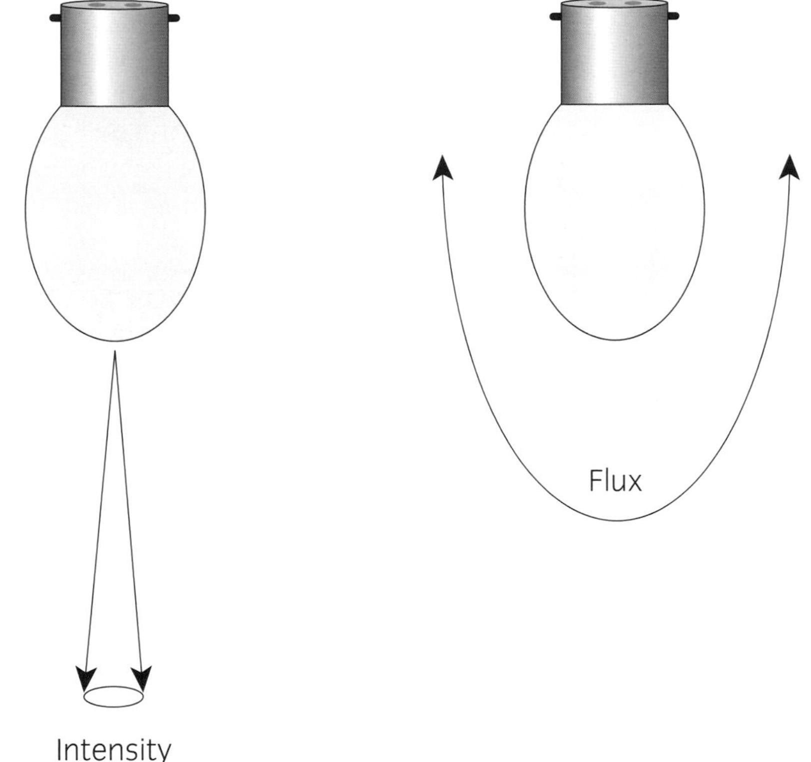


Candela is the rating of intensity; lumens are the rating of total output in all directions

## ACTIVITY

A luminaire emits 1250 candela in all directions. Calculate the illuminance at

a) 2.5 m and b) 5 m directly below the luminaire.

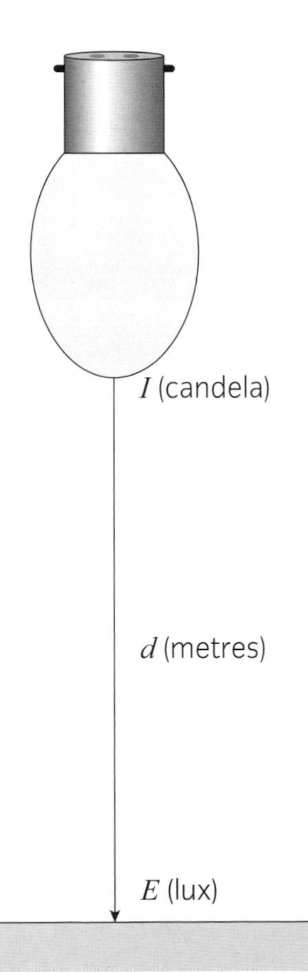


Inverse square law: illuminance on a surface from a light source positioned directly above

## Inverse square law

The amount of light falling onto a surface changes depending on the distance from the light source. If you hold a torch just above a surface, the light falling on the surface is intense. As you move the torch further away, the light directly below the torch becomes less intense because the light spreads. The amount of light on the surface is the illuminance. To determine the amount of light on the surface directly below the source, use the inverse square law.

$$E = \frac{I}{d^2} \text{ (lux)}$$

where:

$I$ = luminous intensity in candela (cd)

$E$ = illuminance on the surface in lux

$d$ = distance between the lamp and surface in metres (m).

**Example**

A 1000 cd light source is suspended above a level plane. Calculate the illuminance of the surface at 2 m and 4 m from the source.

At 2 m:

$$E = \frac{I}{d^2}$$

Therefore:

$$E = \frac{1000}{2^2} = 250 \text{ lux}$$

At 4 m:

$$E = \frac{I}{d^2}$$

Therefore:

$$E = \frac{1000}{4^2} = 62.5 \text{ lux}$$

**ASSESSMENT GUIDANCE**

The inverse square law and cosine law are basically the same. It is just that the cosine of 0° is 1, so it has no effect on the calculation.

## Cosine law of illumination

When light falls obliquely on a surface, not at right angles to it, the light spreads over an increasing area as the angle ($\theta$) between the perpendicular to the surface and the direction of the light increases.

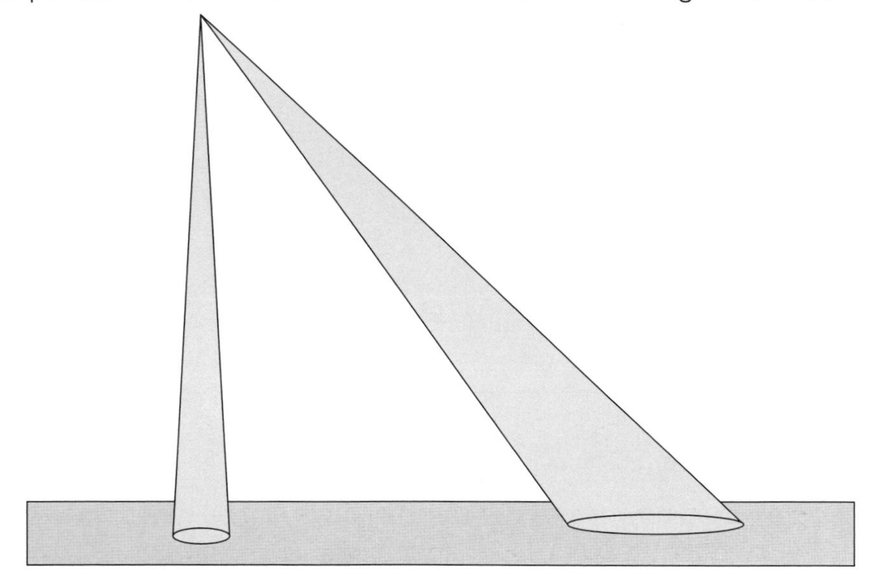

**The spread of light at different angles**

To calculate illumination in such cases, use the cosine law, which takes the additional area illuminated into account. It is expressed as:

$$E = \frac{I}{d^2} \times \cos\theta \text{ (lux)}$$

where:

$I$ = luminous intensity in candela (cd)

$E$ = illuminance on the surface in lux

$d$ = distance between the lamp and surface in metres (m)

cos $\theta$ = cosine of the angle at which the light is emitted from the lamp.

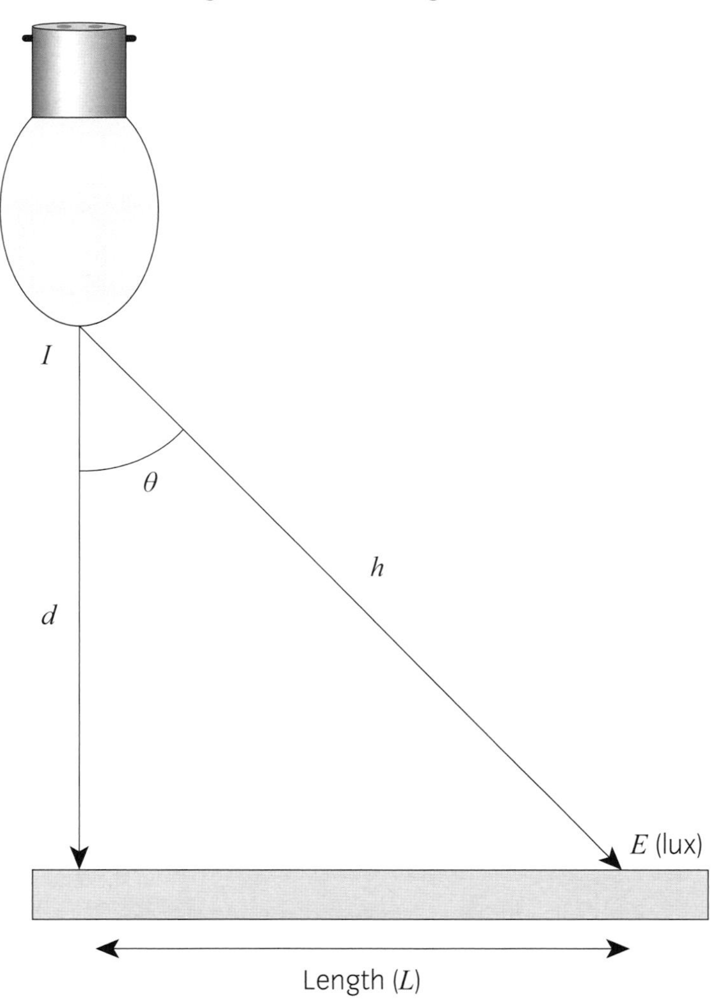


Calculating illumination at different angles

If the angle is unknown, the cosine of the angle can be determined by using Pythagoras' theorem and trigonometry.

As:

$$h = \sqrt{d^2 + L^2}$$

and:

$$\cos\theta = \frac{d}{h}$$

### Example

A 1200 cd light source is suspended above a level plane. Calculate the illuminance of the surface at 2 m directly below the source ($E_1$) and then calculate the illuminance on a surface 4 m away ($E_2$).

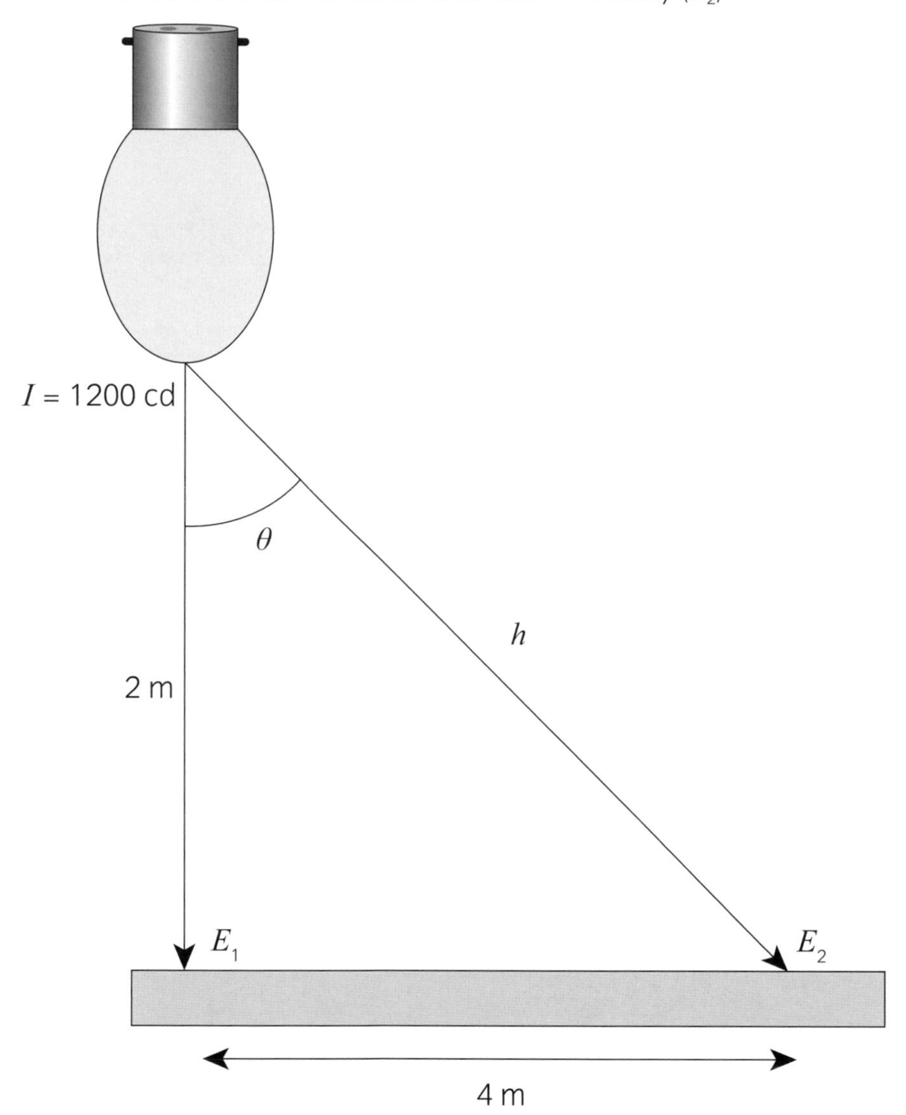


Calculating illuminance directly below the source and 4 m away

$$E_1 = \frac{I}{d^2}$$

Therefore:

$$E_1 = \frac{1200}{2^2} = 300 \text{ lux}$$

To determine $E_2$, find $\cos\theta$.

So:

$$h = \sqrt{d^2 + L^2}$$

Therefore:

$$h = \sqrt{2^2 + 4^2} = 4.47\,\text{m}$$

and

$$\cos\theta = \frac{d}{h}$$

Therefore:

$$\frac{2}{4.47} = 0.44$$

Therefore:

$$E_2 = \frac{1200}{2^2} \times 0.44 = 132\ \text{lux}$$

## Lumen method

As can be seen from the cosine law, the level of light varies across an area due to the distance and angle of the light source. When a desired level of illuminance is specified, an average figure is used across the area or working surface.

Lighting guides give values of illuminance that are suitable for use in various areas. The average value is usually quoted in lux.

The lumen method is used to determine the number of lamps that should be installed for a given area or room to achieve a specific average illuminance level.

**Guided average illuminance levels for different activities or areas**

| Activity or area | Illumination (lux, lumen/m$^2$) |
|---|---|
| Public areas with dark surroundings | 20–50 |
| Working areas where visual tasks are only occasionally performed | 100–150 |
| Warehouses, homes, theatres, archives | 150 |
| Classrooms | 250–350 |
| Normal office work, computer work, study library | 350–450 |
| Supermarkets, mechanical workshops | 750 |
| Normal drawing work, detailed mechanical workshops | 1000 |
| Detailed drawing work, very detailed mechanical works | 1500–2000 |
| Performance of very prolonged and exacting visual tasks | 5000–10 000 |

## Calculating for the lumen method

The lumen method is appropriate for use in lighting design if the luminaires are to be mounted overhead in a regular pattern.

The luminous flux output (lumens) of each lamp needs to be known as well as details of the luminaires and the room surfaces.

Usually, the **illuminance** will already have been specified by the designer, eg office 350–450 lux.

The formula is:

$$N = \frac{E_{\text{average}} \times \text{area}}{\text{Mf} \times \text{Uf} \times F}$$

where:

$E_{\text{average}}$ = average illuminance over the horizontal working plane in lux

$N$ = number of luminaires required

$F$ = luminous flux of each luminaire selected in lumens (as declared by the manufacturer)

Uf = utilisation factor based on the reflectance of the room walls, ceiling and work surface

Mf = maintenance or light loss factor (llf)

$A$ = area to be illuminated.

### Example

Determine the number of luminaires required in an office measuring 18 m by 20 m. The room is to be illuminated using recessed modular luminaires with a luminous flux given as 4800 lumens per fitting. The room has a white ceiling and walls painted a light colour; as a result, the utilisation factor is 0.9. As dust in an office is minimal, the maintenance factor is 0.8. The average illuminace desired is 400 lux.

Using:

$$N = \frac{E_{\text{average}} \times \text{area}}{\text{Mf} \times \text{Uf} \times F}$$

$$N = \frac{400 \times (18 \times 20)}{0.9 \times 0.8 \times 800} = 41.66 \text{ or } 42 \text{ luminaires}$$

**ASSESSMENT GUIDANCE**

The maintenance or light loss factor is composed of three items: dirt on the walls and in the air, dirt on the fittings and aging of the lamps.

**ACTIVITY**

Determine the number of luminaires needed to illuminate your classroom to an average illuminance of 350 lux. Determine the utilisation factor and maintenance factor, based on the brightness and dust accumulation of the room. Use manufacturer's data to obtain the luminous flux of a luminaire.

Assessment criteria

2.3 Explain the operation of luminaires

# OPERATION OF LUMINAIRES

## Space–height ratio

The space–height ratio determines how far apart luminaires should be in relation to their intended mounting height, or visa versa. The ratio depends on the particular luminaire and is determined by the manufacturer.

**ASSESSMENT GUIDANCE**

The actual number of fittings installed may be slightly higher than the calculated number to give equal numbers in each row.

### Example

If the space–height ratio of a luminaire is 3:2 and the mounting height is 2.4 m above the working plane (the area where the maximum illuminance is needed), determine the distance needed between luminaires.

$$\frac{S_r}{H_r} = \frac{S}{H}$$

where:

$S_r$ is the space ratio
$H_r$ is the height ratio
$S$ is the actual spacing between luminaires (centre to centre)
$H$ is the height the luminaires are mounted.

$$\frac{3}{2} = \frac{S}{2.4}$$

So:

$$\frac{3 \times 2.4}{2} = S = 3.6\,\text{m}$$

Assessment criteria 2.3 continues on page 180.

Assessment criteria

2.4 Explain the application of luminaires

# APPLICATION OF LUMINAIRES

## Efficacy

Lamps are given an efficacy rating based on the amount of luminous flux (in lumens) emitted by the lamp for every watt of power consumed by the lamp, including losses (in watts). An efficacy rating is a good indication of a particular lamp's energy efficiency. The higher the efficacy rating, the better the energy efficiency.

$$\text{efficacy}\ \frac{\text{lm}}{\text{W}} = \frac{\text{light output (lm)}}{\text{electrical input (W)}}\ \text{lm/W}$$

The purpose of a lamp is to produce light. Many lamps also produce heat during operation. As energy is required to produce the heat, this counts as a loss of energy because heat is not the intended product of the lamp.

In order for buildings to comply with current Building Regulations or the Code for Sustainable Homes, guidelines are given for the minimum lumens of light per circuit watt consumed.

**ACTIVITY**

Using the internet, look up the current requirements for lighting efficacy according to the Code for Sustainable Homes.

### Example

The light output from a lamp is 8000 lumens and the input power is 150 watts. Calculate the efficacy of the lamp.

$$\text{efficacy} \frac{\text{lm}}{\text{W}} = \frac{8000}{150} = 53.33 \text{ lm/W}$$

**Indication of the efficacy of different lamp types. Consult manufacturer's data for accurate ratings for particular lamps.**

| Lamp type | Efficacy (lm/W) |
|---|---|
| 60 W GLS incandescent lamp | 14 |
| 100 W tungsten halogen | 16 |
| 50 W high-pressure sodium (SON) | 120 |
| 18 W compact fluorescent (CF) | 61 |
| 70 W metal halide | 64 |
| 20 W T5 fluorescent tube | 99 |
| 2 W LED | 100 |
| 90 W low-pressure sodium (SOX) | 180 |

## Colour rendering

The term 'colour rendering' describes the ability of a lamp or luminaire to keep objects looking their true colour. Some lamps emit orange light and so objects lit by the lamp appear orange.

Colour rendering is very important when selecting luminaires, depending on their application. For example, general amenity lighting or street lighting does not require good colour rendering unless closed circuit TV (CCTV) is present, in which case this factor is important. Studies have shown that better colour rendering of street lighting in some areas can reduce crime, as criminals are aware that they can be identified more easily.

**KEY POINT**

Just because a lamp has a good efficacy rating does not necessarily mean it is suitable. For example, the SOX lamp has extremely good efficacy but very poor colour rendering as everything illuminated by it looks orange!

**ASSESSMENT GUIDANCE**

A low-pressure sodium lamp over a dining table would produce a high level of illumination but the food would look terrible.

Colour rendering is also important in shops. Installing the correct type of fluorescent tube can make clothes look vibrant, food look appetising and also reduce eye strain. There are many types of fluorescent tube colours available from warm white to daylight and colouright tubes.

More information relating to the application of different types of luminaires and lamps is given in Assessment criteria 2.3 below.

**Assessment criteria**

2.3 Explain the operation of luminaires

## OPERATION OF LUMINAIRES

There are several types of lamp, which work in different ways:

- incandescent lamps
- discharge lamps
- compact fluorescent lamps
- LED lamps.

**ASSESSMENT GUIDANCE**

General lighting service (GLS) lamps have largely been replaced by compact fluorescent lamps, due to the low efficacy of the GLS types.

### Incandescent lamps

This is the simplest form of lamp, with a current passed through a filament. The filament gets white hot and therefore emits light.

### Tungsten filament lamps

Tungsten has a high melting temperature (3380 °C) and the ability to be drawn out into fine wire.

In order to prevent premature failing through oxidisation, oxygen must be removed from the enclosing glass bulb. Small lamps are evacuated, creating a vacuum in the bulb, but larger lamps are filled with argon, which reduces filament evaporation at high temperatures. The efficacy of filament lamps is relatively low but increases with the larger sizes. Colour rendering is generally very good but does depend on the glass finish of the lamp.

### Tungsten-halogen lamps

Adding a halogen, such as iodine, to the enclosure prevents evaporation and allows the lamp to be run at a higher temperature. Colour rendering is very good with these lamps.

### Halogen cycle

If a halogen gas is present in a lamp with a tungsten filament, the atoms of tungsten that are driven off the filament attach to halogen molecules instead of collecting on the lamp wall. They are eventually returned to the filament and separated. The tungsten is deposited on the filament. The halogen gas molecules are free to circulate again and available to intercept other tungsten atoms.

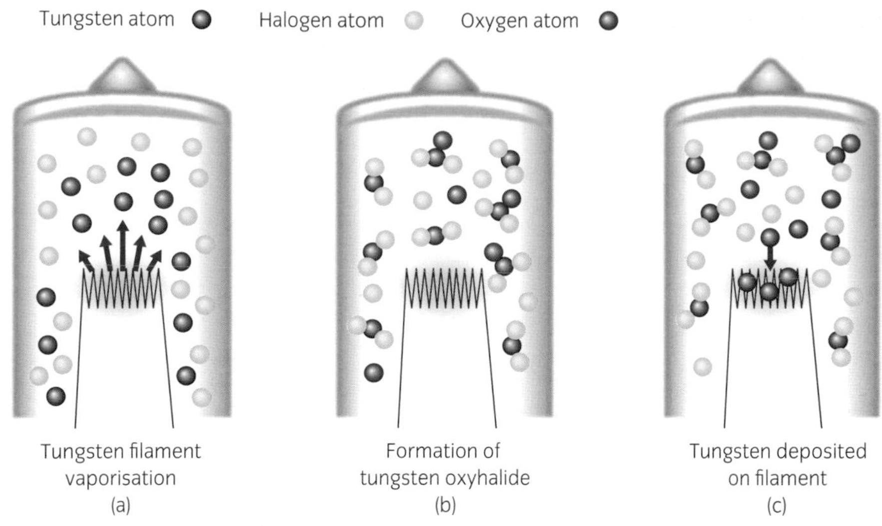


**Halogen regenerative cycle**

Halogens, condense at about 300 °C, so the lamp must be kept above this temperature. The lamp is made of quartz glass, which can be weakened if touched by a person. Oil from the skin can lead to gas leaking from the glass. Handling these lamps without protection can therefore shorten their lifespan.

Tungsten-halogen lamps are widely used for floodlighting and vehicle-lighting applications as well as low-voltage recessed lighting.

## Discharge lamps

The way a heated filament produces light is relatively simple to understand. How light is emitted when an electric current flows in a gas or vapour can be more difficult to understand. Think about how in nature lightning produces large quantities of light when electric current passes through a gas (air).

Several types of lamp produce light by establishing a permanent electric arc in a gas. This process is known as electric discharge or gaseous discharge. It is used to produce light in fluorescent and high-intensity discharge lamps.

## How discharge lamps work

Electrons are driven through the gas or vapour by the tube voltage, colliding with atoms as they go. The collisions are severe enough to break a loosely held electron from an atom, leaving behind a positively charged ion. This type of collision needs a fairly high tube voltage and results in ionisation to produce light.

Different gases and pressures of gas produce light of different wavelengths or colours, which can be further enhanced by using a coating such as phosphor around the inside of the gas tube.

### ACTIVITY

Consult manufacturers' data sheets for circuit diagrams of high- and low-pressure sodium discharge lamps.

### ASSESSMENT GUIDANCE

Some discharge lamps will not restart immediately after being switched off; the pressure has to drop before they will restrike.

## ASSESSMENT GUIDANCE

Electronic starters prevent damage to the tube as they only allow a few attempts at starting, whereas a glow starter will continue to attempt to strike the tube.

When gas is cold, it has a high resistance and therefore requires a high-voltage 'strike' to ionise the gas. Once gas is ionising, its resistance falls, meaning lower voltages can maintain ionisation. However, as proved by Ohm's law, a lower resistance will result in larger currents flowing. In order to create a high-voltage strike and limit running currents, discharge luminaires require control gear.

For a.c. supplies, an electronic control or an iron-cored inductor (also known as a choke or ballast) is used, rather than the resistor used for d.c. supplies.

To understand how the control gear works, look at the process, using the fluorescent luminaire (low-pressure mercury discharge lamp) as an example.

The sequence of events in this discharge luminaire is as follows:

1 When the luminaire is switched on, the starter switch closes and current flowing through the inductor induces a magnetic field within the inductor. As the gas in the tube is cold, the voltage is not enough to break down the resistance.

2 The starter switch opens, which open circuits the inductor. This causes the magnetic field in the inductor suddenly to collapse, creating a high voltage. This voltage strikes across the tube, ionising the gas, which reduces in resistance, and completing the circuit.

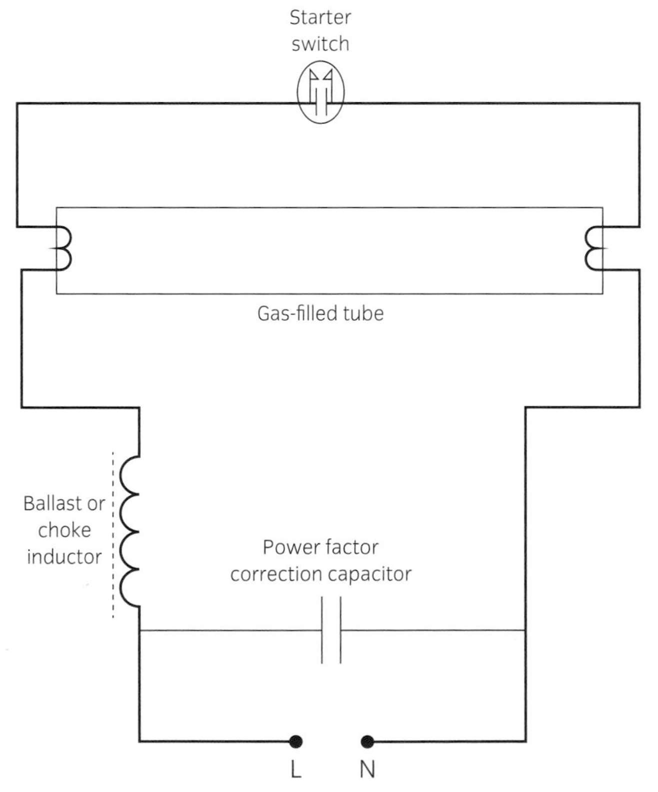


**How control gear works in a discharge lamp**

3 With the circuit once again complete, current flows through the inductor, which re-induces a magnetic field, causing self-induction, which limits the current flow. The starter switch is no longer required as the gas remains ionised with a constant current passing through it.

## The switch starter

The switch starter is enclosed in a small glass tube containing neon gas, which glows and produces heat. The switch contacts are bimetals and the heat causes them to bend so that they touch.

When the fluorescent tube is lit, the current in the tube causes a voltage drop in the choke, so that the lamp voltage across the switch contacts is too low to cause a glow in the neon. As it is open circuit when unheated, the switch starter has no more effect on the luminaire circuit. If the tube fails to strike first time, the process repeats until striking is achieved.

Fluorescent tubes come in many lengths, shapes, colours and ratings.

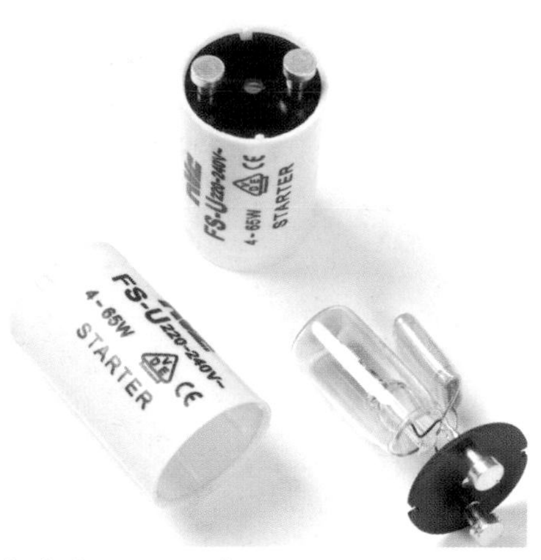


Switch starter unit

**ACTIVITY**

Look on the internet at a wholesaler's website to see the various types of fluorescent tube.

## Other types of discharge lamp

Other discharge lamps that work on a similar principle as the fluorescent tube (low-pressure mercury discharge lamp) above, include:

- high-pressure mercury
- low-pressure sodium
- high-pressure sodium
- metal halide.

### High-pressure mercury (MBFU)

This type of lamp produces a near-white light with a blue tinge. It is commonly used for:

- amenity lighting
- street lighting in residential areas
- bollard lighting.

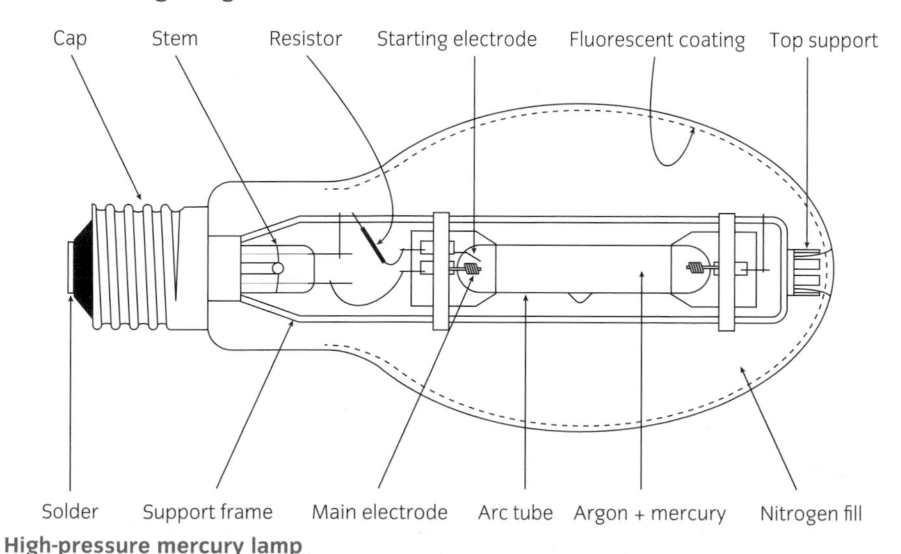


**High-pressure mercury lamp**

Because of its good colour correction, it is good for CCTV applications or areas where coloured objects need identifying.

**ASSESSMENT GUIDANCE**

M = mercury
B = high pressure
SO = sodium
X = low pressure
N = high pressure
F = fluorescent coating
E = external ignitor
U = universal operation

## ASSESSMENT GUIDANCE

Older types of SOX lamp used a step-up auto transformer for starting. Modern types use an ignitor for starting.

## Low-pressure sodium (SOX)

As well as containing a low-pressure sodium gas, this type of lamp will also contain a neon-based gas that ionises at lower temperatures. The neon-based gas, which gives a pink appearance when starting, heats up the sodium gas, which then produces orange light.

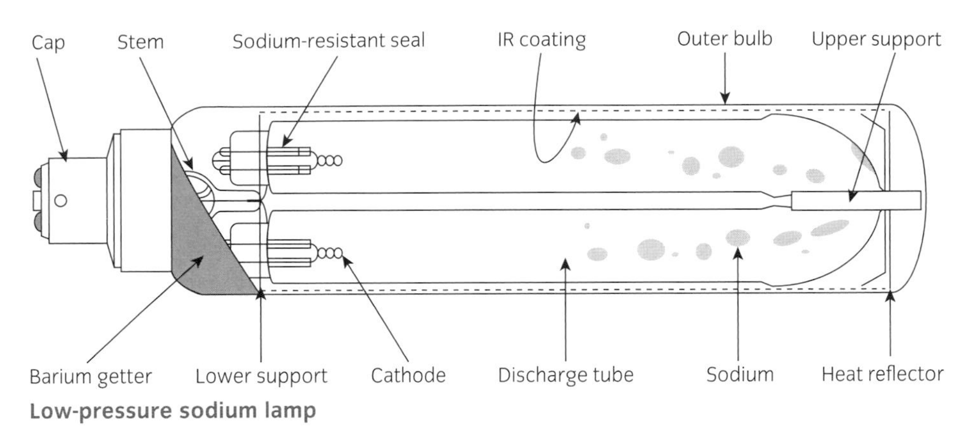


**Low-pressure sodium lamp**

These lamps have poor colour rendering but very good efficacy. They were widely used for most roadway applications but, due to increasing use of CCTV in towns and on roads, they are slowly being phased out and replaced with high-pressure sodium lamps.

**High-pressure sodium lamp**

## High-pressure sodium (SON)

These lamps are commonly used for street and amenity lighting as well as car parks, high-bay lighting and security perimeter lighting. They have reasonably good colour rendering although the light output is light orange. They have a good efficacy despite the colour rendering, which is why they are a common choice of lamp. They come in two varieties: SON-E elliptical and SON-T tubular.

Some lamps contain internal ignition (starter) switches, but others rely on separate starter units within the control gear. If you are replacing one of these lamps, make sure you check the type needed, which should be indicated by the appropriate triangular symbol.

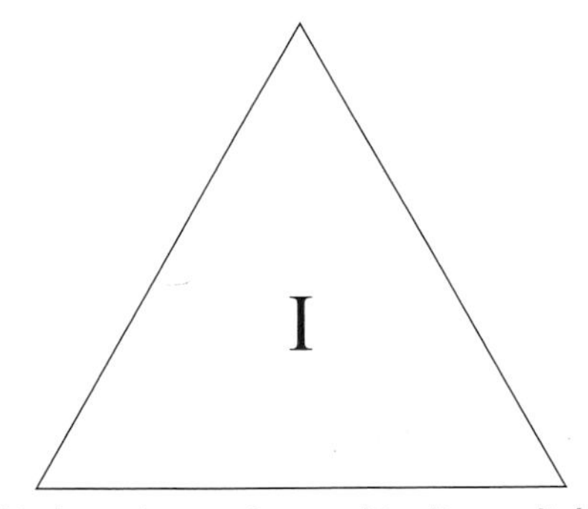


**This lamp has an internal ignitor switch**

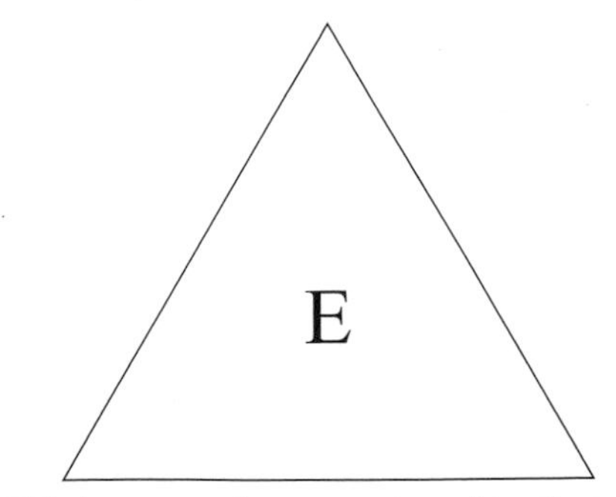


**This lamp requires an external ignitor switch**

## Metal halide (HID)

**Metal halide lamp**

These lamps have excellent colour rendering and good efficacy. They are extensively used for sports arena floodlighting as well as general amenity or security lighting.

The Waste Electrical and Electronic Equipment Regulations 2006 (WEEE) and other environmental legislation place strict controls on the disposal of discharge tubes. Care is required when handling these tubes and lamps as the mercury in the tubes is toxic and the sodium in the lamps burns when in contact with moisture. Therefore it is important that the tubes and lamps remain intact for specialist disposal.

Compact fluorescent lamp

## Compact fluorescent lamps

These are miniature fluorescent tubes compacted into a small space. The control gear is contained in the base of the lamp. They are intended as energy-saving replacements for incandescent lamps although the colour rendering and flickering mean that many people find them difficult to use for reading or close work.

## LED lights

The LED is a light-emitting diode. These lights are usually made from inorganic substances such as gallium indium nitride and gallium phosphide. The colour of the light output depends on the material used for the diode. The main colours are red, orange and green, and a variety of shades of blue.

The light output is usually monochromatic, ie the light emitted is at a single wavelength. The most common way to create a white light is to apply a phosphor-based coating to a blue diode. The phosphor converts the blue light to white light in a range of colour temperatures. The quality of the white light is affected both by the choice of LED and by the properties of the phosphor.

Basic LED colours

LEDs are very small; the active light-emitting surface is no bigger than 1–2 $mm^2$. A single diode can rarely produce enough light for a given lighting situation. For the unit to work, it must be mounted on a circuit board, with multiple LEDs in a cluster to form a LED module.

LEDs can be powered in two ways: with constant current or constant voltage. The ballast, which is referred to as the driver, is the unit that drives an LED array.

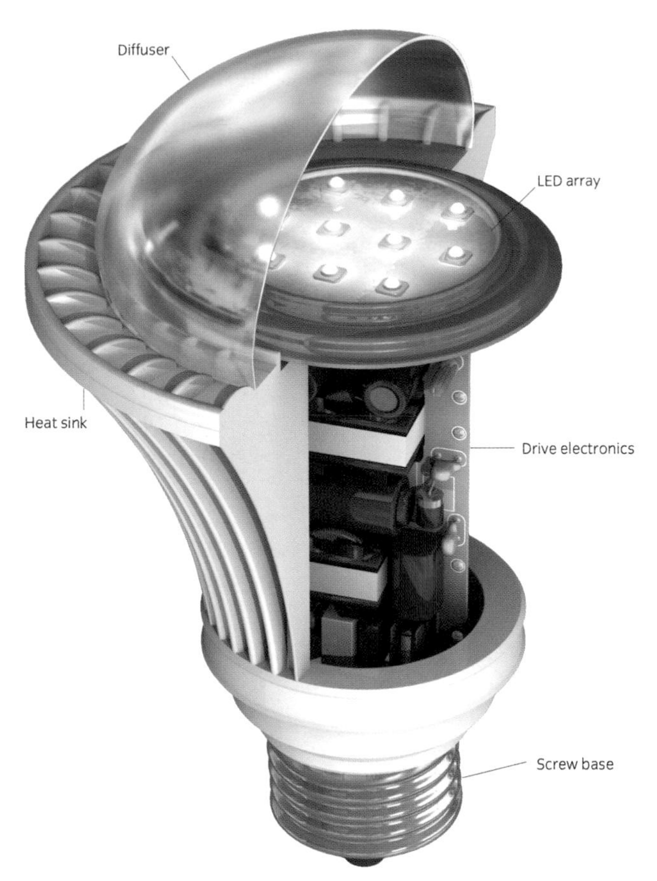


**Retrofit LED lamp with cut-away showing internal parts**

Although some LEDs run on conventional transformers, these can lack certain kinds of safety feature, such as short-circuit protection. When driven correctly, LEDs are claimed to be able to run for 50 000 hours, which is considerably longer than other technologies.

OUTCOME 3

# Understand electrical quantities in star delta configurations

Star (Y) and delta (Δ) configurations are used throughout the building services industry. Each configuration has different characteristics in terms of voltage and current values. Calculation of these values is essential in modern electrical engineering.

## VOLTAGE AND CURRENT IN STAR-CONFIGURED SYSTEMS

In a star (Y) connected load:

- the line current ($I_L$) flows through the cable supplying each load
- the phase current ($I_P$) is the current flowing through each load.

So:

$$I_L = I_P$$

and:

- the voltage between any line conductors is the line voltage ($V_L$)
- the voltage across any one load is the phase voltage ($V_P$)

so:

$$V_P = \frac{V_L}{\sqrt{3}} \text{ or } V_L = V_P \times \sqrt{3}$$

In a balanced three-phase system there is no need to have a star-point connection to neutral as the current drawn by any one phase is taken out equally by the other two. Therefore the star point is naturally at zero.

**Assessment criteria**

**3.1** Calculate values of voltage and current in star-configured systems

**ACTIVITY**

For a star-connected system calculate the phase voltage when the line voltage is a) 400 V, b) 415 V, c) 11 kV, d) 110 V.

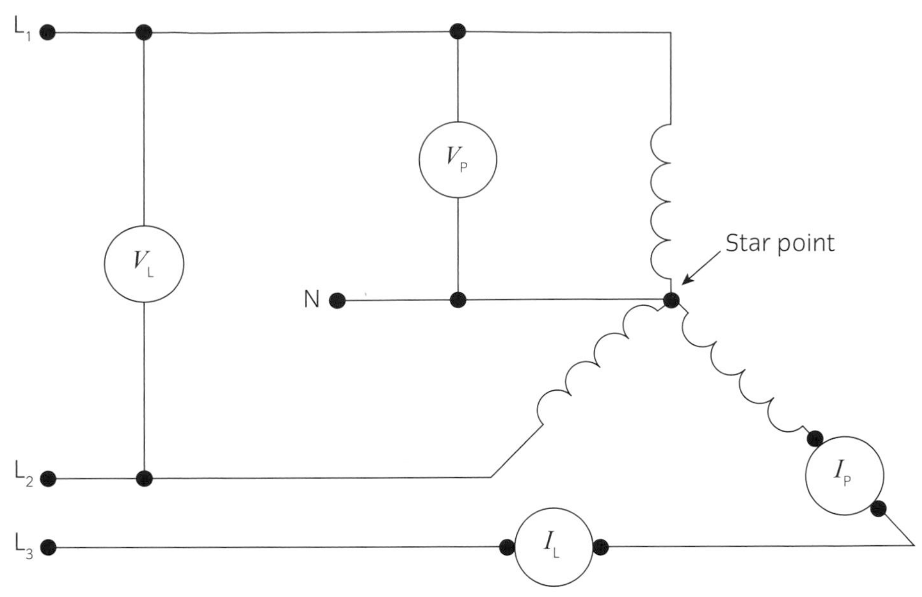


Star-connected load

So if a line current is 10 A, the phase current will also be 10 A. If the line voltage was 400 V, the phase voltage would be:

$$\frac{400}{\sqrt{3}} = 230\,V$$

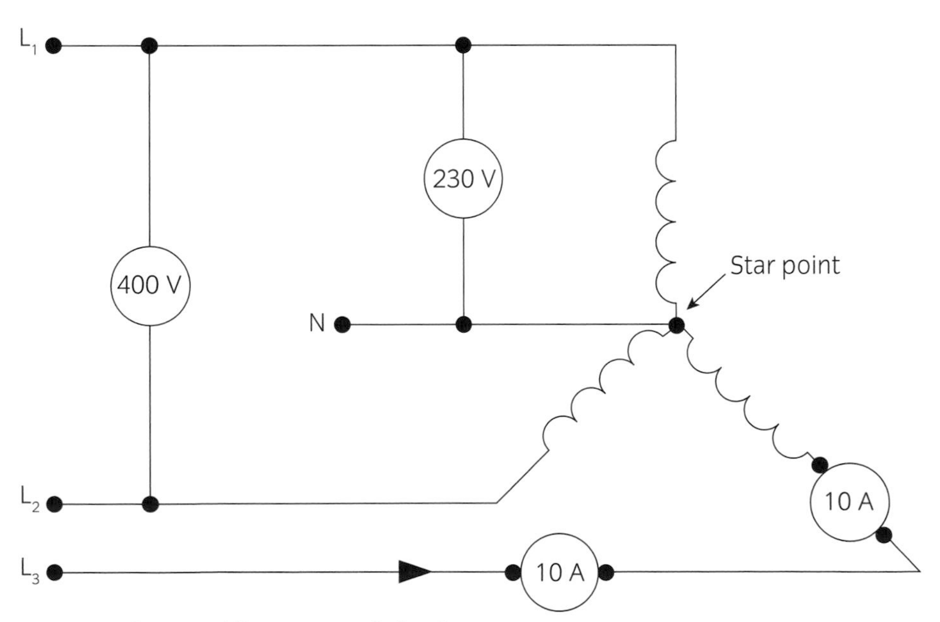


Common phase and line currents in load

# VOLTAGE AND CURRENT IN DELTA-CONFIGURED SYSTEMS

**Assessment criteria**

3.2 Calculate values of voltage and current in delta-configured systems

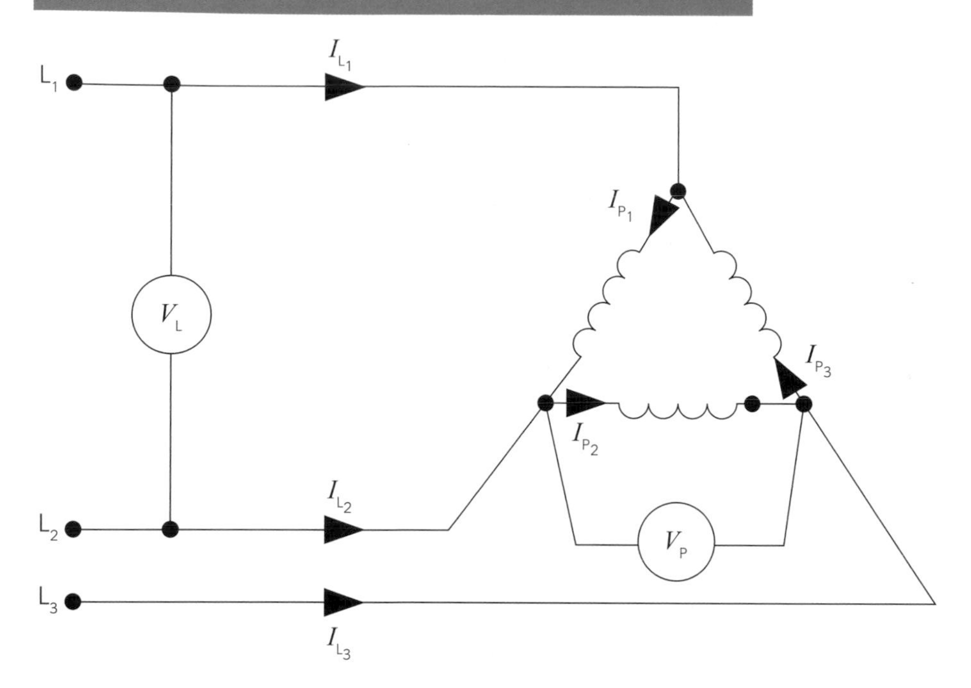


**Delta-connected load similar to delta-connected supply**

In a delta (Δ) connected load:

- the line current ($I_L$) flows through the cable supplying each load
- the phase current ($I_P$) is the current flowing through each load.

So:

$$I_P = \frac{I_L}{\sqrt{3}} \text{ or } I_L = I_P \times \sqrt{3}$$

and:

- the voltage between any line conductors is the line voltage ($V_L$)
- the voltage across any one load is the phase voltage ($V_P$)

so:

$$V_L = V_P$$

As there is no provision for a neutral connection, items such as delta motors would automatically be balanced – but complex loads on transmission systems could be unbalanced.

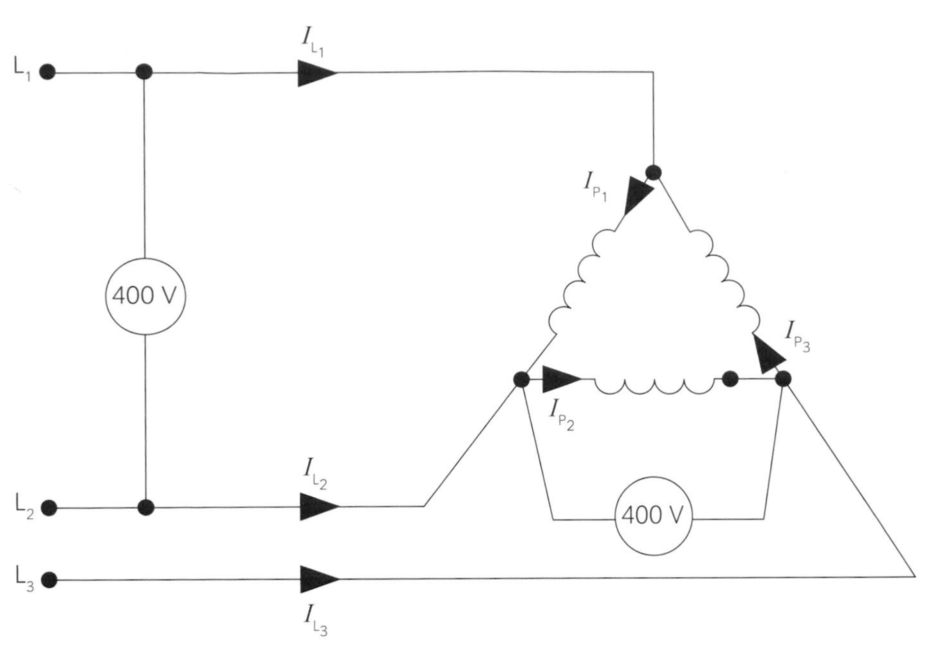


**Phase and line currents in the load**

Therefore if the phase current is 100 A and the phase voltage is 400 V, the line current can be calculated as follows:

$$I_{\text{line}} = 1.732 \times 100 = 173.2\,\text{A}$$

**Assessment criteria**

3.3 Determine the neutral current in a three-phase and neutral supply

**ASSESSMENT GUIDANCE**

A three-phase heater or motor would provide a balanced load. Three identical houses connected to separate single-phase supplies probably wouldn't as each would have a different load.

## NEUTRAL CURRENT IN THREE-PHASE AND NEUTRAL SUPPLIES

In a balanced three-phase system there is no requirement to have a star-point connection as the three phases have a cancellation effect on each other. Therefore the star point is naturally at zero current. While the load is balanced and the waveforms are symmetrical (not containing harmonics or other waveform distorting influences), this statement is relatively accurate. However, in practice, this is sometimes not the case.

Where the load is not in balance, different currents circulate in the load through the source winding and back. This gives rise to a change in star-point voltage, which can result in the system 'floating' away from its earthed reference point. In essence, a current will flow in the neutral. The three ways in which this current value could be determined are:

- by phasor, an accurate method that indicates the angle at which the maximum current occurs
- by calculation, which gives an accurate value
- by equilateral triangle, which gives a good indication of the value.

## Determining current value by phasor

Assuming that a load has the current values $L_1 = 85$ A, $L_2 = 50$ A and $L_3 = 60$ A per phase, the current value can be determined by the following steps.

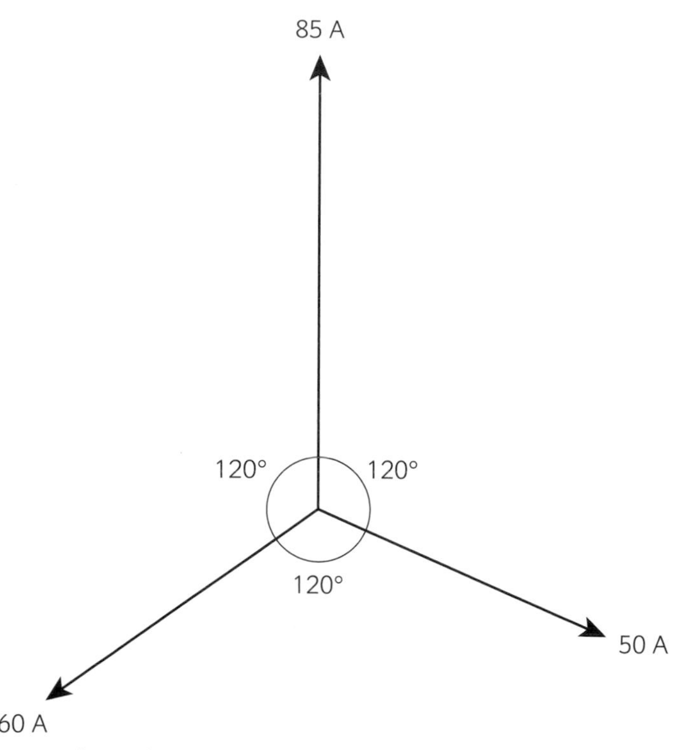


**Step 1 Construct a basic three-phase phasor to a suitable scale, ensuring the phases are 120° apart**

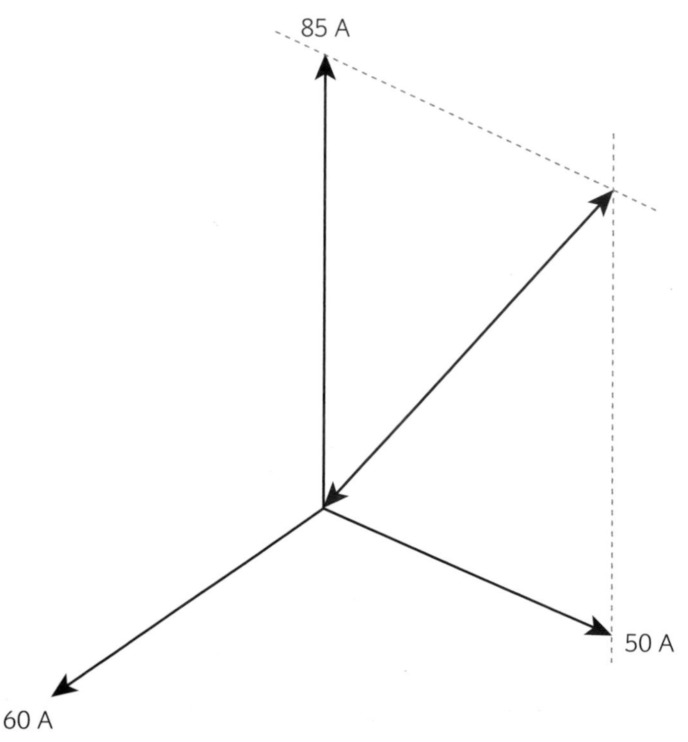


**Step 2 Now construct another parallelogram between the new line and the remaining phase. Where the two new parallelogram lines intersect represents the neutral current value to the scale selected**

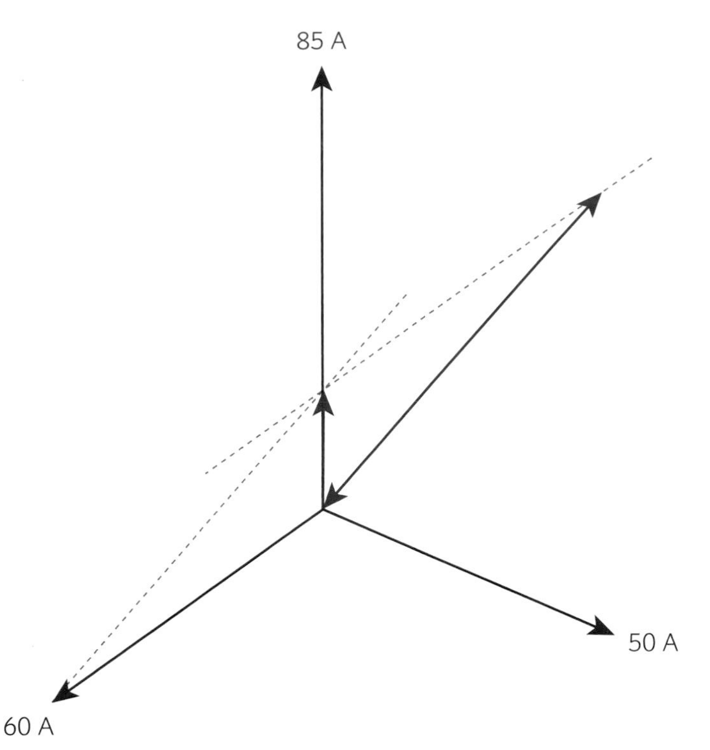

**Step 3 Construct a parallelogram with two of the phases and draw a line from the centre point of the phasor to the point where the two new lines meet**

### ASSESSMENT GUIDANCE

You may find the neutral current by any method you choose.

### ACTIVITY

Three equal 40 A loads are connected in star to a 400 V supply. The neutral current is zero. What will be the neutral current if:

a) one phase is disconnected
b) two phases are disconnected?

## Determining current value by equilateral triangle

The value of the neutral current can be determined using a scale drawing based on an equilateral triangle. If all phases are balanced, and therefore equal, all sides of the triangle are equal in length and meet to give equal angles of 60°.

If the phases are not balanced, there will be a gap at the top of the two sloping sides which represents the neutral current. The triangle here represents a balanced system.

### KEY POINT

Remember, an equilateral triangle is formed by three equal sides and has three equal angles of 60°.

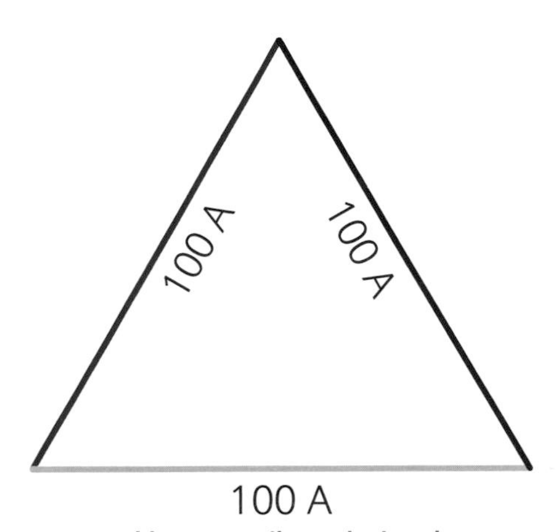

**A balanced system represented by an equilateral triangle**

Now consider an unbalanced system with these values:

- L1 = 70 A
- L2 = 100 A
- L3 = 60 A

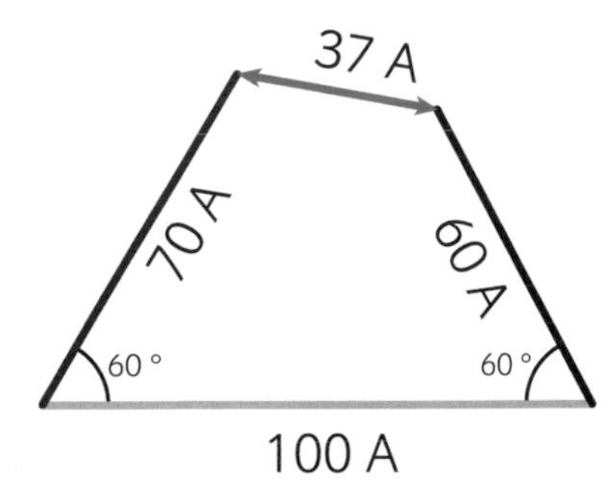


**In an unbalanced system, the gap (shown in light blue) represents the neutral current**

The neutral current is represented by the gap left where the two sides do not meet (shown in light blue). In this example, that gap represents a current of approximately 37 A.

**KEY POINT**

When constructing scaled diagrams, it is crucial that you use a good scale and measure accurately.

# OUTCOME 4

# Understand the principles of electrical machines

It is important to understand the differences in a.c. and d.c. machines and to appreciate where there are similarities in certain machine configurations.

**Assessment criteria**

**4.1** Explain the operating principle of d.c. machines

## HOW D.C. MACHINES OPERATE

Direct current (d.c.) machines were once the most popular type of machine because of the ability to control speed and direction. With advances in cheaper a.c. alternatives, d.c. machines were used less. Now that the parts and control devices for d.c. machines are cheaper, the use of d.c. is on the increase again. The competent electrician must therefore have a knowledge of d.c. machine operating principles.

**ACTIVITY**

Losses occur in d.c. machines which reduce the efficiency. Name three such losses.

There is no difference in the construction of d.c. motors and generators. They are rotating machines with three basic features: a magnetic-field system, a system of conductors and provision for relative movement between the field and the conductors.

The magnetic field in most d.c. machines is set up by the stationary part of the machine, called the field windings. The rotating part, known as the armature, is made up of multiple loops of cable linked to a commutator. Power is either delivered to (motor) or taken from (generator) the armature by brushes in contact with the moving commutator.

## d.c. generators

The d.c. generator is supplied with mechanical energy and gives out most of the energy, less losses, as electrical energy.

The d.c. generator has many loops and a multi-segmented commutator. With electricity flowing in the armature through brushes, the commutator reverses current flow as it passes from one pole to another so that the current in both conductors will always be the same.

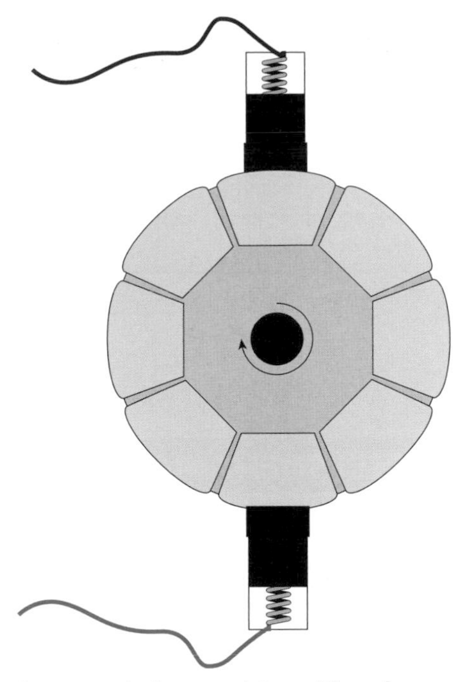

**A segmented commutator with carbon brushes making contact. This commutator would have four loops wound around the armature.**

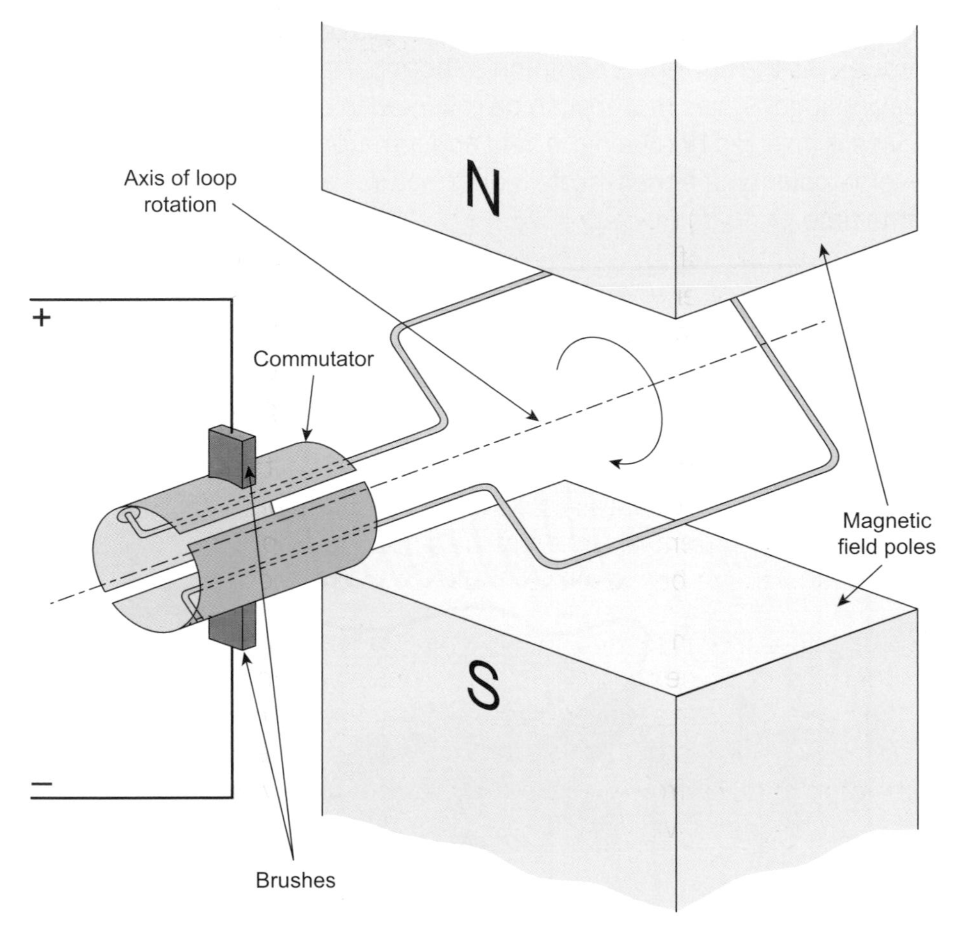


**A simple generator arrangement**

When the loops within the armature are rotated within the magnetic field, an emf is induced into the loop. The commutator ensures that the brushes are always in contact with the loop, which is in the strongest part of the magnetic field, at all times. This ensures a steady flow of direct current.

## ASSESSMENT GUIDANCE

In the past, d.c. shunt generators were used in automobiles to keep the battery charged. In modern cars a three-phase alternator and bridge rectifier are used as this arrangement has a much higher output.

## ASSESSMENT GUIDANCE

Series machines should always be connected to a load, otherwise they will run dangerously fast.

## d.c. motors

The d.c. motor takes in electrical energy and provides mechanical power, less losses.

There are three types of d.c. motor: series, shunt and compound.

### Series motors

Series motors are also known as universal motors as they can also be used on alternating current. The field and armature in a series motor carry the same current and are capable of providing high starting torques. As the current is common to both parts, the windings are heavy gauge. Series motors can be reversed in direction if a switch device is inserted between the field and armature, allowing simple reverse polarity of either the field or armature, but not both at the same time.

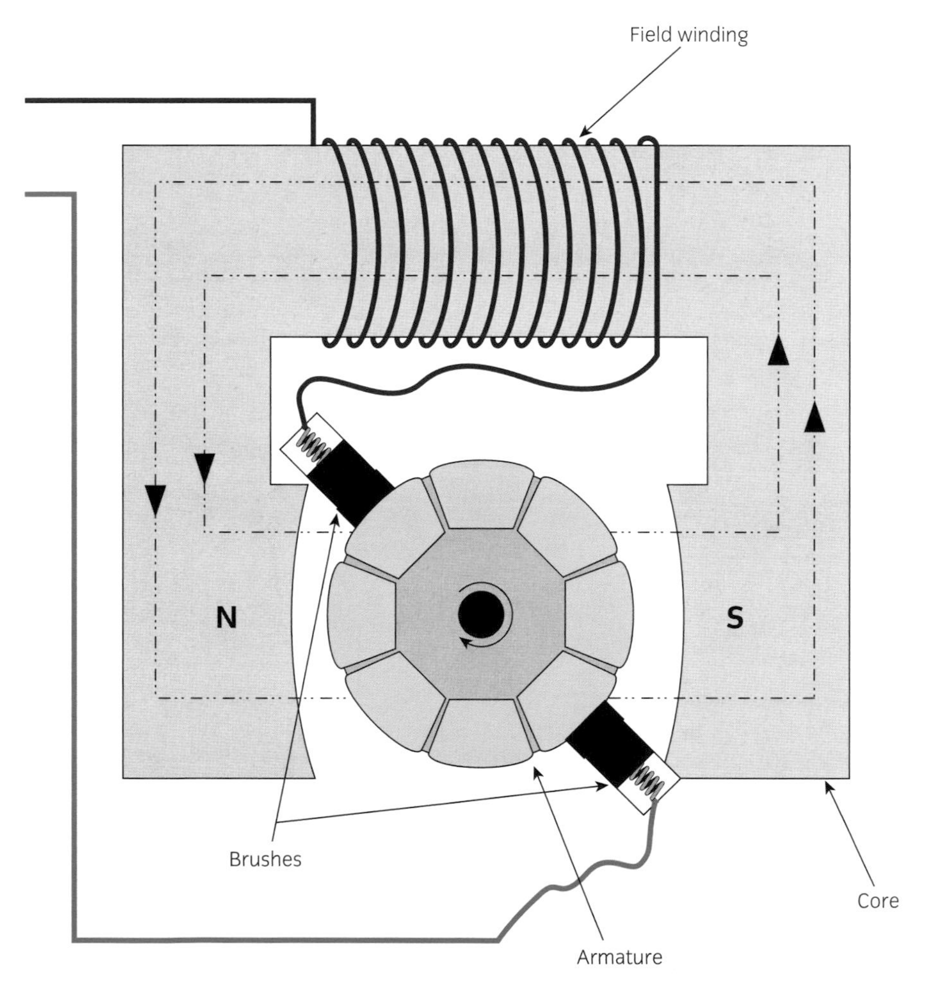


**Simplified series motor arrangement**

## Shunt motors

A shunt-connected d.c. motor consists of a field winding in parallel with the armature. This type of motor does not have the same common current characteristics as the series motor and therefore does not have a high starting torque. However, speed control of the shunt motor is considerably easier than the series motor as the field current can be controlled independently from the armature. Shunt motors can be reversed in direction if the polarity of either the field or armature is reversed, but not both at the same time.

### ACTIVITY

What is the difference between a self-excited and separately excited machine?

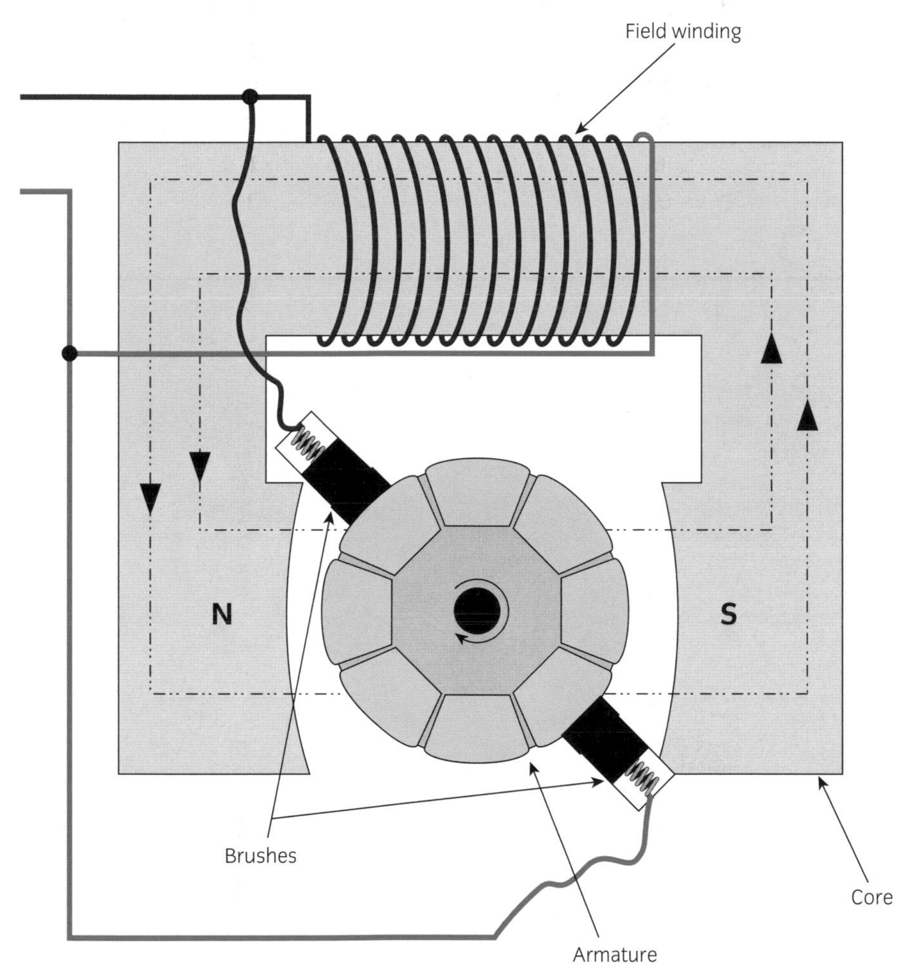


**Simplified shunt motor arrangement**

### ASSESSMENT GUIDANCE

The shunt motor uses a shunt field regulator to control the speed. Increasing resistance decreases the current and flux, causing the motor speed to increase.

## Compound motors

The compound motor is a mixture of the series and shunt motor circuits, offering the benefits of each type of machine, ie high starting torque and good speed control. To reverse a compound motor, the armature field must be reversed.

> **ASSESSMENT GUIDANCE**
>
> Compound machines can be 'long shunt' where the shunt winding is connected across the armature and series winding or 'short shunt' where the shunt winding is connected across the armature only.
>
> A compound motor can be connected as a cumulative machine where the two fields assist each other or a differential machine where the two fields oppose each other.

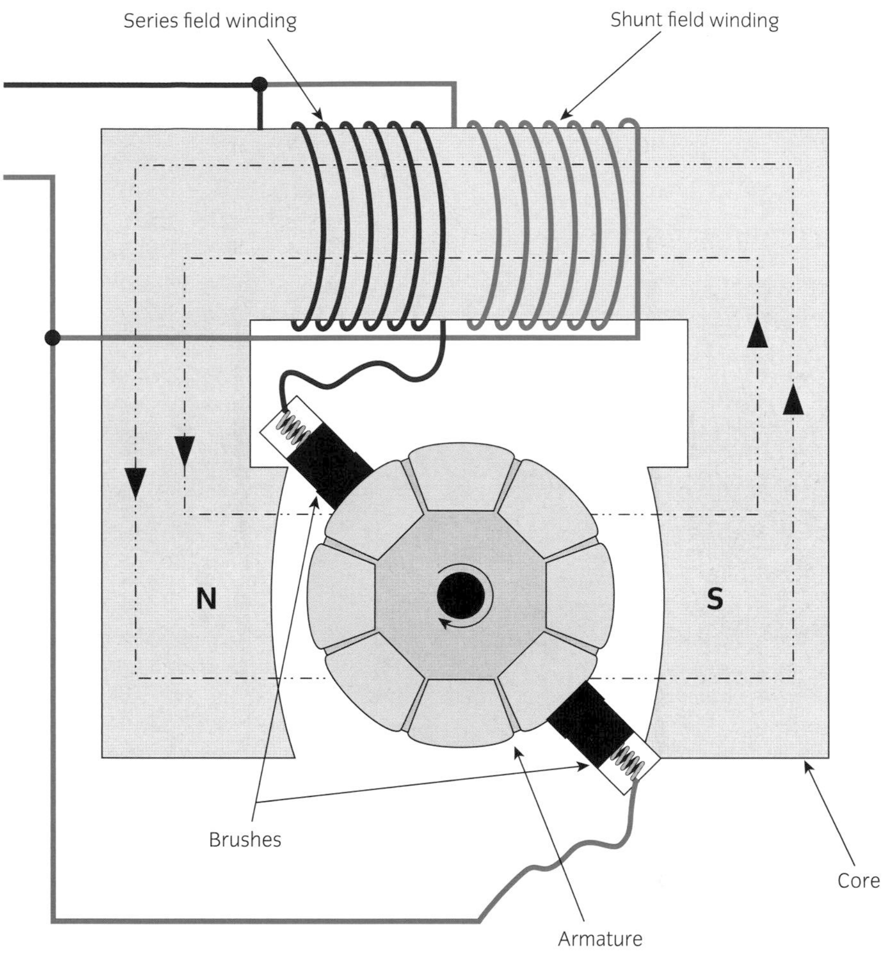


Compound motor arrangement

> **ASSESSMENT GUIDANCE**
>
> Series-biased compound motors are used on older underground trains. This means the series winding is stronger than the shunt winding.

**Assessment criteria**

**4.2** State the applications of d.c. machines

## APPLICATIONS OF D.C. MACHINES

Series-wound motors have excellent torque (load) characteristics and are used for applications such as dragline excavators, where the digging tool moves rapidly when unloaded but slowly when carrying a heavy load.

Shunt motors are best used where constant speed and torque are to be maintained, for example, on a production line, so that items placed on it do not affect the speed.

Compound motors offer the benefits of both series and shunt motors and have been used in older underground trains.

Direct current generators do not have many practical uses in their own right. However, they can provide a reliable energy supply directly into batteries or where a d.c. supply is required.

# HOW THREE-PHASE A.C. MACHINES OPERATE

Three-phase a.c. machines are motors and generators that use or produce three-phase power.

**Assessment criteria**

4.3 Explain the operating principle of three-phase a.c. machines

## Three-phase a.c. generators

A three-phase a.c. generator has a stator with three sets of windings arranged so that there is a phase displacement of 120°. The three-phase output is produced by either star- or delta-connected windings on the stator.

In the UK, three-phase generators are used in power stations.

### ACTIVITY

A three-phase a.c. alternator has four poles per phase. Calculate the speed in revs/sec needed to produce an output of:

a) 40 Hz
b) 50 Hz
c) 60 Hz.

Large-scale a.c. generation

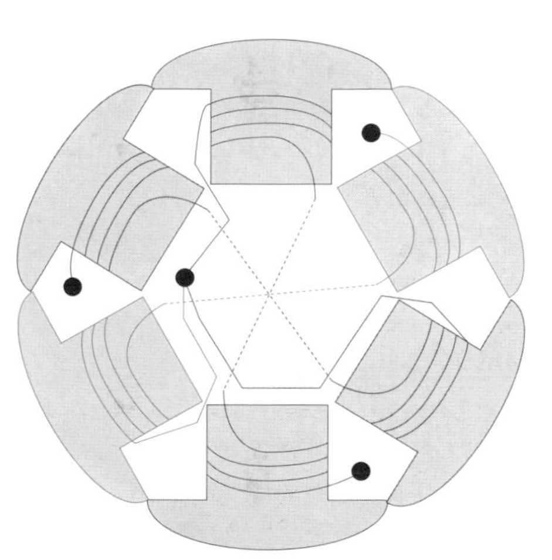

Six-pole (three pairs) salient pole rotor (star wound) for an alternator

## How a.c. generators work

For simplicity, the description below relates to one phase. However, it is important to remember that there are three phases, displaced at 120° from each other.

As each pair of poles passes through the strongest part of the magnetic field at right angles, the maximum electromotive force (emf) is induced into that particular phase. At that point, the other two pairs are in a weaker part of the field and a lower voltage is induced. The moving rotor is connected to the stationary stator by slip rings, which keep each phase in constant contact.

The output of an a.c. system, when measured and tracked, is usually referred to as a waveform. This is because, as the rotating machine induces an emf, the value rises to a peak, falls to zero, then to a negative peak value and then rises back to zero.

### ASSESSMENT GUIDANCE

$N = \frac{f}{p}$ where $N$ = speed in revs/s, $f$ = frequency and $p$ = pairs of poles per phase

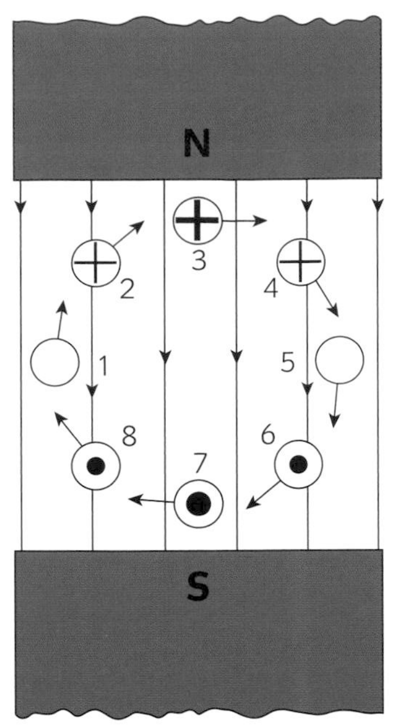


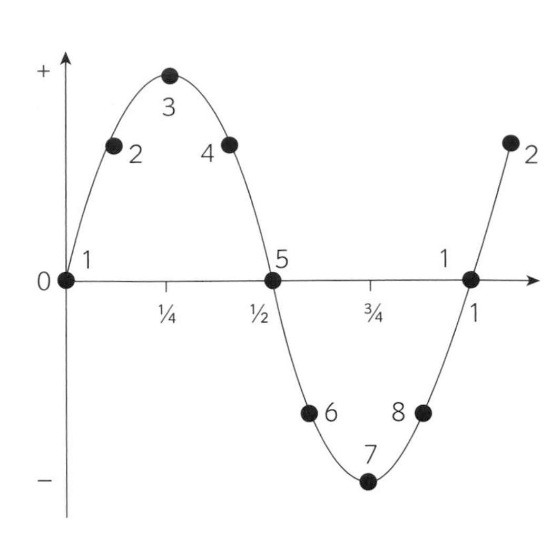


The emf-generated per phase per rotation

**ASSESSMENT GUIDANCE**

Rotor bars are also formed by casting aluminium into the rotor.

The time taken for the cycle to return to its starting position (from position 1 back to 1 in the example above) is the periodic time $t$. This process can be described in terms of Faraday's law because the rotation of the coil continually changes the magnetic flux through the coil and therefore generates an emf.

## Three-phase a.c. motors

Three-phase a.c. motors have a number of advantages over their single-phase equivalents, including:

- smaller physical size for a given output
- steady torque output
- the ability to self-start without additional equipment.

The induction motor is the simplest and most common form of motor. The stator consists of a laminated body with slots for the field windings to pass through. The rotor is a laminated cylinder with conducting bars just below the surface.

In its simplest form, the rotor consists of a number of conductors passing through holes in a rotor drum. The ends are brazed together, causing the formation of a cage, which is why it is called a cage rotor.

Cage induction motors are cheaper and smaller, but produce less torque, than wound rotor induction motors. Wound rotor motors provide speed control via resistors and slip rings, but this is an inefficient method of controlling speed.

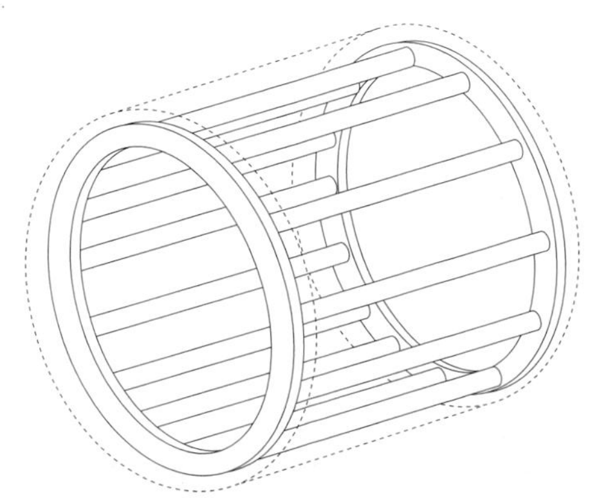

Cage arrangement

## How cage induction motors work

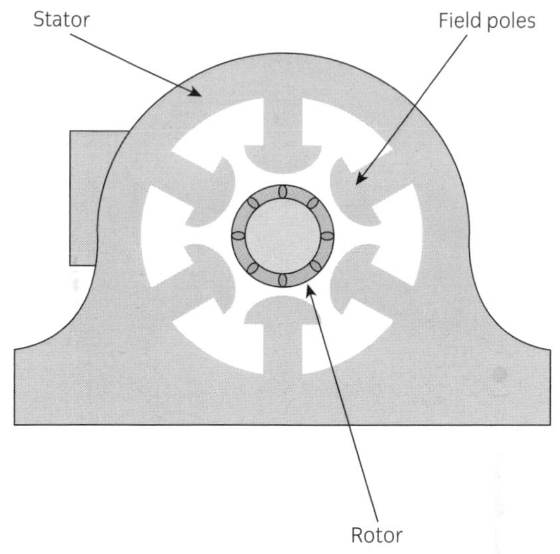


**Component parts of a six-pole cage induction motor**

If a rotor is placed in the field set up by the field windings in the stator, it will be cut by alternate north and south fields as the field rotates through one cycle. This creates an emf in the rotor, which causes current to flow in the conductor bars of the rotor. This current flow sets up a magnetic field, which causes the rotor to move and follow the rotating magnetic field in the stator.

The rotor rotates faster, heading towards **synchronous speed**. However, as the rotor speed increases, the difference between **rotor speed** and stator field speed reduces, causing the emf induced to reduce. The reduction in torque reduces the acceleration/velocity, meaning that synchronous speed cannot be met because no field cuts the rotor and, in turn, no emf or current in the rotor is induced.

This means that induction motors reach an ideal balancing velocity where there is sufficient **slip** to ensure that an emf is generated, resulting in a torque to turn the rotor. The fundamental operating principle of induction motors is that there has to be slip for the motor to work.

Slip may be represented as a percentage or a factor. If a motor with a synchronous speed of 16.67 revolutions/second had a slip of 8%, the rotor speed would be:

$$\text{rotor speed } (n_s) = \text{synchronous speed} - \left(\frac{\text{synchronous} \times \text{slip percentage}}{100}\right)$$

$$= 16.67 - \left(\frac{16.67 \times 8}{100}\right)$$

$$= 15.34 \text{ r/s}$$

### ASSESSMENT GUIDANCE

The three-phase coils set up a rotating magnetic field in the stator. This can be reversed by changing over any two supply phases.

**Synchronous speed**

The speed that the rotating field rotates around the field poles

**Rotor speed**

The actual speed at which the rotor rotates in revolutions/second (r/s)

**Slip**

The difference between the synchronous speed and the rotor speed expressed as a percentage or per unit value

### KEY POINT

Consider that a particular pole is north and that north then moves to the next pole (and the next) until there has been one complete rotation. The synchronous speed is the number of rotations completed in one second.The synchronous speed is affected by the supply frequency and the number of pairs of poles. For example, if a motor had six poles (three pairs) and the supply frequency was 50 Hz, the synchronous speed would be:

$$\text{synchronous speed } (n_s) = \frac{f}{p} = \frac{50}{3} = 16.67 \text{ r/s}$$

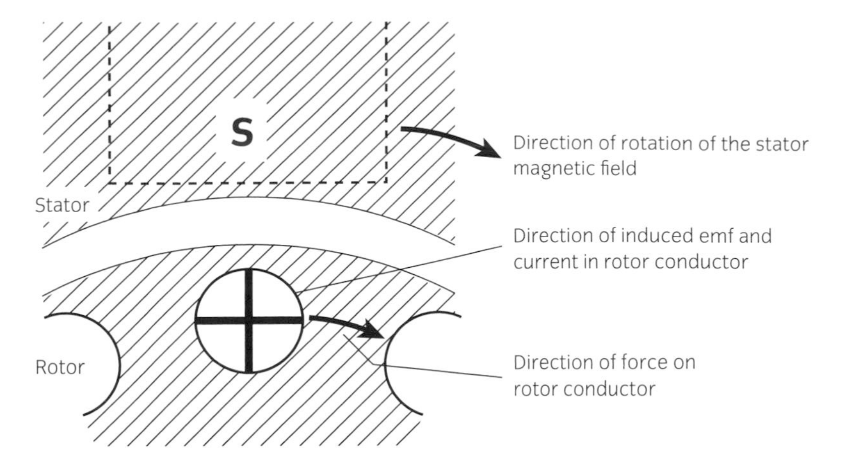


Induction motor principles

The speed of an induction motor can be varied by switching field pole pairs in and out.

### Example

Calculate the synchronous speed of a two-, a four- and a six-pole motor fed from a 50Hz supply.

For a two-pole motor = one-pole pair:

$$n_s = \frac{f}{p} = \frac{50}{1} = 50 \text{ r/s}$$

For a four-pole motor = two-pole pair:

$$n_s = \frac{f}{p} = \frac{50}{2} = 25 \text{ r/s}$$

For a six-pole motor = three-pole pair:

$$n_s = \frac{f}{p} = \frac{50}{3} = 16.66 \text{ r/s}$$

It can be seen that the speed of a motor is varied by the number of poles. However, electronic controls such as inverter drives are more effective and allow the speed to be varied over a wide range, matching it to the load requirements.

## Wound rotor induction motors

In wound rotor induction motors, the rotor windings are connected through slip rings to external resistances. Adjusting the resistance allows control of the speed/torque characteristic of the motor. These motors can be started with low inrush current by inserting high resistance into the rotor circuit; as the motor accelerates, the resistance can be decreased.

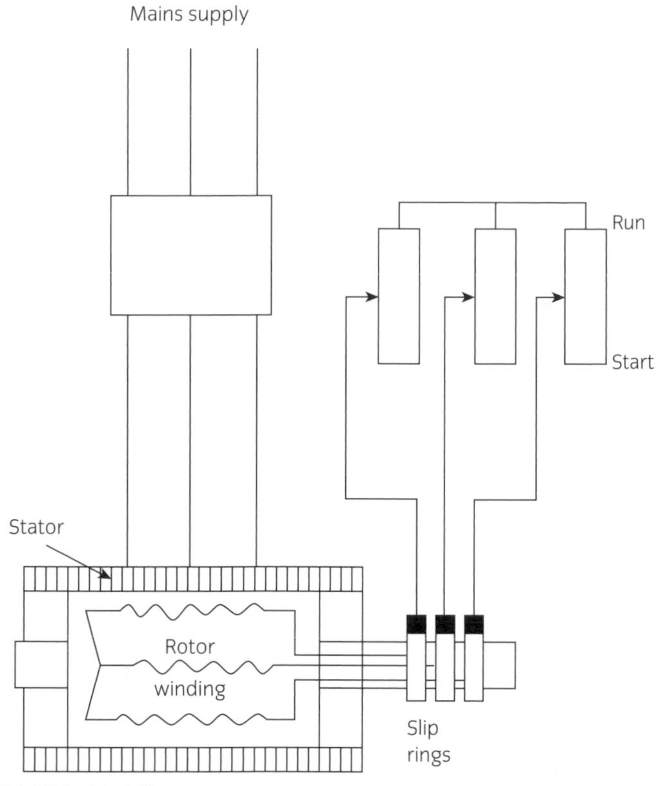


**Wound rotor arrangement**

## ASSESSMENT GUIDANCE

The control circuit needs interlocks to prevent the motor being started with the resistance out of circuit (run position).

The wound rotor has more winding turns than a cage and has a series variable power resistor in the circuit. In start-up, the inrush current is reduced by the resistors with, resulting in a higher torque than the cage motor. However, the ease of speed control has been overtaken by the use of simple induction motors with readily available variable frequency drives, as changing frequency will also affect the synchronous speed and therefore the rotor speed.

# APPLICATIONS OF THREE-PHASE A.C. MACHINES

**Assessment criteria**

**4.4** State the applications of three-phase a.c. machines

## a.c. generators

The a.c. generator or alternator is very widely used, eg in hybrid electrical vehicle drives, small and large-scale power generators, wind turbines or micro-hydro systems (small hydro-electric plant using stream water).

## Induction motors

The cage induction motor is a simple, cost-effective induction motor. Advances in technologies for drive systems (eg Thyristor (silicon-controlled rectifier)) and speed-control drives have enabled simple induction motors to replace more expensive wound rotor induction and some d.c. motors. Such technologies offer simple speed-controlled drives and reduced current starting for 'soft start' systems.

As a result, induction motors are used in a wide range of applications such as pumps, hoists, lifts and many other machines. The cage rotor is particularly useful because it has fewer parts that are subject to wear, having no brushes or slip rings.

**Assessment criteria**

**4.5** Explain the operating principle of single-phase a.c. machines

## HOW SINGLE-PHASE A.C. MACHINES OPERATE

A single-phase generator is composed of a single stator winding with one pair of terminals. With a single pair of rotating poles, the output waveform is as shown below.

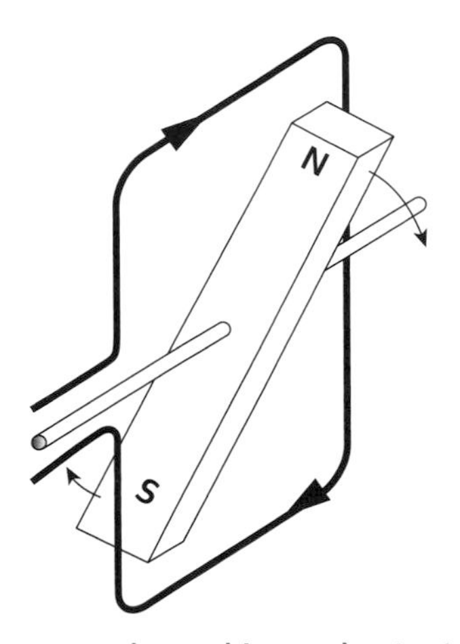


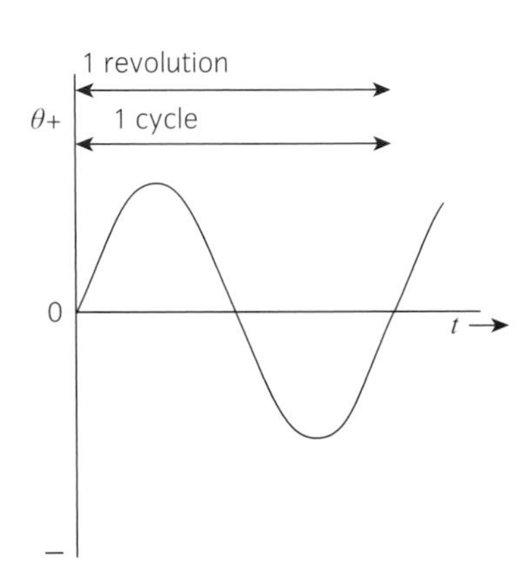


Two-pole machine and output

When a four-pole machine is used, the output waveform is changed as follows.

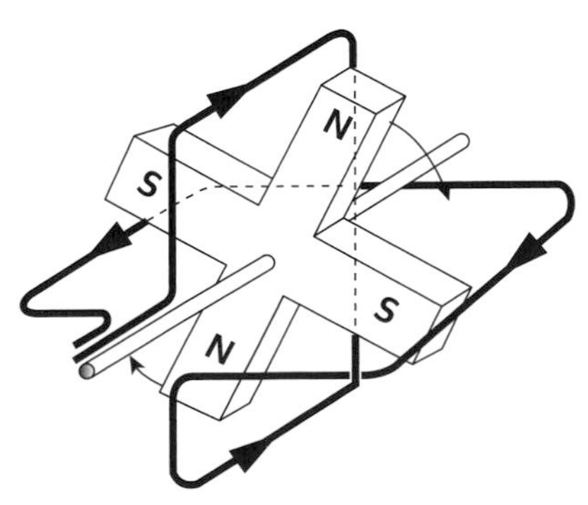


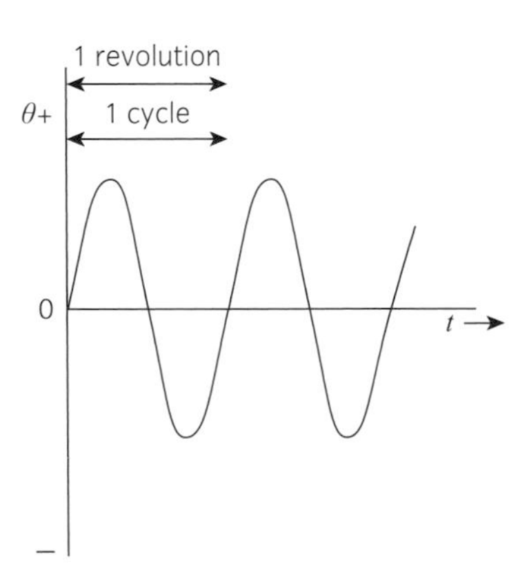

Four-pole machine and output

## Single-phase induction motors

Three-phase motors use one phase to induce current into the rotor bars with another phase, at a different polarity, creating the repulsion/attraction. However, the single-phase motor will not start by itself. This is due to the magnetic flux components being equal and opposite, cancelling out and leaving no torque to turn the rotor. Single-phase motors need to be modified to give the phase shift needed for the motor to start by itself.

Single-phase motors have some form of additional start winding. The current in this winding can be made to lead or lag the main field winding by various methods. We saw on page 162 that a motor winding is an inductor and resistance. The phase angle can be changed by changing resistance values or by adding a capacitor.

As a result, single-phase motors can create a phase-shift in a start winding with:

- split-phase induction motors
- capacitor-start motors
- shaded-pole motors.

**KEY POINT**

A synchronous motor works with a magnet as the rotor. Electromagnet rotors with a field fed from a d.c. supply are more widely used than permanent magnet rotors.

### Split-phase induction motors

The split-phase induction motor has a separate start winding connected to the main supply through a centrifugal switch.

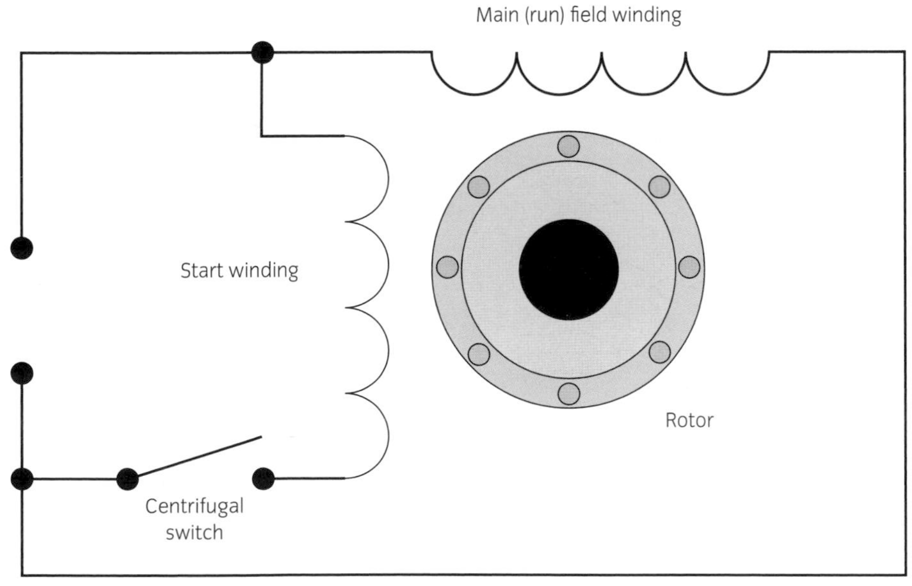


**Split-phase induction motor**

The separate winding causes a slight phase shift by having a different resistance. This causes the rotation. The direction of rotation is determined by the polarity of the start winding, which is switched out by the centrifugal switch once a particular speed is achieved.

## ACTIVITY

Capacitor start-and-run motors have two capacitors. One is electrolytic; the other is a paper type. Why are two types used?

## ASSESSMENT GUIDANCE

There are also capacitor-start/capacitor-run motors. These are easily identifiable as they have two external capacitors.

## Capacitor-start motors

The capacitor-start motor is a variation of the split-phase induction motor. A starting capacitor is in series with the start winding, which creates a phase shift in the circuit due to the inductive and capacitive circuit formed by the winding and the capacitor.

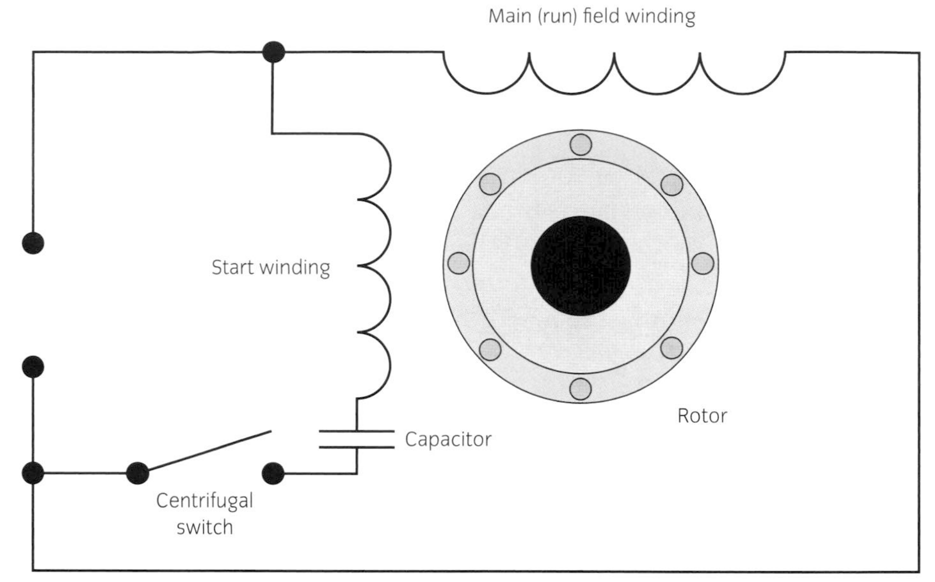


**Capacitor start motor**

In most capacitor-start motors, the capacitor is switched out of circuit by a centrifugal switch. The capacitor-start-and-run motor is a variant that does not switch out the capacitor.

## Shaded-pole motors

The shaded-pole motor is a quite commonly used in devices, such as domestic appliances, that require low starting torque.

The motor is constructed with small single-turn copper shading coils, which create the moving magnetic field required to start a single-phase motor. A small section of each pole is encircled by a copper coil or strap; the induced current in the strap opposes the change of flux through the coil. This causes a time lag in the flux passing through the shading coil, creating an opposing pole to the main part of the pole.

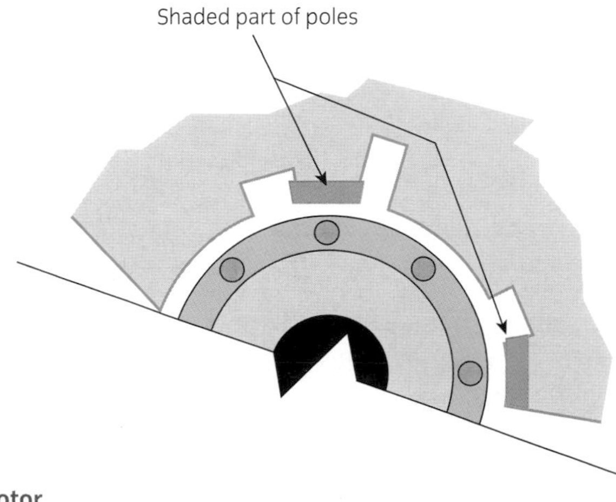


**Shaded-pole motor**

## Universal motors

An a.c. universal motor is very similar in construction to a d.c. series-wound motor. These devices combine the advantages of a.c. machines with some of the characteristics of d.c. machines, including high starting torques.

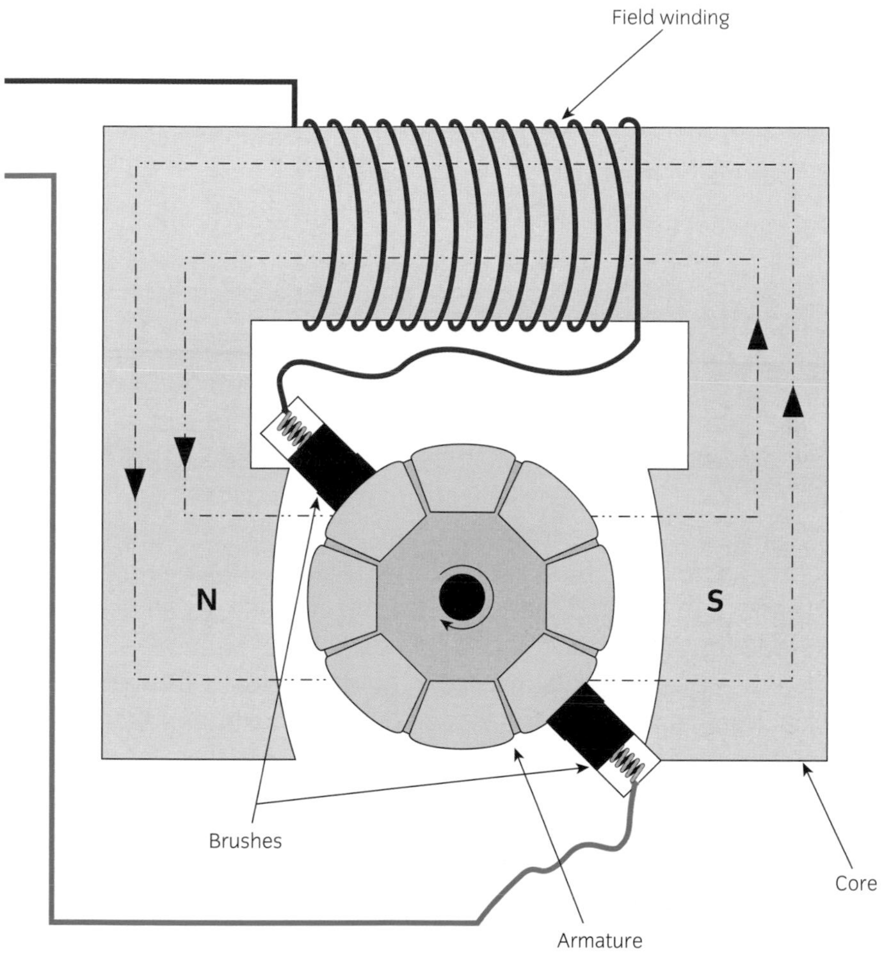


**Section through a universal motor**

### ASSESSMENT GUIDANCE

a.c series universal motors are used in all kinds of domestic appliances such as vacuum cleaners, spin dryers and lawn mowers although some use induction motors, Drills, saws, angle grinders etc all use a.c. series universal motors.

## APPLICATIONS OF SINGLE-PHASE A.C. MACHINES

The commonly used shaded-pole motor is used where a high starting torque is not required, for example, in electric fans or drain pumps for washing machines and dishwashers, and in other small household appliances.

Universal motors are commonly used in small household appliances. such as food blenders, or power tools, such as drills, where smaller motors are beneficial.

**Assessment criteria**

**4.6** State the applications of single-phase a.c. machines

Capacitor-start motors are commonly found in applications such as central-heating circulation pumps.

Split-phase motors are better suited to belt-drive applications due to poorer starting torque.

**Assessment criteria**

**4.7** Explain methods of starting motors

## HOW MOTOR STARTERS OPERATE

Motors are started in various ways. The choice of motor starting/control device depends on:

- available supply (single- or three-phase)
- motor start-up current
- starting speed.

### Direct online starters (DOL)

These motor control devices literally switch a motor on or off with no varying speed or reduction in starting current. They are suitable for small, low-powered motors and can be used for single- or three-phase applications. The unit contains a electromagnetic coil that operates a contactor. The unit also incorporates overload contacts, finely tuned to the motor's start and running current to trip the device should an overcurrent occur. The coil that operates the contactor also provides undervoltage protection, as the contactor will open should a voltage lower than the coil voltage rating occur, eg during loss of supply. This means that the machine will not automatically restart unexpectedly should power resume.

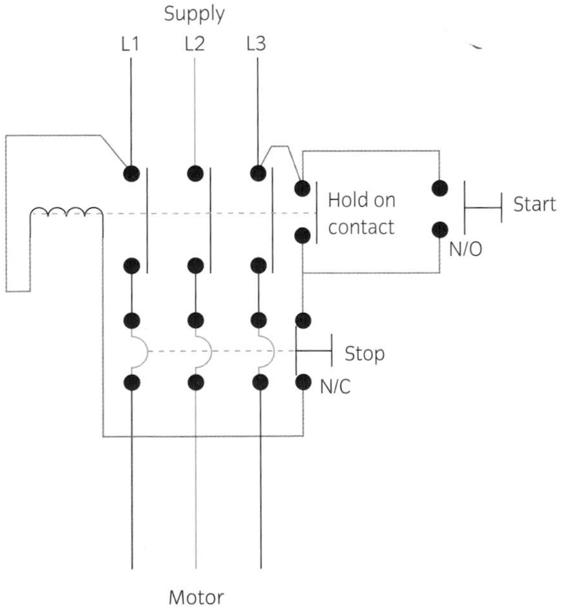

**ASSESSMENT GUIDANCE**

When adding remote controls, remember stop buttons are connected in series with the original stop, and start buttons are connected in parallel with the original start.



**Direct online (DOL) starter**

## Star-delta starters

Star-delta starters must be connected to three-phase motors, which have six connection points at the motor. The motor is started by being put into a star connection, which reduces the starting current. Once the motor reaches a given speed, an electronic time switch switches over the contactors automatically, putting the motor into delta connection and allowing full load current. The purpose of this is to reduce high current on start-up.

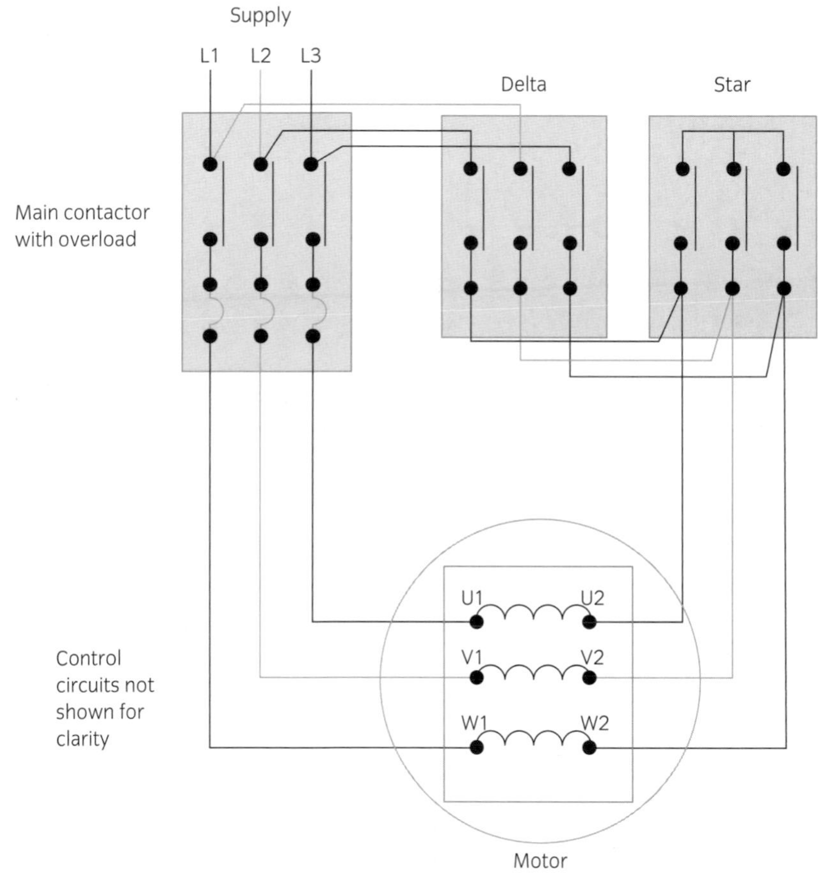


**Star delta starter arrangement**

To avoid short circuits, the star and delta contactors are normally physically interlocked by an electrical interlock and a mechanical connecting rod to prevent both contactors being in circuit at the same time.

### ACTIVITY

What fraction of power is available in a star connection compared with a delta connection?

### ASSESSMENT GUIDANCE

Star-delta starting requires access to all six ends of the motor windings.

## Rotor resistance starters

This type of starting device works by introducing variable resistance to the rotor windings. It requires the motor to be a wound rotor induction motor, not a cage induction motor. The rotor windings are connected to an external variable resistance unit by slip rings and bushes. Variations in resistance can reduce start-up currents in the rotor.

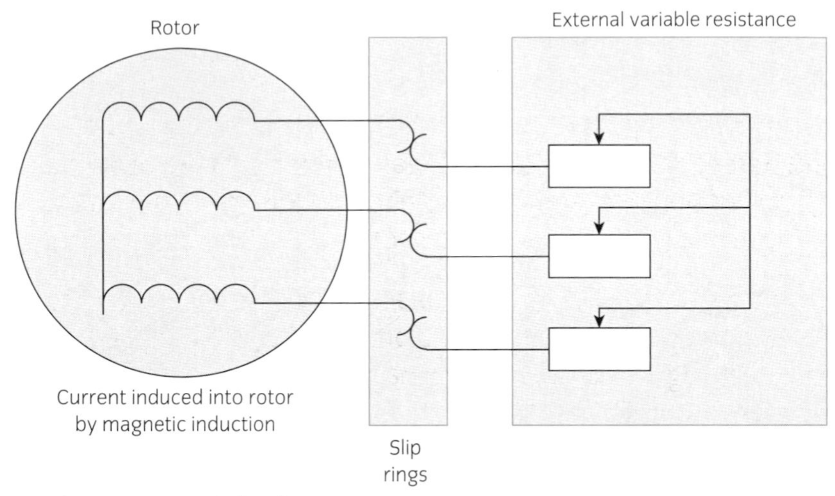


**Rotor resistance control circuit**

## Electronic motor starters

These motor soft-starters are used with a.c. motors to reduce the load and torque in the motor circuit during start-up. This is normally achieved by reducing the voltage; then, as the motor starts to achieve running speed, the voltage is ramped up. The term 'soft-start' also applies to the mechanical stresses placed on the components as they are also not subjected to intense starting forces.

Variable frequency drives (VFD) control the motor speed and torque of a.c. induction motors, by adjusting the frequency and voltage. They save energy and money by adjusting constant-speed devices such as pumps and fans to match the appropriate outputs.

# OUTCOME 5

# Understand the principles of electrical devices

## HOW ELECTRICAL COMPONENTS OPERATE

**Assessment criteria**

**5.1** Explain the operating principle of electrical components

To understand how and why particular components are used in electrical installations, and to be able to make judgements on their suitability, requires some knowledge of how they operate.

Many of the common components used in electrical installations for protection and control rely on electromagnetism. So first, refresh your memory on magnetism and electromagnetism.

Electromagnets are often thought of as large magnets in vehicle salvage yards, lifting scrap metal from one place to another. Although this is one example, electromagnets are also used in lots of electrical installation equipment such as circuit breakers, RCDs, contactors and relays.

Before exploring electromagnets, first consider what a magnet is.

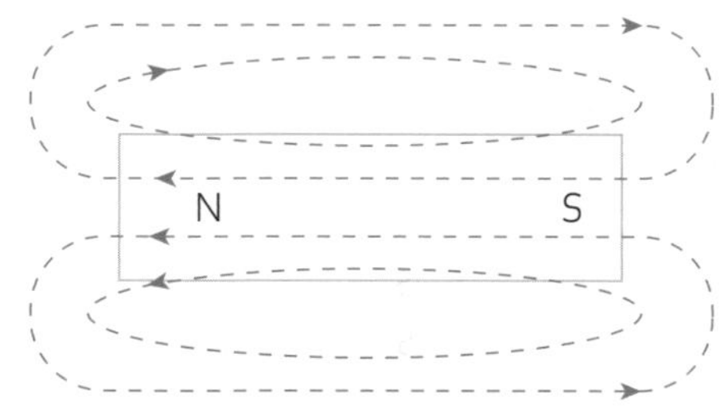


**The pattern of magnetic flux lines that pass through a magnet from south to north**

The bar magnet above shows a magnet and its north and south poles. It also shows the pattern of the lines of magnetic flux that pass through the magnet from south to north, and also outside the magnet from north to south. These flux patterns can be seen when a piece of paper is put over a bar magnet and iron filings are sprinkled over the paper. When the paper is tapped, the iron filings form a pattern because they are drawn into the flux lines.

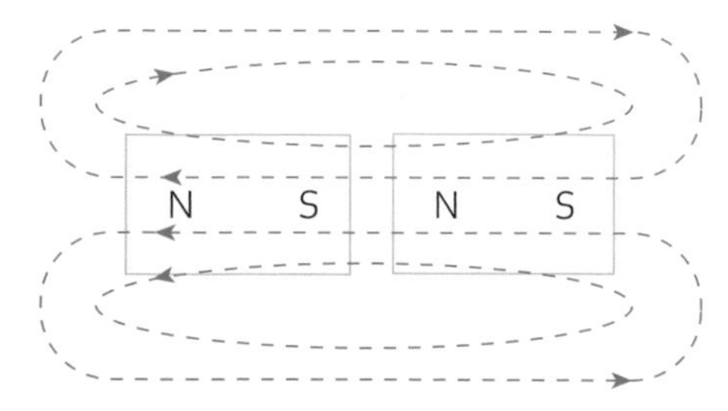


**Two magnets, showing that opposite charges attract**

When two magnets are put together, with a north pole facing a south pole, the lines of flux move together in the same direction. This causes the magnets to attract, pulling together and forming one larger magnet. Opposites attract.

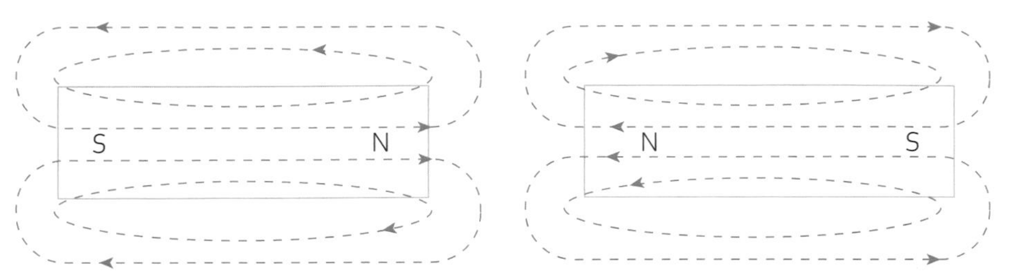


Two magnets, showing that like charges repel

When two magnets are placed with the same poles together, the flux paths move against each other. The force of the magnetic flux causes the magnets to repel and move away from each other.

The planet we live on is a giant magnet with a magnetic field. People navigate around the world, using this magnetic field by placing a small piece of iron on a pivot. Like the iron filings, this small piece of iron follows the flux direction. It is called a compass.

## Magnetic flux patterns of electromagnets

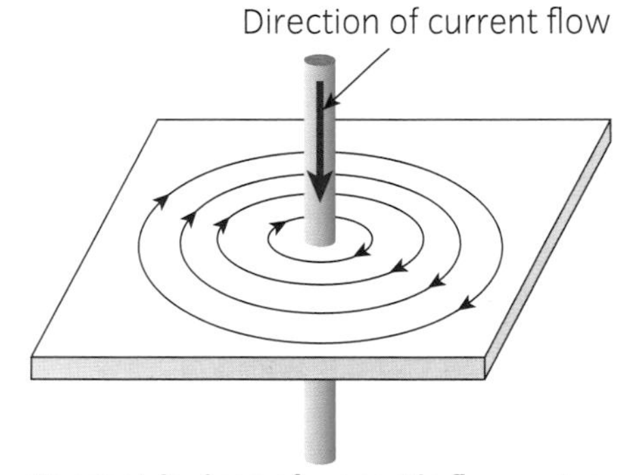


Concentric rings of magnetic flux centre around the conductor

Experimentation with a compass needle or iron filings on a sheet of paper with a conductor passing vertically through it shows that a magnetic field is created around a conductor when current flows through it. If the current is removed, the effect on the compass or iron filings disappears.

This effect occurs throughout the length of a conductor. However, the effect on iron filings on a sheet of paper shows a 'slice' of the field in the plane where the paper is at right angles to the conductor.

## Current and field convention

It is usual to indicate current flow in a conductor because there is a three-dimensional relationship between current flow and magnetic field. Current flowing away from the viewer is shown with a cross, rather like an arrow or dart passing through a tube. Current flowing towards the viewer is shown as a large dot, like an arrow or dart point emerging from a hollow tube.

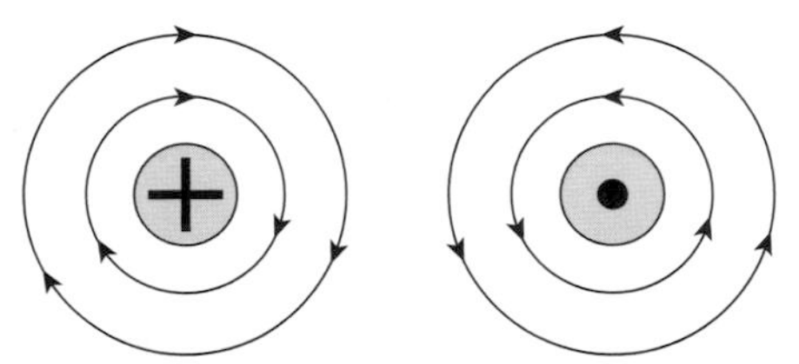

Direction of magnetic field around a conductor (shown in cross-section): current flowing away from view (left) and current flowing into view (right)

The direction of the magnetic field (field rotation) of the concentric rings can be checked with a compass needle. When current flows away from the viewer, the magnetic field rotates clockwise. When current flows towards the viewer, the magnetic field rotates anticlockwise. The magnetic field rotates in the same way as a screw: clockwise to tighten the screw (forcing it away), anti-clockwise to undo it (drawing it closer).

The strength of the magnetic flux is proportional to the current flowing through the conductor. The more current flowing, the stronger the magnetic field will be.

**KEY POINT**

Remember that parallel conductors with currents flowing in opposite directions will push away from each other. Currents flowing in the same direction will cause the conductors to pull towards each other.

Placing two conductors together changes the effects. If two conductors are placed together in a conduit, for example, with the current flowing in opposite directions, there is a cancelling effect between the opposing magnetic fields, as long as the magnetic fields are of equal strength. This arrangement is therefore adopted in electrical installations. Magnetic fields can cause problems in electrical installations and therefore need to be cancelled and minimised as far as is reasonably practicable.

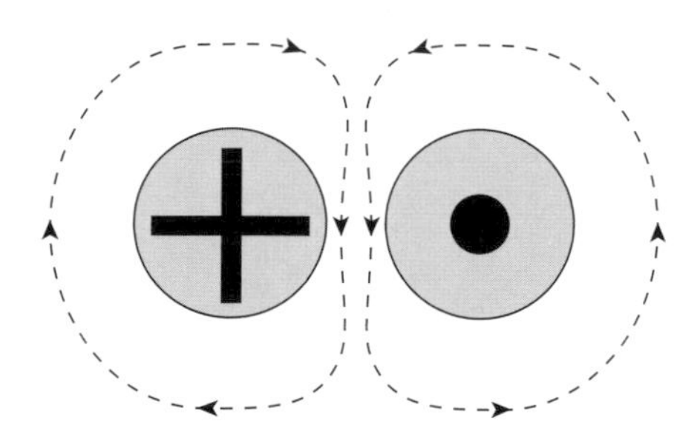

Cancellation effect of opposing conductors

Where conductors are placed together, with the current flowing in the same direction, there is an additional effect. This is undesirable in electrical installations as the increase in the magnetic field will cause additional losses in the circuit and possibly electromagnetic compatibility issues.

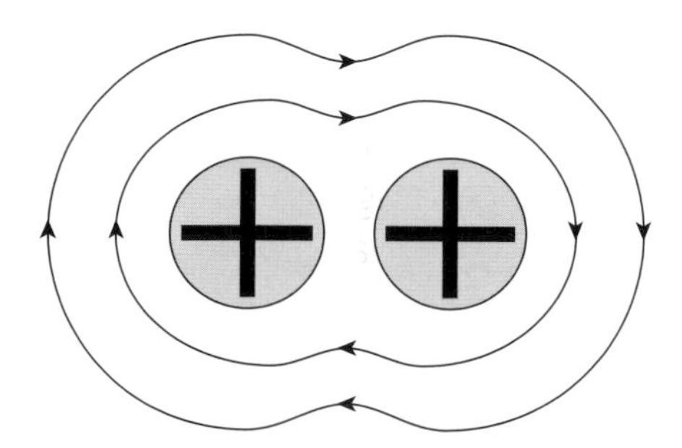

Total effect of magnetic fields

## Solenoids

The strength of a magnetic field is proportional to the current flowing through it. Even with high currents passing through the conductors the field produced is relatively weak, in terms of useful magnetism. To obtain a stronger magnetic field a number of conductors can be added by turning the cable.

The most common form of this is the solenoid, which consists of one long insulated conductor wound to form a coil. The winding of the coil causes the magnetic fields to merge into a stronger field similar to that of a permanent bar magnet. The strength of the field depends on the current and the number of turns.

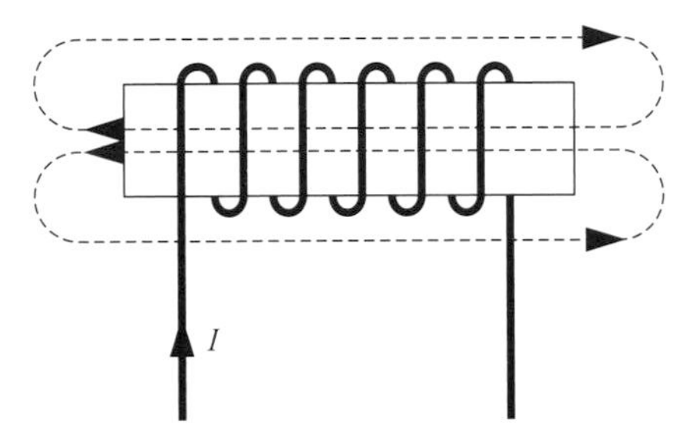


A cable wound around a tube: the current at the top moves away from the viewer and the current at the bottom moves towards the viewer

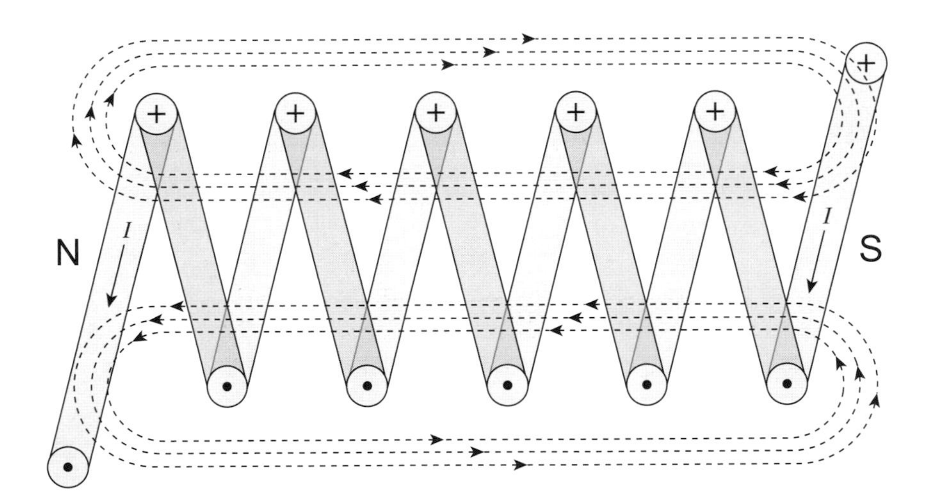


Coiling produces a bar magnet effect

As a solenoid is the electrically powered equivalent of a bar magnet, its field strength is dependent on the current passing through it. The magnetic field can be switched on or off.

The polarity of a solenoid is determined by the current direction. Using the NS rule, the letter N and/or S can be drawn, following the current direction.

The arrow heads on the letters, as shown in this diagram, indicate the direction of the magnetic field rotation.

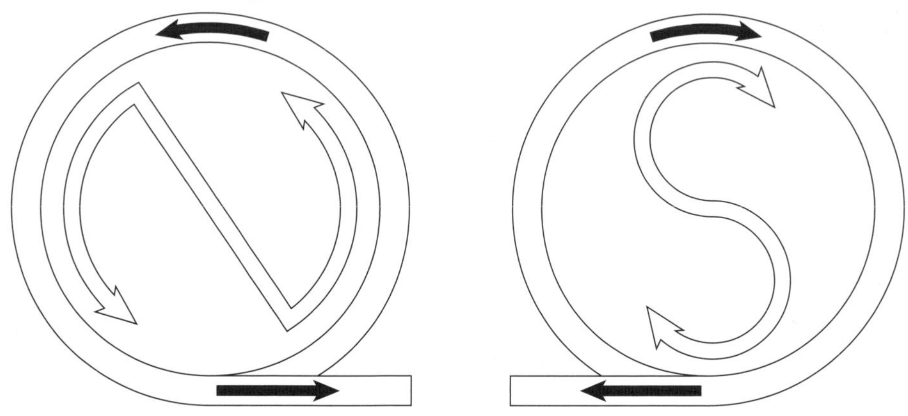

Tracing the letters shows the direction of the magnetic field rotation

The direction of magnetic field rotation can also be determined using the right-hand grip method. If the fingers of the right hand follow the current flow direction, the thumb points to the north pole.

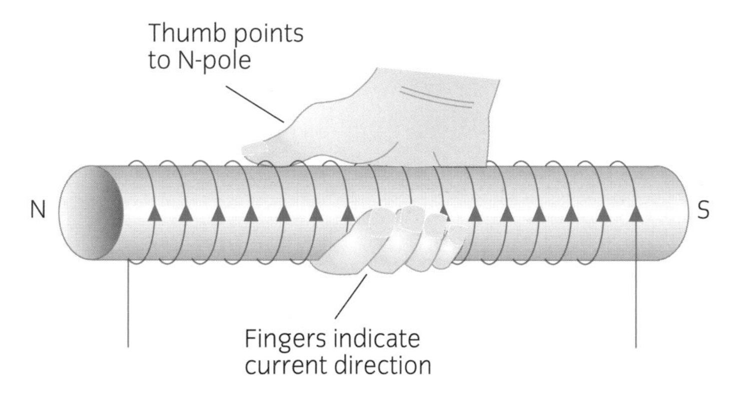


Holding a solenoid in a right-hand grip indicates the direction of magnetic field

## Units of magnetic flux

The unit of magnetic flux is the weber (which is pronounced 'veyber'), abbreviated to Wb. It is represented by the Greek letter phi ($\Phi$). Magnetic flux is a measure of the quantity of magnetic flux not a density.

Flux density (the amount of flux in a given area) is represented by the symbol $B$ which is measured in webers per square metre (Wb/m$^2$), called teslas (T). One weber of flux spread evenly across a square metre of area will give a flux density of 1 tesla. Therefore:

$$B = \frac{\Phi}{A}$$

where:

$B$ = magnetic flux density (T) in webers per square metre (Wb/m$^2$)

$\Phi$ = magnetic flux in webers (Wb)

$A$ = the cross-sectional area of flux path in square metres (m$^2$).

## Relays

A relay is an electrically operated switch, which uses an electromagnet to operate a set of contacts mechanically. This mechanical movement allows complete isolation from the initial signalling.

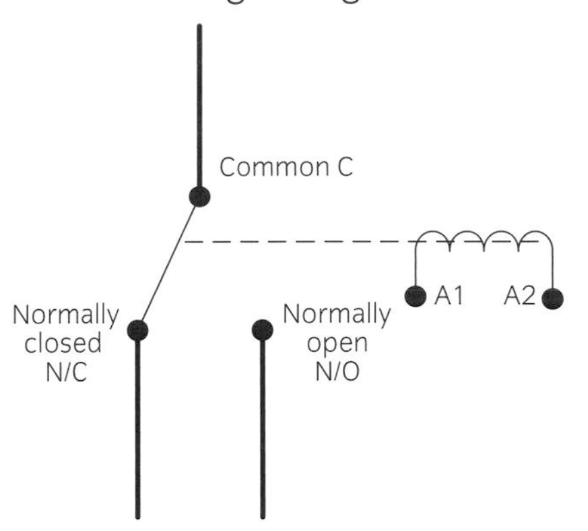


Relay showing contact positions

Relays are used often to control a circuit by a low-power signal in complete isolation from the larger circuit, or multiple circuits, being controlled. The first relays were used in long-distance telegraph circuits, repeating the signal coming in from one circuit and re-transmitting it to another. Relays were then used extensively in telephone exchanges to perform logic functions and operations.

With modern technological advances, not all relays consist of a coil operating a set of magnets. Solid-state relays either replace or are available in conjunction with electromechanical relays. Solid-state relays control electrical circuits despite having no moving parts. Instead, they use a semiconductor device to perform the switching.

**ASSESSMENT GUIDANCE**

A typical application of a contactor is to control a large heating load by the use of a thermostat. The thermostat can operate at extra low voltage but still control a large heating load.

## Contactors

A relay that can handle the high power used to control directly an electric motor or other loads is called a contactor. There is little difference between a relay and a contactor, but generally contactors are devices that switch heavier loads on and off, whereas relays either switch or divert (like a two-way switch) lower current loads.

**Assessment criteria**

5.2 State the application of electrical components in electrical systems

## APPLICATION OF ELECTRICAL COMPONENTS IN ELECTRICAL SYSTEMS

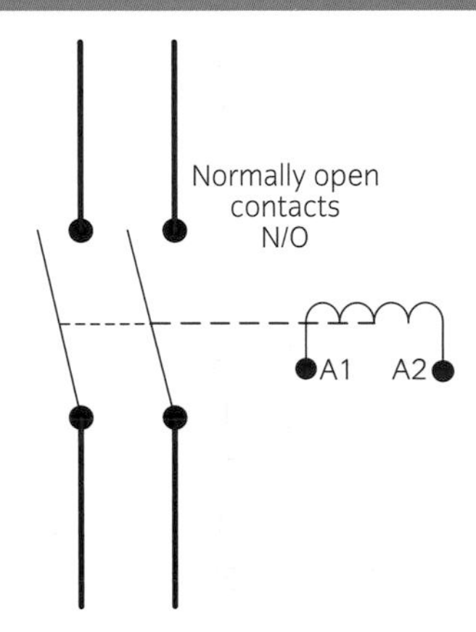


Typical contactor

## Solenoids

Solenoids have a number of functions. They are often used as electrical–mechanical transducers (converters), ie they convert an electrical signal into mechanical action. This can be as some form of limit switch that trips a non-automatically resettable device or, more commonly, to operate a control valve on heating and other similar systems.

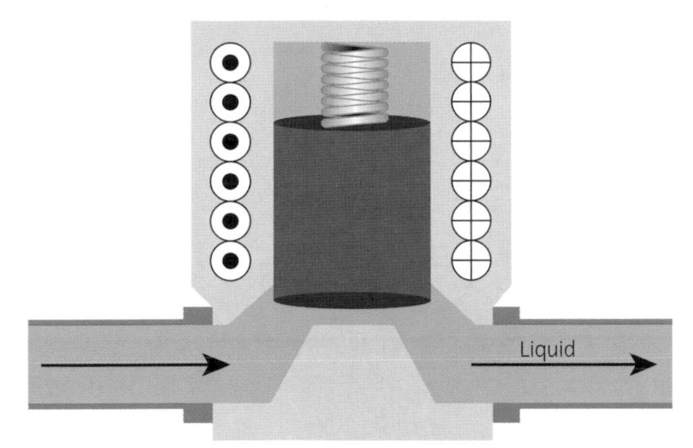


Solenoid valve

Solenoids are also often used in electromagnetic locking devices, either to engage or to retract the locking mechanism. Where safety is essential, such solenoids are usually positioned so that the system drops to a safe position if the electrical circuit fails. For example, a door release magnet will release the door in the event of a fire.

## Relays

A relay is used so that one circuit, normally a low-current circuit, controls another by use of remote contacts. Some relays operate a large number of contacts, switching multiple circuits, with complete electrical isolation of the switching circuit from the operating circuit. Others switch high-current circuits using either low-power circuits or even extra low-voltage circuits, which are in turn controlled by logic devices such as programmable logic controllers (PLCs).

## Contactors

The term 'contactor' is often used instead of 'relay'; however, 'contactor' is more accurately used for a large relay operating large loads, such as a motor.

In the case of motor starters such as the direct online (DOL) starter above, the contactor is also coupled to an overload device. The contactor itself provides control (on and off) as well as undervoltage protection, which is required where the loss of supply and subsequent restoration may cause danger. In the case of a motor or machine, the machine cannot restart after a loss of supply until someone physically pushes the start button on the starter.

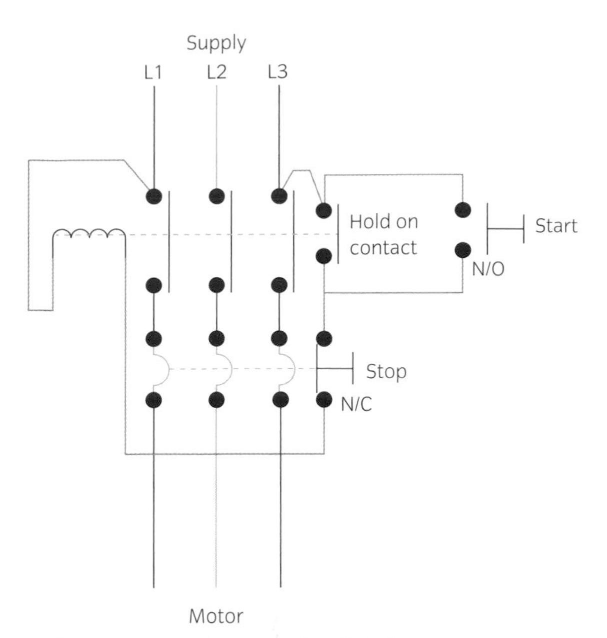


A typical example of a contactor: direct online (DOL) starter arrangement

## ACTIVITY

A direct online motor starter is a type of contactor with built-in start and stop controls. It also has overload protection. Name two types of overload protection that are used.

In the case of motor starters such as the direct online (DOL) starter above, the contactor is also coupled to an overload device. The contactor itself provides control (on and off) as well as undervoltage protection, which is required where the loss of supply and subsequent restoration may cause danger. In the case of a motor or machine, the machine cannot restart following loss of supply until someone physically pushes the start button on the starter.

**5.3** Explain the operating principle of overcurrent protective devices

# PROTECTIVE DEVICES

Protective devices may be one or a combination of:

- fuses
- circuit breakers (CBs)
- residual current devices (RCDs).

## Fuses

Fuses have been a tried and tested method of circuit protection for many years. A fuse is a very basic protection device that melts and breaks the circuit should the current exceed the rating of the fuse. Once the fuse has 'blown' (ie the element in the fuse has melted or ruptured), it needs to be replaced.

Fuses have several ratings.

- $I_n$ is the nominal current rating. This is the current that the fuse can carry, without disconnection, without reducing the expected life of the fuse.
- $I_a$ is the disconnection current rating. This is the value of current that will cause the disconnection of the fuse in a given time.
- Breaking capacity (kA) rating. This is the current up to which the fuse can safely disconnect fault currents. Any fault current above this rating may cause the fuse and carrier to explode.

## BS 3036 rewirable fuses

In older equipment, the fuse may be just a length of appropriate fuse wire fixed between two terminals. There are increasingly fewer of these devices around as electrical installations are rewired or updated.

One of the main problems associated with rewirable fuses is the overall lack of protection, including insufficient breaking capacity ratings. Another major problem is that the incorrect rating of wire can easily be inserted when changing the fuse, leaving the circuit underprotected.

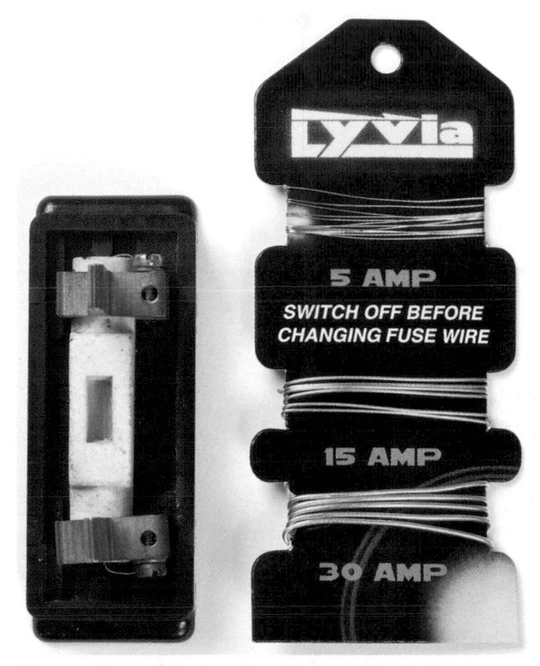


**Rewirable fuse and fuse wire card showing how wrong wire can easily be used**

## BS 88 fuses

These modern fuses are generally incorporated into sealed cylindrical ceramic bodies (or cartridges). If the element inside blows, the whole cartridge needs to be replaced. Although these devices have fixed time current curves, they can be configured to assist discrimination. The benefit of BS 88 and similar fuses is their simplicity and reliability, coupled with high short-circuit breaking capacity.

Within some types of BS 88 fuse, usually the bolted type, there may be more than one element. The purpose of this is to minimise the energy from a single explosion, should the fuse be subjected to high fault currents. Instead there will be several smaller explosions, allowing these devices to handle much higher fault currents of up to 80 kA.

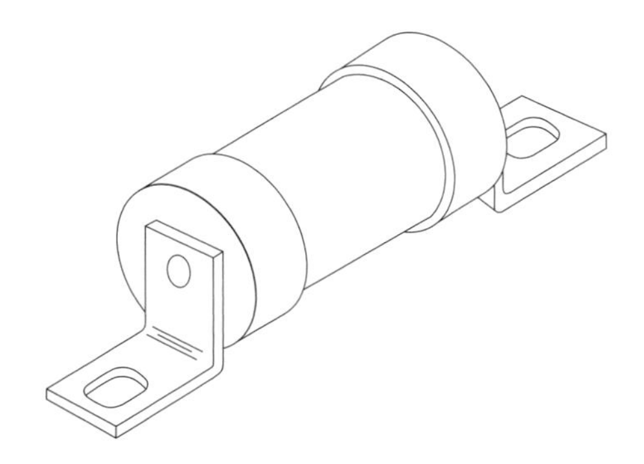

**BS 88 bolted type fuse**

Other BS 88 devices may be the clipped type, which do not have the two bolt tags. They are simply barrel shaped and slot into place in the carrier. They are often called cartridge fuses.

Another type of cartridge fuse is the BS 1362 plug fuse. These are fitted into 13 A plugs and are available in a range of ratings. Typical ratings are 3 A, 5 A and 13 A.

**ASSESSMENT GUIDANCE**

The great advantage of circuit breakers is they do not need replacing in the event of a fault, just resetting.

## Circuit breakers

Circuit breakers (CB) have several ratings.

- $I_n$ is the nominal current rating. This is the current that the device can carry, without disconnection and without reducing the expected life of the device.
- $I_a$ is the disconnection current. This is the value of current that will cause the disconnection of the device in a given time.
- $I_{cn}$ is the value of fault current above which there is a danger of the device exploding or, worse, welding the contacts together.
- $I_{cs}$ is the value of fault current that the device can handle and remain serviceable.

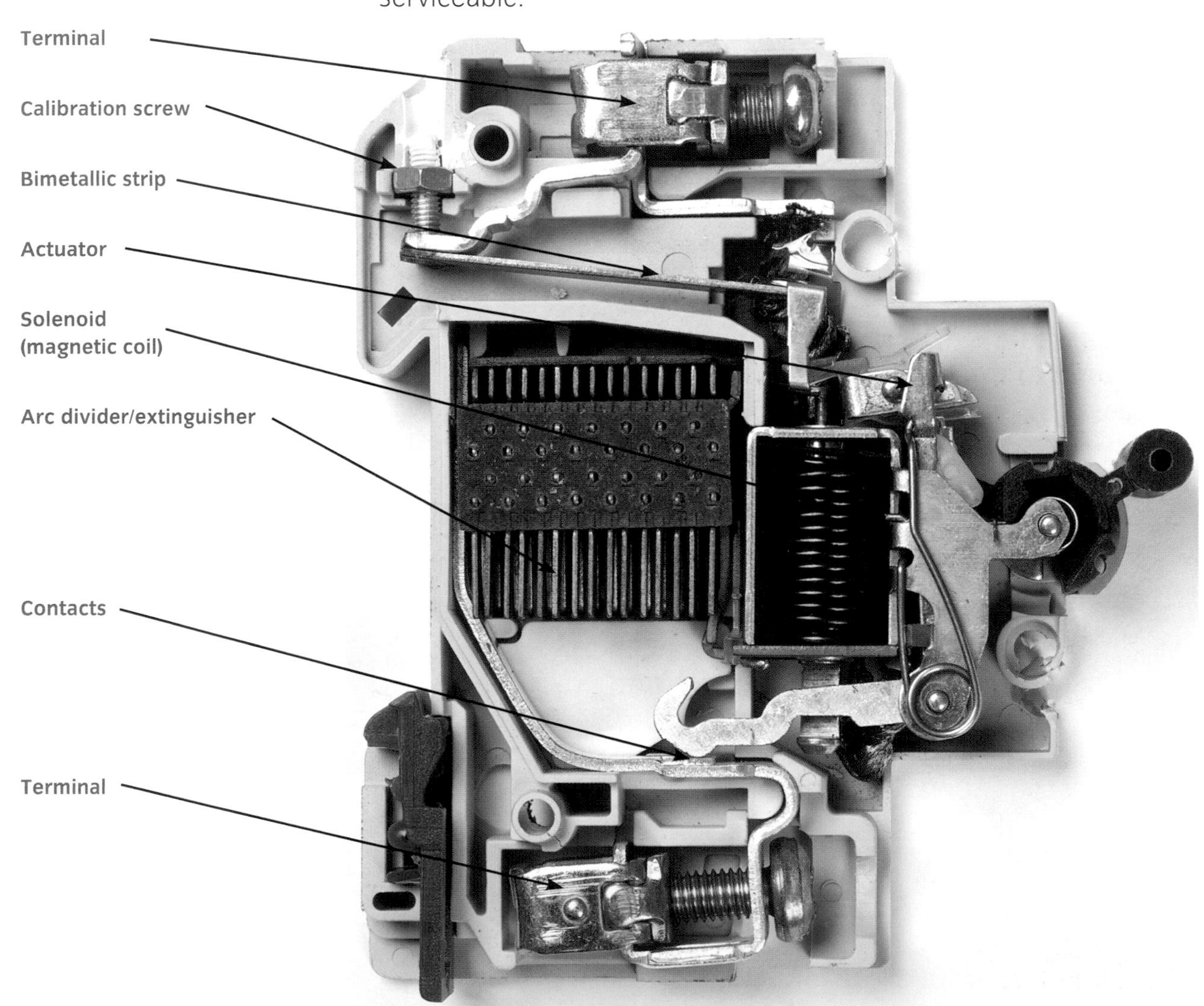


Section through a circuit breaker

Circuit breakers are thermomagnetic devices capable of making, carrying and interrupting currents under normal and abnormal conditions. They fall into two categories: miniature circuit breakers (MCBs), which are common in most installations for the protection of final circuits, and moulded-case circuit breakers (MCCBs), which are normally used for larger distribution circuits.

Both types work on the same principle. They have a magnetic trip and an overload trip, which is usually a bimetallic strip. If a CB is subjected to overload current, the bimetallic strip bends due to the heating effect of the overcurrent. The bent strip eventually trips the switch, although this can take considerable time, depending on the level of overload.

## Miniature circuit breakers (MCBs)

These thermomagnetic devices have different characteristics, depending on their manufacture. They generally have a lower prospective short-circuit current rating than a high-rupturing capacity (HRC) fuse, ranging from approximately 6 kA to 10 kA. Specialist units are available for higher values.

**KEY POINT**

BS 7671 refers to both miniature circuit breakers (MCB) and moulded case circuit breakers (MCCB) as circuit breakers (CB).

The operating characteristics of MCBs can be shown in graphical form by a time–current curve. MCBs are generally faster acting than the standard curve in BS 88 fuses. A CB has a curve, then a straight line, whereas the BS 88 fuse is fully curved. This demonstrates the two tripping mechanisms in a CB. The magnetic trip is represented by the straight line on the graph, indicating that a predetermined value of fault current will disconnect the device rapidly. The curve represents the device's thermal mechanism. Like a fuse, the thermal mechanism reacts within a time specific to the overload current. The bigger the overload, the faster the reaction.

**ASSESSMENT GUIDANCE**

The cost of circuit breakers has come down as their use has become more widespread.

▼ **Table 7.2.7(i)** Rated short-circuit capacities

| Device type | Device designation | Rated short-circuit capacity (kA) | |
|---|---|---|---|
| Semi-enclosed fuse to BS 3036 with category of duty | S1A<br>S2A<br>S4A | 1<br>2<br>4 | |
| Cartridge fuse to BS 1361 type I<br>type II | | 16.5<br>33.0 | |
| General purpose fuse to BS 88-2 | | 50 at 415 V | |
| BS 88-3 type I<br>type II | | 16<br>31.5 | |
| General purpose fuse to BS 88-6 | | 16.5 at 240 V<br>80 at 415 V | |
| Circuit-breakers to BS 3871 (replaced by BS EN 60898) | M1<br>M1.5<br>M3<br>M4.5<br>M6<br>M9 | 1<br>1.5<br>3<br>4.5<br>6<br>9 | |
| Circuit-breakers to BS EN 60898* and RCBOs to BS EN 61009 | | $I_{cn}$<br>1.5<br>3.0<br>6<br>10<br>15<br>20<br>25 | $I_{cs}$<br>(1.5)<br>(3.0)<br>(6.0)<br>(7.5)<br>(7.5)<br>(10.0)<br>(12.5) |

* Two short-circuit capacities are defined in BS EN 60898 and BS EN 61009:

$I_{cn}$ the rated short-circuit capacity (marked on the device).
$I_{cs}$ the in-service short-circuit capacity.

**Rated short circuit capacities of protective devices (from the On-Site Guide, IET)**

## Moulded-case circuit breakers (MCCBs)

Although moulded case circuit breakers (MCCBs) work on the same principle as MCBs, the moulded case construction and physical size of MCCBs gives them much higher breaking capacity ratings than those of MCBs. Many MCCBs have adjustable current settings.

## Residual current devices (RCDs) and residual current circuit breakers with overload (RCBOs)

**ACTIVITY**

What rating of RCD would be used to provide additional protection?

**KEY POINT**

Remember that RCDs only offer earth fault protection not short-circuit (line to neutral) or overload protection. If an RCD is fitted to a circuit, appropriate protective devices must also be installed to offer short-circuit and overload protection.

Residual current devices (RCDs) operate by monitoring the current in both the line and neutral conductors of a circuit. If the circuit is healthy with no earth faults, the toroidal core inside the device remains balanced with no magnetic flux flow. If a residual earth fault occurs in the circuit, slightly more current flows in the line conductor compared to the neutral. If this imbalance exceeds the residual current setting of the device, the flux flowing in the core is sensed by the sensing coil, which induces a current to a solenoid, tripping the device.

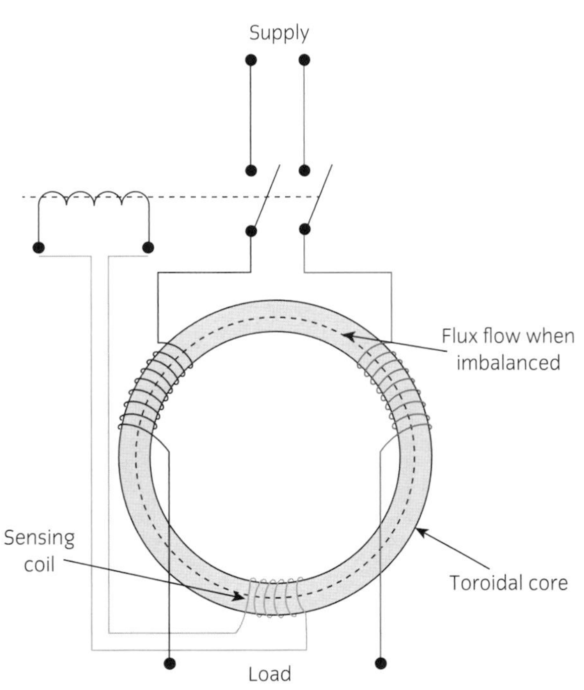


Internal circuit diagram for an RCD

RCBO to BS EN 61009

Residual current breakers with overload (RCBOs) combine an overcurrent protective device with a RCD in the body of the CB.

Unlike CBs, RCDs and RCBOs have a test button, which should be pressed at very regular intervals to keep the mechanical parts working effectively. If the mechanical components in a CB stick, there is not much concern as the energy needed to trip a CB is large enough to unstick any seized parts. As RCDs and RCBOs operate under earth fault conditions, with relatively small residual currents, there may not be enough energy to free any seized parts.

# APPLICATION OF PROTECTIVE DEVICES

Assessment criteria

5.4 State the application of overcurrent protective devices in electrical systems

## BS 3036 rewireable fuses

Unlike most other protective devices, the BS 3036 fuse arrangement does not have a very accurate operating time or current as it is dependent upon factors such as age, level of oxidation on the element and how it has been installed (eg whether it was badly tightened, open to air movement).

The lack of reliability of these fuses is a concern to designers and duty holders. Due to the lack of sensitivity, special factors have been applied to Appendix 4 of BS 7671 Requirements for Electrical Installations (the IET Wiring Regulations) to account for these fuses. This rating factor to be applied ($C_f$) is 0.725. This rating factor is explained in greater detail in Unit 305, pages 195–301.

A range of BS 3036 rewireable fuses: 5 A (white), 15 A (blue) and 20 A (yellow)

## BS 88 fuses

High-rupturing capacity (HRC) or high-breaking capacity (HBC) fuses are common in many industrial installations. They are also very common in switch fuses or fused switches controlling specific items of equipment. They are particularly suited to installations with a high prospective fault current ($I_{pf}$) as they have breaking capacities of up to 80 kA. BS 88 fuses come in two categories:

- gG for general circuit applications, where high inrush currents are not expected
- gM for motor-rated circuits or similar, where high inrush currents are expected.

### ACTIVITY

Identify two other rewireable fuse carrier ratings and colours?

## MCBs

There are three common types of MCB: Type B, Type C and Type D. The difference between the devices is the value of current ($I_a$) at which the magnetic part of the device trips. The different types are selected to suit loads where particular inrush currents are expected.

### ASSESSMENT GUIDANCE

In a fused switch the fuses are mounted on the moving contacts. In a switch fuse, the fuse and switch are in series and the fuse does not move.

**Type B** trips between three and five times the rated current (3 to 5 × $I_n$). These MCBs are normally used for domestic circuits and commercial applications where there is no inrush current to cause it to trip. For example, the magnetic tripping current in a 32 A Type B CB could be 160 A. So $I_a = 5 \times I_n$. These MCBs are used where maximum protection is required and therefore should be the choice for general socket-outlet applications.

**Type C** trips between five and ten times the rated current (5 to 10 × $I_n$). These MCBs are normally used for commercial applications where there are small to medium motors or fluorescent luminaires and where there is some inrush current that would cause the CB to trip. For example, the magnetic tripping current in a 32 A Type C CB could be 320 A. So $I_a = 10 \times I_n$.

**Type D** trips between ten and twenty times the rated current (10 to 20 × $I_n$). These MCBs are for specific industrial applications where there are large inrushes of current for industrial motors, x-ray units, welding equipment, etc. For example, the magnetic trip in a 32 A Type D CB could be 640 A. So $I_a = 20 \times I_n$.

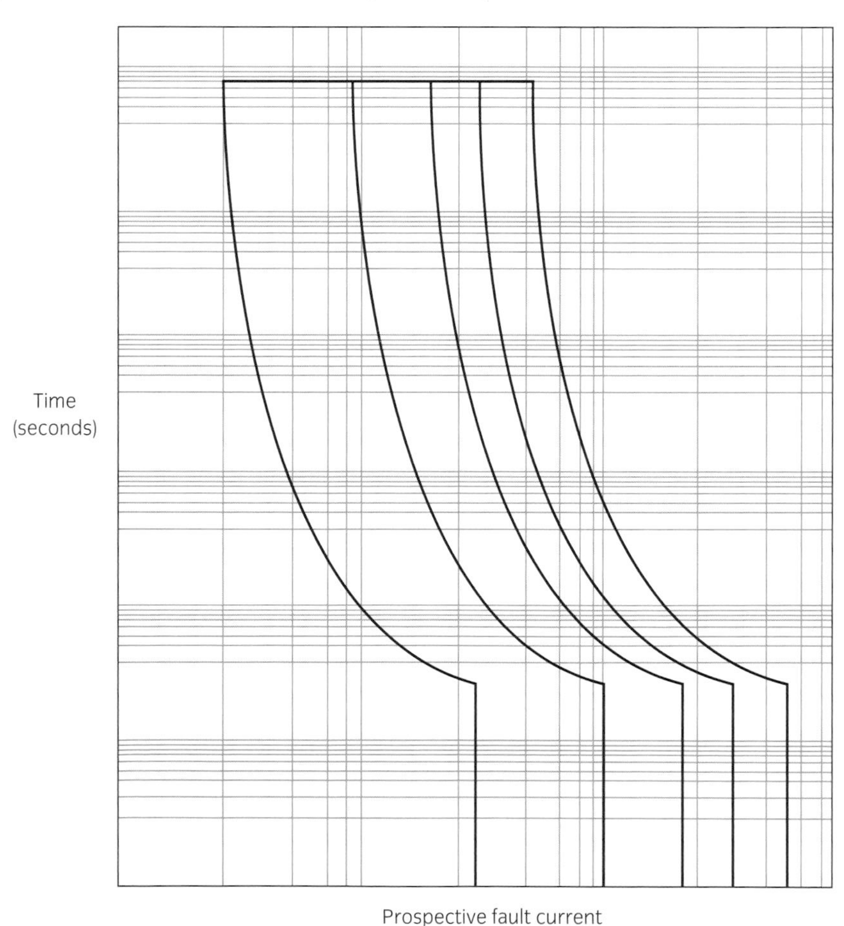


**Sample time–current characteristic graph which is found in Appendix 3 of BS 7671**

## MCCBs

MCCBs are available in various ranges. Lower-cost simpler versions are thermomagnetic with no adjustment. Other devices have electronic trip units and sensitivity settings or the ability to be de-rated.

Most MCCBs are used on larger circuits or distribution circuits where larger prospective short circuits are likely but the flexibility of an electronic trip is also required.

# OUTCOME 6

# Understand the principles of electrical heating systems

Electric heating includes any process in which electrical energy is converted to heat, such as cooking and heating space and water.

**Assessment criteria**

**6.1** Explain the principles of electrical space-heating systems

## HOW ELECTRICAL SPACE-HEATING SYSTEMS WORK

The three ways to transfer heat from one medium to another are:

- convection
- conduction
- radiation.

Many heat sources use a combination of these methods.

### Convection

Hot air rises and colder air falls through the process of convection. A simple convection panel heater mounted on a wall uses this principle to move warm air around a room.

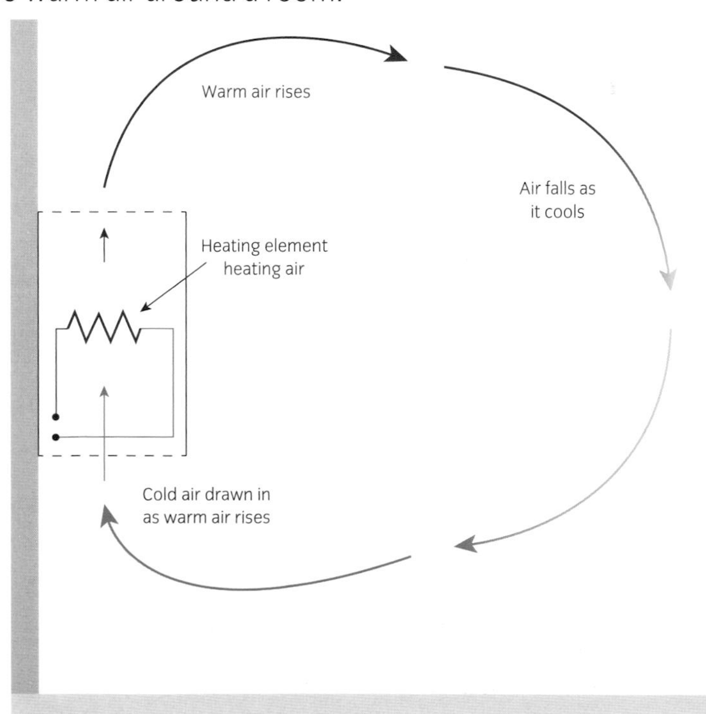


**Convection cycle**

A convection heater usually has a low-temperature 'black-heat' element. Air in contact with the element is warmed and becomes less dense, so that it rises and is replaced by colder air, which is then warmed in turn.

Some convection heaters heat up another medium by conduction. For example, the element can be submersed in oil. The oil transfers heat around the unit, giving a larger body of heat to start the convection cycle.

Traditionally, convection heaters such as central heating emitters (radiators) are positioned where colder air is present, such as under windows, as this produces a larger cycle effect. This is less relevant today as modern windows provide better insulation.

## Conduction

Conduction is the effect of heating something by direct contact. For example, the underfloor heating elements directly below a tiled floor warm the tiles and heat is transferred up through them. Similarly, in an immersion heater, a heating element is placed in water and the heat is transferred from the element directly to the water.

**ASSESSMENT GUIDANCE**

Immersion heaters are usually considered for back-up heating, perhaps in conjunction with solar thermal systems.

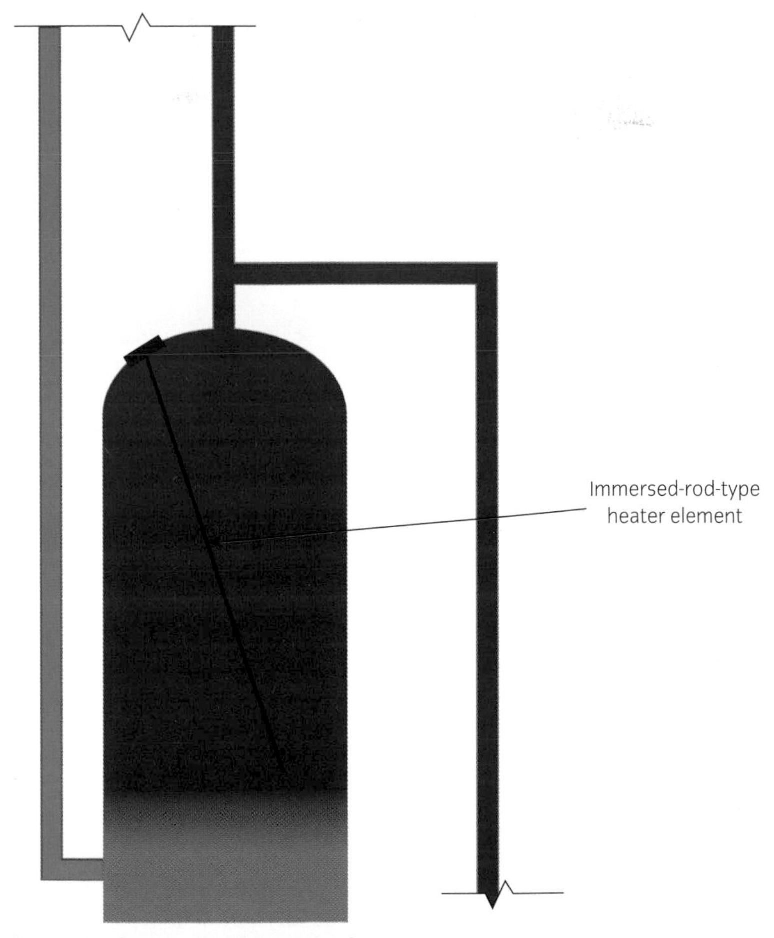


**Immersion water heater heating water by conduction**

## Radiation

In this process heat is radiated (or thrown out) from a source and warms objects nearby. A standard coal fire does this; the heat can be felt on surfaces facing the fire, but not on surfaces facing away from the heat source.

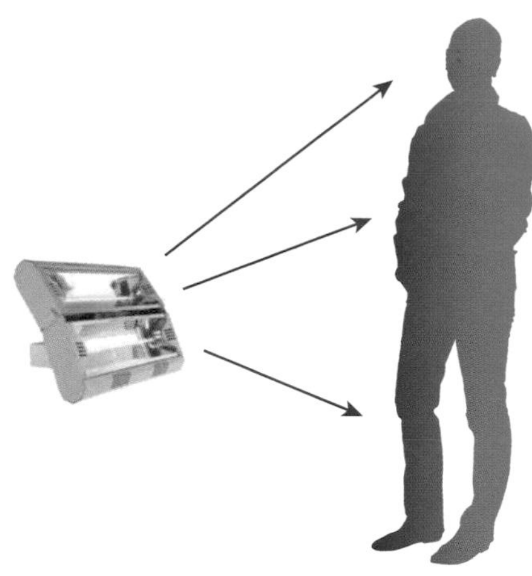

Radiant heat

Radiant heaters include:

- traditional electric fires
- infrared heaters
- oil-filled radiators
- tubular heaters
- underfloor heaters.

Most heating devices use more than one of these methods of heat transfer.

## Heat sources

### Underfloor heating

Underfloor heating systems have been around for centuries, although electrically powered versions have been available for a much shorter period. Since the 1960s the popularity of electric underfloor heating has fluctuated. In more recent times, it has become particularly associated with bathrooms and tiled areas. In many cases, underfloor heating is only installed in new houses or extensions as it is very costly to install into existing floors.

Electric underfloor heating before the final screed is laid

Underfloor heating is an effective system of transferring heat into the floor surface by conduction and then heating the room by radiant heat and convection. It is a very good way of providing a uniform heat in a space.

### Storage heaters

Storage heaters charge up at night when energy is available at low cost on off-peak tariffs. The energy is then stored in the unit in fireclay blocks, which release it slowly during the day. Radiator- and fan-type storage heaters are available.

The radiator type is heated by elements in fireclay blocks. The release of heat is controlled by insulation. The storage heater is sized to store enough heat to last all day, under controlled release conditions.

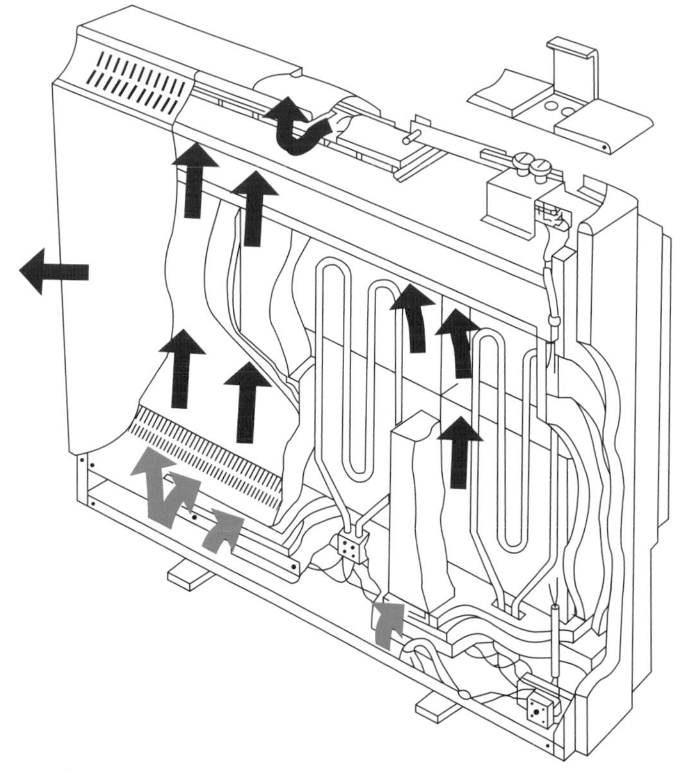

Electric storage heater

Fan-type storage heaters have thicker insulation so that very little heat is lost. A small unit, on a 24-hour supply for the fan, will provide warm air controlled by a thermostat. Because of the fan, it is possible to use up the heating charge, so there is often a short boost option. However, daytime boosting is not an economical way to use storage heaters.

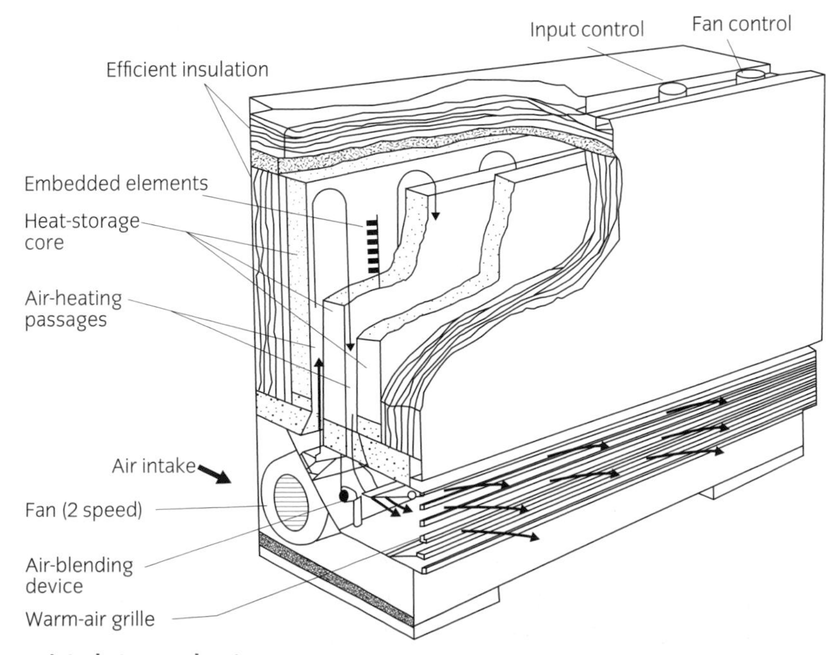


Fan-assisted storage heater

## ACTIVITY

Name the tariff that should be used with storage heaters.

## ASSESSMENT GUIDANCE

Block storage heating tends to be retrofitted where the installation of a wet system could cause considerable disruption.

The fan-assisted storage heater is usually larger and always noisier than the radiator-type equivalent.

## Panel heaters

Panel heaters heat spaces by convection and radiation. They can be slim in design and can even be fitted to the front of storage heaters to provide daytime heating if the stored charge has been lost.

Most panel heaters are provided with thermostatic and time controls. Although they can be fairly expensive to run, they are often chosen as a means of heating reasonably small locations or locations where other services, such as hot-water central-heating systems, are difficult to install.

An infrared heater

## Radiant or infrared heaters

These types of heater are particularly useful in large, cold areas where heating the air is difficult. Examples are garage workshops and warehouses. The heat radiated warms bodies but not the air around them. They would work just as effectively in a vacuum.

**Assessment criteria**

**6.2** Explain the principles of electrical water-heating systems

# HOW ELECTRICAL WATER-HEATING SYSTEMS WORK

There are many different types of electrically powered water heaters, including rod-type immersion heaters and instantaneous water heaters.

**ACTIVITY**

Why are infrared heaters commonly used in bathrooms?

## Immersion heater systems

Immersion heater systems usually contain large amounts of water, that are heated over a period of time. The size of the vessel and limited amount of circulation and mixing allows relatively hot water to be used on demand. Then the heat is replenished over a period of time.

A large copper or stainless steel vessel is filled with water from a separate cold-water tank or sealed pressurisation unit. The vessel has an electric heater element fitted through a screw-threaded fitting, known as a boss.

The heater element can be the length of the cylinder, but in some instances two shorter elements are used, one positioned high and the other positioned low in the vessel. In domestic premises, the two heater elements used to be known as the 'bath and sink' function. By switching on the top element, just a small proportion of the water is heated, saving energy. This relies on the stratification process (ie hot liquids stay at the top) and works if the water is used before significant circulation takes place.

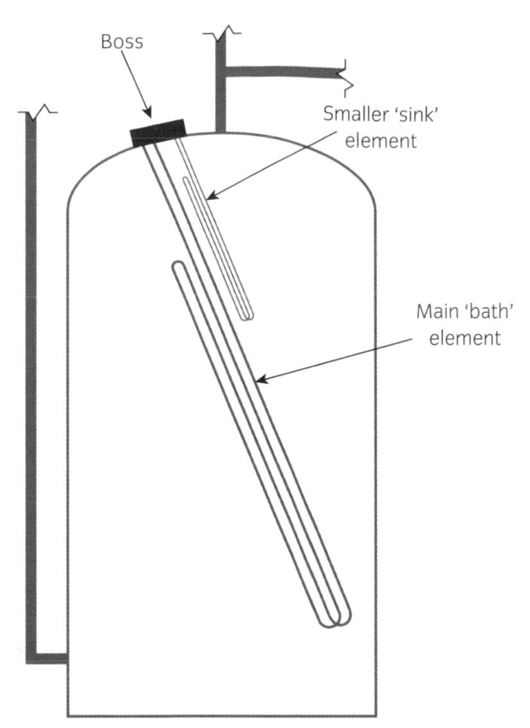


Cylinder element arrangements

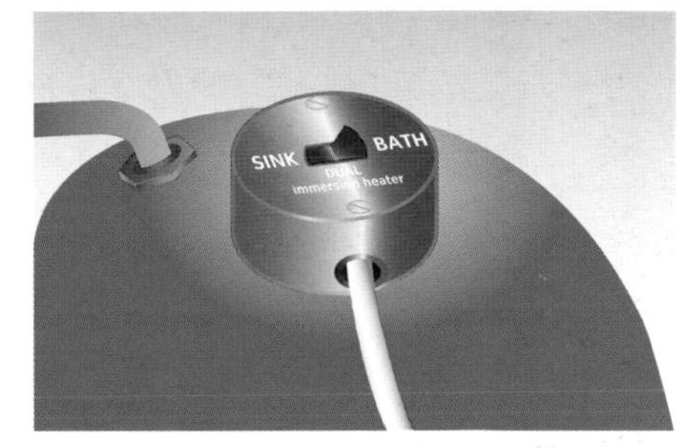


Top entry boss with 'bath and sink' switch

To ensure that there is a full tank of hot water, a longer element is fitted to heat the whole tank. Temperature is controlled by a rod thermostat fitted in a pocket tube in the head of the heater. However, there are also strap-on thermostats that fix to the exterior of the hot water cylinder. Many immersion heaters also have a thermal cut-out to open the element circuit, should the thermostat fail and the water reach dangerously high temperatures, which could lead to enormous pressures in the tank and the venting of scalding water through the overflow or expansion pipe.

In order to save energy, thermal insulation is added to the vessel during manufacture or a thermally insulated jacket is fitted after installation.

Thermostat inside pocket in heater element

## Instantaneous water heaters

There are many examples of instantaneous water heaters; most common are electric showers and point-of-use hand washers, fitted above or below sinks.

With all of these heaters there is a limit to how much water can be raised to a specific temperature in a given time. This depends on the flow rate. The slower the flow rate, the hotter the water will get.

There are two types of instantaneous hot-water system. The tank type is like a miniature hot water cylinder. In the other type, the elements wrap around the water pipe inside the unit.

**ASSESSMENT GUIDANCE**

Immersion heaters fitted with efficient lagging have a very high efficiency.

## ASSESSMENT GUIDANCE

Water heaters should be supplied via a double-pole switch.

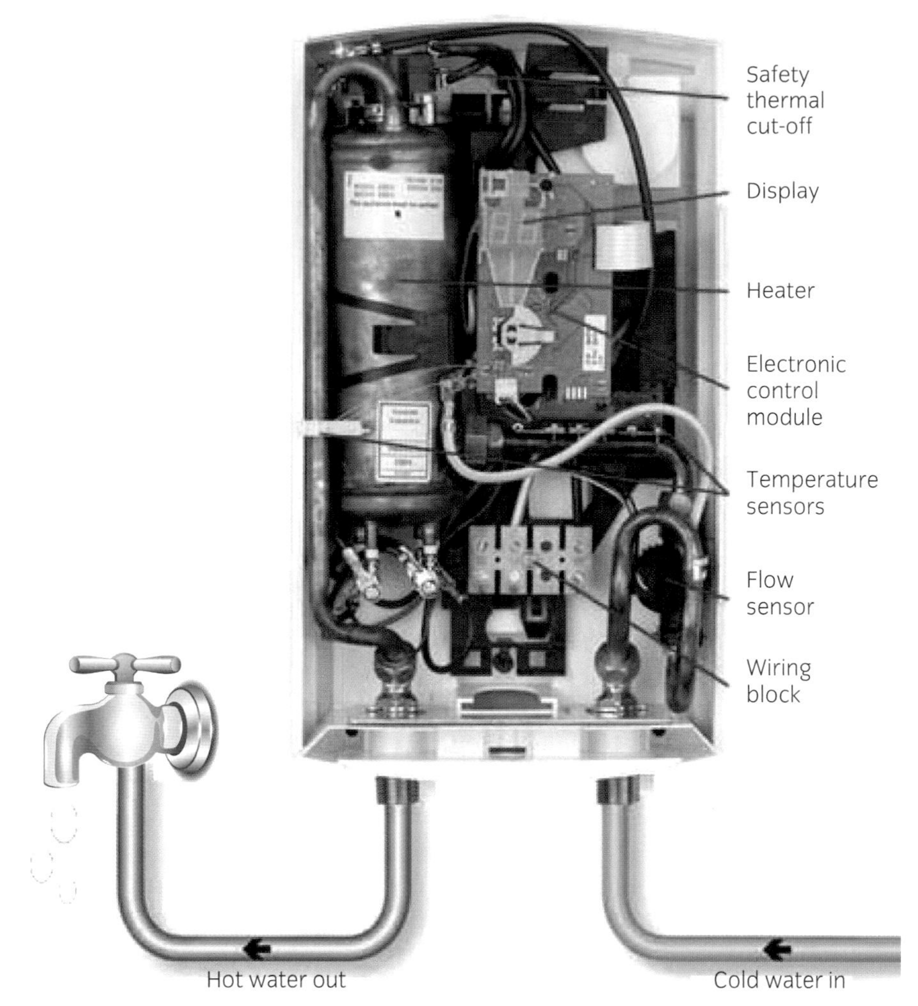


**Over-sink tank-type instantaneous water heater**

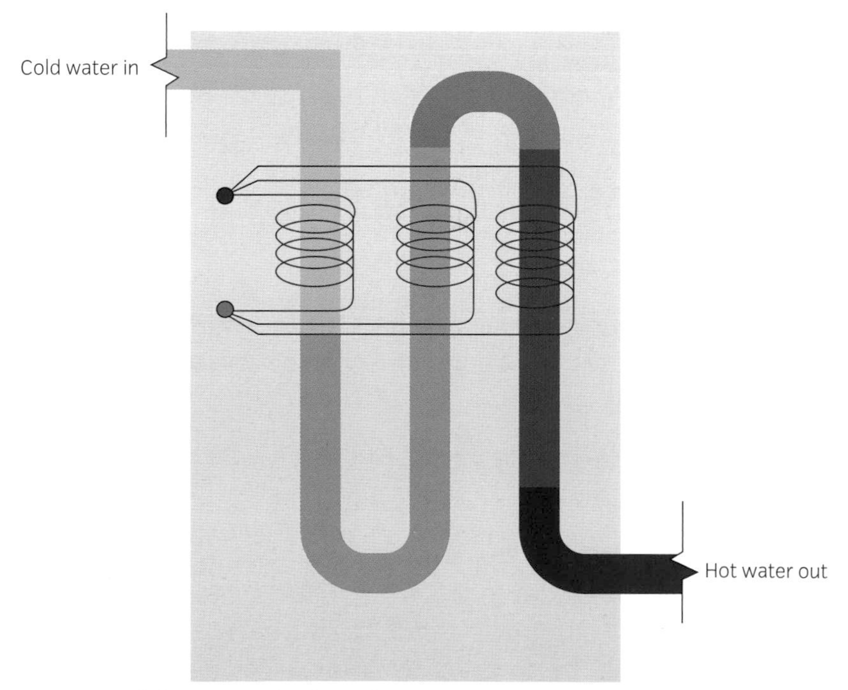


**Elements around the water pipe in an instantaneous water heater**

# HOW TO CONTROL HEATING SYSTEMS

Assessment criteria

6.3 Explain the method of control for heating systems

Heating and hot-water systems are not controlled just for economic reasons. Control is also a legal requirement for safety reasons. Building Regulations demand control.

## Room thermostats and control circuits

Room thermostats are used to provide temperature control when heating spaces. Traditionally, this is done by means of a simple adjustable bimetallic sensor incorporating a set of contacts. When the desired temperature is reached, the contacts open due to the bimetal bending. This stops water-based heating systems pumping heat around or, on more advanced systems, operates a valve that shuts off the water, but still heats water until it reaches the desired temperature, when the pump or boiler/heater will switch off.

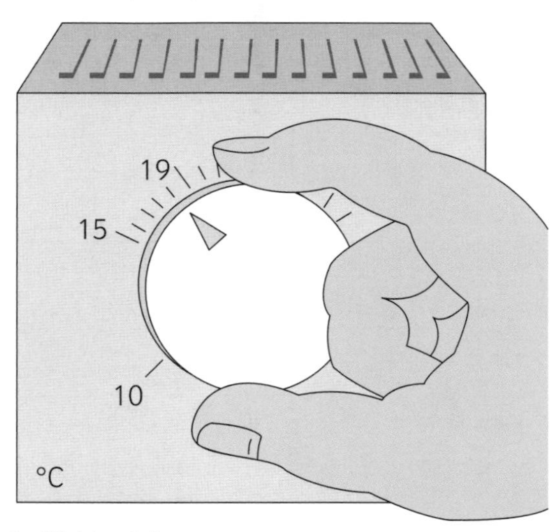


Simple thermostat with bimetal

The valve control arrangement is always used in commercial premises as it gives more accurate and zoned control, as required by Approved document L to the Building Regulations.

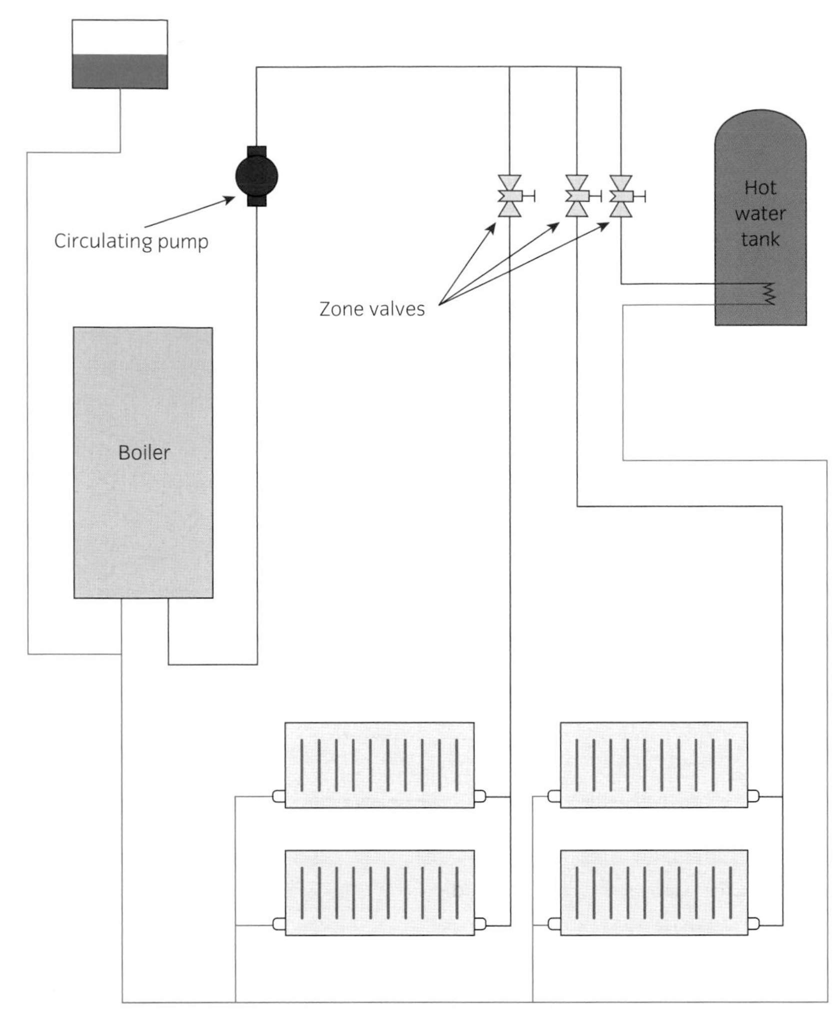

Simplified central heating system

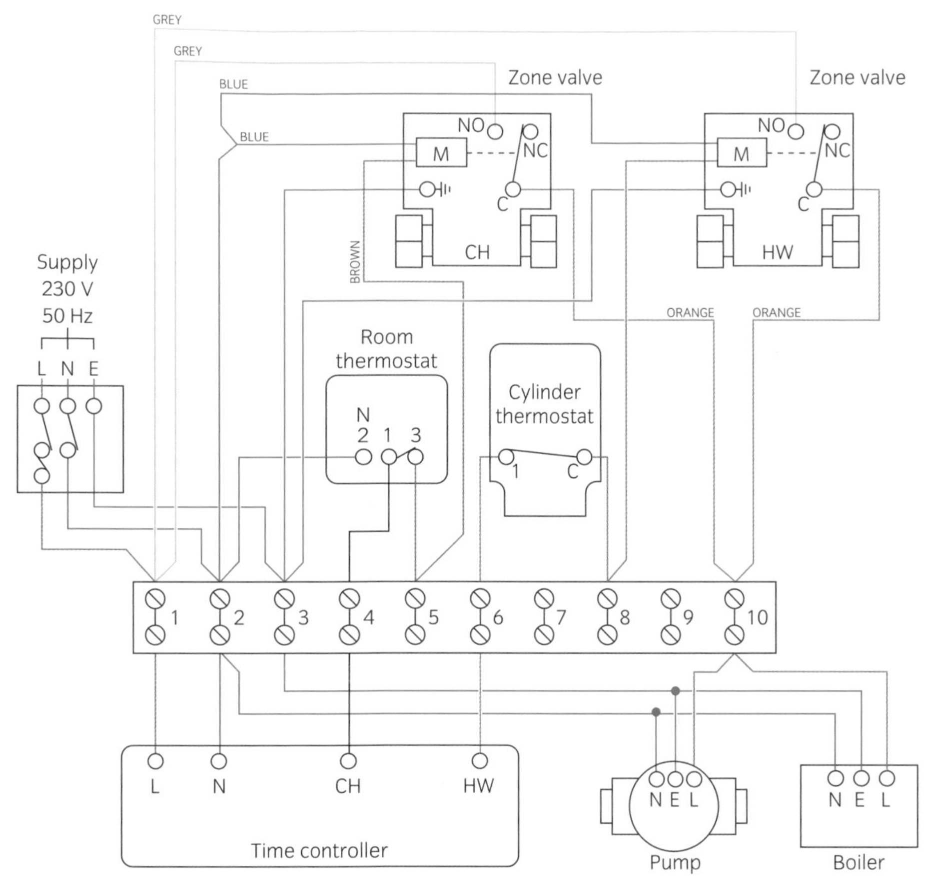

Typical domestic central heating circuit diagram

As the bimetal thermostatic control is accurate to only plus or minus 3 Celsius degrees, many commercial and some domestic systems use digital thermostats containing temperature-sensitive electronic thermistors. These units give a signal that can be converted directly into a temperature, usually with an accuracy of up to 0.1 of a Celsius degree.

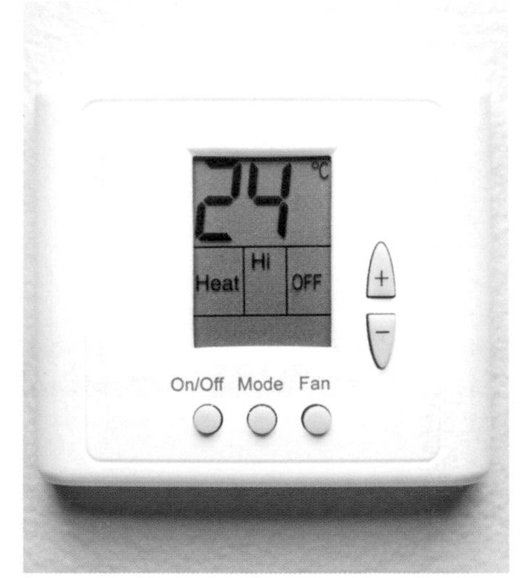

Digital thermostat with temperature reading

## Time switches and programmers

Time switches and similar devices are used to provide energy at the correct time and minimise wastage (eg not heating an office at weekends or a house when everyone is out.

In its simplest form, a 24-hour time switch cannot differentiate between days of the week. This is wasteful and a nuisance on days when the heating is required at different types from the norm. A programmable time controller is therefore normally installed as a minimum requirement for Building Regulations and convenience.

Simple domestic programmers allow the user to select the temperature required at set points on individual days of the week.

Simple daily time switch

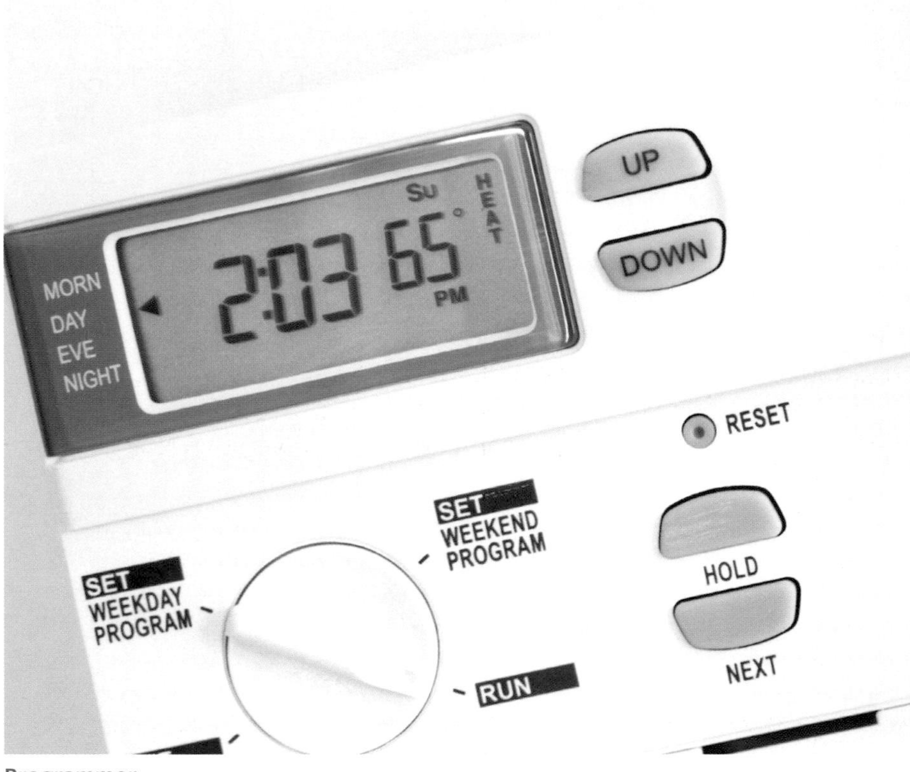


Programmer

## ASSESSMENT GUIDANCE

Larger heating systems should be split into zones to allow accurate control.

## ASSESSMENT GUIDANCE

Most programmers have an over-ride facility to change the settings if the temperature drops suddenly or there is a heavy demand for hot water.

In commercial environments, controllers can be more sophisticated with optimum start functions. The user inputs the time at which they wish a specific temperature to be reached. The programmer uses thermostats to estimate the start-up time in order to reach the specified temperature and the time to switch off again in order to reduce the temperature at the end of the day.

Taking into account the thermal mass of the building, and the actual temperatures inside the building, the system adjusts itself for the next day. As with all systems that use computers to estimate comfort conditions, this type of system is often criticised because British weather is very unpredictable and conditions can change rapidly.

OUTCOME 7

# Understand the principles of electronic components in electrical systems

In today's world of micro-components, it is becoming increasingly common to replace a whole circuit board rather than replacing single components. Nevertheless, it is far easier to diagnose faults in electrical systems if you understand how particular electronic components function.

**KEY POINT**

Always take care when handling or replacing electronic components or circuit boards as many can be damaged by static electricity. It is always important to ensure that static risks are minimised by earthing yourself to the equipment before handling components.

## HOW ELECTRONIC COMPONENTS WORK

**Assessment criteria**

7.1 Describe the operating principle of electronic components

### Diodes

A diode is a silicon P-N junction, which allows current flow in one direction, but not the other. When current flows through a diode, it is called 'forward bias'. When current is restricted, it is called 'reverse bias'. There are several types of diode, from the simple one described above, used for rectification or signalling, to:

- a zener diode, which only allows current flow when a set voltage is reached
- a light-emitting diode (LED), which emits a light when current flows through it
- a photo diode, which allows forward bias current flow when it detects light.

**ACTIVITY**

How could a capacitor, zener diode and resistor be used to smooth the output of a full-wave bridge rectifier?

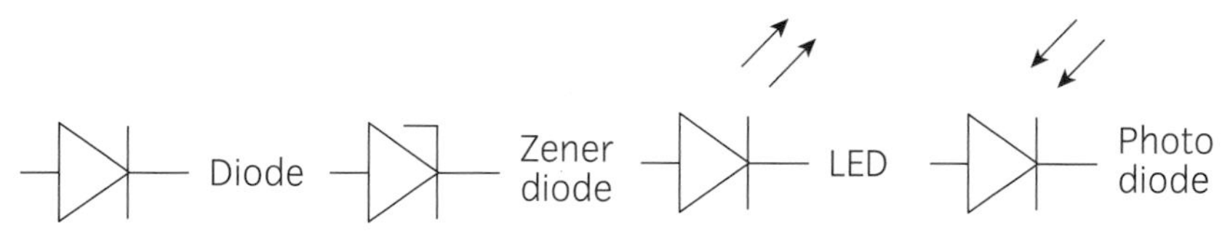


**Symbols for different types of diode**

### Diacs

A diac is a junction of two zener diodes, with two terminals. It works on a.c. circuits, hence the name **di**ode for **a.c.** A diac will not allow current flow unless a pre-set voltage is reached. Once this voltage is reached, current can flow in both directions. Current will continue to flow until the voltage falls below the level set, at which point the diac restricts current flow.

**Symbol for a diac**

## Thyristors

A thyristor is a solid-state switch that allows current flow between two of its terminals if a small current is sensed on the third. There are two types: silicon-controlled rectifiers (SCR) and triacs.

### ACTIVITY

What effect would varying the trigger angle have on a triac?

### SCRs

The SCR is similar to the diode, in that current can only flow between the anode and cathode in one direction. However, it also has a gate terminal, which controls the switch when a small current is sensed on that terminal. Essentially, it allows a large current to be controlled by a small current. The SCR will continue to allow current flow between anode and cathode until this current is stopped. It does not require a constant gate terminal current, except when allowing the main current to pass.

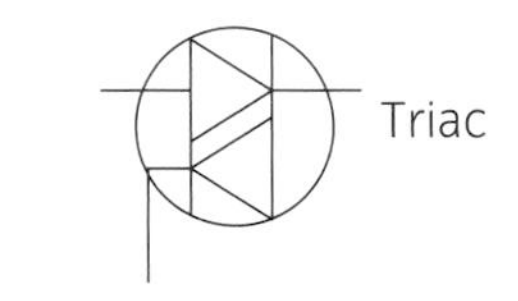

Symbol for a triac

### Triacs

A triac has three terminals, one called the gate. If the gate senses a very small control current, a.c. is allowed to flow between the other two main terminals ($MT_1$ and $MT_2$). If the gate current is removed, the device will stop current flow when the alternating cycle reaches 0 V.

## Transistors

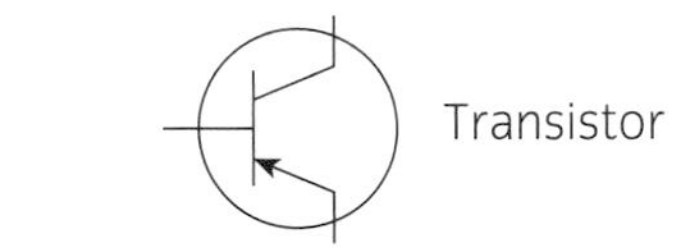

Symbol for a transistor

The transistor is the fundamental building block of modern electronics and the reason why electronic systems are now so affordable.

The three terminals on a bipolar transistor are known as base (B), collector (C) and emitter (E).

A transistor may be used either as a switch or an amplifier. When the base of an NPN transistor is grounded (0 V), no current flows between emitter and collector, so the transistor is off. If the base voltage is increased above 0.6 V, a current will flow from emitter to collector and the transistor is on. If the base current varies in value, the emitter to collector current will follow this pattern of variation with a larger current flow; in this situation, the transistor acts as an amplifier.

A PNP transistor operates in the same way as an NPN transistor but with current flow allowed in the reverse direction.

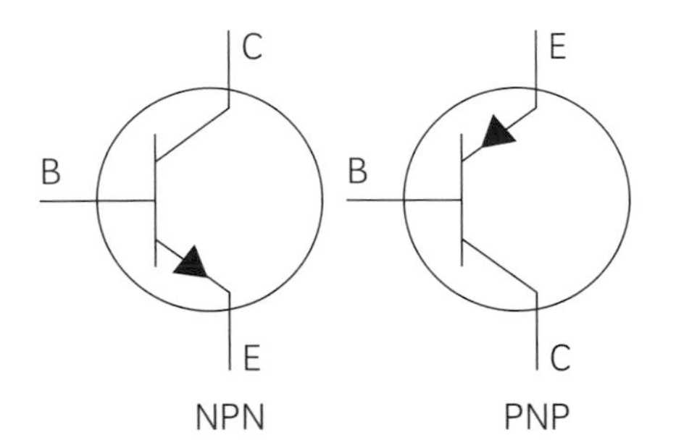

An NPN and a PNP transistor, showing the polarity of each device

Another type of transistor is the field effect transistor (FET), which has terminals marked gate, source and drain. The FET is much cheaper to produce as it requires less silicon. It also has the major advantage of operating at virtually no current on the gate terminal as long as a voltage above 0.6 V is present.

## Resistors

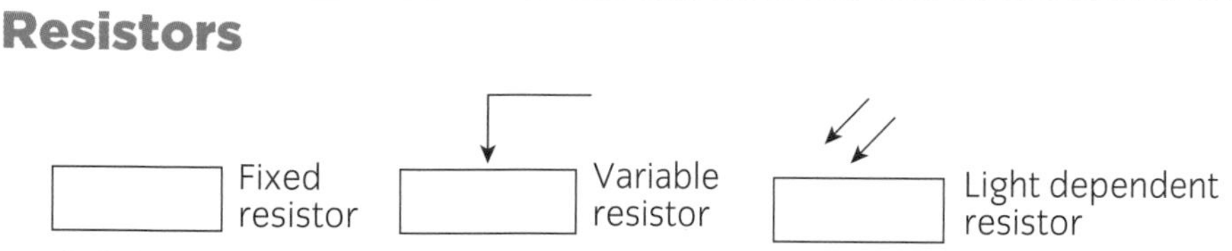


Symbols for different types of resistor

Resistors are used to control or reduce current flow in electronic circuits. With a sufficiently high resistance, they can also be used as voltage dividers on certain circuits to allow a fixed voltage, less than the input voltage, to be obtained. Fixed-value resistors are either made from carbon film with an insulated coating, as shown below, or are wire wound for larger power applications.

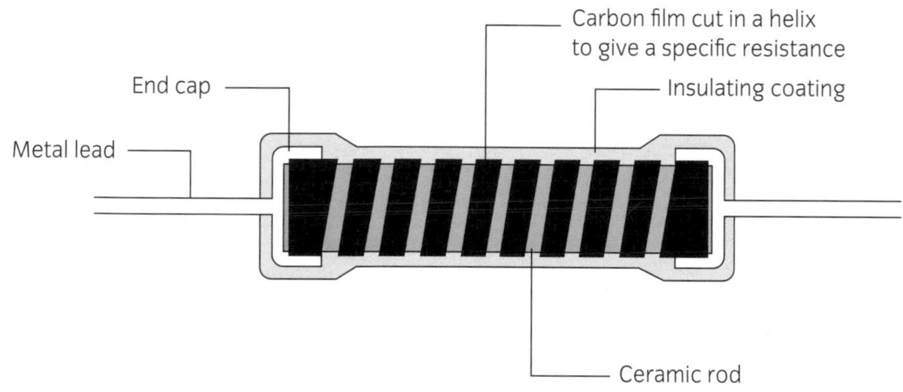


Section through a carbon-film resistor

Carbon-film resistors are colour-coded to indicate their value and tolerance as shown below:

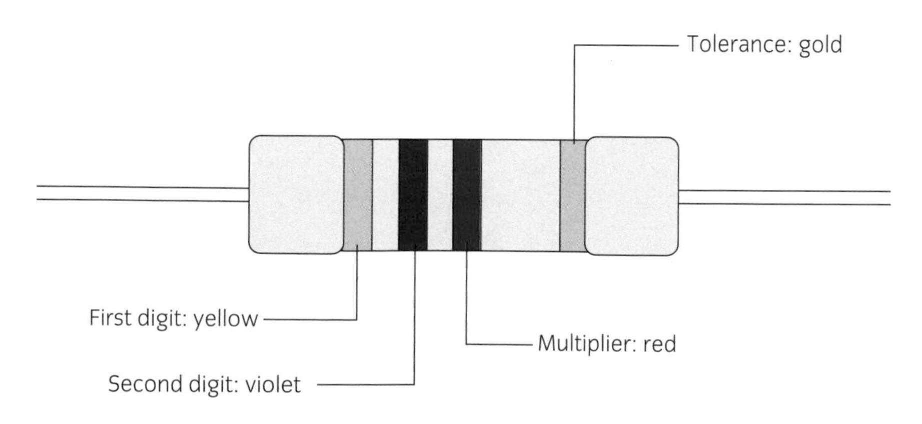


Resistor colour-coding system

Resistor colour values

| Colour | Digit | Multiplier | Tolerance |
|---|---|---|---|
| Black | 0 | 1 | |
| Brown | 1 | 10 | 1.0% |
| Red | 2 | 100 | 2.0% |
| Orange | 3 | 1 000 | |
| Yellow | 4 | 10 000 | |
| Green | 5 | 100 000 | 0.5% |
| Blue | 6 | 1 000 000 | 0.25% |
| Violet | 7 | 10 000 000 | 0.1% |
| Grey | 8 | | 0.05% |
| White | 9 | | |
| Gold | | 0.10 | 5.0% |
| Silver | | 0.01 | 10.0% |

**KEY POINT**

A resistor is identified by the colour bands red, violet, orange and gold. What is its value?

Wire-wound resistors are normally coded in order to establish the value, eg a 2R resistor is 2 Ω, whereas 2R2 is 2.2 Ω.

## Thermistors

T Thermistor

**Symbol for a thermistor**

A thermistor is a type of resistor in which the resistance varies significantly with temperature. This variation is so defined that there is a definite temperature-related use for them. Thermistors typically achieve high precision within a limited temperature range, typically –90 °C to 130 °C.

Thermistors are widely used as temperature sensors, self-resetting overcurrent protectors for self-regulating heating elements and current inrush limiters.

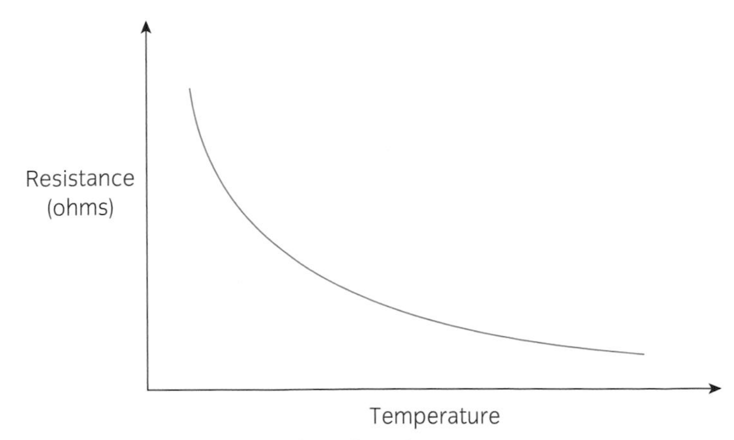


**Effect of temperature on resistance in a thermistor**

## Photoresistors

These are resistors that vary in resistance depending on the amount of light falling on them. They are often referred to as photocells and used to control lighting as day/night switches.

## Variable resistors or potentiometers

These resistors are used to vary resistance in a circuit manually. Their applications are wide, including use as sound volume controllers and speed controllers.

## Capacitors

Capacitors are widely used in electrical circuits in many common electrical devices. A capacitor is a passive two-terminal electrical component used to store energy electrostatically in an electric field, rather than by chemical reaction as in a battery. (Originally capacitors were known as condensers, but the original term has now been widely superceded.)

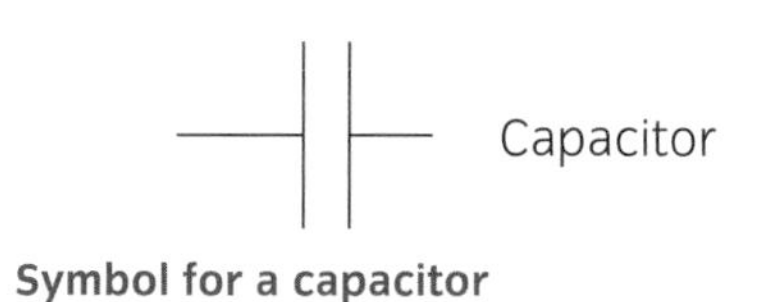


**Symbol for a capacitor**

Capacitors vary widely, but all contain at least two electrical conductors separated by a dielectric (insulating layer), which acts as an insulator between the conducting plates. The plates are usually made from foils. The capacitance is varied by the area of the plates and the size of gap between the plates. The narrow gaps that are used require a very high dielectric strength.

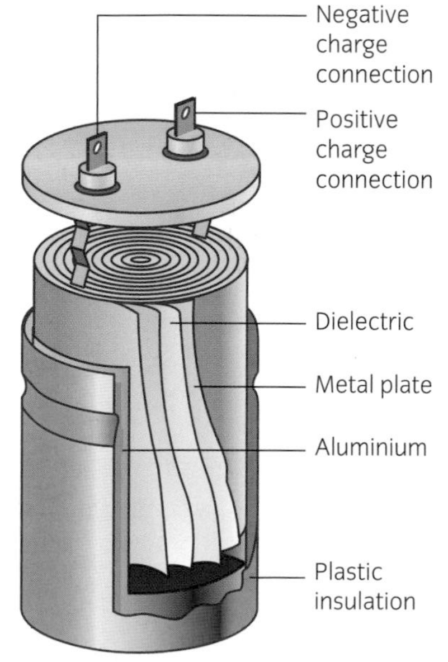


**Construction of a typical capacitor**

### ACTIVITY

Name five different types of capacitor.

## Rectifiers

A rectifier is an electrical device that uses diodes to convert alternating current (a.c.), which periodically reverses direction as it cycles, to direct current (d.c.), which flows in only one direction.

Half-wave and full-wave rectifiers are available.

**ASSESSMENT GUIDANCE**

Three-phase rectification requires three diodes for half-wave and six diodes for full-wave rectification.

### Half-wave rectification

In half-wave rectification of a single-phase supply, either the positive or negative half of the a.c. wave is passed, while the other half is blocked. Half-wave rectification requires a single diode in a single-phase supply, or three in a three-phase supply.

As only one half of the input waveform reaches the output, the mean voltage is lower than full-wave rectification.

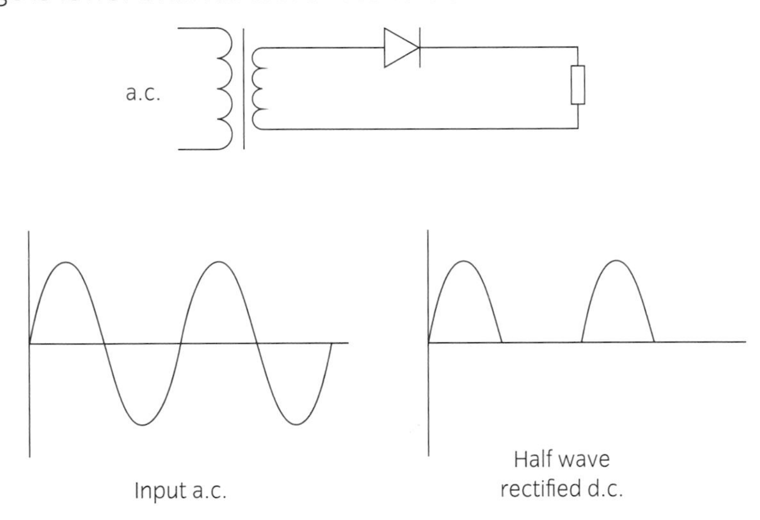


**Half-wave rectification**

### Full-wave rectification

A full-wave rectifier converts the whole of the input, both positive and negative components, of the waveform to one of constant polarity at its output. Full-wave rectification output gives a pulsating d.c waveform with a higher average output voltage than its half-wave counterpart.

The unit works with two diodes and a centre-tapped transformer, or four diodes in a bridge configuration, as shown opposite.

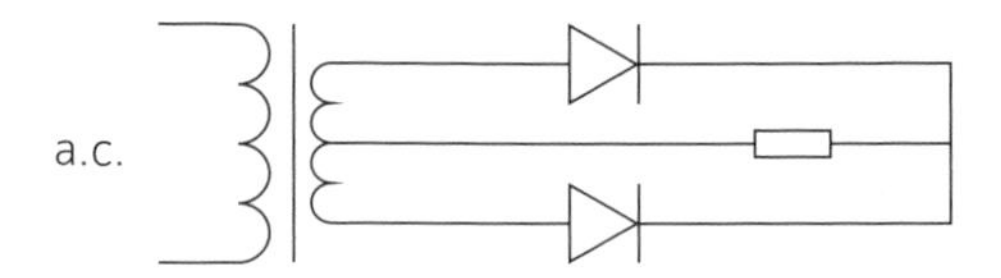


Two-diode method with centre-tapped transformer

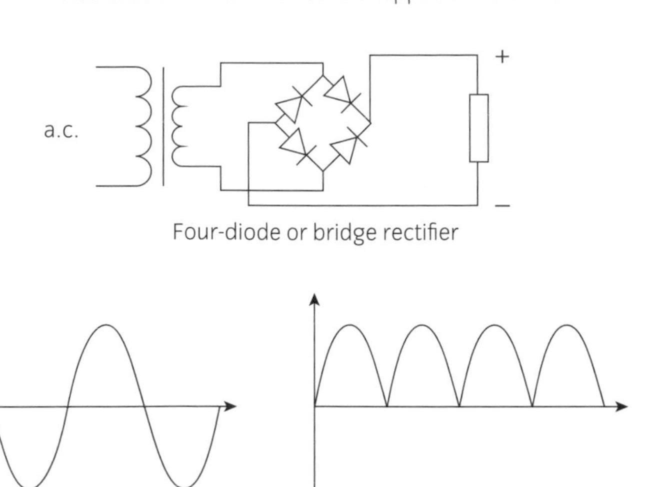


Four-diode or bridge rectifier

**Full-wave rectification using two diode and bridge arrangements**

## THE FUNCTION OF ELECTRONIC COMPONENTS IN ELECTRICAL SYSTEMS

**Assessment criteria**

**7.2** Describe the function of electronic components in electrical systems

Most electronic systems use many different electrical components in their power supply and in their operational systems.

Security alarm systems use full-wave rectification and smoothing through capacitors to supply a 12 V operating system. In addition, the closed-loop system uses transistor and similar technology to convert low-level signals from components such as passive infrared (PIR) detectors into an alarm output signal to components that operate the alarm.

Thyristors or SCRs are used extensively in motor speed-control circuits for heating and other applications, where motors and pumps require variable output. The ability to trigger a high-power switch is essential for controlling the output waveform, which in turn controls the motor speed.

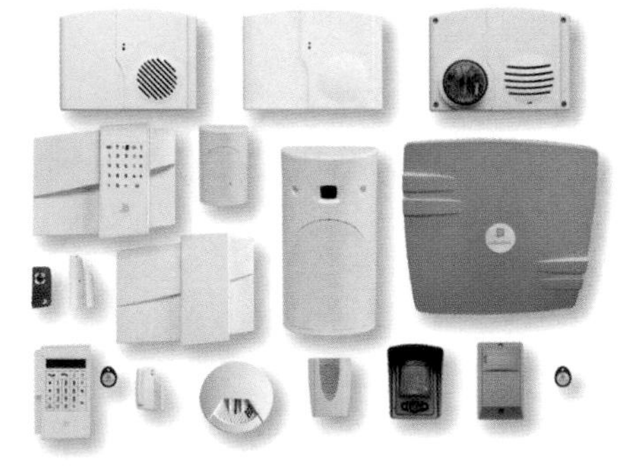

**A range of different detection devices using different technologies**

Heating control systems use a number of components. The most important element of any form of heating control is probably the ability to sense temperature in the airspace or water systems that are being heated. A thermistor is used to determine accurately the temperature of the space or heating medium. The heating control system then uses feedback from the sensor to determine how much heat needs to be passed. The temperature is controlled via valves and/or variable speed pumps.

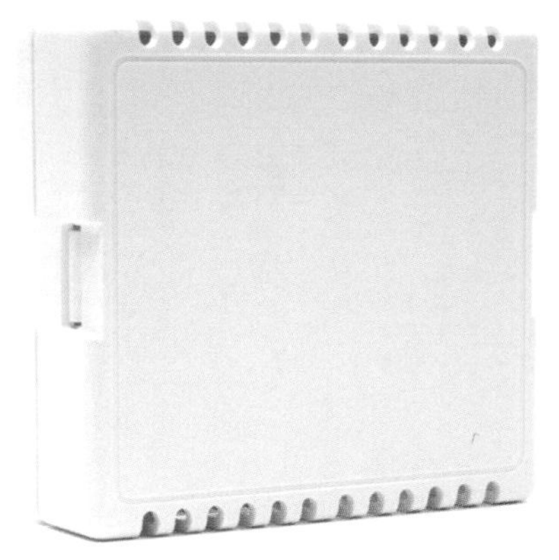
Typical thermistor-based space temperature detector

Diacs and triacs are used in lighting control circuits. The ability to trigger the device through a separate voltage allows dimming to be provided through proprietary dimming systems.

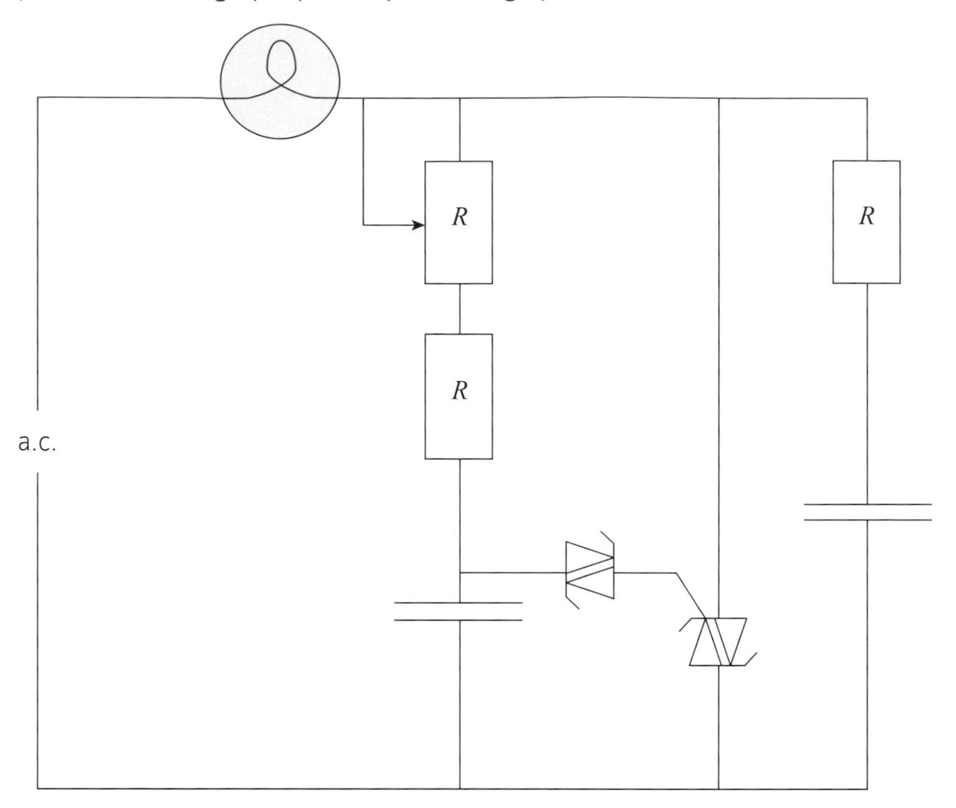

Typical diac-control dimming arrangement

# ASSESSMENT CHECKLIST

## WHAT YOU NOW KNOW/CAN DO

| Learning outcome | Assessment criteria | Page number |
|---|---|---|
| 1 Understand the principles of a.c. theory | *The learner can:* | |
| | 1 Explain the effects of components in a.c. circuits | 148 |
| | 2 Calculate quantities in a.c. circuits | 156 |
| | 3 Explain the relationship of power quantities in a.c. circuits | 166 |
| | 4 Calculate power factor | 163 |
| | 5 Explain power factor correction | 169 |
| | 6 Compare methods of power factor correction. | 169 |
| 2 Understand the principles of lighting systems | *The learner can:* | |
| | 1 Explain the laws of illumination | 171 |
| | 2 Calculate illumination quantities | 171 |
| | 3 Explain the operation of luminaires | 178, 180 |
| | 4 Explain the application of luminaires. | 178 |
| 3 Understand electrical quantities in star delta configurations | *The learner can:* | |
| | 1 Calculate values of voltage and current in star configured systems | 187 |
| | 2 Calculate values of voltage and current in delta configured systems | 189 |
| | 3 Determine the neutral current in a three-phase and neutral supply. | 190 |

| Learning outcome | Assessment criteria | Page number |
|---|---|---|
| 4 Understand the principles of electrical machines | *The learner can:* | |
| | 1 Explain the operating principle of d.c. machines | 194 |
| | 2 State the applications of d.c. machines | 198 |
| | 3 Explain the operating principle of three-phase a.c. machines | 199 |
| | 4 State the applications of three-phase a.c. machines | 203 |
| | 5 Explain the operating principle of single-phase a.c. machines | 204 |
| | 6 State the applications of single-phase a.c. machines | 207 |
| | 7 Explain methods of starting motors. | 208 |
| 5 Understand the principles of electrical devices | *The learner can:* | |
| | 1 Explain the operating principle of electrical components | 211 |
| | 2 State the application of electrical components in electrical systems | 216 |
| | 3 Explain the operating principle of overcurrent protective devices | 218 |
| | 4 State the application of overcurrent protective devices in electrical systems. | 223 |
| 6 Understand the principles of electrical heating systems | *The learner can:* | |
| | 1 Explain the principles of electrical space-heating systems | 226 |
| | 2 Explain the principles of electrical water-heating systems | 230 |
| | 3 Explain the method of control for heating systems. | 233 |
| 7 Understand the principles of electronic components in electrical systems | 1 Describe the operating principle of electronic components | 237 |
| | 2 Describe the function of electronic components in electrical systems. | 243 |

## ASSESSMENT GUIDANCE

The assignment is in two parts and two hours is allowed.

- Part A is a short-answer paper with 25 questions.
- Part B is a practical exercise with back-up questions
- You must use a pen with black or blue ink to complete your answers.
- You may use a scientific calculator (graphical and programmable calculators are not permitted).
- Remember to write your name and candidate details at the top of your answer sheet.
- Any calculations or rough working can be done on the paper.
- Attempt all questions. If you find a question difficult, leave it and return to it later.

## KNOWLEDGE CHECK

1 Explain how current flow is restricted in a 50 Hz a.c. circuit when connected to:

**a)** a pure resistor

**b)** a pure inductor

**c)** a pure capacitor.

2 For each of the following series combinations, state whether the overall power factor will be in phase, leading or lagging.

**a)** 20 Ω resistor with 30 Ω inductive reactance.

**b)** Capacitor $X_c$ = 80 Ω and resistor = 69 Ω.

**c)** $X_l$ = 314 Ω and capacitor = 150 Ω.

**d)** $R$ = 10 Ω, $X_l$ = 87 Ω, $X_c$ = 87 Ω.

**e)** $R$ = 45 Ω and $X_L$ = 45 Ω.

3 State the effect on the supply and lamp current of removing the capacitor from a 58 W fluorescent circuit.

4 A single-phase motor operates at 0.8 PF lagging when connected to a 230 V supply. The input power is recorded as 5.69 kW. What would be the input current?

5 A 400 V single-phase welding unit draws a current of 20 A at a power factor of 0.5 lagging.

Determine the kW, kVA and $kVA_r$ values.

6 Explain why it is necessary to have neutral conductor in an unbalanced three-phase and neutral system.

7 The star-connected secondary of an 11 kV /400 V transformer supplies a 24 kW delta-connected load. Determine the:

**a)** transformer secondary-phase and line voltage

**b)** load phase voltage.

8 State how the following motors may be reversed.

**a)** Series d.c.

**b)** Series universal a.c.

**c)** Three-phase induction.

**d)** Single-phase capacitor start.

**e)** Shaded pole.

9 A lamp emits 2000 cd in all directions. It is suspended 1.8 m immediately above the centre of a table that is 2 m in diameter. Calculate the illuminance immediately below the lamp and at the edge of the table.

10 **a)** Explain how heat is transferred by the following methods.

**i)** Radiation.

**ii)** Convection.

**iii)** Conduction.

**b)** Name one type of heater that uses each method of heat transfer.

# UNIT 305
# Electrical systems design

This unit provides an introduction to electrical installation design. Designing electrical installations is a complex task in which the designer has to interpret a great deal of information and ideas from the client, work those ideas into the regulatory requirements and produce a sound electrical design scheme.

Good electrical installation design begins by taking in and sorting a great deal of information. This may be documented, verbal or physical, such as the actual building.

This unit includes advice on:

- the law
- types of electrical system earthing
- collation of design data
- reading plans
- design of the installation
- estimating (the cost of the works)
- management of the on-site work
- handover procedures.

## LEARNING OUTCOMES

There are four learning outcomes to this unit. The learner will:

1. Understand how to interpret design information
2. Understand principles for designing electrical systems
3. Understand procedures for connecting complex electrical systems
4. Understand how to plan work schedules for electrical systems installation.

This unit will be assessed by:

- assignment projects based around drawings and scenarios
- a written question paper (open book).

# OUTCOME 1

# Understand how to interpret design information

On completion of this learning outcome, you will understand what information you need to bring together before starting an electrical installation design. This includes information on:

- the availability and type of electricity supply
- the extent of the work
- the environment
- building dimensions
- the location and nature of the electrical **loads**.

**KEY POINT**

The British Standard detailing the requirements for electrical installations is BS 7671:2008 Requirements for Electrical Installations (the IET Wiring Regulations 17th edition).

**Load**

A device intended to absorb power, including active power and apparent power.

**Assessment criteria**

1.1 State the criteria used when selecting electrical systems

## ELECTRICAL SYSTEMS

In this section you will learn about the supply arrangements that may be available for electrical systems. You will consider the type of system earthing that may be adopted and the most suitable arrangement for different types of installation.

**ACTIVITY**

Look up the conductor identification colour in Table 51 of BS 7671.

What colour should the **protective conductor**, line conductor and **neutral conductor** be?

### Electrical systems

An electrical system consists of:

- the source or sources (where the electricity is generated), for example, a generator or more commonly a transformer
- the distribution (the means of moving the energy to where it is needed), such as distribution cables or overhead lines
- the installation, for instance, the electrical wiring and equipment in a dwelling or shop.

There are a number of types of electrical system. The type or name is determined by the earthing arrangements of the particular system.

**Protective conductor (PE)**

A conductor used for protection against electric shock; it is often called the 'earth wire' or 'earth conductor'.

**Neutral conductor**

A conductor connected to the neutral point of a system that contributes to the transmission of energy.

**SmartScreen Unit 305**

Additional resources to support this unit are available on SmartScreen.

## Earthing arrangements of electrical systems

You need to be familiar with both the conductor arrangements and the earthing arrangements adopted by common supply systems.

The systems commonly adopted in the United Kingdom are TN-S, TN-C-S and TT, as shown in the diagrams below and in the following pages.

The first letter of the system type describes the earthing of the source:

- T for 'terre' (French for 'earth') when there is an earth connection of the neutral point
- I for 'isolated' when the source is not earthed.

The second group of letters describes the nature of the installation connection with earth via protective conductors:

- N-C indicates that the neutral and protective conductors are combined
- N-S indicates that the neutral and protective conductors are separated
- T is used when the installation has its own earth connection.

### ACTIVITY

Look up conductor arrangements and types of system earthing in Regulation 312 of BS 7671. What do the letters PME stand for?

### TN-S earthing systems

TN-S systems often use the armouring of cables as the protective conductor. The risk of corrosion of the armouring causes unreliability in old systems and is the reason for electricity distribution companies converting TN-S distribution systems to TN-C-S.

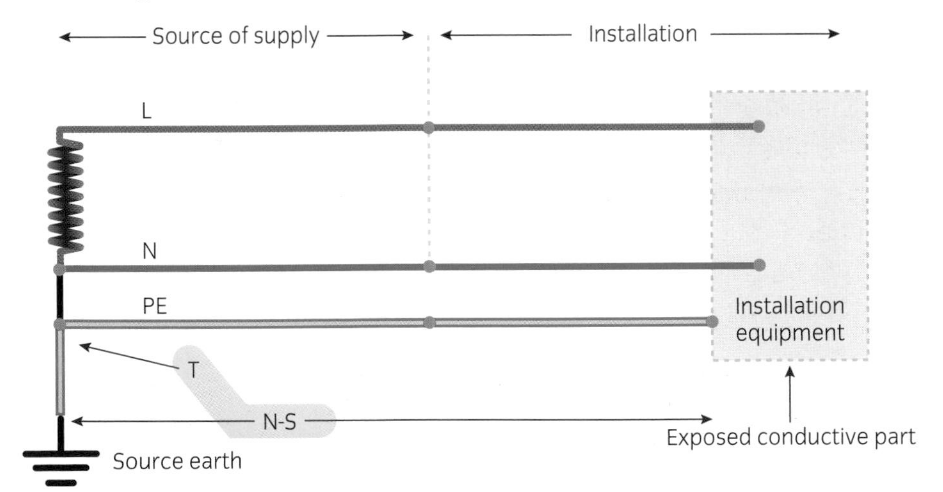


**TN-S earthing system**

TN-S earthing systems are often adopted for large installations supplied by their own transformer. They are also found in old (circa pre-1960) domestic installations. The diagram on the next page shows a domestic TN-S supply position arrangement.

### KEY POINT

In TN-S systems, the neutral and protective conductors are separated at the system supply, for example, the transformer to and throughout the installation.

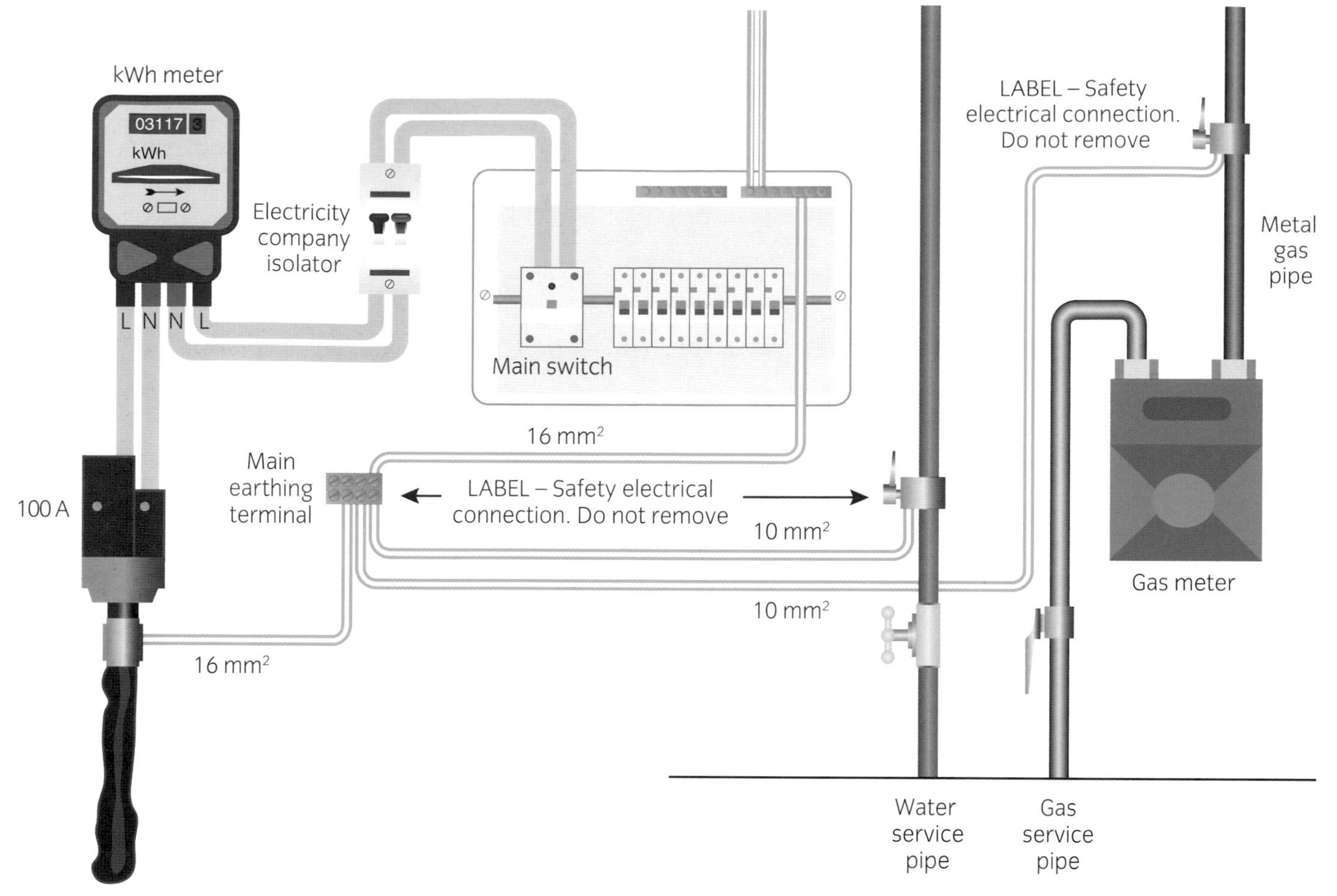


Domestic TN-S earthing system

## TN-C-S earthing systems

TN-C-S is the most common system. It is used in the great majority of installations supplied from the public supply system, particularly new installations.

**PEN**

Protective earthed neutral. The PEN conductor is both a live conductor and a combined earthing conductor.

**PME**

Protective multiple earthing. This term is commonly used to describe a TN-C-S system.

**ASSESSMENT GUIDANCE**

The TN-C-S system is not suitable for all installations, caravan parks and petrol stations being two of these. Where an installation is supplied by TN-C-S it must not be extended outside the building to an outbuilding, but must be converted to TT for that part of the installation.

L
PEN
N
PE
Link
Installation equipment
Additional source electrode
Source earth
T
N-C
S
(Combined neutral and protective conductor (**PEN**) in source of supply, with **PME** applied)
(Separate neutral (N) and protective conductors (PE) within installation)

TN-C-S earthing system

Most electricity distribution systems in the UK have a combined neutral and protective conductor, called a PEN conductor. Such systems are required by the Electricity Safety, Quality and Continuity (ESQC) Regulations to be multiply earthed to provide protection to the user in the event of one or more earth connections failing. Electricity suppliers (distribution network operators or DNOs) will usually offer protective multiple earthing (PME) to someone requesting a new supply.

The Electricity Safety, Quality and Continuity (ESQC) Regulations prohibit the use of protective multiple earthing in caravans and boats. PME is also prohibited in agricultural and horticultural installations by Section 705 of BS 7671.

If the PEN conductor supplying the caravan or boat goes open circuit (breaks), all the metalwork connected to the earthing terminal may rise in potential relative to true earth. This could be a particular risk, for instance, for a person standing outside a metal caravan with bare feet in contact with wet grass.

**KEY POINT**

TN-C-S system: the neutral and protective conductors are combined in the supplier's network then separated at the supplier's fused cut-out. They remain separated throughout the installation.

**ACTIVITY**

Look at the supply position in your home. What system is it part of?

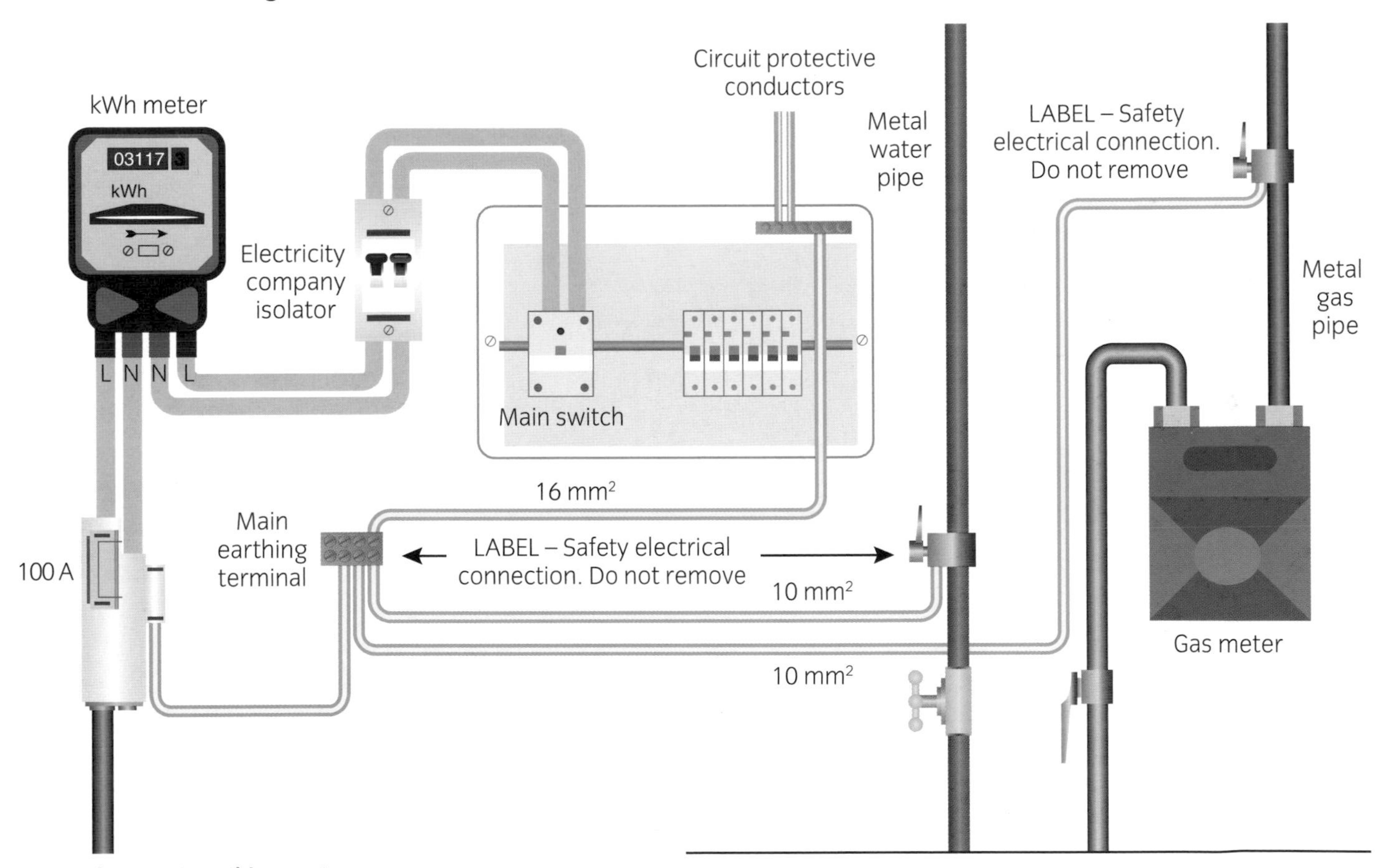


Domestic TN-C-S earthing system

**KEY POINT**

A TT system has its own connection with earth. The system's protective conductors are connected to this earth and are not connected to an earth terminal provided by the electricity supplier.

## TT earthing systems

TT systems are common in rural areas, particularly to farms and similar premises.

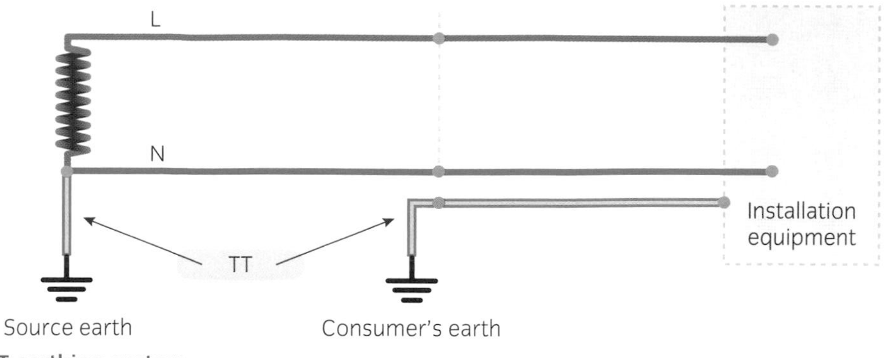


**TT earthing system**

**ACTIVITY**

Identify three items that may be used as an earth electrode in accordance with BS 7671.

Electricity suppliers will often refuse to provide protective multiple earthing (PME) to farms. This is because of the difficulty of bonding all connections with earth together in a farm and the risks of an open circuit PEN conductor. The installation must then have its own earth, not connected to the supply, and the installation becomes TT.

**KEY POINT**

A TN supply can be converted to a TT system by not using the earth connection provided by the supplier. Rather, the consumer uses their own earth electrode.

The connection to earth for an installation forming part of a TT system will use an earth electrode as a means of earthing. A complete list of what is accepted as an earth electrode can be found in Chapter 54 of BS 7671.

## TN-C earthing systems

TN-C systems (within buildings) require an exemption from the Electricity Safety, Quality and Continuity (ESQC) Regulations and are, as a result, generally not permitted in the UK. The problem is that normal **load currents** will flow in earth conductors and all metal connected with the earth. This can cause corrosion and significant interference with information technology (IT) systems, particularly communications.

**KEY POINT**

TN-C systems are not permitted in the UK for general use.

**Load current**

The amount of current an item of equipment or a circuit draws under full load conditions.

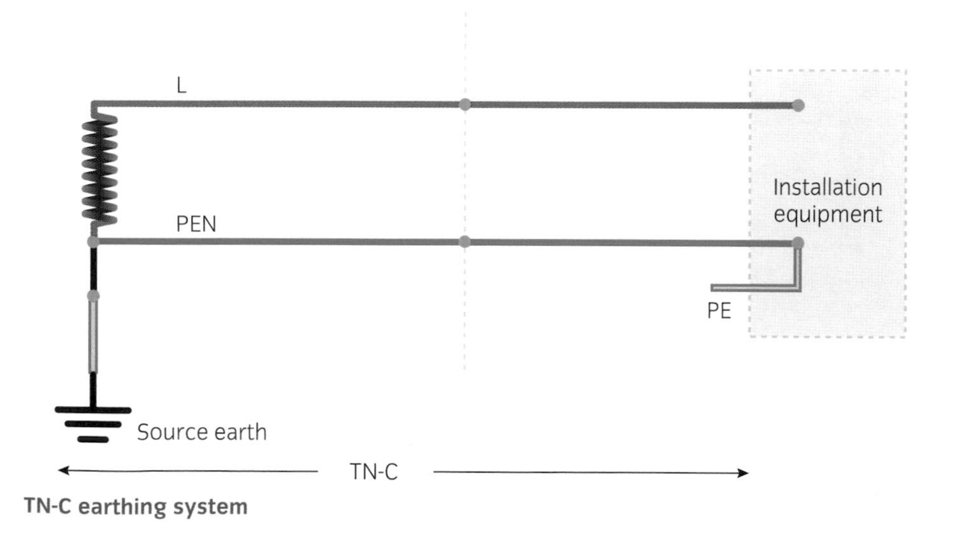


**TN-C earthing system**

## IT earthing system

Isolated earthing (IT) is used in parts of electrical installations where interruptions to the supply can have serious consequences, for example, life-support systems in a hospital operating theatre. There is monitoring of the installation for faults without automatic disconnection. This monitoring is by the use of residual current monitors (**RCM**s) or insulation monitoring devices (**IMD**s).

**RCM**

Residual current monitor. This is much like an RCD but does not disconnect the circuit. Instead it gives a warning that a fault is present.

**IMD**

Insulation monitoring device. An end-of-line resistor together with a fire detection system is an example.

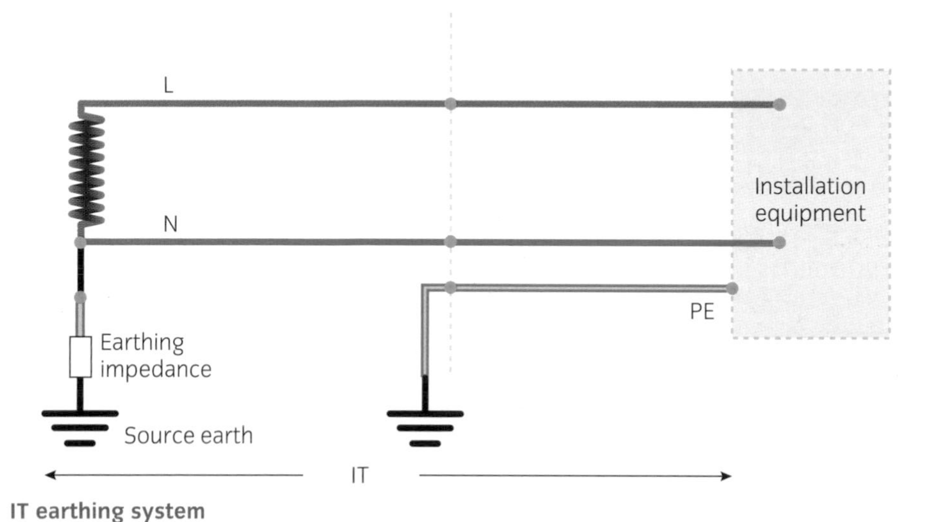


**IT earthing system**

## Earth fault loop impedance

The earthing system adopted will determine the **earth fault loop impedance**, and this will determine the method of protection against electric shock.

- TN-C-S systems tend to have low earth fault loop impedances external to the installation, of the order of 0.35 Ω.
- TN-S systems tend to have higher earth fault loop impedances compared to TN-C-S systems. The typical maximum declared value is 0.8 Ω.
- When TT systems are adopted, the resistance of the installation earth will be high (of the order of a 100 Ω). This means that residual current devices (RCDs) will need to be adopted for protection against electric shock as they operate at much lower earth fault currents than standard protective devices.

The earth fault loop comprises the following, starting at the point of fault:

- the circuit protective conductor
- the consumer's earthing terminal and earthing conductor
- (for TN systems) the metallic return path
- (for TT and IT systems) the earth return path
- the path through the earthed neutral point of the transformer
- the transformer winding
- the line conductor from the transformer to the point of fault.

**Earth fault loop impedance**

The impedance of the earth fault current loop starting and ending at the point of earth fault. This impedance is denoted by the symbol $Z_s$.

**KEY POINT**

For reasons of safety, the distributor must not provide caravan parks, marinas, swimming pools, farms and horticultural establishments with an earth terminal. Rather, a TT system must be adopted.

**ACTIVITY**

Draw the earth fault loop path for a fault on a TT installation.

**ACTIVITY**

Look up Regulation 24.4 of the Electricity Safety, Quality and Continuity (ESQC) Regulations. Is the supplier required to provide an earth connection to an existing supply?

## Other considerations

Generally for new installations, a TN-C-S system is adopted.

Unless it is inappropriate for reasons of safety, the Electricity Safety, Quality and Continuity (ESQC) Regulations require the distributor to provide an earth terminal for new connections. (This is not retrospective; there are many TT installations where in the past the distributor has not provided an earth.)

**Assessment criteria**

**1.2** Explain positioning requirements when designing electrical systems

# ACCESSIBILITY AND SUITABILITY OF SYSTEMS

There are various considerations regarding the position of components in electrical installations.

## Mains position

The position of the main incoming supply terminal (for example, fused cut out) will need to be agreed with the electrical supplier. This is necessary:

- to provide easy access to metering equipment for meter reading
- to facilitate installation of the supply cable and equipment.

For installations with demands exceeding 100 A, for instance, the incoming supply position will have to be determined so as to minimise voltage drop and energy loss (copper loss).

**ASSESSMENT GUIDANCE**

Most domestic installations will never draw anything near the maximum current. If you consider the average consumption over a month it is probably closer to 5–10 A.

## Space and access

Electrical equipment must be installed and arranged so as to provide sufficient space for:

- initial installation
- later repair and replacement
- accessibility for safe operation, inspection and testing, and maintenance.

## Access to and use of buildings

Approved document M to the Building Regulations (see Learning outcome 3, pages 335–347) includes requirements that reasonable provision be made for disabled people to gain access to and use the building. Approved document M provides specific guidance on the accessibility of wiring accessories, including switches and socket outlets in new dwellings. Note this is not retrospective: a re-wire or addition to an existing property that does not undergo major building works will not need to comply with Part M.

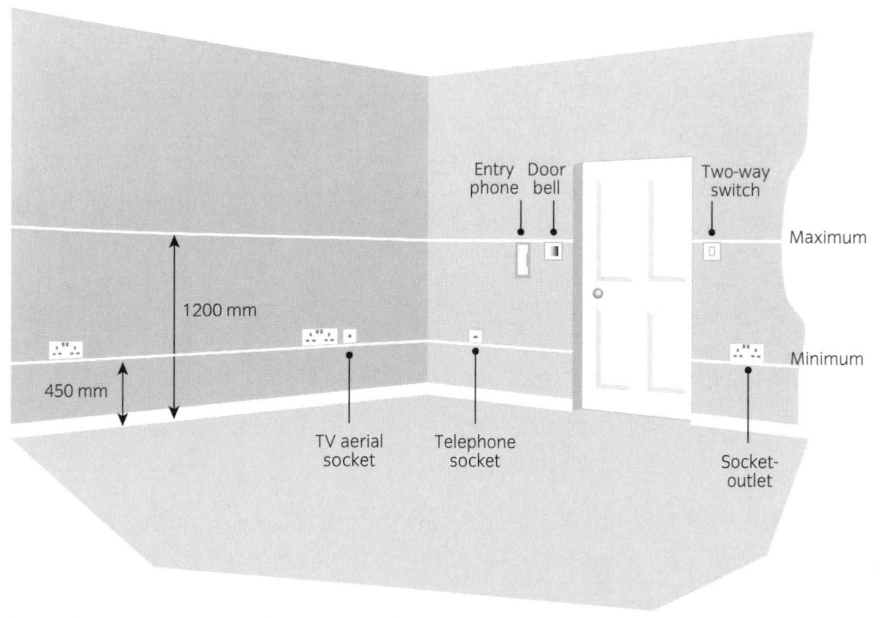


Height of switches and sockets in dwellings

## Environment

Electrical equipment must be suitable for the environment and, where practicable, should be installed where it will not be subjected to harsh conditions. Locations with extreme environments, such as explosive, corrosive, flammable, very wet, hot or cold conditions, should be avoided where possible to reduce the costs of suitable equipment and to provide a more reliable installation. An example of this could be installing a switch outside a location with an explosive or wet environment rather than within the location.

# SUSTAINABLE DESIGN

Electrical equipment and installations should be energy efficient, long-lasting and functional. They should also minimise use of materials and not pollute the environment.

Humankind's impact on the natural world is now so great that the quality of life of future generations will be damaged. The Earth's resources are limited and its environment fragile. Electrical installations, like all new projects, should preferably be designed and installed so they use the minimum of the Earth's resources over the period of their life. They should be designed to last a long time (if this is appropriate) and to be energy efficient – so long as the resources used in making the project energy efficient do not negate the energy savings.

**Assessment criteria**

1.3 Describe the importance of **sustainable design**

**Sustainable design**

Designing equipment, the built environment and services to comply with the principles of social, economic and ecological sustainability.

Electrical installation designers will consider:

- low-loss transformers
- energy-efficient lighting
- efficient motors
- power factor correction
- occupancy controls
- voltage optimisation (reduction of iron losses)
- increasing conductor cross-sectional area (reduction of copper losses)
- renewable sources of energy.

Sustainable designs may be able to adopt solar, wind, hydro or other renewable sources of energy. However, designs must be appropriate to the circumstances. For example, large cross-sectional area cables will reduce voltage drop and energy losses, but will cost more and use more of a scarce resource – copper.

### Voltage optimisation

Voltage optimisation usually involves installing an automatic transformer at the origin to control the voltage. Typically, this will only be carried out on very large installations. Voltage optimisation will limit the voltage rise at low load. Higher voltage will increase current consumed and will then increase energy consumed unless automatic controls switch off the load, for instance, when the required temperature is reached.

Voltage optimisation is an effective way of reducing the iron losses associated with inductive loads such as motors or control gear in luminaires (with the exception of high-frequency luminaires).

**Assessment criteria**

**1.4** Identify information required for electrical systems design

## ELECTRICAL INSTALLATION DESIGN

A certain amount of information is required when designing an electrical installation.

### Design sequence

To carry out an electrical design, the work has to be completed in a particular sequence. For example, cables cannot be selected until design load currents are calculated; voltage drop cannot be calculated until the current is known, a cable size has been selected and the cable length measured.

The design sequence required is described on the next page.

1. Determine load characteristics, load centres, maximum current demands of load centres and standby generation if required.
2. Determine supply characteristics and agree tariffs with electricity company.
3. Position main and sub-distribution boards.
4. Prepare installation outline of a distribution system.
5. Design distribution system:
   - **a)** Estimate maximum current ($I_b$) demands of each distribution board.
   - **b)** Select protective devices and rating ($I_n$).
   - **c)** Select cable type and size (S).
   - **d)** Calculate the voltage drop.
   - **e)** Calculate prospective fault currents.
   - **f)** Calculate earth fault loop impedances.
   - **g)** Select earthing and protective conductor size (S).
6. Size final circuits:
   - **a)** Determine circuit design current ($I_b$).
   - **b)** Select protective devices and rating ($I_n$).
   - **c)** Select cable type and size (S).
   - **d)** Calculate the voltage drop.
   - **e)** Calculate prospective fault currents.
   - **f)** Calculate earth fault loop impedances.
   - **g)** Select earthing and protective conductor size (S).
7. Check isolation and switching requirements.
8. Carry out final assessment.

**KEY POINT**

Circuit design involves many categories of current values. Common symbols used include:

| Symbol | Meaning |
|---|---|
| $I_a$ | Current causing automatic operation of protective device within the time stated |
| $I_b$ | Design current of circuit A in rated current or current setting of protective device |
| $I_n$ | Rated current or current setting of protective device |
| $I_{\Delta n}$ | Rated residual operating current of the protective device |
| $I_{pf}$ | Prospective fault current |
| $I_{sc}$ | Short-circuit current |
| $I_t$ | Tabulated current-carrying capacity of a cable |
| $I_z$ | Current-carrying capacity of a cable for continuous service under the particular installation conditions concerned |

## ASSESSMENT GUIDANCE

The power factor of a domestic or small commercial installation is not normally a concern as most of the equipment is purely resistive or has power-factor correction equipment pre-fitted.

### KEY POINT

Inrush and starting currents are common with inductive loads such as fluorescent luminaires. When the luminaires are first switched on, the current surge can be much greater than the normal running current.

## ACTIVITY

Calculate the design current $I_b$ for a 10 kW three-phase motor with a power factor of 0.8 where $U$ is 400 V.

# Design sequence step 1: Determine load characteristics

The first step is to determine the characteristics of the major loads, including:

- location
- load current $I_b$ / time profiles with maximum demand
- power factors ($\cos\theta$)
- inrush and starting currents
- reliability requirements
- need for standby supplies.

(See Assessment criteria 2.3.)

Information on the need for, and nature of, standby supplies is obtained through meetings with the client and their advisors.

### KEY POINT

**Electrical power** is calculated using the following equations.

For single phase a.c. equipment:

$$\text{active power } (P) \text{ in kW} = \frac{U_0 I_b \cos\theta}{1000}$$

For three-phase a.c. equipment:

$$\text{active power (P) in kW} = \frac{\sqrt{3} U I_b \cos\theta}{1000} = \frac{3 U_0 I_b \cos\theta}{1000}$$

and:

$$\text{power factor} = \frac{\text{power in watts}}{\text{product of rms values of voltage and current}}$$

where:
$P$ is active power, a measure of ability to carry out work, sometimes called active power
$U$ is the rms voltage between lines
$U_0$ is the rms voltage to earth
$I$ is the rms load current
$\cos\theta$ is the power factor of the equipment
$\theta$ is the phase angle between the voltage and the current
rms is the root mean square value of an alternating current or voltage.

**Apparent power** is calculated using the following equations.

For single phase a.c. equipment:

$$\text{apparent power (S) in kVA} = \frac{U_0 I_b}{1000}$$

For three-phase a.c. equipment:

$$\text{apparent power (S) in kVA} = \frac{\sqrt{3} U I}{1000} = \frac{3 U_0 I}{1000}$$

## Design sequence step 2: Electricity supply information

The Electricity Safety, Quality and Continuity (ESQC) Regulations place requirements upon the distributor to provide suitable equipment in suitable locations; see Regulation 24 below.

**Regulation 24: Equipment on a consumer's premises**

**1** A distributor or meter operator shall ensure that each item of his equipment which is on a consumer's premises but which is not under the control of the consumer (whether forming part of the consumer's installation or not) is:

**a)** suitable for its purpose;

**b)** installed and, so far as is reasonably practicable, maintained so as to prevent danger; and

**c)** protected by a suitable fusible cut-out or circuit breaker which is situated as close as is reasonably practicable to the supply terminals.

**2** Every circuit breaker or cut-out fuse forming part of the fusible cut-out mentioned in paragraph (1)(c) shall be enclosed in a locked or sealed container as appropriate.

**3** Where they form part of his equipment which is on a consumer's premises but which is not under the control of the consumer, a distributor or meter operator (as appropriate) shall mark permanently, so as clearly to identify the polarity of each of them, the separate conductors of low-voltage electric lines which are connected to supply terminals and such markings shall be made at a point which is as close as is practicable to the supply terminals in question.

**4** Unless he can reasonably conclude that it is inappropriate for reasons of safety, a distributor shall, when providing a new connection at low voltage, make available his supply neutral conductor or, if appropriate, the protective conductor of his network for connection to the protective conductor of the consumer's installation.

**5** In this regulation the expression 'new connection' means the first electric line, or the replacement of an existing electric line, to one or more consumer's installations.

The designer needs to have information from the supplier in order to carry out a design. The Electricity Safety, Quality and Continuity (ESQC) Regulations require the distributer to supply information on request; see Regulation 28 on the next page.

**Regulation 28: Information to be provided on request**

A distributor shall provide, in respect of any existing or proposed consumer's installation that is connected or is to be connected to his network, to any person who can show a reasonable cause for requiring the information, a written statement of:

**a)** the maximum prospective short-circuit current at the supply terminals;

**b)** for low-voltage connections, the maximum earth loop impedance of the earth fault path outside the installation;

**c)** the type and rating of the distributor's protective device or devices nearest to the supply terminals;

**d)** the type of earthing system applicable to the connection; and

**e)** the information specified in regulation 27(1),

which apply, or will apply, to that installation.

**ASSESSMENT GUIDANCE**

The tariff for small installations is charged on energy use (kWh) not current demand.

In reality, the supplier will not provide the specific values for the installation in question. Suppliers are more likely to specify the maximum, depending on the system arrangement.

## Design sequence step 3: Positioning of equipment

**ACTIVITY**

A load centre or local distribution board needs to be positioned as close as possible to the centre of the loads it supplies. What advantage is there in this?

The mechanical services consultant and the client generally determine the location of major items of plant. However, the location of distribution boards may be at the discretion of the electrical designer, and the number and positions of socket outlets, light switches and luminaires, etc, usually is. These will need to be divided into circuits. Knowledge of the maximum circuit cable lengths will greatly help in allocating sockets and luminaires to particular circuits. The designer will need to consider accessibility and ease of use, as well as balancing of loads and reducing voltage drop.

## Design sequence step 4: Prepare distribution outline

With the major loads identified and marked on a site drawing, the location of the main distribution board and sub-distribution boards can then be decided and the distribution cable routes selected. A distribution outline can be prepared as a schematic drawing and on the site drawing, as shown on the next page.

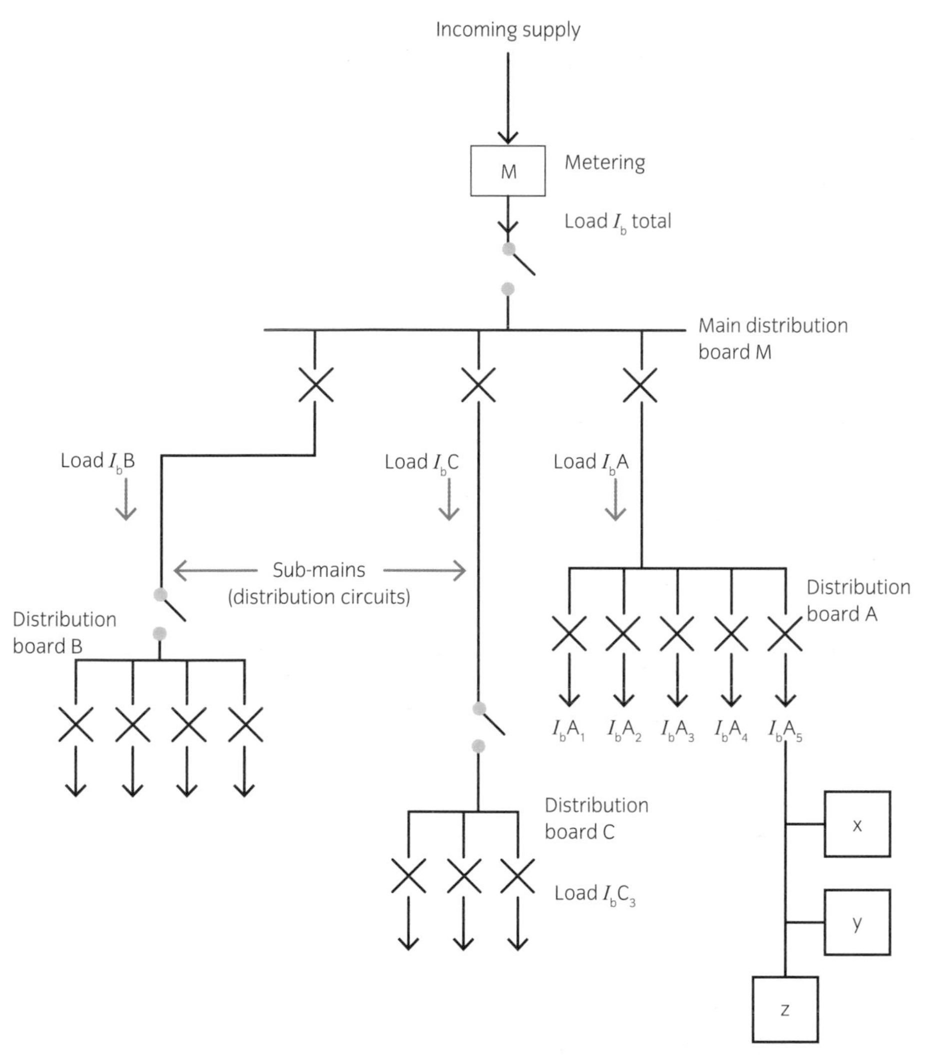


Installation outline

## Design sequence step 5: Carry out detailed design calculations for the distribution circuits

For more information on this step, see coverage of Assessment criteria 2.5 to 2.9 on pages 295–326.

## Design sequence step 6: Carry out detailed design calculations for the final circuits

The process is the same as for distribution circuits in step 5 above.

## Design sequence step 7: Isolation and switching

For more information on this step, see page 332.

## Design sequence step 8: Final assessment

Following completion of the design work, the proposals should be reviewed with the client to check that all the client's requirements have been met in an efficient design.

## SPECIAL INSTALLATIONS OR LOCATIONS

**Special locations in BS 7671:2008 Requirements for Electrical Installations (the IET Wiring Regulations 17th edition)**

| Section number | Installation or location | Particular risk |
|---|---|---|
| 701 | Locations containing a bath or shower | Shock risk increased as low body resistance from immersion in water; wet environment for equipment |
| 702 | Swimming pools and other basin | As 701 |
| 703 | Rooms and cabins containing sauna heaters | As 701 |
| 704 | Construction and demolition site installations | Difficulty in main bonding all metal in contact with earth; wet environment |
| 705 | Agricultural and horticultural premises | Difficulty in main bonding all metal in contact with earth; wet environment |
| 706 | Conducting locations with restricted movement | Good contact with earth; low body resistance due to sweat; difficulty escaping from electric shock |
| 708 | Electrical installations in caravan/camping parks and similar locations | Metal constructions might become raised in potential if there is an open circuit fault on the electricity supply; dangerous for persons outside the caravan or boat with easy access to earth; reduced body resistance due to wet grass/water |
| 709 | Marinas and similar locations | As 708 |
| 710 | Medical locations | Body resistance reduced by invasive surgery, etc; dependence upon maintenance of electricity supply to life-support equipment; unconscious |
| 711 | Exhibitions, shows and stands | Temporary supplies; outside locations with difficulty in main bonding |
| 712 | Solar photovoltaic (PV) power supply systems | Specialist equipment; difficulty in making the installation dead |
| 717 | Mobile or transportable units | Similar problems to caravans |
| 721 | Electrical installations in caravans and motor caravans | Vibration |
| 729 | Operating and maintenance gangways | Restricted space for operating equipment and escape |
| 740 | Temporary electrical installations for structures, amusement devices and booths at fairgrounds, amusement parks and circuses | As 711 |
| 753 | Floor and ceiling heating systems | Risk of damage to heating elements; risk of shock should ceilings/floors be penetrated |

Assessment criteria

1.5 State additional considerations for special locations

Certain installations such as construction-site installations and those in locations such as bathrooms require additional precautions to be taken to ensure the safety of people using them.

Part 7 of BS 7671 describes particular requirements that supplement or modify the general requirements that apply to all installations.

The table on page 264 lists the special installations or locations described in BS 7671; the number of special installations or locations grows with each edition.

It is important to note that many installations, such as installations in explosive atmospheres, require reference to other standards as well as BS 7671. Some installations are outside the scope (not covered at all) by BS 7671 (for instance, see Regulations 110.1.3 and 110.2).

Some installations that require further precautions are excluded from the scope of BS 7671. In these cases, reference must be made to other documents, such as the APEA/IP publication Guidance for Design, Construction, Modification, Maintenance and Decommissioning of Filling Stations (Revised June 2011). See Regulation 110.2 for other examples; most if not all of these require special precautions to be taken.

A list of some of other standards that are commonly referred to is given in the table below.

**Locations and installations requiring supplementary reference to another standard**

| Location or installation | Supplementary reference |
|---|---|
| Electric signs and high voltage luminous discharge tube installations | BS 559 and BS EN 50107 |
| Emergency lighting | BS 5266 |
| Electrical apparatus for explosive gas atmospheres | BS EN 60079 |
| Electrical apparatus for use in the presence of combustible dust | BS EN 50281 and BS EN 61241 |
| Fire detection and fire-alarm systems in buildings | BS 5839 |
| Telecommunications systems | BS 6701 |
| Electric surface heating systems | BS EN 60335-2-96 |
| Electrical installations for open-cast mines and quarries | BS 6907 |

Care has to be taken with some installations that fall outside the scope of BS 7671. These include:

- 'systems for the distribution of electricity to the public
- railway traction equipment, rolling stock and signalling equipment
- equipment of motor vehicles, except those to which the requirements of the regulations concerning caravans or mobile units are applicable
- equipment on board ships covered by BS 8450
- equipment of mobile and fixed offshore installations
- equipment of aircraft
- those aspects of mines and quarries specifically covered by statutory regulations
- radio interference suppression equipment, except so far as it affects safety of the electrical installation
- lightning protection systems for buildings and structures covered by BS EN 62305
- those aspects of lift installations covered by relevant parts of BS 5655 and BS EN 81-1
- electrical equipment of machines covered by BS EN 60204
- electric fences covered by BS EN 60335-2-76.'

**ASSESSMENT GUIDANCE**

Petrol filling stations obviously need specialist cables and equipment because of the petrol in the pumps and the risk of spillage on the fore court.

**ACTIVITY**

Look up the BS EN 62305, for lightning conductors and name the parts of the system.

Even if a location or installation is not listed, the designer must consider whether the particular design will require additional precautions to be taken. This may include other statutory documents or local authority requirements.

**Assessment criteria**

**1.6** Interpret measurements from design plans

## ESTIMATING FROM DESIGN PLANS

Designers will be invited to prepare designs and cost installations when the building/construction is in the design stage. Therefore all information including measurements will have to be taken from plans.

### Plans and scales

Large items can be represented on plans by scaling down the object. For example, Ordnance Survey 'Explorer' maps adopt a scale of 1:25 000 whereby every 1 cm on the plan represents 25 000 cm = 250 m.

For building plans the scales used are typically 'larger scale', such as 1:20 or 1:200. These are called larger scale than 1:25 000, as the resultant drawing on the plan is larger than one using a 1:25 000 scale.

## Scale rules

A scale rule is used to measure directly from plans, as shown in the photograph.

The scale ruler will read the dimension directly off the plan; only the scale used when drawing the plan is needed.

Triangular scale rule

**KEY POINT**

It is important when scaling off plans to check that the drawing you are measuring from has not been reduced or magnified through photocopying or printing and that it is to a proper scale.

# COSTING

**Assessment criteria**

1.7 Research cost of equipment used in electrical systems using different sources

The main cost elements of an electrical installation are:

- company overheads
- labour
- materials
- sub-contract items (eg parts of the work involving specialists who carry out the work on your behalf)
- prime cost sums
- main contractor's discounts.

## Company overheads

The indirect costs of carrying out an electrical installation need to be estimated so that they can be recovered from all the work carried out, on a reasonable basis. Indirect costs will include:

- premises (rent and rates)
- supervisors'/managers' salaries
- vehicles
- company tax
- dividends to the owners.

**KEY POINT**

**Direct and indirect costs**

Direct costs are those costs directly associated with the work, eg:

- the direct labour costs of electricians installing equipment
- the direct material cost of the equipment installed.

Indirect costs are those costs not directly related to a particular job of work, eg:

- costs of premises
- costs of vehicles
- labour costs of estimators, accountants, quantity surveyors.

## Labour costs

For many installations, particularly small installations such as in dwellings, labour costs are the most significant. Labour costs include:

- company National Insurance contributions
- employee salaries.

From each employee's salary will be deducted:

- employee National Insurance contributions
- tax.

From these figures a labour cost per hour can and must be calculated. A labour cost estimate can then be prepared from an estimate of the time to carry out each unit of the work (in hours).

There are publications that provide both estimates of labour costs based on nationally agreed rates and estimates of time to carry out units of work. These are usually referred to as a 'schedule of rates'.

## Material costs

An estimate of the cost of equipment (eg cables and switchgear) is usually obtained from potential suppliers. Equipment is commonly purchased:

- from electrical wholesalers
- directly from manufacturers
- from builders' merchants
- from DIY stores.

### Electrical wholesalers

Electrical wholesalers will provide a price for a complete range of equipment, and it is not usually necessary to specify the manufacturer. However, with experience, an electrical contractor will start to specify particular manufacturers that are known to produce goods that are:

- easier to install (reducing labour costs)
- more reliable (leading to customer satisfaction and reduced after-sale costs)
- easier to operate
- preferred by customers (as the client may wish to standardise equipment across a range of installations).

## Direct from manufacturers

Manufacturers will provide competitive estimates if the quantities are large or the cost of the item is high.

## Builders' merchants

Builders' merchants may supply a limited range of electrical accessories and the items can be inspected on the shelves. However, costs are likely to be relatively high compared to the discounts often given by wholesalers.

## DIY stores

There are many DIY stores that are readily accessible, with extended opening hours. However, they generally stock only a small range of electrical items from just a few suppliers and costs are high.

Note:

- It is important to remember that buying from wholesalers also gives an element of product expertise. Equipment bought from the internet or through DIY-type outlets can often lead to products being mis-sold. As an example, a device labelled a 'shower switch' may only be rated as an isolator (off-load switching) device where an on-load rated switch is required.
- Ensure your supply terms cover you against the terms of the contract. You do not want to be in a position where the supplier can increase their charges to you but you cannot increase your charges to your customer or the main contractor. Always check the time limit of any quotations given to you by a supplier.
- Visit the site, if possible, so you are fully aware of any practical difficulties there may be, such as dimensions and space constraints.

### ACTIVITY

Electrical wholesalers normally offer a monthly account, which means you do not have to pay until a month after the statement date. Look on the internet for local wholesalers close to your home address.

## Sub-contractors

Your quotation or tender may need to include work to be carried out by others, for example, the lightning protection may be included in your contract but you need a specialist sub-contractor to undertake this work. You must ensure the sub-contract includes the terms of your contract, otherwise the subcontractor may be able to levy charges against you that you cannot recover from the client.

If you need to supervise this work, or if you have to pay this contractor before you are paid, you will need to add these costs to the tender price.

## Prime cost sums (PCs)

Prime costs are an allowance to be made for equipment or installations that cannot be determined at the time of tendering. The tender documents will usually advise the sum to be allowed. The contractor is not allowed to add profit to these items.

## Builders' discount

The builder may be entitled to take a discount from any claim for payment you make, to cover their costs in supervising the work. This is often 2.5%. You will need to add this cost to your tender.

# ESTIMATING FOR AN INSTALLATION

**Assessment criteria**

**1.8** Calculate requirements for electrical systems for a domestic property

After a design has been carried out (see Learning outcome 2, pages 276–281), an estimate of the cost of carrying out an installation can be made.

**ASSESSMENT GUIDANCE**

The profit margin is up to you. It depends on the competition, as well as on your trading history with the client. Some profit is necessary for future expansion of your company.

## Estimating tools

There are many tools available such as:

- estimating software
- estimating guides, eg Luckins
- simple schedules.

The essence of all estimating is to divide the installations into appropriately small units so that the cost can be estimated, with sufficient accuracy, in terms of:

- the labour (time to complete)
- the materials
- subcontracted items.

This may be achieved by dividing the estimate into smaller chunks such as control equipment at the intake position, distribution circuits and final circuits.

## Simple schedules

Below is an example of a simple estimating schedule.

**Company name:** A. N. Contracts Ltd

| **Client:** A House builder | **Job no and description:** 1874 |
|---|---|
| **Return date** 31March | **Estimator:** PNB |

| | Item | | Cat. No | quantity | Unit | Price per unit £ | Discount % | Net £ |
|---|---|---|---|---|---|---|---|---|
| 1 | Consumer unit 8-way c/w 2 x RCDs | | 12345 | 1 | each | 59.90 | 15 | 50.92 |
| 2 | 2-gang socket outlets | | 12346 | 20 (4 packs) | Pack of 5 | 18.26 | 30 | 51.13 |
| 3 | 2-gang surface boxes | | 13458 | 20 | each | 0.96 | 30 | 13.44 |
| 4 | Lighting pendant sets | | 15677 | 15 | each | 2.98 | 30 | 31.29 |
| 5 | 1-gang 1-way switches | | R678 | 15 | each | 2.24 | 30 | 23.52 |
| 6 | 1-gang 2-way switches | | R679 | 6 | each | 2.24 | 30 | 9.41 |
| 7 | Cable 2.5/1.5 mm$^2$ 6242Y | | C77877 | 160 m | 100 m | 62.00 | 40 | 59.52 |
| 8 | Cable 1.5/1.0 mm$^2$ 6242Y | | C77978 | 150 m | 100 m | 47.00 | 40 | 42.30 |
| 9 | 45 A cooker switch | | 45689-1 | 1 | each | 6.89 | 30 | 4.83 |
| 10 | Cooker switch surface box | | 45690-2 | 14 | each | 1.65 | 30 | 16.17 |
| 11 | Cable 4/1.5 mm$^2$ 6242Y | | C8458 | 8 | 100 m | 72.00 | 40 | 3.47 |
| 12 | Tails 16 mm$^2$ | | 66666 | 1 m | 1 m | 12.00 | 30 | 8.40 |
| 13 | Recessed light fittings | | 6784 | 30 | Pack of 3 | 27.00 | 30 | 189.00 |
| 14 | Towel rail | | -- | 1 | each | 24.00 | 30 | 16.80 |
| 15 | Sundries | | | | | | | 20.00 |
| Total materials | | 540.20 | | | | | | |
| On cost % | 30 | 162.06 | | | | | | |
| Total net labour | 80 hours at £15.25 per hour | 1241.60 | | | | | | |
| On cost % | 40 | 496.64 | | | | | | |
| Sub total | | 2440.50 | | | | | | |
| VAT | | 488.10 | | | | | | |
| Total | | 2928.60 | | | | | | |

**Simple estimating schedule**

**Assessment criteria**

**1.9** Explain implications for tender content

# CONTRACTS

A tender is an offer to carry out work for a certain amount of money, in accordance with a particular contract. There are legal implications that you need to bear in mind when tendering for work.

**ACTIVITY**

Make a table similar to the estimating schedule on page 271. Go round your own home and estimate all the equipment that would have been required.

## Contract law

English contract law regulates contracts in England and Wales. A contract generally forms when the contractor makes an offer and another person accepts it. Once a contract is made, it is enforceable in court.

If problems are to be avoided, the offer must be made clearly, in sufficient detail, including its terms and conditions. Obviously, it is preferable if the offer is made and accepted in writing.

## Small works

Small works such as house rewires or minor works will usually be carried under the contractor's own terms, perhaps amended by the client.

## Tender procedure

When a company receives an invitation to tender for a particular contract, the normal process to follow would be:

- record all documentation
- gather information to form the estimate
- determine prime costs
- determine provisional sum
- make final tender.

### Record all documentation

All documents received in the invitation will be recorded and checked against the list of documents, which should be included. This ensures that all relevant documents are present and the most up-to-date revisions of drawings are used. If, by mistake, an older version of drawings is used, this will reflect in the overall cost submitted.

### Gather information to form the estimate

All documents and drawings will be carefully scrutinised and lists of materials will be constructed. These may be regular materials, such as switches and socket outlets, that are listed on a 'take-off sheet'.

| Area | DB reference | Drawing number | [symbol] | [symbol] | [symbol] | [symbol] | [symbol] | [symbol] | [symbol] |
|---|---|---|---|---|---|---|---|---|---|
| Ground floor | DB-2 | A-2011-3-R1 | 12 | 3 | 18 | 1 | 6 | 2 | 3 |
| First floor | DB-3 | A-2011-4-R1 | 6 | 4 | 15 | 1 | 12 | 2 | 0 |

**Sample take-off sheet**

The sample take-off sheet shows the number of particular items needed for a particular area of the installation. The symbols would be used from the drawing, to minimise any confusion, and the specification would be checked for any specific manufacturer or finish required. At the bottom of the sheet, totals would indicate the quantity of each product required. The advantage of showing the quantity of items per area is that, should a contract be won, this can be used to order materials at suitable installation stages.

Other major items of specialist equipment, such as large motors and generators, will be put out for quotation to major suppliers or manufacturers.

In some situations the client may provide a Bill of Quantities. This would be a ready-made list of specific materials and would normally be made by a quantity surveyor. This would replace the take-off sheet.

Once the material costs, including any specialist plant, have been obtained, the cost estimate can be calculated. This will involve total material costs and estimated labour costs against each specific area. In some instances, a **schedule of rates** may be used to assist in this process.

**Schedule of rates**

This is a document that lists a labour time/cost for installing and connecting an individual item of equipment. Examples may be £2.50 to fix a ceiling rose securely to a plaster ceiling and £1.50 to connect the cable to it. Running the cable to the ceiling rose could be charged at a rate of £2.00 per metre.

It is always wise, if possible, to visit the site at this point, to see whether there are any issues that may affect the installation process. These may be issues, such as space constraints, giving delivery and storage problems (which would require a lot of labour time spent moving materials around from storage to site).

## Prime cost

Prime cost is an allowance for the supply of a particular item of equipment where the equipment has been specified by the client. For example, if the client specifically requires a particular type of decorative chandelier to be installed, the prime cost would be for the contractor to purchase the chandelier. This would not include the cost of installation. In some circumstances, prime cost may be a cost for supplying a service by a nominated sub-contractor specified by the client. An example of this would be the prime cost for a specialist sub-contractor to create and install a particular piece of decorative artwork, such as a statue or feature within a building.

### Provisional sum

This is a sum of money that will cover the work required to install a particular item of equipment, when the actual cost is yet unknown. An example could be the installation of a decorative chandelier where the detail of its full size or weight is not yet known. In this case a provisional sum would be given to the client, which can be re-negotiated later in the contract.

### Final tender

Once all the costs of labour and materials, including prime costs and provisional sums, are determined, the company overheads will then be added to make the final tender. Overheads include the hidden costs of a business, such as the costs of:

- van insurances
- office staff
- holiday pay
- National Insurance contributions.

**ASSESSMENT GUIDANCE**

National insurance contributions have to be paid by everyone in work, whether employed or self-employed.

Overheads include many factors but the list above gives some examples of what should be allowed for.

Finally, the tender is completed by a senior manager, using the benefit of their own experience to account for further costs, such as profit and other considerations that may arise. This involves a very careful balancing act to ensure that the company benefits from the work while keeping the overall cost competitive.

## EXPLAIN IMPLICATIONS FOR TENDER CONTENT

This section briefly looks at the contractual procedures that are followed, should a tender be accepted.

### Alterations to a tender

If a tender is successful, a contract will be agreed and signed, based on the work contained in the tender and the agreed price. Rarely will an installation follow the tender exactly, as the various factors affecting the overall installation are likely to change during the work. Examples of this could be socket outlets needing to be re-sited, as a radiator needs to be fitted in a particular location and this not having been anticipated.

Where these instances occur, the installer must obtain a variation order (VO), which is a declared change to the original contract. At this stage, an additional sum for the alterations may be agreed or the work could be undertaken on a rate per day. Day work sheets will be used to record the additional work and, at various points in the contract, the costs of these additional jobs will be agreed. Some major contracts require the day rates to be agreed prior to work starting.

## Contract progress

The progress of a contract is set and monitored using a variety of tools such as critical path networks and Gantt charts. These are covered in detail on page 351 and in Unit 210 of the 2365 Level 2 book.

Should a contract fall behind expected progress, due to factors under the control of the electrical sub-contractor, it is likely that penalty clauses would take effect. These clauses are imposed as part of a contract agreement to ensure timely completion. If completion is delayed, the client may need compensating due to loss of facilities. These penalties are set, normally at a single day rate, at the point where contracts are agreed. The longer the delay in completion, the larger the penalty imposed. For this reason, using tools such as critical path networks can assist in avoiding delays.

Companies that have been awarded ISO 9000 have demonstrated that they:

- perform effectively, using these management systems
- meet regulatory requirements
- work to ensure they meet their clients' needs.

# OUTCOME 2

# Understand principles for designing electrical systems

**Assessment criteria**

2.1 Explain how regulatory requirements impact upon the (design and) installation of electrical systems

## REGULATORY REQUIREMENTS

There is a great deal of legislation and regulation to ensure that those involved with electrical systems work safely and without risk, not only during the construction phase but also during the system's subsequent use and demolition. Legislation also covers electrical installation design.

### The risks from electrical installations

The objective of good design is to ensure the safety of people, livestock and property and to provide against dangers and damage that may arise in the reasonable use of electrical installations.

In electrical installations, risk of injury may result from:

1. shock currents
2. excessive temperatures likely to cause burns, fires and other injuries
3. ignition of a potentially explosive atmosphere
4. undervoltages, overvoltages and electromagnetic disturbances likely to cause or result in injury or damage
5. mechanical movement of electrically actuated equipment, in so far as providing a suitable means of preventing injury by emergency switching or a suitable means of switching equipment off for the mechanical maintenance of non-electrical parts
6. power supply interruptions and/or interruption of safety services
7. arcing or burning likely to cause blinding effects, excessive pressure and/or toxic gases.

**ASSESSMENT GUIDANCE**

An undervoltage may be caused by excessive loading on the system. Motor circuits are protected by undervoltage or no-volt coils, which disconnect the motor should the voltage drop below a predetermined voltage. The motor cannot restart until the starter is reset.

## Protective provisions

The risks outlined above are described in electrical terms as protection against the risks listed in the table below.

| Risk | Chapter in BS 7671:2008 Requirements for Electrical Installations |
|---|---|
| 1 Electric shock | 41 |
| 2 Thermal effects | 42 |
| 3 Overcurrent | 43 |
| 4 Fault current | 43 |
| 5 Voltage disturbances | 44 |
| 6 Electromagnetic disturbances | 44 |
| 7 Power supply interruptions | 56 |

The protective provisions listed above must be considered when assessing those factors affecting the installation, such as:

- the use of the building
- the environmental conditions
- the building construction material and methods.

(See Appendix 5 of BS 7671.) When carrying out a design and subsequent installation, those factors involved will need to meet these protective provisions.

### ACTIVITY

Look at Chapter 13 of BS 7671. This chapter covers an important area for a designer of an electrical installation.

Explain in detail what this chapter covers.

## Safety

The key piece of legislation is the Health and Safety at Work etc Act 1974 (HSW Act). This is a comprehensive piece of legislation that covers all aspects of safety at work. Importantly, this Act empowers the Secretary of State to make regulations that provide more detail to the general requirements of the Health and Safety at Work etc Act 1974. The regulations most relevant to electrical installation work are:

- Electricity at Work Regulations 1989 (EAWR)
- Construction (Design and Management) Regulations 2007 (CDM 2007)
- Provision and Use of Work Equipment Regulations 1998 (PUWER)
- Personal Protective Equipment at Work Regulations 1992
- Electricity Safety, Quality and Continuity (ESQC) Regulations 2002.

### KEY POINT

It is important to be aware that, in the event of any accident or 'dangerous occurrence', ignorance or lack of understanding of the laws and regulations is no defence for those responsible, if they are prosecuted. This is the case, no matter how difficult or complicated the legislation appears to be.

It is also essential to note that, while the legislation imposes requirements on the employer, designer, builder/constructor, etc, it also imposes requirements on everyone involved in the work to ensure that the work is carried out without danger to anyone during the installation of the system, while it is in use and during subsequent dismantling.

## The IET Wiring Regulations 17th edition

The IET Wiring Regulations are not legislation; they are a British Standard: BS 7671. However, in the view of the Health and Safety Executive (HSE), compliance with BS 7671 is likely to lead to compliance with the Electricity at Work Regulations 1989 (EAWR). The IET Regulations are also referred to as a standard in the Electricity Safety Quality and Continuity (ESQC) Regulations 2002 and as guidance in Part P to the Building Regulations.

**Assessment criteria**

**2.2** Describe factors affecting selection of electrical systems

# ASSESSMENT OF GENERAL CHARACTERISTICS

You need to consider a variety of factors when selecting an electrical system.

## General characteristics

An assessment must be made of the general characteristics of the installation, as detailed in the table below. The relevant chapter of BS 7671 is listed against each characteristic.

| Characteristic | Chapter |
|---|---|
| The purpose for which the installation is intended to be used, its general structure and its supplies | 31 |
| The external influences to which it is to be exposed | 32 |
| The compatibility of its equipment | 33 |
| Its maintainability | 34 |
| Recognised safety services | 35 |
| Assessment for continuity of service | 36 |

## Purpose, structure and supplies

Within Part 3 of BS 7671, Chapter 31 requires a designer of an electrical installation to assess, during the design stage:

- maximum demand and diversity
- conductor arrangement and system earthing
- types of system earthing
- supplies
- division of installation
- external influences
- compatibility of characteristics and electromagnetic compatibility
- maintainability
- safety services
- continuity of service (need for back-up supplies).

While others may give the designer some of the above detail, such as supplies, it is still the designer's responsibility to assess suitability.

**ASSESSMENT GUIDANCE**

When choosing and installing equipment it is essential that it can be maintained easily. Access is an important issue.

## External influences

According to Regulation 512.2 of BS 7671, all equipment must be suitable for the situation in which it is to be installed and used, in terms of temperature, exposure to water, dust, mechanical impact and use, as well as other situations in which the equipment is to be used. These factors are listed in full in Appendix 5 of BS 7671, with each influence given a specific code. Designers must be aware of all the influences to which the installation is likely to be subjected.

**ACTIVITY**

Identify the three zones associated with bathrooms.

## Cable installation method

Like all equipment, the cabling system adopted for a location must be suitable for the environment. Table 42A of Appendix 4 of BS 7671 lists 120 cable installation methods. The common systems are briefly described in the table on the next page.

(See also Assessment criteria 2.4 on pages 290–295.)

**Common cable installation methods (Source: Table 42A of Appendix 4 of BS 7671:2008 Requirements for Electrical Installations (the IET Wiring Regulations 17th edition), 2011)**

| Table 42A reference number | Method | Description | Application |
|---|---|---|---|
| 100–103 | Flat twin with cpc | An insulated and sheathed cable with protective conductor. Installed in insulation | Very limited mechanical protection – plastic sheath. Generally used to install in walls, ceilings, floors, etc where building structure provides limited protection. Cheapest method |
| C | Steel wire armoured cable | | Clipped directly to walls |
| D | Insulated and sheathed multicore cables with armouring | | Installed underground |
| E | | | On cable tray |
| E | Non-armoured cable | Insulated and sheathed multicore cables without armouring including fire-retardant cables | For use on cable tray |
| C | Armoured or non-armoured cables with sheath including fire-retardant cables | | Clipped directly to a wood or masonry surface |
| B | Steel conduit (black enamelled) and trunking | Insulated single or multicore cables with or without separate protective conductor.<br>Also includes multi-compartment trunking such as dado trunking | Good mechanical protection, for use indoors. Training on threading and bending required |
| B | Steel conduit (galvanised) | Insulated single or multicore cables with or without separate protective conductor | Good mechanical protection, for use outdoors and indoors. Training on threading and bending required |
| B | Surface-mounted PVC conduit and trunking | Single-core insulated cables with separate protective conductor or multicore cable.<br>Also includes multi-compartment trunking such as dado trunking | Very suitable for outdoor and corrosive environments, eg agricultural. Cheap, easy to install |
| A | Conduit in a thermally insulated wall | Single-core insulated cables with separate protective conductor or multicore cable | Steel or PVC |
| B | Floor trunking | Insulated single or multicore cables with or without separate protective conductor | Provides accessible sockets in very large open-plan offices. Expensive |
| Not classified in table | Busbar trunking | Fitted with busbars | For machine shops and similar. Allows flexibility if machines are to be moved. Expensive |

## Design sequence

As described in Assessment criteria 1.4 (see pages 258–264), the designer will prepare an installation outline, having made appropriate allowances for diversity and determined the design current $I_b$ of final circuits, distribution boards, sub-mains, main distribution board, etc.

For each circuit the following proceedure can be followed.

1 Determine design current $I_b$.
2 Select overcurrent device type and rating $I_n$.
3 Select cable type and minimum required current-carrying capacity in the particular installation conditions $I_z$.
4 Calculate the cable-tabulated current $I_t$ and select cable nominal cross-sectional area (csa) $S$.
5 Knowing the cable length, calculate the voltage drop and check that it is satisfactory.
6 Check that the disconnection time meets the requirements of BS7671 Regulations 411.3.2.1 to 411.3.2.4 and Table 41.1.
7 Check that the protective conductor cross-sectional area (csa) is adequate.

**Assessment criteria**

**2.3** Explain how a circuit's maximum demand is established after diversity factors are applied

**KEY POINT**

**BS 7671 Regulation 132.3: Nature of demand**

The number and type of circuits required for lighting, heating, power, control, signalling, communication and information technology, etc, shall be determined from knowledge of:

- location of points of power demand
- loads to be expected on the various circuits
- daily and yearly variation of demand
- any special conditions such as harmonics
- requirements for control, signalling, communication and information technology, etc
- anticipated future demand if specified.

**ACTIVITY**

Calculating the size of each main and distribution cable takes a long time. Any changes may mean that you have to recalculate the whole lot. Cable design packages are readily available and will recalculate the whole installation in a few seconds.

Look on the internet for companies offering such packages.

**ASSESSMENT GUIDANCE**

On large installations the metering will be via current transformers rather than direct connection.

# ESTIMATING MAXIMUM DEMAND

## Installation outline

When the position and demand, in amperes (A) and kilovolt-amperes (kVA), of the installation have been ascertained (from drawings, specifications, discussion with the client, etc), an installation outline can be prepared. This will show the main and sub-distribution boards as well as the sub-mains that link the sub-distribution boards with the main board, as illustrated below.

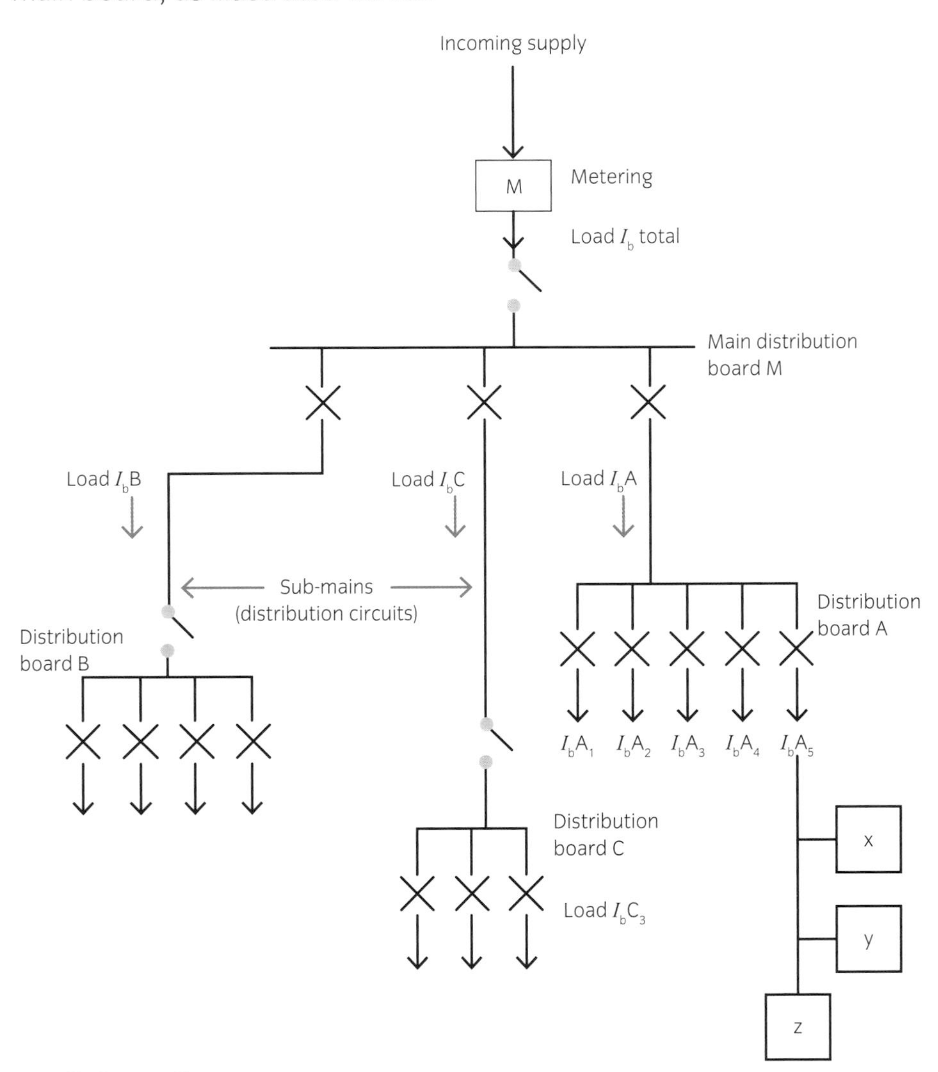


**Installation outline**

For ease of drawing, this is normally a schematic. For an existing building, cable route lengths will be estimated by scaling off a map or even by direct measurement with a tape measure.

For a domestic installation the outline will be brief, as shown below.

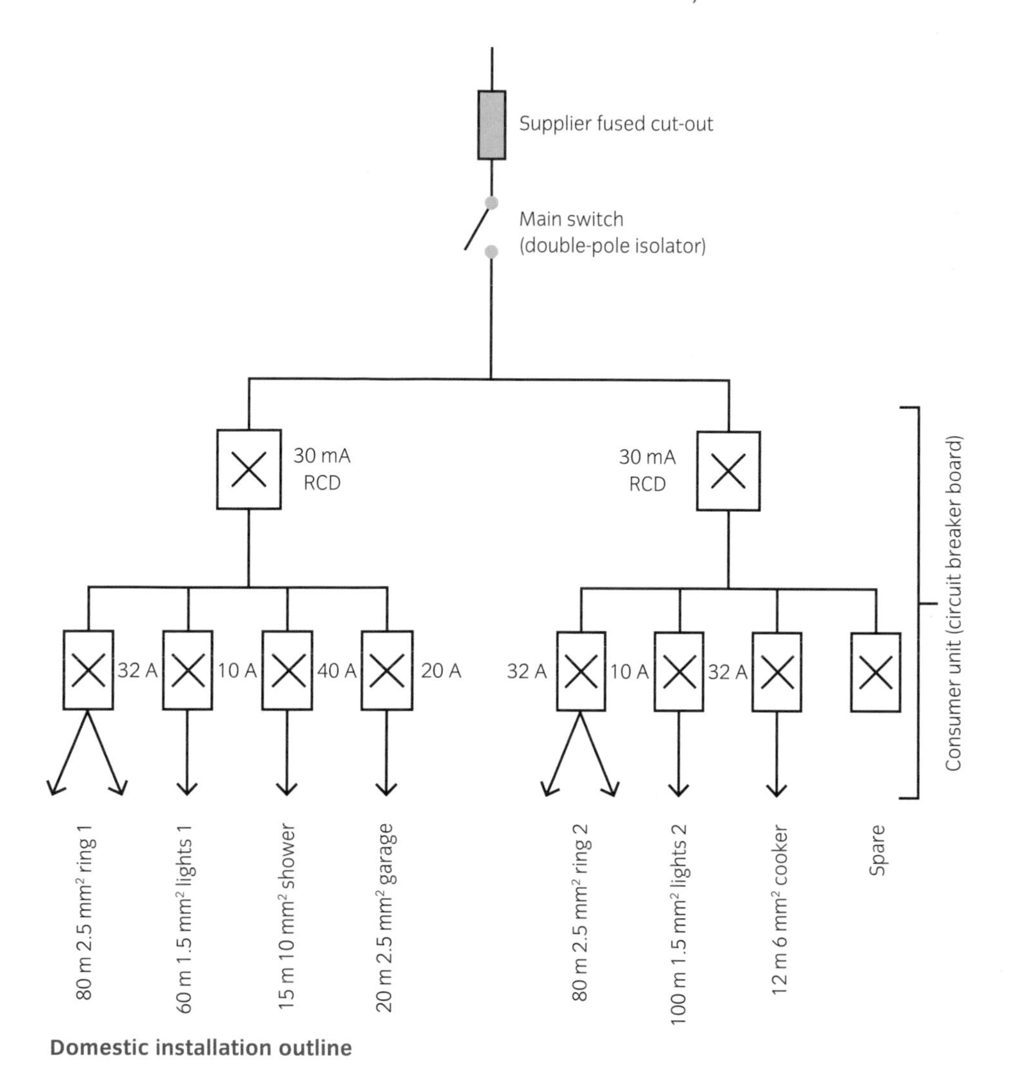


**Domestic installation outline**

## Calculating current from kVA and kW

The current demand for an item of equipment for a given kVA rating can be calculated as follows.

For a single-phase supply the current I is given by:

$$I = \frac{\text{kVA} \times 1000}{U_0}$$

where:

$I$ is current in amperes
kVA is the rating of the equipment
$U_0$ is the nominal voltage (line to earth) of the equipment.

When the rating of the equipment is in kW, the current $I$ is given by:

$$I = \frac{\text{kW} \times 1000}{U_0 \times \cos\theta}$$

where:

$I$ is current in amperes
kW is the rating of the equipment in kW
$U_0$ is the nominal voltage (line to earth) of the equipment
$\cos\theta$ is the power factor.

For three-phase equipment, there are two methods for finding the current $I$.

**Method 1:**

$$I = \frac{\text{kVA} \times 1000}{3U_0}$$

where:

$I$ is current in amperes
kVA is the rating of the equipment
$U_0$ is the nominal voltage (line-to-earth) of the equipment.

**Method 2:**

$$I = \frac{\text{kVA} \times 1000}{\sqrt{3}U}$$

where:

$I$ is current in amperes
kVA is the rating of the equipment
$U$ is the nominal voltage (line-to-line) of the equipment.

Where the design current for discharge lighting is determined, a factor of 1.8 is applied if particular detail is not available (such as power factor ratings or particular starting current demands).

## Final circuit current demand

The current demand of a final circuit is estimated by adding the current demands of all points of utilisation (eg socket outlets) and items of equipment connected to the circuit and, where appropriate, making allowances for diversity.

For domestic and similar small installations, the table below gives current demands to be used for final circuits. The rated current given in item 1 relates to the nominal rating of the protective device selected ($I_n$).

**Final circuit current demand to be assumed for points of utilisation and current-using equipment (IET On-Site Guide, Table A1)**

| | Point of utilisation or current-using equipment | Current demand to be assumed |
|---|---|---|
| 1 | Socket outlets other than 2 A socket outlets and 13 A socket outlets (see Note 1 below) | Rated current |
| 2 | 2 A socket outlets | At least 0.5 A |
| 3 | Lighting outlet (see Note 2 below) | Current equivalent to the connected load, with a minimum of 100 W per lampholder |
| 4 | Electric clock, shaver supply unit (complying with BS 3535), shaver socket outlet (complying with BS 4573), bell transformer and current-using equipment of a rating not greater than 5 VA | May be neglected |
| 5 | Household cooking appliance | The first 10 A of the rated current plus 30% of the remainder of the rated current plus 5 A if a socket outlet is incorporated in the control unit |
| 6 | All other stationary equipment | Normal current or British Standard rated current |

Note 1: See Appendix H (of the IET On-Site Guide) for the design of standard circuits using socket outlets to BS 1363-2 and BS 4343.

Note 2: Final circuits for discharge lighting must be arranged so as to be capable of carrying the total steady current, namely that of the lamp(s) and any associated gear and also their harmonic currents. Where more exact information is not available, the demand, in volt-amperes (VA), is taken as the rated lamp watts multiplied by not less than 1.8. This multiplier is based upon the assumption that the circuit is corrected to a power factor of not less than 0.85 lagging, and takes into account control gear losses and harmonic current.

**Allowances for diversity between final circuits for sizing distribution circuits (IET On-Site Guide, Table A3)**

| | | Type of premises | Type of premises | Type of premises |
|---|---|---|---|---|
| | Purpose of final circuit fed from conductors or switchgear to which diversity applies | Individual household installations including individual dwellings of a block | Small shops, stores, offices and business premises | Small hotels, boarding houses, guest houses, etc |
| 1 | Lighting+ | 66% of total current demand | 90% of total current demand | 7% of total current demand |
| 2 | Heating and power (but see 3 to 8 below)+ | 100% of total current demand + up to 10 A +50% of any current demand in excess of 10 A | 100% full load (f.l.) of largest appliance +75% f.l. of remaining appliances | 100% f.l. of largest appliance +80% f.l. of second largest appliance +60% f.l. of remaining appliances |
| 3 | Cooking appliances | 10 A +30% f.l. of connected cooking appliances in excess of 10 A +5 A if socket outlet incorporated in control unit | 100% f.l. of largest appliance +80% f.l. of second largest appliance +60% f.l. of remaining appliances | 100% f.l. of largest appliance +80% f.l. of second largest appliance +60% f.l. of remaining appliances |
| 4 | Motors (other than lift motors that are subject to special consideration) | Not applicable | 100% f.l. of largest motor +80% f.l. of second largest motor +60% f.l. of remaining motors | 100% f.l. of largest motor +50% f.l. of remaining motors |
| 5 | Water-heaters (instantaneous type)* | 100% f.l. of largest appliance +100% f.l. of second largest appliance +25% f.l. of remaining appliances | 100% f.l. of largest appliance +100% f.l. of second largest appliance +25% f.l. of remaining appliances | 100% f.l. of largest appliance +100% f.l. of second largest appliance +25% f.l. of remaining appliances |
| 6 | Water-heaters (thermostatically controlled) | No diversity allowable† | No diversity allowable† | No diversity allowable† |
| 7 | Floor warming installations | No diversity allowable† | No diversity allowable† | No diversity allowable† |
| 8 | Thermal storage space heating installations | No diversity allowable† | No diversity allowable† | No diversity allowable† |
| 9 | Standard arrangement of household and similar final circuits (in accordance with Appendix 8 of the IET On-Site Guide)+ | 100% of current demand of largest circuit +40% of current demand of every other circuit | 100% of current demand of largest circuit +50% of current demand of every other circuit | 100% of current demand of largest circuit +50% of current demand of every other circuit |

| | | Type of premises | Type of premises | Type of premises |
|---|---|---|---|---|
| 10 | Socket outlets other than those included in 9 above and stationary equipment other than those listed above | 100% of current demand of largest point of utilisation +40% of current demand of every other point of utilisation | 100% of current demand of largest point of utilisation +70% of current demand of every other point of utilisation | 100% of current demand of largest point of utilisation +75% of current demand of every other point in main rooms (dining rooms, etc) +40% of current demand of every other point of utilisation |

* For the purpose of this table, an instantaneous water-heater is deemed to be a water-heater of any loading that heats water only while the tap is turned on and therefore uses electricity intermittently.

† It is important to ensure that the distribution boards and consumer units are of sufficient rating to take the total load connected to them without the application of any diversity.

+ The current demand may be that estimated, for example in accordance with Table A1 IET On-Site Guide. Where the circuit is a standard circuit for household or similar installations, the current demand is the rated current of the overcurrent protective device of the circuit.

**KEY POINT**

The first table may be applied to final circuits but the second table is only applied to a collection of final circuits in order to size a distribution circuit or overall maximum demand including diversity.

**ACTIVITY**

2 A sockets were often used to connect table lamps or similar, the socket being controlled from a wall switch. The current equivalent was based on a 100 W lamp. Tungsten lamps of that rating are no longer available, and have typically been replaced by 18 W fluorescent lamps. What would be the current demand at 230 V of:

**a)** a 100 W tungsten lamp

**b)** a 18 W compact fluorescent?

## Examples of circuit current demand

### Shower circuit

(See row 6 of the table on the previous page.)

Rating of shower (P) from technical specification 7.9 kW at 230 V:

$$\text{Current demand } I = \frac{\text{kW} \times 1000}{U_0 \times \cos\theta}$$

$\cos\theta = 1$ for a resistive load

$$\text{current demand } I = \frac{7.9 \times 1000}{230} = 34.3\text{ A}$$

**KEY POINT**

Certain loads have no diversity because they are either 100% off or 100% on.

### Cooker circuit

(See row 3 of the table on the previous page.)

Consider an electric cooker with:

- hob comprising 4 of 2 kW elements
- main oven 2 kW
- grill/top oven 2 kW
- total installed capacity of 12 kW at 230 V.

$$I = \frac{\text{kW} \times 1000}{U_o \times \cos\theta} = \frac{12 \times 1000}{230 \times 1}$$

So the short-time peak demand = 52.2 A

From row 3 of the table on page 286, the circuit design current ($I_b$) applying diversity is the first 10 A of the rated current plus 30% of the remainder of the rated current plus 5 A if a socket outlet is incorporated in the control unit.

$$\text{After diversity demand } I = 10 + \left(\frac{30}{100}\right) \times (52.2 - 10) = 22.7 \text{ A}$$

Or as an alternative method:

$$I_b = 52.2 - 10 = 42.2 \times \left(\frac{30}{100}\right) + 10 = 22.7 \text{ A}$$

After diversity is applied:

$$I_b = 22.7 \text{ A (or +5 A if a socket outlet is included = 27.7 A)}$$

### Lighting circuit

(See row 1 of the table on the page 286 of allowances for diversity between final circuits for sizing distribution circuits.)

Consider 10 downstairs lights with standard brass bayonet cap (BC) lampholders. In line with the table of final circuit current demand to be assumed for points of utilisation and current-using equipment on page 285, assume 100 W demand per lighting point. Therefore:

$$\text{Circuit demand } I_b = \frac{\text{total load in watts}}{\text{voltage}} = \frac{10 \times 100}{230} = 4.3 \text{ A}$$

A 6 A circuit would be suitable for tungsten lamps. If extra low voltage or discharge lighting is to be supplied by a type B circuit breaker, you would need to specify a 10 A circuit to reduce unwanted tripping on switch on, or a 6 A type C circuit breaker due to starting surge.

For all but the simplest circuits the load characteristics should be assessed and the manufacturer's data applied for the selection of circuit breakers. Manufacturer's data should be consulted in particular for tungsten flood lamps, heat lamps, discharge lighting or transformers.

**KEY POINT**

Tungsten lamps exhibit a brief inrush current when initially switched on. This inrush current is normally ignored for multiple incandescent GLS lamps that are not switched in large groups. For tungsten flood lamps and heat lamps, account should be taken of the inrush current.

### Immersion heater circuit

(See row 6 of the table on page 286.)

Consider a single-phase 230 V, 3 kW immersion heater circuit:

$$I = \frac{3 \times 1000}{230 \times 1} = 13.04\text{ A}$$

### Motor circuit

(See row 4 of the table on page 286.)

Consider a single-phase 230 V, 3 kW motor with power factor $\cos\theta = 0.8$:

$$I_b = \frac{\text{kW} \times 1000}{U_o \times \cos\theta} = \frac{3 \times 1000}{230 \times 0.8} = 16.3\text{ A}$$

For three-phase equipment, the design current $I_b$ is given by:

$$I_b = \frac{\text{kVA} \times 1000}{3U_o}$$

Where:

$I_b$ is design current (amperes)
kVA is the rating of the equipment
$U_o$ is the nominal voltage (line to earth) of the equipment.

Or more commonly:

$$I_b = \frac{\text{kVA} \times 1000}{\sqrt{3}U}$$

Where:

$I_b$ is design current (amperes)
kVA is the rating of the equipment
$U$ is the nominal voltage (line to line) of the equipment.

The above calculation is for balanced three-phase loads. The design current determined is per line; in other words, the current that each cable is sized for.

Where three-phase loads are unbalanced, they should be treated as three individual single-phase loads and the circuit cables sized on the largest of the three loads determined.

## Estimating the demand of a distribution circuit supplying a number of final circuits

The current demand of a distribution circuit that supplies a number of final circuits may be assessed by using the allowances for diversity given in the table of allowances for diversity between final circuits for sizing distribution circuits on pages 286–287 (IET On-Site Guide, Table A3); you apply these allowances to the total current demand of all the equipment supplied by that circuit.

## ACTIVITY

Using BS 7671 and the On-Site Guide, fill in the blanks in the table as best you can. Use Table 4D5, etc and Method C.

| $I_b$ load | $I_n$ BS EN 60898 | $I_t$ cable pvc/pvc |
|---|---|---|
| 15 | | |
| 25 | 32 | |
| 3 | 6 | |
| 60 | | |
| 41 | | |

You do not assess the current demand of the distribution circuit by adding the current demands of the individual final circuits (as outlined above). In the table on pages 286–287 (IET On-Site Guide, Table A3), the allowances are expressed either as percentages of the current demand or – where followed by the letters f.l. (full load) – as percentages of the rated full-load current of the current-using equipment.

The current demand for any final circuit that is a conventional circuit arrangement, and which complies with Appendix H of the Institution of Engineering and Technology (IET) On-Site Guide, is the rated current of the overcurrent protective device of that circuit. For example, a 20 A radial circuit that supplies socket outlets would be protected by a 20 A circuit breaker, so the design current would be assessed as being 20 A.

An alternative method of assessing the current demand of a circuit that supplies a number of final circuits is to add the diversified current demands of the individual circuits and then apply a further allowance for diversity, in the form of a factor. In this method, the allowances given in Table A2 of the IET On-Site Guide are not to be used; rather, the values to be chosen are the responsibility of the installation designer. This may be particularly useful when assessing the total demand for the supply to a building that may contain many circuits, but consumption is likely to be low. This method is only recommended for experienced designers.

As the value of supply voltage has an effect on the design current, some situations may require the designer to use actual values of voltage relevant to the installation instead of the standard 230 V.

**Assessment criteria**

**2.4** Assess suitability of types of wiring systems for different environments

## SELECTING AND ERECTING A WIRING SYSTEM

The type of wiring system you choose must be suitable to the environment in which it will be used. This is covered in BS 7671, Chapter 52: Selection and erection of wiring systems.

There is a variety of factors to consider, as described below and in the following pages.

### KEY POINT

**Regulation 132.5 (BS 7671): Environmental conditions**

Regulation 132.5.1 (BS 7671): The design of the electrical installation shall take into account the environmental conditions to which it will be subjected.

Regulation 132.5.2 (BS 7671): Equipment in surroundings susceptible to risk of fire or explosion shall be so constructed or protected, and such other special precautions shall be taken, as to prevent danger.

### Cable insulation

#### Bare cables

Bare cables (that is, cables without **basic insulation**) are only allowed when installed on insulators (see the table on the next page). They must also be placed out of reach, unless they are supplied by a safety source at a voltage not exceeding 25 V a.c. or 60 V d.c. in normal dry conditions (ie indoors away from water, and not near bathrooms, swimming pools, etc).

The requirements of the Electricity Safety, Quality and Continuity (ESQC) Regulations 2002 must also be complied with.

**Basic insulation**

Insulation applied to live parts to provide basic protection. It does not necessarily include insulation used exclusively for functional purposes.

## Non-sheathed cables

Non-sheathed live cables (insulated but unsheathed) must be installed in conduit, trunking or ducting. They may also be installed on insulators.

**KEY POINT**

**Regulation 417.3 (BS 7671): Placing out of reach**

Note: Protection by placing out of reach is intended only to prevent unintentional contact with live parts.

A bare or insulated overhead line for distribution between buildings and structures shall be installed to the standard required by the Electricity Safety, Quality and Continuity (ESQC) Regulations 2002.

**ASSESSMENT GUIDANCE**

Non-sheathed cables (singles code 6491X) are normally installed in conduit but bonding conductors of 4 mm$^2$ or above may be installed without protection.

**Schedule of installation methods in relation to conductors and cables installation method**

| Conductors and cables | Without fixings | Clipped direct | Conduit systems | Cable trunking systems* | Cable ducting systems | Cable ladder, cable tray, cable brackets | On insulators | Support wire |
|---|---|---|---|---|---|---|---|---|
| Bare conductors | NP | NP | NP | NP | NP | NP | P | NP |
| Non-sheathed cables | NP | NP | P[1] | P[1,2] | P[1] | NP[1] | P | NP |
| Multicore non-sheathed cable[3] | P | P | P | P | P | P | N/A | P |
| Single-core sheathed cables[3] | N/A | P | P | P | P | P | N/A | P |

Notes:

P = Permitted

NP = Not permitted

N/A = Not applicable, or not normally used in practice

* including skirting trunking and flush floor trunking

1 = Non-sheathed cables that are used a protective conductors or protective bonding conductors need not be laid in conduits or ducts

2 = Non-sheathed cables are acceptable if the trunking system provides at least the degree of protection IPXXD or IP4X and if the cover can only be removed by means of a tool or a deliberate action

3 = Sheathed cables including mineral insulated and armored cables

## Type of wiring and method of installation

Regulation 132.7 of BS 7671 requires the following to be considered when selecting the wiring system and method of installation:

- the nature of the location including the structure that supports the wiring (large cables can pull walls down)
- vulnerability to damage by persons, vermin, livestock (vermin will eat insulation; people and livestock may cause mechanically damage to it)
- voltage
- electromagnetic stresses (which can force parallel cables apart under fault conditions)
- electromagnetic interference (caused by or from the installation)
- all external influences.

## ACTIVITY

Look up Section 522 of BS 7671 and research the factors that need to be considered when selecting a wiring system (main regulation headings).

### KEY POINT

**BS 7671 numbering system**

In the numbering system used, the first digit signifies a Part, the second digit a Chapter, the third digit a Section and the subsequent digits the Regulation number. For example, the Section number 413 is made up as follows:

- Part 4 – Protection for safety
- Chapter 41 (first chapter of Part 4) – Protection against electric shock
- Section 413 (third section of Chapter 41) – Protective measure: electrical separation.

## External influences

The wiring system must be able to withstand all the external influences it will be subjected to, including those listed in the table below. Particular care must be taken when a system enters a building or changes direction.

**External influences on a wiring system**

| External influence | Example |
|---|---|
| Ambient temperature | The general constant temperature of the environment |
| External heat sources | Proximity to hot-water pipes, heaters |
| Water or high humidity | Immersed cables in a marina, condensation in an unheated building |
| Solid foreign bodies | Dust |
| Corrosive or polluting substances | Substances found in a tannery, battery room, plating plant, cowshed |
| Impact | Vehicles in a garage or car park |
| Vibration | Connections to motors |
| Other mechanical stresses | Bends too tight to pull in cables, insufficient supports |
| Presence of flora or mould | Plants or moss |
| Presence of fauna | Vermin, livestock |
| Seismic effects | In locations susceptible to earthquakes |
| Movement of air | Sufficient to stress mountings |
| Nature of processed or stored materials | Risk of fire or degradation of insulation, for example, from oil spills, acid in plating shops |
| Building design | From building movement |

## Which wiring system to use?

General uses of the more common wiring systems are described in the table opposite.

Common wiring systems

| Wiring system | Requirements | Common use |
|---|---|---|
| Clipped direct | Sheathed without armour | General application in domestic and commercial installations. Not usually acceptable to the client in new installations. Mechanical protection may be required in some locations by the installation of guards or the use of armoured cable |
| Clipped direct | Sheathed with armour | Commonly adopted in industrial premises where presence of cables is acceptable |
| Installed in the building structure | If using insulated and sheathed cables (that is, not armoured or enclosed in conduit, etc), precautions must be taken to prevent damage | Insulated and sheathed cables installed in metal-framed walls will probably require additional protection |
| Buried cables | Buried cables must be enclosed in a conduit or duct or be armoured. A sufficient depth is required to prevent damage by reasonably foreseeable disturbance to the ground | This is the most practical solution for large-sized cables for power distribution in a supply company network or between buildings |
| Plastic conduit systems | A very wide range of plastic conduits is available. Suitable for almost every location | Manufacturer's instructions must be read to ensure the conduit is suitable for the particular environment. Expansion needs to be allowed for |
| Steel conduit systems | Common types are black enamel and galvanised. Stainless steel conduits may be specified for onerous locations | Galvanised is suitable for onerous environments and may be selected where the good mechanical properties are necessary |
| Cable trunking systems* | Insulted cables may be used (without sheath) | Used where a number of small cables need to be run |
| Cable ducting systems | The most practical system for distributing large cables through a building or site | Normally formed by concrete for large cables or circular pipe for smaller cables |
| Cable ladder, cable tray, cable brackets | Cables need further mechanical protection by a sheath as this is not a containment system | Practical solution for distributing large cables around a building or site |
| Mineral-insulated | Particular skills are required to install these cables | They have exceptional fire-resistant properties and so are used for circuits requiring high integrity under fire conditions |
| On insulators | Widely adopted for electricity supply distribution in rural areas | Cheaper than laying cables in the ground, but may be considered unsightly and are susceptible to damage by high vehicles. Care necessary in some locations, eg caravan sites, riverbanks, farms, etc |
| Support wire | Low cost and may avoid considerable disturbance | Check that the height is sufficient for all vehicle movements, etc |

## Application of cable types

The table below provides guidance on the use of cables for particular cable systems.

**Cable use for particular cable systems**

| Type of cable | Uses | Comments |
|---|---|---|
| Thermoplastic, thermosetting or rubber-insulated non-sheathed | In conduit, cable ducting or trunking | ▪ Intermediate support may be required on long, vertical runs<br>▪ 70 °C maximum conductor temperature for normal wiring grades including thermosetting types<br>▪ Cables run in PVC conduit shall not operate with a conductor temperature greater than 70 °C |
| Flat thermoplastic or thermosetting insulated and sheathed | ▪ General indoor use in dry or damp locations; may be embedded in plaster<br>▪ On exterior surface walls, boundary walls, etc<br>▪ Overhead wiring between buildings<br>▪ Underground in conduits or pipes<br>▪ In building voids or ducts formed in-situ | ▪ Additional protection may be necessary where exposed to mechanical stresses<br>▪ Protection from direct sunlight may be necessary<br>▪ Black sheath colour is better for cables in sunlight<br>▪ Unsuitable for embedding directly in concrete<br>▪ May need to use hard-drawn (HD) copper conductors for overhead wiring |
| Split-concentric thermoplastic insulated and sheathed | General | ▪ Additional protection may be necessary where exposed to mechanical stresses<br>▪ Protection from direct sunlight may be necessary<br>▪ Black sheath colour is better for cables in sunlight |
| Mineral-insulated | General | With overall PVC covering where exposed to the weather or risk of corrosion, or where installed underground or in concrete ducts |
| Thermoplastic or XLPE-insulated, armoured, thermoplastic sheathed | General | ▪ Additional protection may be necessary where exposed to mechanical stresses<br>▪ Protection from direct sunlight may be necessary<br>▪ Black sheath colour is better for cables in sunlight |

Notes:

1 The use of cable covers or equivalent mechanical protection is desirable for all underground cables that might otherwise subsequently be disturbed. Route-marker tape should also be installed, buried just below ground level.

2 Cables with thermoplastic insulation or sheath should preferably be installed only when the ambient temperature is above 0 °C and has been for the preceding 24 hours. Where they are to be installed during a period of low temperature, precautions should be taken to avoid risk of mechanical damage during handling. A minimum ambient temperature of 5 °C is advised for some types of thermoplastic insulated and sheathed cables. Manufacturer's information must be followed.

3 Cables should be suitable for the maximum ambient temperature and should be protected.

4 Thermosetting cable types (to BS 7211 or BS 5467) can operate with a conductor temperature of 90 °C. This must be limited to 70 °C when drawn into a conduit, etc, with thermoplastic insulated conductors or connected to electrical equipment (Regulation 512.1.5 and Table 52.1), or when such cables are installed in plastic conduit or trunking.

5 For cables to BS 6004, BS 6007, BS 7211, BS 6346, BS 5467 and BS 6724, further guidance may be obtained from those standards. Additional advice is given in BS 7540-2:2005 Guide to use for cables with a rated voltage not exceeding 450/750 V for cables to BS 6004, BS 6007 and BS 7211.

6 Cables for overhead wiring between buildings must be able to support their self-weight and any imposed wind or ice/snow loading. A catenary support is usual but hard drawn copper types may be used.

## CABLE SELECTION (FOR CURRENT-CARRYING CAPACITY)

### Symbols

The symbols used in this section are those used in BS 7671.

- $I_z$ is the current-carrying capacity of a cable for continuous service, under the particular installation conditions concerned.
- $I_t$ is the value of current tabulated in Appendix 4 of BS 7671 for the type of cable and installation method concerned, for a single circuit in an ambient temperature of 30 °C.
- $I_b$ is the design current of the circuit, ie the current intended to be carried by the circuit in normal service.
- $I_n$ is the nominal current or current setting of the device protecting the circuit against overcurrent.
- $I_2$ is the operating current (ie the fusing current or tripping current for the conventional operating time) of the device protecting the circuit against overload.
- $C$ is a range of rating factors to be applied where the installation conditions differ from those for which values of current-carrying capacity are tabulated (in Appendix 4 of BS 7671). The various rating factors are identified as:
  - $C_a$ for ambient temperature
  - $C_g$ for grouping
  - $C_i$ for thermal insulation
  - $C_t$ for operating temperature of conductor.

**Assessment criteria**

**2.5** Determine minimum current-carrying capacity of live conductors for given installation conditions

**ASSESSMENT GUIDANCE**

It should be obvious that a cable has to carry the current of the load but this is actually determined by the protective device rating, which in turn is governed by the load current.

**KEY POINT**

All these correction factors reduce the current rating of a cable or, to put it another way, increase the csa or current rating required.

## Overcurrent protection

Each installation and every circuit within an installation must be protected against overcurrent by devices that will operate automatically to prevent injury to persons and livestock and damage to the installation, including the cables. The overcurrent devices must be of adequate **breaking capacity** and be so constructed that they will interrupt the supply without danger. Cables must be able to carry these overcurrents without damage. Overcurrents may be fault currents or overload currents.

**KEY POINT**

Chapter 43 of BS 7671 provides the requirements for overcurrent protection.

**Breaking capacity**

This is the maximum value of current that a protective device can safely interrupt.

### Fault currents

Fault currents arise as a result of a fault in the cables or the equipment. There is a sudden increase in current, perhaps 10 or 20 times the cable rating, the current being limited by the **impedance** of the supply, the impedance of the cables, the impedance of the fault and the impedance of the return path. The current should be of short duration, as the overcurrent device should operate.

**KEY POINT**

Section 434 of BS 7671 provides the requirements for protection against fault currents.

**Impedance**

Impedance (symbol $Z$) is a measure of the opposition that a piece of electrical equipment or cable makes to the flow of electric current when a voltage is applied. Impedance is a term used for alternating current circuits. The elements of impedance are:

- resistance
- inductance
- capacitance.

(The effects of impedance on current such as leading and lagging are covered in Unit 302, see page 154.)

### Overload currents

Overload currents do not arise as a result of a fault in the cable or equipment. They arise because the current has been increased by the addition of further load. Overload protection is only required if overloading is possible. It would not be required for a circuit supplying a fixed load, although fault protection would be required.

The load on a circuit supplying, for example, a 7.2 kW shower will not increase unless the shower is replaced, when the adequacy of the circuit must be checked against the new load criteria.

A distribution circuit supplying a number of buildings could be overloaded by additional machinery being installed in one of the buildings supplied.

Overload currents are likely to be of the order of 1.5 to 2 times the rating of the cable, whereas fault currents may be of the order of 10 to 20 times the rating.

Overloads of less than 1.2 to 1.6 times the device rating are unlikely to result in operation of the device. Regulation 433.1 of BS 7671 requires that every circuit be designed so that small overloads of long duration are unlikely to occur.

It is usual for one device in the circuit to provide both fault protection and overload protection. A common exception is the overcurrent devices to motor circuits, where the overcurrent device at the origin of the circuit provides protection against fault currents and the motor starter will be providing protection against overload.

**KEY POINT**

Section 433 of BS 7671 provides the requirements for protection against overload currents.

## Design current $I_b$ and device rating $I_n$

Once the design current has been determined (see pages 284–285), the device rating can be selected. The device rating (or current setting) In must be greater than or equal to the design current $I_b$: $I_n \geq I_b$.

For example, if the design current $I_b$ is 29 A and a type B device is to be selected, $I_n$ must be 32A.

As protective device nominal ratings ($I_n$) are the values that the device is rated for in normal service, then $I_b$ can equal $I_n$.

Types of protective device are chosen, based on the type of load. Below is a guide to the type of protective device suitable for particular loads.

| Type of load | Suitable type of device |
|---|---|
| Resistive, such as heating elements, incandescent lighting, etc | Type B circuit breakers<br>gG BS 88 devices |
| Inductive, such as discharge lighting, small motors or ELV lighting transformers | Type C circuit breakers<br>gM BS 88 devices |
| High inductive or surging, such as welding equipment, X-ray machines, large motors without soft starting | Type D circuit breakers |

Protective devices are also selected for suitability against prospective fault current (PFC).

## Protection against overload and short circuit

For circuits where overload and short circuit are likely, as described above, the following method is used to select the correct cross-sectional area (csa) of conductor.

When the device rating is known, the tabulated cable rating can be calculated. Where the overcurrent device (fuse or circuit breaker) provides protection against overload and short circuit, as is usual, then the tabulated current rating $I_t$:

$$I_t \geqslant \frac{I_n}{C_g C_a C_s C_d C_i C_f C_c}$$

**KEY POINT**

Regulation 432.1 of BS 7671 gives a requirement for protection against overload current and fault current.

where:

$I_t$ is the tabulated current-carrying capacity of a cable, found in Appendix 6 of the IET On-Site Guide and Appendix 4 of BS 7671
$I_n$ is the rated current or setting of a protective device
$C_g$ is the rating factor for the grouping (see Table 4C1 of BS 7671 or F3 of the IET On-Site Guide)
$C_a$ is the rating factor for the ambient temperature (see Table 4B of BS 7671 or $F$1 of the IET On-Site Guide)
$C_s$ is the rating factor for the thermal resistivity of the soil
$C_d$ is the rating factor for the depth of burial of a cable in the soil
$C_i$ is the rating factor for conductors surrounded by thermal insulation
$C_f$ is a rating factor applied when overload protection is being provided by overcurrent devices with fusing factors greater than 1.45 (eg $C_f$ = 0.725 for semi-enclosed fuses to BS 3036)
$C_c$ is the rating factor for circuits buried in the ground.

For cables not laid in the ground this simplifies to:

$$I_t \geqslant \frac{I}{C_g C_a C_i C_f}$$

## ASSESSMENT GUIDANCE

A short circuit is a fault of negligible impedance between live conductors of the same circuit.

### KEY POINT

Remember to use brackets on your calculator when you input the bottom line. Also, any value multiplied or divided by 1 remains unchanged, so the 1s are not actually required. For example, if brackets are not used on the bottom line of a calculation, the formula:

$\frac{20}{0.89 \times 0.9}$ becomes:

$\frac{20}{0.89} \times 0.9 = 20.22$ and is incorrect.

$\frac{20}{(0.89 \times 0.9)} = 24.96$, which is the correct answer

## ACTIVITY

Determine the csa for a flat 70 °C thermoplastic cable supplying a 7 kW 230 V cooker circuit where the cooker control incorporates a socket outlet. The single circuit is to be wired clipped directly to a surface over a distance of 15 m.

Ambient temperature is 35 °C. Protection is by a circuit breaker to BS EN 60898.

**Example:**

A circuit is to be installed in a surface-mounted trunking, which contains four other circuits. The circuit is to be wired using single-core 70 °C thermoplastic-insulated cable with copper conductors. The design current for the single-phase circuit ($I_b$) is 17 A, and the ambient temperature is 25 °C. Protection against overload and short circuit is by a Type B circuit breaker to BS EN 60898.

As $I_n \geqslant I_b$ then a 20 A device is selected.

Using the information contained in Appendix 4 of BS 7671, the method of installation is B from Table 4A2.

$C_a$= 1.03 from Table 4B1 for 25 °C and thermoplastic cable
$C_g$= 0.6 from Table 4C1, given the number of circuits is 5 in total, installed as method B
$C_i$= 1 as no thermal insulation exists
$C_f$= 1 as the device is *not* to BS 3036.

$$I_t \geqslant \frac{I}{C_g C_a C_i C_f}$$

Therefore:

$$I_t \geqslant \frac{20}{1.03 \times 0.6 \times 1 \times 1} = 32.36 \text{ A}$$

So, using the correct cable selection table for single-core 70 °C thermoplastic cable from Appendix 4 of BS 7671, we can select a conductor cross-sectional area (csa). Using Table 4D1A and, in particular, column 4, as the circuit is single-phase and the installation method is B, the value of $I_t$ suitable for the current calculated is 41 A, which references a 6 mm² conductor size.

## Protection against short circuit only

Where the overcurrent device (fuse or circuit breaker) provides protection against short circuit only as the load is fixed, the following formula is used:

$$I_t \geqslant \frac{I_b}{C_g C_a C_s C_d C_i C_f C_c}$$

where:

$I_t$ is the tabulated current-carrying capacity of a cable, found in Appendix 6 of the IET On-Site Guide and Appendix 4 of BS 7671

$I_b$ is the design current of the circuit

$C_g$ is the rating factor for grouping (see Table 4C1 of BS 7671 or F3 of the IET On-Site Guide)

$C_a$ is the rating factor for ambient temperature (see Table 4B of BS 7671 or F1 of the IET On-Site Guide)

$C_s$ is the rating factor for the thermal resistivity of the soil

$C_d$ is the rating factor for the depth of burial of a cable in the soil

$C_i$ is the rating factor for conductors surrounded by thermal insulation

$C_f$ is a rating factor applied when overload protection is being provided by overcurrent devices with fusing factors greater than 1.45 (eg $C_f$ = 0.725 for semi-enclosed fuses to BS 3036)

$C_c$ is the rating factor for circuits buried in the ground.

$$I_t \geqslant \frac{I_b}{C_g C_a C_i C_f}$$

**Example:**

A circuit is to be installed in a surface-mounted trunking, which contains four other circuits. The circuit is to be wired using single-core 70 °C thermoplastic-insulated cable with copper conductors. The design current for the single-phase circuit ($I_b$) is 17 A and ambient temperature is 25 °C. Protection against short circuit only is by a type B circuit breaker to BS EN 60898.

As $I_n \geqslant I_b$ then a 20 A device is selected.

Using the information contained in Appendix 4 of BS 7671, the method of installation is B from Table 4A2.

$C_a$= 1.03 from Table 4B1 for 25 °C and thermoplastic cable

$C_g$= 0.6 from Table 4C1 given the number of circuits is 5 in total installed as method B

$C_i$= 1 as no thermal insulation exists

$C_f$= 1 as the device is *not* to BS 3036.

$$I_t \geqslant \frac{I_n}{C_g C_a C_i C_f}$$

Therefore:

$$I_t \geqslant \frac{17}{1.03 \times 0.6 \times 1 \times 1} = 27.5 \text{ A}$$

**KEY POINT**

There are many circumstances where the overcurrent device provides fault protection but not overload protection. The most common is when the load is fixed or when a motor starter provides overload protection. In these circumstances the cable rating must exceed the load but not necessarily the overcurrent device rating.

So, using the correct cable selection table for single-core 70 °C thermoplastic cable in Appendix 4 of BS 7671, we can select a conductor cross-sectional area (csa). Using Table 4D1A, and in particular column 4 as the circuit is single-phase and the installation method is B, the value of $I_t$ suitable for the current calculated is 32 A, which references a 4 mm² conductor size.

The equation for protection against short circuit only (when overload protection is not required) is appropriate for motor circuits where the motor starter provides overload protection and for circuits supplying fixed loads. BS 7671 allows its use for any fixed loads, for example, water heaters. However, it is usual to provide overload protection unless it is impracticable, as for motor circuits (see BS 7671, Regulation 552.1.2). The omission of overload protection is also allowed where unexpected disconnection would cause danger, as stated in BS 7671, Regulation 433.3.3.

If overload protection is not provided then the **adiabatic** check (see page 327) of BS 7671, Regulation 434.5.2 is carried out to ensure the cable is suitably protected.

**Adiabatic**

The adiabatic equation can be used in several ways. It can be used to determine the time taken for a given cable to exceed its final limiting temperature. Alternatively, it can be used to determine the minimum csa to be able to withstand a fault current for a given duration without exceeding the final limiting temperature.

## ACTIVITY

Look up the equation of Regulation 434.5.2 and Table 43.1 in BS 7671:

$$t = \frac{k^2 S^2}{I^2}$$

where :

$S$ is the nominal cross-sectional area of conductor in mm²

$I$ is the value of fault current in amperes

$t$ is the time in seconds taken for the cable to exceed its final limiting temperature

$k$ is a factor taking account of the resistivity, temperature coefficient and heat capacity of the conductor material.

This equation is often called the adiabatic equation; see the note to BS 7671, Regulation 543.1.3.

Using the above formula calculate the value of $t$ if:

$S$ = 10 mm²

$I$ = 2000 A

$K$ = 115.

The value $I$ can be taken as the value of fault current that causes rapid disconnection of a device; it can be found in Appendix 3 of BS 7671.

Using the example on pages 299–300, the required information is:

$k$ = 115 from Table 43.1 of BS 7671
$S$ = 4 mm$^2$
$I$ = 100 A for a 20 A type B circuit breaker (as Appendix 3 of BS 7671) for a disconnection time of 0.1 seconds

Therefore:

$$t = \frac{115^2 \times 4^2}{100^2} = 21.16 \text{ seconds}$$

As the device will disconnect within 0.1 seconds, and the cable can safely carry 100 A fault current for 21.16 seconds before its final limiting temperature is reached, the circuit is acceptable.

However, the equation can be rearranged to demonstrate its objective more clearly, as follows.

$$I^2t = k^2S^2$$

$I^2t$ is proportional to the thermal energy let through by the protective device under fault conditions, and $k^2S^2$ is the thermal capacity.

## Current-carrying capacity tables

### Tabulated current-carrying capacity $I_t$

Appendix 4 of BS 7671 provides current-carrying capacities of cables in certain defined conditions. Each table specifies:

- the cable type
- the ambient temperature
- the conductor operating temperature
- the reference method of installation.

The correct table for any particular cable is found by reference to Table 4A3 of BS 7671.

The tabulated cable current rating $I_t$ is the current that will increase the temperature of the conductor of the cable from the tabulated ambient temperature, usually 30 °C, to the tabulated maximum conductor operating temperature (for example, 70 °C for thermoplastic cables and 90 °C for thermosetting) under the defined conditions.

For example, for a 4 mm$^2$ single-core 70 °C thermosetting cable, at an ambient temperature of 30 °C, enclosed in conduit in an insulating wall, we can see from Table 4D1A of BS 7671 that columns 2 and 3 are for method A (enclosed in a thermally insulated wall).

**ASSESSMENT GUIDANCE**

A BS 3036 fuse is generally given a fusing factor of 2. This means it needs twice the current rating:

$$\frac{1.45}{2} = 0.725$$

Therefore, $I_t = 26$ A for two cables single-phase a.c. or d.c. and 24 A for three or four cables three-phase a.c. or d.c.

Each table has a corresponding voltage drop table, with the exception of Table 4D5 of BS 7671, where the voltage drop is incorporated in the current-carrying capacity table.

## Rating factors

If the actual conditions of installation do not meet the reference conditions, then rating factors are applied to the tabulated rating $I_t$ as multipliers to obtain the cable's current rating ($I_z$) in the actual installation conditions.

$$I_z = I_t C_a C_g C_i C_f$$

where:

$I_z$ is the current-carrying capacity of a cable for continuous service under the particular installation conditions concerned

$I_t$ is the tabulated current-carrying capacity of a cable found in Appendix F of the IET On-Site Guide and Appendix 4 of BS 7671

$C_a$ is the rating factor for the ambient temperature (see Table 4B of BS 7671 or Table F1 of the IET On-Site Guide)

$C_g$ is the rating factor for grouping (see Table 4C of BS 7671 or Table F3 of the IET On-Site Guide)

$C_i$ is the rating factor for conductors surrounded by thermal insulation

**KEY POINT**

Note: $C_f$ is a rating factor applied when overload protection is being provided by overcurrent devices with fusing factors greater than 1.45, eg $C_f = 0.725$ for semi-enclosed fuses to BS 3036.

**KEY POINT**

Where a formula has no symbols showing, it means the values should be multiplied.

As long as the calculated value of $I_z$ exceeds the value of $I_n$ then the circuit is acceptable.

As we have seen before, $I_t$ is the current-carrying capacity for specific cables, as given in BS 7671, Table 4D1A onwards. This method may be used where voltage drop is a greater consideration. For cables selected by voltage drop first, this formula will be used to verify that the selected cables are suitable for current capacity. This method requires practice and experience but can prove to be more efficient, with fewer calculations needed.

**ACTIVITY**

State why aluminium conductors are not used for sizes below 16 mm$^2$.

## Ambient temperature rating factor $C_a$

$C_a$ from 4B1 of BS 7671 is based on ambient air temperatures around the wiring system. The worst-case temperature must be used. For example, if the cable passes through an airing cupboard in a house, the ambient temperature is likely to be 35 °C and that must be applied. The factor selected is also based on the type of cable insulation selected.

Ambient ground temperatures for cables buried in the ground are found in Table 4B2 of BS 7671.

## Grouping rating factor $C_g$

Where more than one circuit is contained in a wiring system, or where cables on a tray or wall are close enough together to be classed as touching, rating factors for grouping must be applied. In order to reduce complicated calculations, circuits of similar size and type should be grouped together and kept to a minimum. If cables of different type and size are enclosed together, further calculations, based on single or poly-phase circuits and additional factors, need to be applied. This is beyond the scope of this course.

Rating factors for grouping are based on the *total* number of circuits as well as the method of installation. Where cables are, for example, clipped to a wall, they are classed as spaced if the distance between them is more than twice the diameter of the cable. If not, they must be classed as grouped.

## Thermal insulation factor $C_i$

$C_i$ is a factor for thermal insulation.

For a single cable surrounded by thermally insulating material over a length of more than 0.5 m, the current-carrying capacity should be taken, in the absence of more precise information, as 0.5 times the current-carrying capacity for that cable clipped direct to a surface and open (Reference Method C). This means that cables in thermal insulation for this length will only be able to carry half the current of one clipped to a wall, so a cable will theoretically need to be twice as big.

Where a cable is to be totally surrounded by thermal insulation for less than 0.5 m, the current-carrying capacity of the cable should be reduced appropriately, depending on the size of cable, length in insulation and thermal properties of the insulation. The rating factors in Table 52.2 of BS 7671 are appropriate to conductor sizes up to 10 $mm^2$ installed and surrounded by thermal insulation having a thermal conductivity ($\lambda$) greater than 0.0625 $Wm^{-1}K^{-1}$. (Manufacturer's data relating to the thermal insulation needs to be checked.)

**KEY POINT**

Cables having cross-sectional area exceeding 10 $mm^2$ are large enough to conduct heat to a cooler environment, where the distance covered by thermal insulation is no greater than 0.5 m.

**Cable surrounded by thermal insulation**

| Length in insulation (mm) | Derating factor |
|---|---|
| 50 | 0.89 |
| 100 | 0.81 |
| 200 | 0.68 |
| 400 | 0.55 |

**Extract from Table 52.2 of BS 7671:2008 Requirements for Electrical Installations (the IET Wiring Regulations 17th edition) (2011)**

If a cable is installed in a thermally insulated wall or ceiling, in line with one of the recognised methods indicated in the cable selection tables, the rating factors for thermal insulation, as shown on the previous page, do not need to be applied. For example:

- Method 100: flat profile cable above a ceiling covered on one side only by thermal insulation not exceeding 100 mm in depth
- Method 101: flat profile cable above a ceiling covered on one side only by thermal insulation exceeding 100 mm in depth
- Method 102: flat profile cable in a thermally insulated stud wall with one side touching the inner wall surface
- Method 103: cable in a thermally insulated stud wall not touching the inner wall surface
- Method A: cable in conduit in an insulated wall.

**ASSESSMENT GUIDANCE**

Method 103 covers cable in a studding wall surrounded by insulation (such as fibre glass or rockwall, etc) not touching the inner wall surface. Studding walls are normally constructed of timber or metal with plasterboard covering.

The effect of the thermal insulation, as detailed above, is taken into consideration by the cable selection tables.

In situations not given above, it is often better to avoid thermal insulation or, if it is unavoidable, to sleeve the cable as it passes through any insulation, thus providing an air gap. This is only possible where protection against the spread of fire is not required.

**Example:**

Confirm the suitability for a 6 mm$^2$ 70 °C thermoplastic, flat profile twin and cpc cable to be installed, enclosed in conduit within a thermally insulated wall. The ambient temperature is 25 °C and the conduit contains one other circuit. The circuit is protected by a 20 A type B circuit breaker to BS EN 60898.

Using the information contained in Appendix 4 of BS 7671, the method of installation is A from Table 4A2.

$C_a$ = 1.03 from Table 4B1 for 25 °C and thermoplastic cable

$C_g$ = 0.8 from Table 4C1, given the number of circuits is 2 in total installed as Method A

$C_i$ = 1 as no further thermal insulation rating factor is required

$C_f$ = 1 as the device is *not* to BS 3036

$I_t$ = 32 A from Table 4D5 for flat profile cable

As we are confirming the cables suitability, we could use:

$$I_z = I_t C_a C_g C_i C_f$$

Therefore:

$$I_z = 32 \times 1.03 \times 0.8 \times 1 \times 1 = 26.4 \text{ A}$$

As the rating of the protective device ($I_n$) is 20 A and the resulting $I_z$ is greater than this, the cable is suitably protected.

## Overcurrent device fuse rating factor $C_f$

Points i) and ii) of Regulation 433.1.1 in BS 7671 require that, where overload protection is necessary for any circuit:

$$I_n \geqslant I_b \qquad\qquad I_z \geqslant I_n$$

This is something we have already seen. Point iii) of Regulation 433.1.1 also states:

$$I_2 \leqslant 1.45 \times I_z$$

This means that the cable rating after factors have been applied, ($I_z$), must be capable of taking 45% more current, thereby ensuring effective disconnection of the protective device under overload ($I_2$) to guarantee the cable is suitable for any overload.

Equally, by transposing the formula, we can determine that the cable should be de-rated by:

$$\frac{1.45}{I_2}$$

Regulation 433.1.100 of BS 7671 recognises that the common protective devices used in installations will satisfy this requirement, with the exception of the semi-enclosed (re-wireable) BS 3036 fuses. BS 3036 fuses will not effectively disconnect under overload until a current ($I_2$) of approximately twice (× 2) the rating of the device is reached. This means that where a BS 3036 device is used, a further rating factor ($C_f$) is applied; this factor is:

$$\frac{1.45}{2} = 0.725$$

This means that, as a minimum, the cable must be suitably rated to withstand overloads of:

$$\frac{I_n}{0.725}$$

**ACTIVITY**

Look at the time–current graphs in Appendix 3 of BS 7671. Compare the minimum values of current where the different devices disconnect. As an example, from the graphs, a 5 A BS 88-3 device will disconnect with a minimum current of 8 A, whereas a 5 A BS 3036 fuse will disconnect at 9.5 A.

## Cable selection for voltage drop

As well as selecting a cable for current-carrying capacity, a cable cross-sectional area (csa) must also be suitably sized to minimise voltage drop.

### Voltage drop limit

BS 7671 requires that, under normal service conditions, the voltage at the terminals of any fixed current-using equipment shall be greater than that of the product standard.

Furthermore, in the absence of any product standard, the voltage drop should not exceed that for the proper working of the equipment.

As a result of this, the following guidance is given in Appendix 4 of BS 7671:

**Voltage drop in consumers' installations**

The voltage drop between the origin of an installation and any load point should not be greater than the values in the table below expressed with respect to the value of the nominal voltage of the installation.

| Type of installation | Lighting | Other uses |
|---|---|---|
| Low-voltage installations supplied directly from a public low-voltage distribution system | 3% | 5% |
| Low-voltage installation supplied from private LV supply | 6% | 8% |

**Extract from Table 4Ab in Appendix 4 of BS 7671:2008 Requirements for Electrical Installations (the IET Wiring Regulations 17th edition) (2011)**

Although experienced designers may make different tolerances to the above values, compliance with the table above will satisfy most installations.

## Basic voltage-drop calculation

### Single-phase circuits

In order to calculate the voltage drop in a single-phase circuit, the following information is required regarding the circuit:

- the value of milli-volts per ampere per metre (mV/A/m) for the cable selected for current-carrying capacity (found in the voltage-drop table associated with the cable selection tables in Appendix 4 of BS 7671)
- the design current for the circuit ($I_b$) in amperes
- the total circuit length in metres.

Then we apply the equation:

$$\text{voltage drop (V)} = \frac{\text{mV/A/m} \times I_b \times L}{1000}$$

This will determine the voltage drop at the maximum operating temperature of the circuit.

To verify the voltage drop determined, the figure must be compared to the maximum permitted for the supply voltage (assumed to be 230 V) depending on the type of load.

For lighting circuits, the maximum would be:

$$230 \times \frac{3}{100} = 6.9\ \text{V}$$

For power circuits, the maximum would be:

$$230 \times \frac{5}{100} = 11.5\ \text{V}$$

**Example:**

A single-phase cooker is wired using 70 °C thermoplastic 6 mm$^2$ single-core cable in surface-mounted conduit (Method B) to a length of 10 m. The design current ($I_b$) is 29 A. Determine the suitability of the cable for voltage drop.

From Table 4D1B (voltage-drop values) in Appendix 4 of BS 7671, the value of mV/A/m for a 6 mm$^2$ cable as column 3 (single-phase, method B) is 7.3 mV/A/m. Therefore:

$$\text{voltage drop} = \frac{7.3 \times 29 \times 10}{1000} = 2.1\ \text{V}$$

As the maximum permitted for a power circuit is 11.5 V, this is acceptable.

### Three-phase voltage drop

The voltage-drop tables in Appendix 4 of BS 7671 also include values of voltage drop in mV/A/m for three-phase circuits. These values assume that the loads are balanced, resulting in no neutral current. The permitted value of voltage drop now relates to the line-to-line voltage of 400 V ($U$) instead of the line-to-neutral value.

## ACTIVITY

The supply voltage, nominally 230 V, will be divided between the load and the (L & N) conductors. The greater the voltage drop in the cables, the less voltage there is for the load. Less voltage; less current; less power. A pure resistive load takes a current of 20 A at 230 V (4600 W). What would be the power output at 220 V assuming the load resistance remains the same?

## ASSESSMENT GUIDANCE

Three-phase loads connected to a three-line supply, such as motors and water-heating equipment, would normally be balanced. Groups of single-phase loads connected to a three-line supply should be balanced across the three supply cables as evenly as possible.

Voltage drop (per ampere per metre): Conductor operating temperature: 70° C

| Conductor cross-sectional area | Two-core cable d.c. | Two-core cable, single-phase a.c. | | | Three- or four-core cable, three-phase a.c. | | |
|---|---|---|---|---|---|---|---|
| 1 | 2 | 3 | | | 4 | | |
| (mm²) | (mV/A/m) | (mV/A/m) | | | (mV/A/m) | | |
| 1.5 | 29 | 29 | | | 25 | | |
| 2.5 | 18 | 18 | | | 15 | | |
| 4 | 11 | 11 | | | 9.5 | | |
| 6 | 7.3 | 7.3 | | | 6.4 | | |
| 10 | 4.4 | 4.4 | | | 3.8 | | |
| 16 | 2.8 | 2.8 | | | 2.4 | | |
| | | r | x | z | r | x | z |
| 25 | 1.75 | 1.75 | 0.170 | 1.75 | 1.50 | 0.145 | 1.50 |
| 35 | 1.25 | 1.25 | 0.165 | 1.25 | 1.10 | 0.145 | 1.10 |
| 50 | 0.93 | 0.93 | 0.165 | 0.94 | 0.80 | 0.140 | 0.81 |
| 70 | 0.63 | 0.63 | 0.160 | 0.65 | 0.55 | 0.140 | 0.57 |
| 95 | 0.46 | 0.47 | 0.155 | 0.50 | 0.41 | 0.135 | 0.43 |

**Extract from Table 4D4B of BS 7671:2008 Requirements for Electrical Installations (the IET Wiring Regulations 17th edition) (2011)**

**Example:**

Verify the voltage drop for a three-phase 400 V motor with a design current ($I_b$) of 17.2 A. The circuit is wired using 4 mm² 70 °C thermoplastic insulated steel-wire armoured cable to a length of 24 m clipped directly to a wall.

Using Table 4D1B (voltage-drop values) in Appendix 4 of BS 7671, for a 4 mm² steel-wire armoured cable, the value of voltage drop in mV/A/m, from column 4 (three or four-core cable, three-phase), is 9.5 mV/A/m. Therefore:

$$\frac{9.5 \times 17.2 \times 24}{1000} = 3.92\text{ V}$$

The maximum permitted value of voltage drop permitted is:

$$400 \times \frac{5}{100} = 20\text{ V}$$

Therefore the value of 3.92 is well within the permitted value.

### Determining cable size by voltage drop

In some situations, voltage drop rather than current-carrying capacity ultimately affects the size of a conductor. This is common for circuits having long lengths but relatively small design currents. In this situation, the minimum value of mV/A/m can be determined by:

$$\text{max mV/A/m} = \frac{\text{max permitted voltage drop} \times 1000}{I_b \times L}$$

The value of mV/A/m is compared to the voltage-drop tables in Appendix 4 of BS 7671, in order to select a cable with a value lower than that determined. Following this, the cable may be proved for adequate capacity by selecting the particular cable-current capacity ($I_t$) and applying:

$$I_z = I_t C_a C_g C_i C_f$$

As long as the calculated value of $I_z$ exceeds $I_n$, the circuit is satisfactory for both current capacity and voltage drop.

**KEY POINT**

The following information regarding load power factor is beyond the requirements of this course but is included here to provide further explanation.

### Correction for load power factor

For cables with conductors of cross-sectional area (csa) of 16 mm$^2$ or less, their inductances are not significant and only $(\text{mV/A/m})_r$ values are tabulated.

For cables with conductors of cross-sectional area greater than 16 mm$^2$, the impedance values are given as $(\text{mV/A/m})_z$ together with the resistive component $(\text{mV/A/m})_r$ and the reactive component $(\text{mV/A/m})_x$. This can be seen in Table 4D4B from BS 7671, on page 308.

If the power factor of the load is not known, the $(\text{mV/A/m}_z)$ value of voltage drop is used. This would include circuits such as distribution circuits, where the final circuits have multiple items in which power factor may affect the circuit but a fixed value is not known.

Where a more accurate assessment of voltage drop is required and the power factor ($\cos \theta$) is known, such as motor circuits, the following methods may be used.

For cables with conductors of cross-sectional area of 16 mm$^2$ or less, the design value voltage drop is determined approximately by multiplying the tabulated value of mV/A/m by the power factor of the load, $\cos \theta$:

$$\text{voltage drop} = \frac{\text{mV/A/m} \times \cos\theta \times I_b \times L}{1000}$$

For cables with conductors of cross-sectional area greater than 16 mm$^2$, the design value of mV/A/m is determined approximately as $\cos\theta$ (tabulated $(\text{mV/A/m})_r$) + $\sin\theta$ (tabulated $(\text{mV/A/m})_x$), and then this value of mV/A/m is used in the formula:

$$\frac{\text{mV/A/m} \times I_b \times L}{1000}$$

**KEY POINT**

To determine $\sin\theta$ from the given power factor ($\cos\theta$), the angle must be found. This is done on a calculator by using $\cos^{-1}\theta$ to get an angle, then $\sin\theta$ to determine the sin value.

**Assessment criteria**

**2.6** Verify circuit disconnection time is achieved

**KEY POINT**

Protection against electric shock is one of the fundamental principles of BS 7671. Look up Chapter 13 and Regulation 131.2.

**ACTIVITY**

Double-insulated equipment does not have an earth connection and must not be connected to earth. Name three power tools that use double-insulated protection.

# SHOCK PROTECTION

Chapter 41 of BS 7671 details the requirements for protection against electric shock and provides many ways to provide protection; these are known as 'protective measures' (see the table below). Each protective measure comprises two protective provisions.

## Protective provisions

BS 7671 requires two lines of defence against electric shock. These are:

- a basic protective provision (eg basic insulation of live parts)
- a fault protective provision (eg automatic disconnection of supply).

The combination of the two is a protective measure.

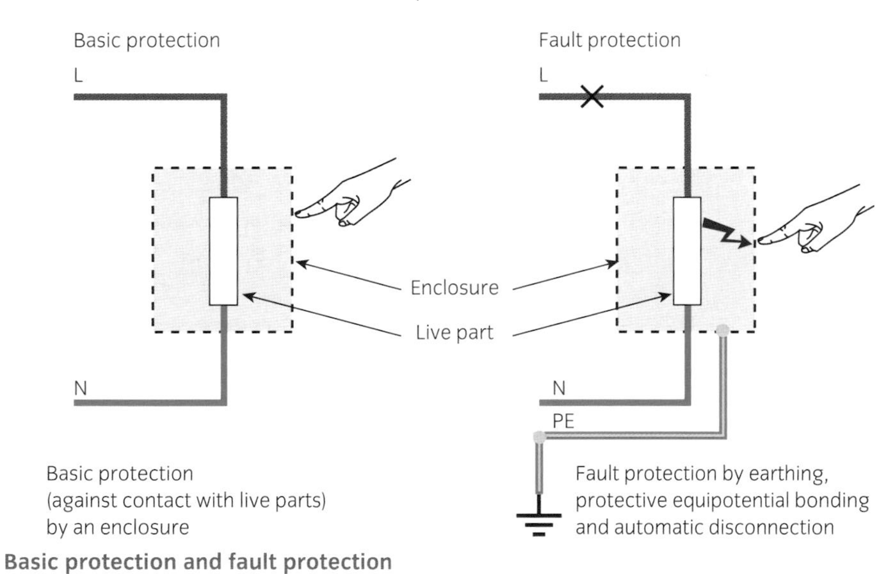


**Basic protection and fault protection**

**Protective measures**

| Protective measures | Protective provisions | |
|---|---|---|
| Relevant BS 7671 sections shown in brackets | Basic protective provision | Fault protective provision |
| Automatic disconnection of supply (411) | Insulation of live parts<br>Barriers or enclosures | Protective earthing<br>Automatic disconnection<br>Protective bonding |
| Double insulation (412) | Basic insulation | Supplementary insulation |
| Reinforced insulation (412) | Reinforced insulation | Reinforced insulation |

| Protective measures | Protective provisions | |
|---|---|---|
| Electrical separation for one item of equipment (413) | Insulation of live parts | One item of equipment, simple separation from other circuits and earth |
| Extra-low voltage (SELV and PELV) (414) | Limitation of voltage, protective separation, basic insulation | |
| **For supervised installations** | | |
| Non-conducting location (418.1) | Insulation of live parts, barriers or enclosures | No protective conductor; insulating floor and walls, spacings/obstacles between exposed conductive parts and extraneous conductive parts |
| Earth-free local equipotential bonding (418.2) | Insulation of live parts, barriers or enclosures | Protective bonding, notices, etc |
| Electrical separation with more than one item of equipment (418.3) | Insulation of live parts | Simple separation from other circuits and earth, separated protective bonding, etc |

## Protective measure: automatic disconnection of supply

The protective measure automatic disconnection of supply is the most commonly used protective measure in electrical installations. It requires:

1 basic protection provided by insulation of live parts or by barriers or enclosures

2 fault protection provided by:
   a) earthing
   b) protective equipotential bonding
   c) automatic disconnection in case of a fault.

## Maximum disconnection times

Chapter 41 of BS 7671 sets maximum disconnection times for earth faults that, if met, will result in the circuit meeting the fault protection requirements for automatic disconnection of supply. The table on the next page, from BS 7671, gives disconnection times for all circuits rated up to and including 32 A.

**Maximum disconnection times for TN and TT systems (Table 41.1 of BS 7671:2008 Requirements for Electrical Installations (the IET Wiring Regulations 17th edition) (2011))**

| System | $50\ \text{V} < U_0 \leq 120\ \text{V}$ seconds | | $120\ \text{V} < U_0 \leq 230\ \text{V}$ seconds | | $230\ \text{V} < U_0 \leq 400\ \text{V}$ | | $U_0 > 400\ \text{V}$ | |
|---|---|---|---|---|---|---|---|---|
| | a.c. | d.c. | a.c. | d.c. | a.c. | d.c. | a.c. | d.c. |
| TN | 0.8 | NOTE | 0.4 | 5 | 0.2 | 0.4 | 0.1 | 0.1 |
| TT | 0.3 | NOTE | 0.2 | 0.4 | 0.07 | 0.2 | 0.04 | 0.1 |

Where, in a TT system, disconnection is achieved by an overcurrent protective device and protective equipotential bonding is connected to all the extraneous conductive parts within the installation in accordance with Regulation 411.3.1.2, the maximum disconnection times applicable to a TN system may be used.

- $U_0$ is the nominal a.c. rms or d.c. line voltage to Earth.

Where compliance with this Regulation is provided by an RCD, the disconnection times in accordance with Table 41.1 relate to prospective residual fault currents significantly higher than the rated residual operating current of the RCD.

Note: disconnection is not required for protection against electric shock but may be required for other reasons, such as protection against thermal effects.

BS 7671 relaxes the disconnection time to:

- 5 seconds for:
  - distribution circuits within installations forming part of a TN system
  - final circuits exceeding 32 A within installations forming part of a TN system
- 1 second for:
  - distribution circuits within installations forming part of a TT system
  - final circuits exceeding 32 A within installations forming part of a TT system.

**ASSESSMENT GUIDANCE**

An earth fault could be a direct fault between line and earth, or a leakage (which could be in the milliamp range) through faulty insulation.

## Fault current required

Protective devices require a particular amount of fault current in order to disconnect in the required time. The higher the fault current is, the quicker the disconnection time.

If the minimum value of fault current is not reached during a fault to earth, the circuit will not disconnect effectively. For a circuit to disconnect effectively, the maximum earth fault loop impedance should not be exceeded as:

$$Z_s \leq \frac{U_0}{I_a}$$

where:

$Z_s$ is the total earth fault loop impedance of the supply and installation

$U_0$ is the nominal voltage to earth (typically 230 V)

$I_a$ is the value of current required to cause the protective device to disconnect in the specified time.

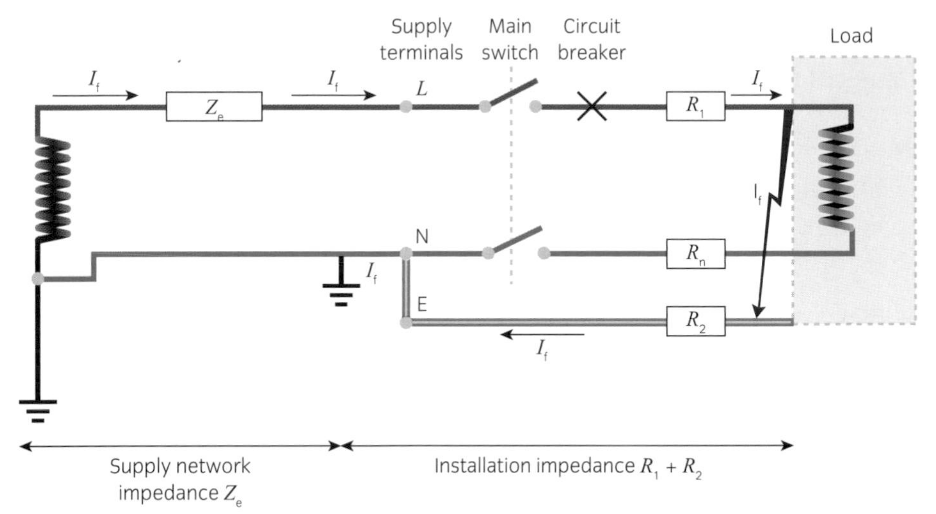


**Fault current ($I_f$) path in a TN-C-S system**

As the total earth fault loop impedance directly affects the value of earth fault current, it is the responsibility of the designer to select the parts of the earth fault path, within the installation, to ensure disconnection times are met. This includes the values of resistance $R_1 + R_2$ for any distribution circuit and to the extremities of all final circuits.

Remember:

$$I_f = \frac{U_0}{Z_e + (R_1 + R_2)}$$

where:

$I_f$ is the fault current

$U_0$ is the nominal a.c. rms line voltage to earth

$Z_e$ is that part of the earth fault loop impedance which is external to the installation

$R_1$ is the resistance of the line conductor of the circuit at conductor operating temperature

$R_2$ is the resistance of the circuit protective conductor of the circuit at conductor operating temperature.

## ACTIVITY

Using the time–current curves in Appendix 3 of BS 7671, determine the disconnection times for a fault current of 100 A for BS EN 60898 circuit breakers rated at 32 A, types B, C and D.

## Determining values of $R_1 + R_2$

$(R_1 + R_2)$ in ohms (Ω) is calculated by multiplying the resistances in milliohms per metre (mΩ/m) at 20 °C, from the table of values of resistance/metre below, by the length of the circuit $L$ and a resistance correction factor. For the purpose of this publication, we will refer to this correction factor for temperature as $C_r$ (see the table on page 316 for values of this factor). Also, to avoid confusion, we will refer to the values of resistance for a line and cpc in mΩ/m at 20 °C as $r_1 + r_2$.

To calculate circuit $(R_1 + R_2)$, we apply the following formula:

$$(R_1 + R_2)\ \Omega = \frac{LC_r(r_1 + r_2)}{1000}\ \Omega$$

where:

$C_r$ is a factor to correct resistances at 20 °C to the conductor maximum operating temperature (see the table on page 316 for factor $C_r$)

$L$ is the circuit cable length

$(r_1 + r_2)$ is the resistance of the line conductor plus the protective conductor, in mΩ/m at 20 °C.

**ACTIVITY**

Calculate the resistance of 50 m of 6 $mm^2$ line and 1.5 $mm^2$ cpc at 20 °C.

**Values of resistance/metre for copper and aluminium conductors at 20 °C in milliohms/metre $(r_1 + r_2)$ (Table I1 of the On-Site Guide)**

| Cross-sectional area $mm^2$ | | Resistance $(r_1 + r_2)$ mΩ/m |
|---|---|---|
| **Line conductor** | **Protective conductor** | **Copper** |
| 1.0 | – | 18.1 |
| 1.0 | 1.0 | 36.2 |
| 1.5 | – | 12.1 |
| 1.5 | 1.0 | 30.2 |
| 1.5 | 1.5 | 24.2 |
| 2.5 | – | 7.41 |
| 2.5 | 1.0 | 25.51 |
| 2.5 | 1.5 | 19.51 |
| 2.5 | 2.5 | 14.82 |
| 4.0 | – | 4.61 |
| 4.0 | 1.5 | 16.71 |
| 4.0 | 2.5 | 12.02 |
| 4.0 | 4.0 | 9.22 |
| 6.0 | – | 3.08 |

| Cross-sectional area $mm^2$ | | Resistance $(r_1+r_2)$ mΩ/m |
|---|---|---|
| Line conductor | Protective conductor | Copper |
| 6.0 | 2.5 | 10.49 |
| 6.0 | 4.0 | 7.69 |
| 6.0 | 6.0 | 6.16 |
| 10.0 | – | 1.83 |
| 10.0 | 4.0 | 6.44 |
| 10.0 | 6.0 | 4.91 |
| 10.0 | 10.0 | 3.66 |
| 16.0 | – | 1.15 |
| 16.0 | 6.0 | 4.23 |
| 16.0 | 10.0 | 2.98 |
| 16.0 | 16.0 | 2.30 |
| 25.0 | – | 0.727 |
| 25.0 | 10.0 | 2.557 |
| 25.0 | 16.0 | 1.877 |
| 25.0 | 25.0 | 1.454 |
| 35.0 | – | 0.524 |
| 35.0 | 16.0 | 1.674 |
| 35.0 | 25.0 | 1.251 |
| 35.0 | 35.0 | 1.048 |
| 50.0 | – | 0.387 |
| 50.0 | 25.0 | 1.114 |
| 50.0 | 35.0 | 0.911 |
| 50.0 | 50.0 | 0.774 |

## ASSESSMENT GUIDANCE

Composite cables such as twin and cpc tend to come with specific live-to-cpc cable ratios. Single cables may be installed in a mix that complies with the adiabatic equation.

**Factor $C_r$ to be applied to table of values of resistance/metre (pages 314–315) in order to calculate conductor resistance at maximum operating temperature for standard devices (Source: IET On-Site Guide)**

| Conductor installation | Conductor insulation | | |
|---|---|---|---|
| | 70 °C thermoplastic (pvc) | 85 °C thermoplastic (rubber) | 90 °C thermosetting |
| Not incorporated in a cable and not bunched (Note 1) | 1.04 | 1.04 | 1.04 |
| Incorporated in a cable or bunched (Note 2) | 1.20 | 1.26 | 1.28 |

Note 1: See Table 54.2 of BS 7671:2008 Requirements for Electrical Installations (the IET Wiring Regulations 17th edition), which applies where the protective conductor is not incorporated or bunched with cables and for bare protective conductors in contact with cable covering.

Note 2: See Table 54.3 of BS 7671, which applies where the protective conductor is a core in a cable or is bunched with cables.

Essentially, the correction factors in the above table raise the resistance of the conductor by 2% for every 5 Celsius degrees change in temperature.

**Example:**

Determine the total earth fault loop impedance ($Z_s$) for a radial final circuit, protected by a 20 A type B circuit breaker of length 25 m, wired in 70 °C thermoplastic insulated and sheathed cable, with 2.5 mm$^2$ line conductors and 1.5 mm$^2$ protective conductor. The external loop impedance $Z_e$ is 0.35 Ω.

Using the cable resistance table, we can see that a 2.5/1.5 mm$^2$ combination gives a $r_1+r_2$ value of 19.51 mΩ/m at 20 °C and the correction factor is 1.20, as the cpc is incorporated in a sheath with the line conductor. So, using:

$$(R_1 + R_2) = \frac{L \times C_r \times (r_1 + r_2)}{1000} \Omega$$

We can determine:

$$(R_1 + R_2) = \frac{25 \times 1.20 \times 19.51}{1000} = 0.585 \ \Omega \text{ at } 20\ °\text{C}$$

The fault current $I_f$ is calculated using:

$$I_f = \frac{U_o}{Z_e + (R_1 + R_2)}$$

Therefore:

$$I_f = \frac{230}{0.35 + (0.585)} = 246 \text{ A}$$

When $I_f$ is known, reference to the device characteristic will give the disconnection time. (See the type B circuit breaker characteristics from Appendix 3 of BS 7671 below.)

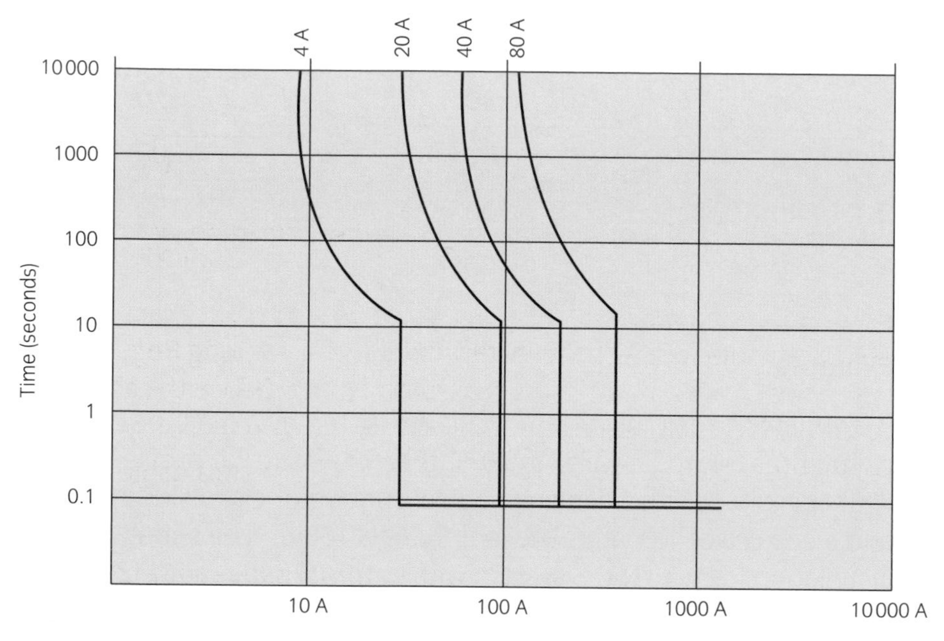


**Type B circuit breaker time–current characteristic (Source: Fig 3A4 of Appendix 3 to BS 7671:2008 Requirements for Electrical Installations (the IET Wiring Regulations 17th edition))**

In the example above, with a 20 A type B circuit breaker with a fault current $I_f$ of 245A, the circuit breaker will operate in 0.1 seconds. This is well within the requirement of 0.4 seconds of Table 41.1 of BS 7671 (see table of maximum disconnection times for TN and TT systems on page 312).

## Maximum earth fault loop impedances

For TN systems, maximum values of earth fault loop impedance are given in Table 41.2, 41.3 and 41.4 of BS 77671:2008 Requirements for Electrical Installations, depending on the type of protective device used. These tables are used where the nominal voltage to earth ($U_0$) is 230 V, and it removes the need to calculate the earth fault current.

As long as the value of earth fault loop impedance ($Z_s$) determined is within the values in the table, disconnection will be achieved.

This maximum fault loop impedance ($Z_s$) to achieve disconnection in the required time in the tables has been determined from:

$$Z_s \leqslant \frac{230}{I_a}$$

## ACTIVITY

Using the current characteristic tables in Appendix 3 of BS 7671, determine the maximum permitted values of $Z_s$ for each of the following circuit breakers. Once determined, compare these values with Table 41.3 of BS 7671:

- 32 A type C
- 100 A type B
- 32 A type D.

### KEY POINT

Remember: types of circuit breaker are selected to suit a particular load, not to suit earth fault loop impedance. If the earth fault loop impedance is too high, changing the rating or type of circuit breaker is not a suitable option.

## ASSESSMENT GUIDANCE

Time–current curves are available from BS 7671 Appendix 3 or from fuse and switchgear manufacturers.

Hence, for the type B circuit breakers above, we can calculate the $Z_s$ values, as shown in this table.

| Rated current $I_n$ | Fault current for 0.1 disconnection $I_a$ | $Z_s = \frac{230}{I_a}$ |
|---|---|---|
| 6 | 30 | 7.670 |
| 20 | 100 | 2.300 |
| 40 | 200 | 1.150 |
| 80 | 400 | 0.575 |

**Example:**

Determine the suitability of a circuit protected by a 6 A type B circuit breaker, to be wired in 70 °C thermoplastic flat profile twin with cpc cable, with a line and neutral conductor csa of 1.5 mm$^2$ and a cpc csa of 1.0 mm$^2$. The length ($L$) is 100 m. The supply and installation form a TN-C-S system with external impedance ($Z_e$) of 0.35 Ω.

First we consider:

$$Z_e + (R_1 + R_2) \leqslant Z_{s\,max}$$

So:

from Table 41.3 of BS 7671 or the table above, $Z_{s\,max}$ is 7.67 Ω

from the table on pages 314–315 (also Table I1 of the On-Site Guide and Table 13.4a of BS 7671), $(r_1 + r_2)$ = 30.2 mΩ/m at 20 °C

from the table on page 316 (Table I3 of the On-Site Guide and Table 13.4b of BS 7671), $C_r$ is 1.20 as the cpc is incorporated in the cable, and:

$$(R_1 + R_2) = \frac{L \times C_r \times (r_1 + r_2)}{1000}\ \Omega$$

Therefore:

$$(R_1 + R_2) = \frac{100 \times 1.20 \times 30.2}{1000} = 3.62\ \Omega$$

As:

$$Z_s = Z_e + (R_1 + R_2)$$

then:

$$0.35 + 3.62 = 3.97\ \Omega$$

As $Z_{s\,max}$ is 7.67 Ω this is satisfactory.

## Where RCDs are used to meet disconnection times

In some cases, earth fault loop impedance values may be too high to satisfy disconnection times of standard protective devices. A solution to this is to install an RCD or RCBO.

As these devices disconnect at much lower fault currents, much higher earth fault loop impedance values are permitted.

Table 41.5 in BS 7671 gives maximum earth fault loop impedance values where earth fault protection is provided by an RCD. There is no maximum residual current rating for an RCD providing automatic disconnection of supply (ADS), but, in most installations, devices with residual current ratings ($I_{\Delta n}$) of 30 mA are commonly installed to provide additional protection.

When a circuit relies on an RCD as fault protection, it must be verified that the circuit still has suitable short circuit and, where required, overload protection.

**KEY POINT**

Additional protection should not be confused with ADS. Additional protection is required where there is greater risk of electric shock through the failure of basic protection, such as insulation leading to direct contact with live parts, whereas ADS provides fault current protection.

**ACTIVITY**

Look in the index of BS 7671. List a sample of the locations where additional protection is required. From this list, work out the common risk and the reason for additional protection.

# COORDINATION OF PROTECTIVE DEVICES

**Assessment criteria**

**2.7** Verify discrimination of protective devices

If there is a fault in an installation, ideally only the equipment or cable that is faulty should be disconnected so that the remainder of the installation can continue to operate normally. It may be necessary to ensure continued functioning of the unaffected parts of the installation to prevent inconvenience or to prevent danger, such as a fault in a circuit causing lighting to be lost.

The first stage of the design process to allow only disconnection of the faulty equipment is division of the installation into circuits.

## Division of an installation

Section 314 of BS 7671 gives reasons why installations should be divided into circuits. These include:

- to avoid danger
- to minimise inconvenience in the event of a fault.

There are further reasons, including:

- to allow inspection and testing
- to reduce unwanted tripping of RCDs
- to reduce electromagnetic interference.

By dividing the installation into circuits, including distribution circuits, we can minimise the chance of losing more than one section of an installation.

## Discrimination between devices

**ASSESSMENT GUIDANCE**

When a fuse blows or a circuit breaker trips, the action is divided into two parts. The pre-arcing energy occurs before the element melts (or the contacts open). The remainder of the energy let through occurs during the arcing period.

Effective discrimination is achieved when a designer ensures that local protective devices disconnect before others located closer to the origin of the installation do. Discrimination is required:

- under normal load conditions
- under overload conditions.

In the event of a fault, discrimination is desirable, but if no danger and no inconvenience arise it may not be necessary.

### Fuse-to-fuse discrimination

Two fuses in series will discriminate between one another under overload and fault conditions if the maximum pre-arcing characteristic of the upstream device exceeds the maximum operating characteristic of the downstream device.

Discrimination will be achieved for fuses if an upstream device is more than twice the rating of any downstream device. For example, if the upstream fuse (A) has a rating of 80 A and the local downstream fuse (B) is 32 A, discrimination would be achieved as neither of the characteristic curves cross one another.

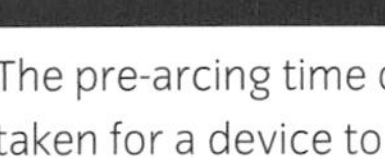

**KEY POINT**

The pre-arcing time of a device is the time taken for a device to react to an overcurrent, up to the point where the device breaks and arcs.

The maximum operating characteristic is the time a device takes to complete the operation of disconnection – including pre-arcing time, arcing and disconnecting – leaving a gap big enough to prevent current flow. Manufacturers provide detailed graphs showing pre-arcing characteristics. Graphs in BS 7671 only show maximum operating characteristics.

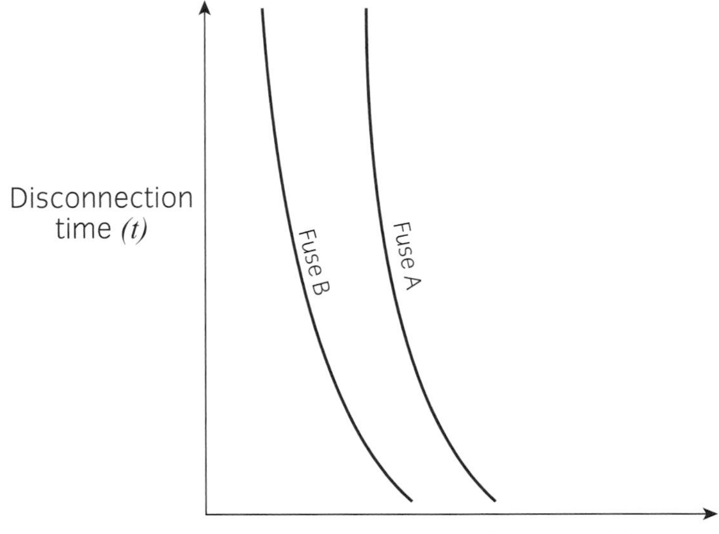


**Fuse characteristics**

### Circuit breaker to circuit breaker

Circuit breaker characteristics are different to those of fuses. The circuit breaker has two characteristic features:

- a thermal characteristic similar to a fuse
- a magnetic characteristic when at a specified current that causes the circuit breaker to operate instantaneously (see diagram opposite).

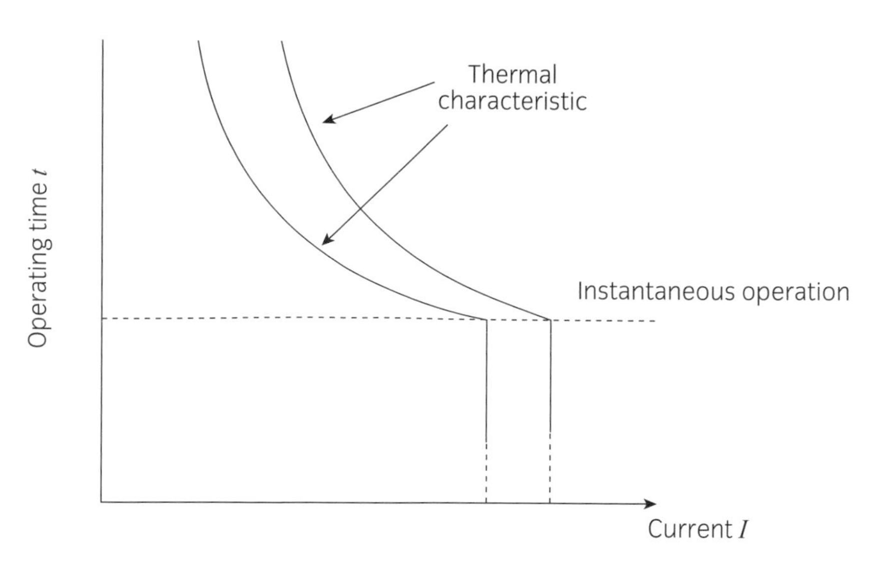


**Circuit breaker characteristics**

If the fault current exceeds the instantaneous operating current of the upstream device, there may not be discrimination if the two devices have the same frame type, whatever their ratings, as both devices may operate at the same time.

Manufacturers will provide information on circuit-breaker discrimination. It will normally be achieved by selecting different frame types for the upstream and downstream circuit breakers.

In some situations a local circuit breaker may be of a different type to one installed nearer the origin. Even though the local device is only rated at 32 A (type D) and the distribution device is rated at 100 A (type B), discrimination will not be achieved.

Looking at Appendix 3 of BS 7671, we can see that a 32 A type D circuit breaker will disconnect at a current of 640 A whereas a 100 A type B will disconnect at 500 A fault current. This means that the 100 A device will disconnect before the 32 A device.

## ACTIVITY

Check Appendix 3 of BS 7671 and determine whether discrimination is achieved between each pair of devices:

| Local device | Distribution device |
|---|---|
| 16 A type C | 40 A type B |
| 40 A type B | 63 A type C |
| 50 A type D | 100 A type C |

## Selection of circuit breaker type

The characteristic tripping current of type B, C and D circuit breakers is specified in Standard BSEN 60898 and Standard BSEN 61009-1. For example, type B circuit breakers will trip instantaneously at 5 $I_n$ and type C at 10 $I_n$ (see the table on the next page and Tables 3A4 to 3A6 of Appendix 3 of BS 7671).

While type B will be suitable for general purposes, loads that surge at switch-on will require a type C or D to be selected. These loads are typically **inductive loads**.

**Inductive loads**

Loads that involve magnetism, such as motors, discharge lighting and (although not a load) transformers.

Application of circuit breakers where $I_n$ is the nominal current rating of the circuit breaker (Source: IET On-Site Guide)

| Circuit-breaker type | Trip current (0.1 s to 5 s) | Application |
|---|---|---|
| 1<br>B | 2.7 to 4 $I_n$<br>3 to 5 $I_n$ | Domestic and commercial installations with little or no switching surge |
| 2<br>C<br>3 | 4.0 to 7.0 $I_n$<br>5 to 10 $I_n$<br>7 to 10 $I_n$ | General use in commercial and/or industrial installations where the use of fluorescent lighting, small motors, etc, can produce switching surges that would operate a type 1 or type B circuit breaker. Type C or type 3 circuit breakers may be necessary in highly inductive circuits such as banks of fluorescent lighting |
| 4<br>D | 10 to 50 $I_n$<br>10 to 20 $I_n$ | Suitable for transformers, X-ray machines, industrial welding equipment, etc, where high inrush currents may occur |

## Moulded case circuit breakers

For moulded case circuit breakers to BS EN 60947-21, time–current characteristics are not specified in BS 7671. The time–current characteristics, voltage rating and making and breaking capacities will all be as specified by the manufacturer. High-current and fault-breaking characteristics are available in devices to this standard, as the moulded case offers much better shielding to high-fault current than standard miniature-type circuit breakers. This gives them a much higher breaking capacity.

**Assessment criteria**

**2.8** Verify breaking capacities of protective devices

# VERIFY BREAKING CAPACITIES OF PROTECTIVE DEVICES

## Breaking capacity

There is a limit to the maximum current that an overcurrent protective device (fuse or circuit breaker) can interrupt. This is called the rated short-circuit capacity or breaking capacity. The table on the next page shows the rated short-circuit capacities of the most common devices used in the UK.

Regulation 434.02.01 in BS 7671 requires the prospective fault current under both short-circuit and earth-fault conditions to be determined at every relevant point of the complete installation.

This means that at every point where switchgear is installed, the maximum fault current must be determined to ensure that the switchgear is adequately rated to interrupt the fault currents.

**Rated short-circuit capacities of protective devices (Source: IET On-Site Guide)**

| Device type | Device designation | Rated short-circuit capacity (kA) |
|---|---|---|
| Semi-enclosed fuse to BS 3036 with category of duty | S1A<br>S2A<br>S4A | 1.0<br>2.0<br>4.0 |
| Cartridge fuse to BS 1361 type I<br>Cartridge fuse to BS 1361 type II<br>Cartridge fuse to BS 88-3 type I<br>Cartridge fuse to BS 88-3 type II | | 16.5<br>33.0<br>16.0<br>31.5 |
| General purpose fuse to BS 88-6 | | 50 at 415 V<br>16.5 at 240 V<br>80.0 at 415 V |
| Circuit breakers to BS 3871 | M1<br>M1.5<br>M3<br>M4.5<br>M6<br>M9 | 1.0<br>1.5<br>3.0<br>4.5<br>6.0<br>9.0 |
| Circuit breakers to BS EN 60898* | | $I_{cn}$ $I_{cs}$<br>1.5 (1.5)<br>3.0 (3.0)<br>4.5 (4.5)<br>6.0 (6.0)<br>10.0 (7.5)<br>15.0 (7.5)<br>20.0 (10.0)<br>25.0 (12.5) |

Notes:

*Two rated short-circuit rating are defined in BS EN 60898 and BS EN 61009

(a) $I_{cn}$ is the rated short-circuit capacity (marked on the device)

(b) $I_{cs}$ is the service short-circuit capacity

The difference between the two is the condition of the circuit breaker after manufacturer's testing.

- $I_{cn}$ is the maximum fault current the breaker can interrupt safely, although the breaker may no longer be usable.
- $I_{cs}$ is the maximum fault current the breaker can interrupt safely without loss of performance.

The $I_{cn}$ value is normally marked on the device in a rectangle, eg: 6000

For the majority of applications the prospective fault current at the terminals of the circuit breaker should not exceed this value.

For domestic installations the prospective fault current is unlikely to exceed 6 kA, up to which value the $I_{cn}$ and $I_{cs}$ values are the same.

From the table on page 323 it can be seen that BS 3036 devices have a very low breaking capacity, compared to other devices. This is the main reason why these devices are no longer suitable for many installations.

Circuit breakers have two short-circuit capacity ratings.

- $I_{cs}$ is the value of fault current up to which the device can operate safely and remain suitable and serviceable after the fault.
- $I_{cn}$ is the value above which the device would not be able to interrupt faults safely. This could lead to the danger of explosion during faults of this magnitude or, even worse, the contacts welding and not interrupting the fault.

Any faults that occur between these two ratings will be interrupted safely but the device will probably require replacement.

**ACTIVITY**

BS 3871 circuit breakers and BS 1361 fuses are no longer listed in BS 7671. Why is it necessary to include them here?

## Determining prospective fault current $I_{pf}$

Prospective fault currents at the origin of an installation may be determined by:

- enquiry of the electricity supplier
- calculation
- measurement.

**ASSESSMENT GUIDANCE**

It is probably sensible to assume the worst case situation with short-circuit currents as it is not usually possible to record the actual value of current.

### Enquiry

For most simple installations, such as domestic installations, an enquiry to the supplier will result in the advice that the prospective fault current will not exceed 16 kA. The supplier is required by the Electricity Safety, Quality and Continuity (ESQC) Regulations 2002 to supply this information.

### Calculation

For installations taking a supply at 11 kV from a private sub-station transformer, the designer will need to calculate the prospective fault current. This is outside the scope of this book.

### Measurement

Instruments are available for measuring prospective fault current at the relatively low fault levels found in domestic installations. They are not accurate for high fault levels which typically exceed 16 kA.

In some situations, the value of short-circuit current may need to be determined within the installation.

This process is similar to the calculation for earth fault loop impedance and earth fault current, but instead of using the cpc ($R_2$) in the calculation, we use the neutral ($R_n$) instead.

## Example:

Determine the short-circuit current at the terminals of the 400 V three-phase motor in the diagram below.

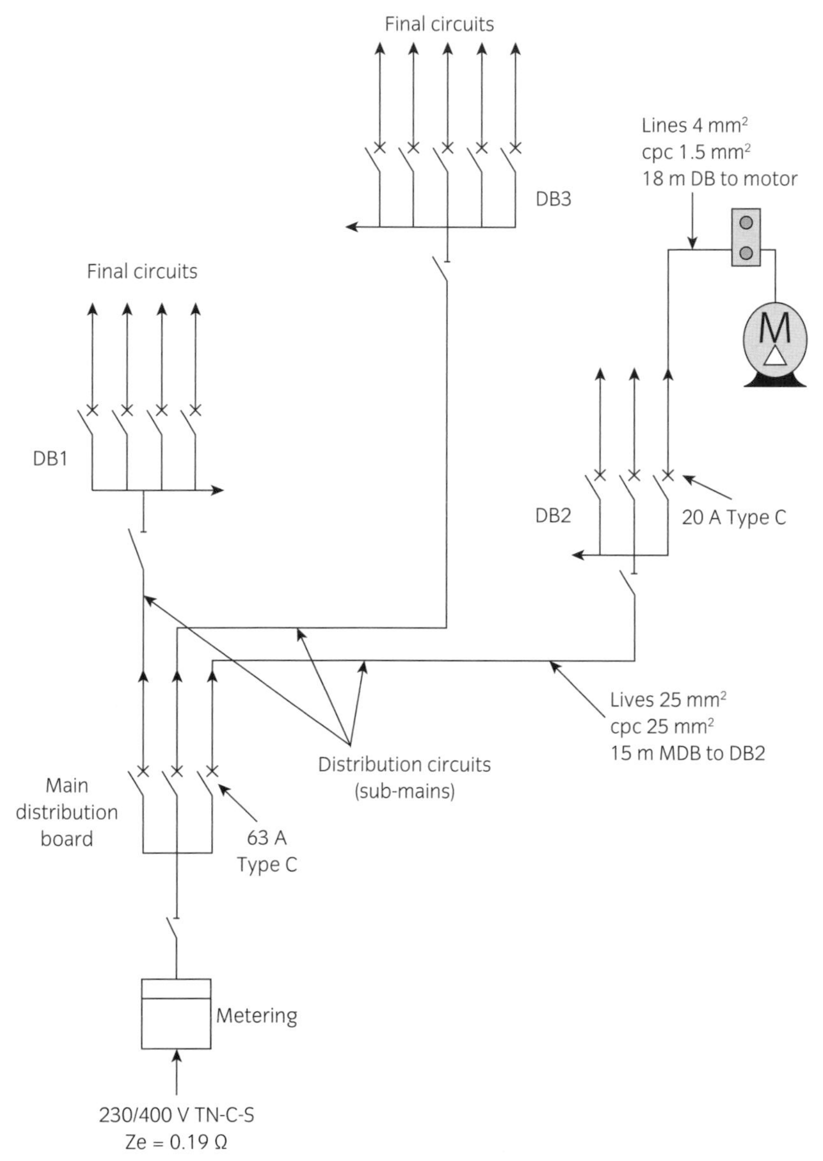


**The sequence of control for an installation**

The maximum short-circuit current within this installation would be between line and line as the voltage is 400 V. As the installation forms part of a TN-C-S earthing arrangement, the $Z_e$ can be used to determine the $I_{pf}$ at the origin of the installation:

$$\frac{400\text{ V}}{0.19\ \Omega} = 2105\text{ A or }2.1\text{ kA}$$

We need to determine the resistance of the distribution circuit line conductors to calculate the $I_{pf}$ at DB2. From Table I1 of the IET On-Site Guide, the value of resistance, in mΩ/m for a 25 mm² line-to-line loop, is 1.45 mΩ/m at 20 °C.

As the line conductors are always enclosed or bunched, the temperature correction factor needed for to adjust the resistance, from Table I3 of the IET On-Site Guide, is 1.20.

Therefore the resistance of the distribution circuit can be determined by:

$$\text{line-to-line } R = \frac{1.45\ \text{m}\Omega/\text{m} \times 15\ \text{m} \times 1.20}{1000} = 0.026\ \Omega$$

The $I_{pf}$ at DB2 will be:

$$\frac{400\ \text{V}}{0.19\ \Omega + 0.026\ \Omega} = 1851\ \text{A or } 1.85\ \text{kA}$$

We now need to determine the resistance of the line-to-line loop of the final circuit that supplies the motor. This calculation will not allow for any resistance within the motor control/starter. From Table I1 of the IET On-Site Guide, the resistance for a 4 mm$^2$ line-to-line loop is 9.22 mΩ/m at 20 °C, and once again, the factor for operating temperature is 1.20. Therefore:

$$\frac{9.22\ \text{m}\Omega/\text{m} \times 18\ \text{m} \times 1.20}{1000} = 0.2\ \Omega$$

By adding all the values together, we can determine the $I_{pf}$ at the motor:

$$\frac{400\ \text{V}}{0.19\ \Omega + 0.26\ \Omega + 0.2\ \Omega} = 615\ \text{A or } 0.6\ \text{kA}$$

So we can see from the example that, due to circuit resistances, the short-circuit current reduces within the installation.

**KEY POINT**

Where single-phase values of short-circuit current are required, the resistance loops will be based on line and neutral values and the voltage applied will be 230 V. Where an installation is not part of a TN-C-S arrangement, the line-to-line (for three-phase) or line-to-neutral (for single-phase) loop impedance values need to be used and the voltage applied is 230 V for single-phase and 400 V for three-phase.

**Assessment criteria**

2.9 Verify thermal constraints

# SELECTION OF PROTECTIVE CONDUCTORS

The selection of protective conductors is covered in Chapter 54 of BS 7671.

## Types of protective conductor

Earthing conductors and circuit protective conductors are both protective earthing conductors. This means they are intended to carry a current under earth fault conditions.

BS 7671 states that a protective conductor may consist of one or more of the following:

- a single-core cable
- a conductor in a cable
- an insulated or bare conductor in a common enclosure with insulated live conductors
- a fixed bare or insulated conductor
- a metal covering, for example, the sheath, screen or armouring of a cable
- a metal conduit, metallic cable management system or other enclosure or electrically continuous support system for conductors
- an extraneous conductive part complying with Regulation 543.2.6.

## ACTIVITY

How would you determine the ambient temperature along the cable run?

## The adiabatic equation

Protective conductors carry very little current in normal operating circumstances (when there is no fault to earth). However, under earth fault conditions, they have to carry the fault currents for the duration of the fault. Fault currents may be very high but last for very short periods (until the overcurrent device operates). Consequently the csa of the protective conductor is not determined from the continuous ratings in Appendix 4 of BS 7671 but by the adiabatic equation of BS7671 Regulation 543.1.3 (see below). This allows a much smaller cable to be used, hence the flat twin cables with reduced protective conductor, commonly used in the UK in domestic and similar installations, and reduced cpc sizes when wiring single-core cable in conduit or trunking.

The adiabatic equation is:

$$S = \frac{\sqrt{I^2 t}}{k}$$

where:

$S$ is the minimum cross-sectional area (csa) of the conductor in $mm^2$

$I$ is the value in amperes (rms for a.c.) of fault current for a fault of negligible impedance that can flow through the associated protective device, due to account being taken of the current-limiting effect of the circuit impedances

$t$ is the operating time of the disconnecting device in seconds, corresponding to the fault current $I$ in amperes

$k$ is a factor taking account of the resistivity, temperature coefficient and heat capacity of the conductor material, and the appropriate initial and final temperatures.

## KEY POINT

The adiabatic equation is applicable for disconnection times not exceeding 5 s. It assumes no heat loss from the cable for the duration of the fault.

Values of $k$ for protective conductors, depending on their type, material and method of installation, are as given in Tables 54.2 to 54.6 of BS 7671. Each table represents the different types of protective conductor or the different methods of installing it, as shown on the next page.

**Values of $k$ for protective conductors in various situations**

| | |
|---|---|
| Table 54.2 | For protective conductors not bunched or incorporated in a cable. Examples include a separate earthing conductor at the intake or 'tails' position in a domestic installation or a separate 'overlay' cpc run with an armoured cable |
| Table 54.3 | For protective conductors bunched or incorporated in a cable. This is the most common table as it covers all multicore composite cables, single-core cables in conduit and trunking |
| Table 54.4 | For protective conductors that form sheaths or armours of a cable. Examples include the use of a steel-wire armouring as a cpc. However, this does not include MICC cable as a copper value is not given for this, so Table 54.2 is used |
| Table 54.5 | For situations where a conduit or trunking is used as a common cpc. Although rare today, this is still an accepted method |
| Table 54.6 | For bare conductors, such as metallic earth strapping, commonly used in main switch rooms (similar to that seen on lightning protection systems) |

**ASSESSMENT GUIDANCE**

PVC conduit must always contain a separate cpc. Table 54.5 refers to metallic conduit and trunking.

Where the application of the formula produces a non-standard size, a conductor with the nearest larger standard cross-sectional area should be used.

**Example:**

Verify the suitability of the cpc for thermal constraints where a circuit has a total earth fault loop impedance of 2.3 Ω and is wired using 2.5/1.5 mm² 70 °C thermoplastic flat profile twin and cpc cable. A 20 A BS 88-3 device protects the circuit.

First we need to determine the earth fault current:

$$I_f = \frac{U_o}{Z_s} \text{(A)}$$

Therefore:

$$I_f = \frac{230 \text{ v } \Omega}{2.3} \text{(A)}$$

Using the graphs in Appendix 3 of BS 7671 (Figure 3A1), we can see that the device will disconnect in 0.7 seconds with a fault current of 100 A.

From Table 54.3 in BS 7671 (cpc incorporated in the cable), the value of $k$ is 115.

The minimum size for thermal constraint ($S$) is determined using:

$$S = \frac{\sqrt{I^2 t}}{k}$$

Therefore:

$$S = \frac{\sqrt{100^2 \times 0.7}}{115} = 0.727 \text{ mm}^2$$

In this particular situation, the minimum csa for the cpc required is 0.727 mm$^2$, which can be rounded to 1 mm$^2$, so the 1.5 mm$^2$ conductor is suitable under fault conditions.

The conductor csa can be calculated as per the adiabatic equation or selected in accordance with Table 54.7 of BS 7671 shown below. This method is simpler than the use of the adiabatic equation but it may not give the most economic results.

**Minimum cross-sectional area of protective conductor in relation to the cross-sectional area of associated line conductor (Source: Table 54.7 of BS 7671:2008 Requirements for Electrical Installations (the IET Wiring Regulations 17th edition), 2011)**

| Cross-sectional area of line conductor $S$ | Minimum cross-sectional area of the corresponding protective conductor | Minimum cross-sectional area of the corresponding protective conductor |
|---|---|---|
| | If the protective conductor is of the same material as the line conductor | If the protective conductor is not of the same material as the line conductor |
| (mm$^2$) | (mm$^2$) | (mm$^2$) |
| $S \leqslant 16$ | $S$ | $\frac{k_1}{k_2} \times S$ |
| $16 < S \leqslant 35$ | 16 | $\frac{k_1}{k_2} \times 16$ |
| $S > 35$ | $\frac{S}{2}$ | $\frac{k_1}{k_2} \times \frac{S}{2}$ |

where:

$k_1$ is the value of $k$ for the line conductor, selected from Table 43.1 in Chapter 43 of BS 76761 according to the materials of both conductor and insulation

$k_2$ is the value of $k$ for the protective conductor, selected from Tables 54.2 to 54.6 of BS 76761 (see above), as applicable.

According to Table 54.7 of BS 7671, the cpc needs to be the same size as the line conductor, for any circuit where the line conductor csa is up to 16 $mm^2$. This means that all twin and cpc cables with a reduced cpc size need to be verified by the adiabatic equation during the design process.

Where high fault levels in the region of 3 kA and above are expected, the minimum protective conductor csa must be increased because of the increased energy let-through of the circuit breakers. This can be seen on Table B7 of the IET On-Site Guide, which gives recommended minimum cpc cross-sectional areas relating to circuit breaker types and ratings.

## Bonding conductors

There are two types of bonding conductor:

- main protective bonding conductor (MPB)
- supplementary bonding conductors.

### Main protective bonding conductors (MPB)

Regulation 411.3.1.2 of BS 7671 requires an MPB to be installed that links the main earthing terminal (MET) to extraneous conductive parts, including:

- water installation pipes
- gas installation pipes
- other installation pipework and ducting (eg oil pipes)
- central heating and air-conditioning systems
- exposed metallic structural parts of the building.

The purpose of the MPB is not only to carry fault current but to ensure that the MET, under earth fault conditions, and the accessible exposed conductive parts are at substantially the same potential (voltage). This is known as 'equipotential bonding' (equal potential). Equipotential bonding reduces the risk of electric shock as current cannot flow if an equal voltage is present. For example, if a person was in contact with a live metal case of a kettle which rose to 230 V due to a fault and was also in contact with a metal water pipe that was connected to the MET by a MPB conductor, this too would be at a similar voltage. The potential difference across the person would be low and the current through the person's body would also be very low.

Due to this, the MPB conductor is sized to decrease resistance and therefore voltage loss, instead of short-duration fault currents. Section 544 of BS 7671 gives the minimum size of MPB required depending on the system earthing arrangement.

**ACTIVITY**

With reference to BS 7671, determine the minimum permissible size of main protective bonding conductor:

- where PME conditions apply and the supply neutral conductor is 35 $mm^2$
- for a TT installation where the earthing conductor is 6 $mm^2$.

## Supplementary bonding

The increased use of RCDs in installations reduces the need for supplementary bonding. Supplementary bonding is intended as a local means of providing equal potential in the event of a fault where a risk of shock exists.

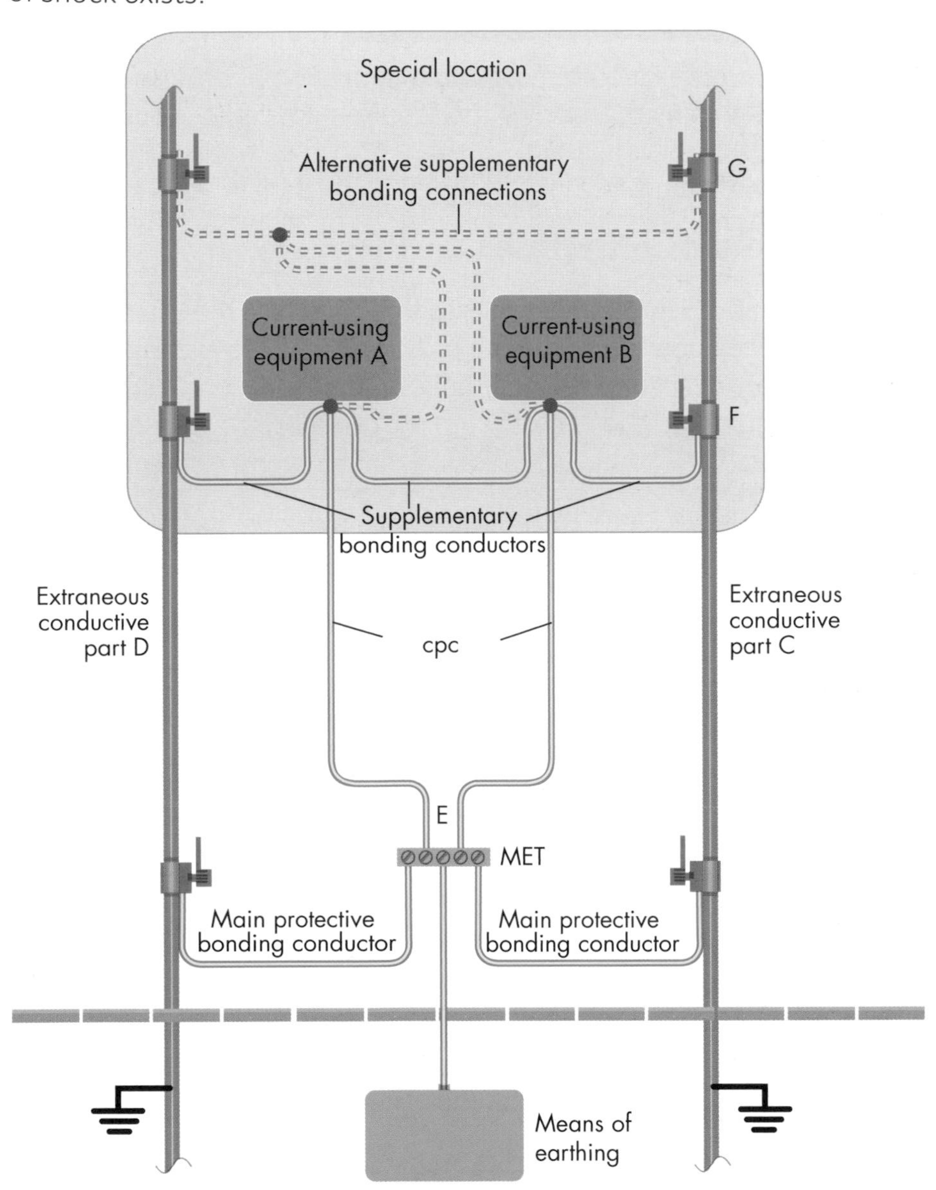


As an example, consider circuits in a kitchen without RCD protection. If a fault to earth occurred on a metallic electric cooker, and someone was in contact with the cooker and a metal sink fed by metal pipes, that person would probably receive an electric shock. This is because an earth fault current could flow through the person to earth.

However, if there was a supplementary bonding conductor linking the cooker to the sink or pipes and the cooker became live, then so would the sink and pipes. This would create an equal potential (or equipotential zone). As the voltage is equal across the person, current cannot flow and the risk of shock is reduced.

As many supplementary bonding conductors are single-core cables that are not contained in wiring systems, Section 544 of 7671 requires that they have a minimum csa of 4 $mm^2$. If they do have further mechanical protection, they may be reduced to 2.5 $mm^2$.

## ISOLATION AND SWITCHING

Section 537 includes requirements for the location of devices for isolation and switching, including the requirements for the actual devices used.

### Fundamental principles

Regulation 132.1.5 in BS 7671 requires that 'Effective means, suitably placed for ready operation, shall be provided so that all voltage may be cut off from every installation, from every circuit thereof and from all equipment, as may be necessary to prevent or remove danger'.

### Switches and isolators

#### Switch

A switch is a device capable of making, carrying and breaking current under normal circuit conditions. The specification of the switch may give the period of time for which it will handle overload conditions and short-circuit currents. Although a switch may be able to connect a short-circuit current, it may not be capable of breaking the circuit under such conditions.

A switch may be an isolator, though not necessarily. A switch may be used to control a piece of equipment such as a light or a motor, but it will not necessarily enable the equipment or circuit to be worked on safely.

#### Isolator

An isolator is a device that cuts off the installation, or part of an installation, from every source of electrical energy. This ensures that work can be carried out safely on the installation, or part of the installation, that has been isolated.

Equipment standards have particular requirements for the effectiveness and reliability of the separation, such as an air gap provided by the isolator. Isolators must also be able to be made secure in the open position so that people working on the installation are not in danger from re-closure.

Isolators are unlikely to be used as a switch unless suitably rated. This must be checked if an isolator is to be located in an area where it may be used to disconnect loads under loaded conditions. In some situations isolators rated at, for example, 100 A may only be capable of switching 60 A.

**ACTIVITY**

Using a wholesaler's catalogue, identify switches suitable for:

a) isolation
b) mechanical maintenance
c) emergency switching
d) functional switching.

A common fault in industrial sites occurs when isolators are placed in locations where ordinary persons use them as a convenient switch. This causes a heat build-up on the switch contacts and eventually failure of the switch or damage to the cable where it is terminated into the isolator.

## Switching for mechanical maintenance

Where plant and machinery requires work to be carried out by a non-electrically skilled person and the work does not involve any form of electrical work, a device must be provided locally to the machinery that is capable of being secured in the off position (or at least supervised) and is capable of switching full-load current.

## Functional switches

Functional switches can a switch on or off or control equipment; the switching may be effected by semi conductors. Examples of functional switches are light switches in a room.

## Emergency switches

Emergency switching is an operation intended to remove, as quickly as possible, danger that may have occurred unexpectedly. Devices with semiconductors cannot be emergency switches. An example of an emergency switch is a stop button in a workshop.

## ASSESSMENT GUIDANCE

Lighting switches are generally micro-gap switches that do not provide an adequate contact gap to be used as an isolator. It is not safe to turn the light switch off and then work on the luminaire.

## Undervoltage protection

Undervoltage protection is required where a reduction in voltage, or loss of voltage, could cause danger and the subsequent restoration of voltage might cause danger.

A common example of this protection is a starter contactor for a motor. If the supply fails, the contactor 'drops out' and opens the circuit. Should the supply resume, the motor will not automatically re-start until someone has pushed the start button after assessing for danger.

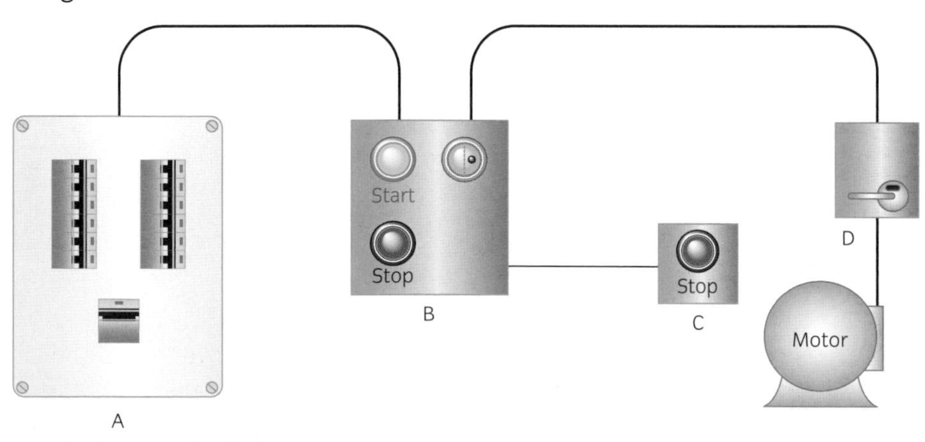


**A typical motor circuit**

The diagram on page 333 shows a typical motor circuit, for example, that might be used to drive a mechanical pump. Each device labelled A to D performs the function as set out in the table below.

| Device | Description | Function |
|---|---|---|
| A | Distribution board | ■ Full isolation (main switch)<br>■ Circuit isolation (circuit breaker)<br>■ Protection (overcurrent) |
| B | Start/stop contactor with switch | ■ Functional switching<br>■ Partial isolation<br>■ Emergency switching<br>■ Undervoltage protection<br>■ Overload protection |
| C | Remote stop button | ■ Emergency switching |
| D | Local switch | ■ Switching for mechanical maintenance<br>■ Isolation (as long as it is lockable) |

Where a switch for mechanical maintenance is installed between a starter and motor, assessment should be made as to the likelihood of the switch being used for functional purposes. In such situations, interlocks between the starter and switch may be required.

# OUTCOME 3

# Understand procedures for connecting complex electrical systems

In this outcome you will look at the methods of connecting and terminating installations as well as the procedures for obtaining a new supply.

## CONNECTION OF SUPPLY

### Applications for supply

For new installations an application must be made for a new electricity supply. This will usually be the responsibility of the electrical design team or consultants, although the installer will also need to provide details. An example of a distribution network operator's (DNO) application form is shown on page 336.

### Compliance with BS 7671

The DNO will expect to receive a statement from the applicant that the installation complies with BS 7671.

The electrical contractor is required by BS 7671 to supply the person ordering the work with an Electrical Installation Certificate, complete with schedules of inspections and schedules of test results. The work is not completed until these have been given to the person ordering the work.

**Assessment criteria**

3.1 Interpret sources of information for the connection of complex electrical systems

**ACTIVITY**

**Building Regulations approved documents**

Approved documents are available free to download from the Planning Portal (www.planningportal.gov.uk/buildingregulations/approveddocuments/). Download Approved document P (Electrical Safety). A copy is essential for installers involved in dwellings.

Electricity North West Electricity Connections

electricity north west

## Application for standard domestic single phase electricity connection

**Maximum supply for a standard domestic property is 20kVA (restrictions may apply), with no generation.**
**Please note all relevant sections of this form must be completed or your application will be on hold or returned to you.**

**Type of supply you are applying for:**

☐ New connection
☐ Temporary supply
☐ Re-energise existing supply

**Section 1 - Applicant** *Please provide applicant details*

| Company name/contact name | | | |
|---|---|---|---|
| Address: | | | |
| | | Postcode: | |
| Landline number: | | Mobile number: | |
| Fax number: | | Email address: | |

**Section 2 - Site contact details**

Are you the site manager/representative? ☐ Yes ☐ No

*If you are not the site manager/representative or will have a builder to manage your on-site activities, including the final connection please provide their details below:*

| Name: | | Contact number: | |
|---|---|---|---|
| Fax number: | | Email address: | |

**Section 3 - Site information** *Please provide site address and details*

| Proposed number of dwellings/connections: | | Type i.e flat, detached house, industrial unit: | |
|---|---|---|---|
| Address: | | | |
| | | Postcode: | |
| Site Telephone number: | | | |
| Does the site have planning permission? | ☐ Yes ☐ No | Does the site have existing connections? | ☐ Yes ☐ No |

**Section 4: Existing supplies information - Re-energising applications only**
***If your site has existing supplies please provide the MPAN (Meter Point Administration Number) of each connection***

***PLEASE NOTE:*** *Your MPAN number should begin with the number 16 if you are within our network area. If your MPAN number does not begin with 16 then your supply may not be part of our network area. If you are within our network area and your MPAN number does not begin with 16 you will need to contact your electricity supplier to confirm who can deal with your connection. Please refer to section 11 of your Application booklet for more information. (Please see map on the back of your booklet, which outlines the areas we cover for Electricity Connections)*

| 1 | 6 | | | | | | | | | | | |
|---|---|---|---|---|---|---|---|---|---|---|---|---|

If you have existing connections do these require disconnection and removal from site? ☐ Yes ☐ No

**Section 5 - New supply information**

Please indicate the preferred position of where you will be having your meter ☐ Internal ☐ External

**Section 6 - Heating type** *if applicable*

☐ Ground source heat pump ☐ Air source heat pump ☐ PV Cell ☐ Not applicable

**If yes to any of the above please ask your consultant or electrician to complete the rest of this section**

| Total single phase motors (in kW) | | | | | |
|---|---|---|---|---|---|
| Largest single phase motor (in kW) | | Type of starting | | | |
| | Frequency of starting (per hour) | | | Starting current (in Amps) | |
| Total 3 phase motors (in kW) | | Type of starting | | | |
| | Frequency of starting (per hour) | | | Starting current (in Amps) | |
| Largest 3 phase motor (in kW) | | Type of starting | | | |
| | Frequency of starting (per hour) | | | Starting current (in Amps) | |

**Section 7 - Acceptance & payment details**

| Name: | | | |
|---|---|---|---|
| Address: | | | |
| | | Postcode: | |
| Landline number: | | Mobile number: | |
| Fax number: | | Email address: | |

**Section 8: Declaration**

| | |
|---|---|
| I acknowledge that in making this application I will be liable for any charges in respect of the connection(s) to which this application relates. | |
| I confirm the following are included with this application where applicable (please tick): | ✔ |
| I confirm I am applying for a standard housing connection with no generation connected | |
| Site location plan identifying the properties requiring connection i.e. Ordnance Survey site location plan | |
| Site layout plan (at a scale of not less than 1:500) which should indicate: | |
| Preferred point for the cable to enter the property | |
| The external or internal meter point location marked with an X | |
| All highways and footpaths | |
| The Boundaries of your property | |
| Letter of authority from your client if you are acting as an agent or consultant | |
| Please indicate a preferred date for your connection | D D M M Y Y |
| Are you the person who will be responsible for accepting and paying for the quotation? *If **No** please provide a letter of authority* | ☐ Yes ☐ No |
| I confirm I have completed all sections which are relevant to my connection. I acknowledge that if I have not supplied the required information my application form will be returned to me and will not be processed until I supply the required information. | |

| **Print name:** | | **Signature:** | |
|---|---|---|---|
| **Company:** | | **Date:** | |

On completion, the application forms and plans should be sent to:
**Electricity Connections, Electricity North West, Frederick Road, Salford M6 6QH**
**Telephone: 0800 048 1820**

www.enwl.co.uk
Electricity Connections • Electricity North West • Frederick Road • Salford M6 6QH
Registered in England and Wales • Registered Number 2366949

06/12/SD/5156

Example of a distribution network operator's application form

## The Building Regulations

The construction of a dwelling must be in accordance with the Building Regulations. Guidance on these requirements is given in a series of 'approved documents'. Those carrying out electrical installations must comply not only with the requirements for electrical installations (see Approved document P of the Building Regulations), but also with all the other requirements, including those listed in the chart below.

**Approved documents to the Building Regulations and what they relate to**

| | |
|---|---|
| Approved document A | Structure (depth of chases in walls, and size of holes and notches in floor and roof joists) |
| Approved document B | Fire safety, Volume 1: dwellings (fire safety of certain electrical installations; provision of fire alarm and fire detection systems; fire resistance of penetrations through floors and walls) |
| Approved document C | Site preparation and resistance to contaminants and moisture (moisture resistance of cable penetrations through external walls) |
| Approved document E | Resistance to the passage of sound (penetrations through floors and walls) |
| Approved document F | Ventilation (ventilation rates for dwellings) |
| Approved document L1 | Conservation of fuel and power (energy-efficient lighting)<br>L1A: new dwellings<br>L1B: existing dwellings |
| Approved document M | Access to and use of buildings (heights of switches and socket outlets) |
| Approved document P | Electrical safety: Dwellings |

When work is carried out on an existing building, any new work must comply with the current building regulations. Other features covered by the approved documents, if unaltered, must be in no worse compliance than before the work commenced.

## Notifying Building Control

Installation of certain types of electrical systems must be notified to the relevant Building Control body.

### Notifiable work

Part P of the Building Regulations applies to low voltage (less than 1000 V) electrical installations in:

- dwellings or buildings attached to a dwelling
- the common parts of a building serving one or more dwellings
- a building receiving its electricity from a dwelling
- a garden, etc, associated with a dwelling.

Importantly, Part P of the Building Regulations does not apply to commercial or industrial dwellings, which must abide by the terms of the Electricity at Work Regulations 1989 (EAWR) enforced by the Health and Safety Executive (HSE).

The detailed requirements in England, Wales, Scotland and Northern Ireland are different.

In England and Wales, notifiable work under the requirements of Part P comprises:

- the installation of a new circuit
- the replacement of a consumer unit
- any addition or alteration to existing circuits in a special location (see diagram opposite).

In this Regulation, the term 'special location' means:

**a)** within a room containing a bath or shower, the space surrounding a bath tap or shower head, where the space extends:

**i** vertically from the finished floor level to **(aa)** a height of 2.25 metres; or **(bb)** the position of the shower head where it is attached to a wall or ceiling at a point higher than 2.25 metres from that level; and

**ii** horizontally **(aa)** where there is a bath tub or shower tray, from the edge of the bath tub or shower tray to a distance of 0.6 metres; or **(bb)** where there is no bath tub or shower tray, from the centre point of the shower head where it is attached to the wall or ceiling to a distance of 1.2 metres; or

**b)** a room containing a swimming pool or sauna heater.

(Source: The Building Regulations 2010: Electrical Safety – Dwellings (Approved Document P))

See the diagram on the page opposite.

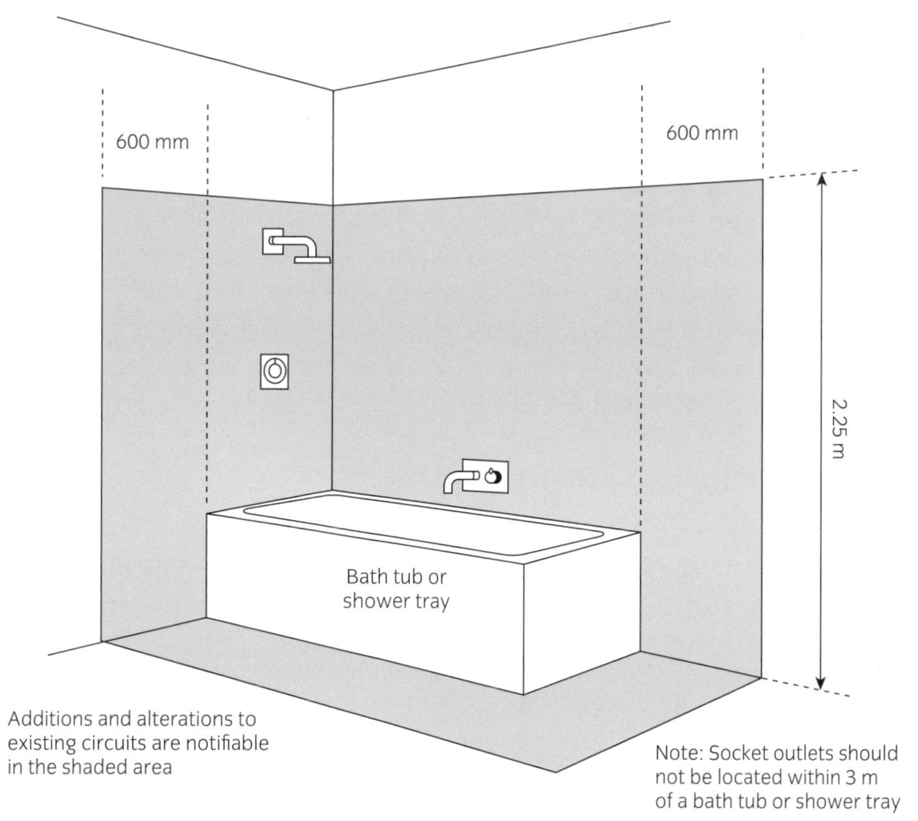


Notifiable work in rooms containing a bath or shower

### Prior notification

In England and Wales, except as below, the relevant Building Control body must be notified of all proposals to carry out electrical installation work in dwellings, etc, before the work begins. It is not necessary to give prior notification of proposals to carry out electrical installation work in dwellings, or in buildings attached to dwellings, if the work is either carried out and certified by a registered competent enterprise, or certified by a registered third-party certifier, or is non-notifiable work.

The technical requirements of Part P of the Building Regulations apply to all electrical installation work in dwellings, whether the work needs to be notified to Building Control or not.

**ASSESSMENT GUIDANCE**

The requirements for Part P have recently been modified and new requirements for a qualified supervisor have been introduced.

## Scottish Building Regulations

The requirements for electrical installations in Scotland are given in the Building (Scotland) Act 2003 and associated legislation. Detailed information on the Scottish system, including Building Regulations, can be found at the Scottish Government Building Standards Division (BSD) website (www.scotland.gov.uk/bsd).

Requirements for electrical installations in Scotland are addressed by:

- Standard 4.5: Electrical safety for all buildings
- Standard 4.6: Electrical fixtures for domestic buildings only
- Standard 4.8: Danger from accidents for all buildings.

There are no significant differences in general installation requirements for electrical work, with Scotland, England and Wales citing BS 7671:2008 Requirements for Electrical Installations (the IET Wiring Regulations 17th edition) (as amended) as the recommended means of satisfying Building Standards requirements. However, Part P electrical self-certification schemes in England and Wales do not apply to work in Scotland. In Scotland, qualified and experienced electricians can certify that their installation work meets the requirements of Building Regulations under the Certification of Construction (Electrical Installations to BS 7671) scheme, which is approved under Section 7(2) of the Building (Scotland) Act 2003.

**ACTIVITY**

Why are bathrooms and other areas where water and steam are present considered to be so dangerous?

## Welsh Building Regulations

The Welsh Ministers are now responsible for remaining functions under the Building Act 1984, including making Building Regulations. The functions transferred are limited under The Welsh Ministers (Transfer of Functions) (No 2) Order 2009 (SI 2009/3019).

The Building Regulations in Wales continue to follow the model used in England but the Welsh Government is responsible for the detail contained in each part of the Regulations.

## Building Regulations in Northern Ireland

The Northern Ireland Building Regulations (see www.buildingcontrol-ni.com) are legal requirements made by the Department of Finance and Personnel that are administered by 26 District Councils.

Guidance on how to meet the requirements is found in Technical Booklets prepared by the Department of Finance and Personnel. Adherence to the methods and standards detailed in the Technical Booklets means that the work will be 'deemed-to-satisfy' and must be accepted by Building Control as complying with the relevant Regulations.

A designer or builder can use other methods, provided it can be demonstrated that the requirements of the Regulations have been met.

## Electrotechnical Assessment Schemes (EAS)

To be a registered competent enterprise, an enterprise or an individual must be registered with an electrical installation competent-person-scheme certification body, such as NICEIC, ELECSA, NAPIT, OFTEC, Benchmark, BSI, BESCA or Stroma.

The Electrotechnical Assessment Specification (EAS) describes the minimum requirements for an enterprise (eg a contractor) to be recognised by a certification body as competent to undertake electrical installation work (design, construction, installation and verification) in England and Wales. It includes the minimum technical competence requirements for enterprises to be considered competent to carry out electrical installation work in dwellings.

# ELECTRICAL CONNECTIONS

The connections are the weak point of any installation, whether electrical or plumbing. According to BS 7671, Regulation 526.1, every connection must be:

- low resistance
- durable in terms of electrical resistance and strength
- mechanically strong
- protected against the environment.

If the connection becomes loose or corroded so that the resistance increases, the connection will rise in temperature, with the heat generated being proportional to the square of the current. Poor electrical connections are often recorded as the source of ignition of fires in buildings, in particular in dwellings.

Assessment criteria

3.2 Assess connection methods for given situations
3.3 Describe procedures for proving that terminations and connections are sound
3.4 Explain consequences of unsound terminations

KEY POINT

**Regulation 526.1 of BS 7671**

Every connection between conductors or between a conductor and other equipment shall provide durable electrical continuity and adequate mechanical strength and protection. (Note: see Regulation 522.8 – Other mechanical stresses)

## Connection methods

The common connection methods are described below.

- **Screw clamping terminals** – only for use with copper cables, and commonly used in wiring accessories. These are the most common of termination methods. Some screw terminations may require tightening to a specific torque, as specified by the manufacturer.
- **Screwless terminals provided with clamping nuts** – used in maintenance-free accessories to BS 5733.
- **Compression terminations such as cable lugs or crimps** – suitable lugs can be used for aluminium cables and to terminate larger csa cables in switchgear. For terminating aluminium cables, the special lugs and compound must be used to reduce the electrolytic effect that occurs when jointing aluminium to other metals, such as brass.
- **Soldered lugs** – with the correct technique and materials, such as fluxes, and solder, these terminals can be used for jointing cables or terminating cables to components. However, mostly they have been replaced by compression-type terminations.
- **Cable resin joints** – used for cables, particularly those laid in the ground, and usually resin-filled to prevent the ingress of moisture. The connection method for the cores within the joint is typically by compression. The resin acts as protection against external influences once it sets hard.

KEY POINT

The solder often used for electrical joints is known as Tinman's solder, an alloy of tin, lead and antimony.

## Selection of connection method

BS 7671 Regulation 526.2 lists factors to be taken into account when selecting a connector.

> The selection of the means of connection shall take account of, as appropriate:
>
> **i** the material of the conductor and its insulation
>
> **ii** the number and shape of the wires forming the conductor
>
> **iii** the cross-sectional area of the conductor
>
> **iv** the number of conductors to be connected together
>
> **v** the temperature attained at the terminals in normal service such that the effectiveness of the insulation of the conductors connected to them is not impaired
>
> **vi** the provision of adequate locking arrangements in situations subject to vibration or thermal cycling (heating and cooling which can put stress on a termination).
>
> Where a soldered connection is used the design shall take account of creep, mechanical stress and temperature rise under fault conditions.
>
> (Source: BS 7671:2008 (2011))

**KEY POINT**

Terminals without the marking R (only rigid conductor), F (only flexible conductor), S or Sol (only solid conductor) are suitable for the connection of all types of conductors. Care must be taken to check the connection for its suitability.

Generally the connector used will be that provided in the equipment to be connected. However, care must be taken to ensure compliance with the equipment manufacturer's instructions. It must be confirmed that the cable is appropriate, neither too small nor too large, is suitable solid or stranded cable of adequate current-carrying capacity and correctly tightened. A torque wrench may be required for some connections.

It is important to note the following points:

- If the conductor current rating results in a conductor operating temperature exceeding 70°C, a cable with a larger cross-sectional area will be required. Few equipment cable connectors are suitable for a cable conductor temperature greater than 70°C. Cable loading should not lead to a conductor operating temperature exceeding 70°C at the connection.
- If aluminium conductor cables are used, special connections will be necessary. Before aluminium cables can be connected to equipment, aluminium/copper lugs need to be compression jointed to the aluminium conductors.
- Checks must be made if the ambient temperature exceeds 30°C.

## The requirements

BS 7671 Regulations 526.3 to 526.9 set forward the requirements for connections. These include:

- accessibility for inspection testing and maintenance (except for certain cable joints and for BS 5733 connectors)
- the temperature rise must not damage the insulation
- the termination must be made in suitable accessories or enclosures
- there shall be no mechanical strain
- enclosures shall be mechanically adequate and suitable for the environment
- the basic insulation of conductors (unsheathed or where the sheath has been removed for terminating the cable) shall be enclosed
- multi-wire and **fine-wire** conductors must be terminated in appropriate terminations, or conductor ends suitably prepared.

### ACTIVITY

Soldering is rarely used for general installation connections. It is far simpler and quicker to use crimp connections. Suggest two possible reasons (apart from those given) not to use soldered joints.

**Fine-wire cable**

Cable with many strands, which gives it much more flexibility. The conductors may need preparing by solder or tinning before they are terminated.

## Basic checks of terminations

According to Part 6 of BS 7671, all terminations should be checked, particularly by:

- tightening after first connecting
- checking again at inspection
- testing for continuity and earth fault loop impedance.

Checks will also confirm:

- the connections are under no undue strain (526.6)
- there is no basic insulation of a conductor visible outside the enclosure (526.8)
- the connections of live conductors are adequately enclosed (526.5)
- the connection is adequately connected at point of entry to the enclosure (glands, bushes, etc) (522.8.5).

## High current connections

High-current connections, particularly those on **busbar trunking**, may need to be checked using microhm meters, which are often called 'ductors'. The manufacturer's instructions for the busbar trunking will need to be followed. The test measures the resistance at the microhm or milliohm level, and is used primarily to verify that electrical connections are made properly and can detect:

- loose connections
- adequate tension on bolted joints
- eroded contact surfaces
- contaminated or corroded contacts.

**Busbar trunking system**

This is a type-tested assembly in the form of an enclosed conductor system that comprises solid conductors separated by insulating material. The assembly may consist of units such as:

- busbar trunking units, with or without tap-off facilities
- tap-off units, where applicable
- phase-transposition, expansion, building-movement, flexible, end-feeder and adaptor units.

Other system components may include tap-off units.

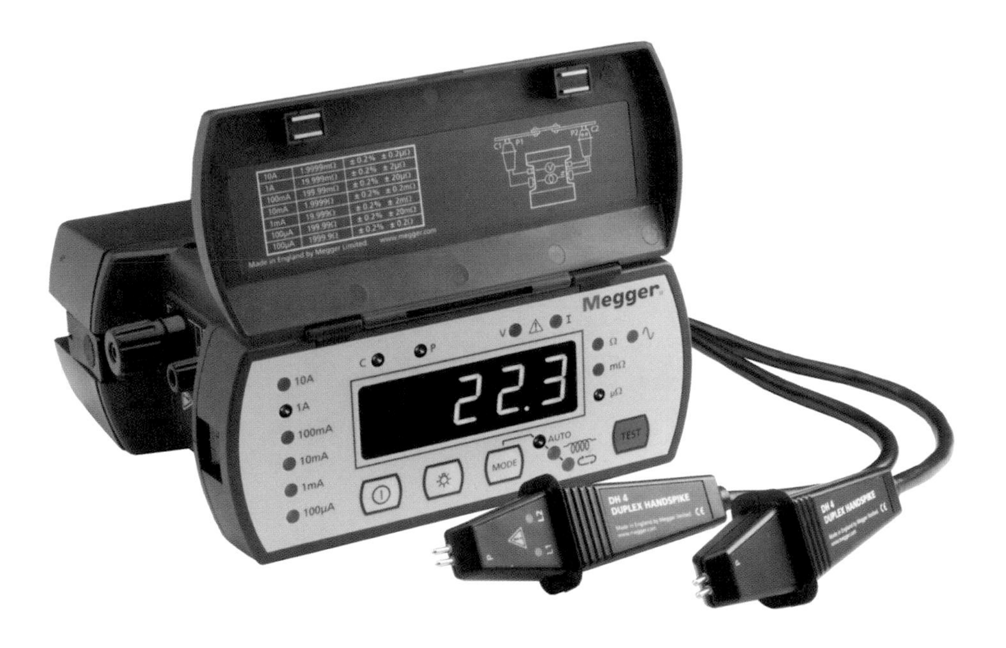


Ductor (microhmmeter)

## Thermal imaging cameras

A thermal imaging camera is a very useful aid when you need to check the condition of electrical installations without isolating the supply. Poor connections overheat and can often be identified by such cameras.

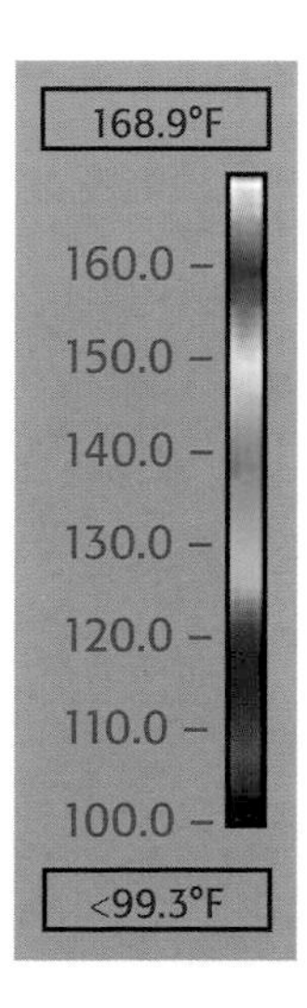


Colour to temperature correlation

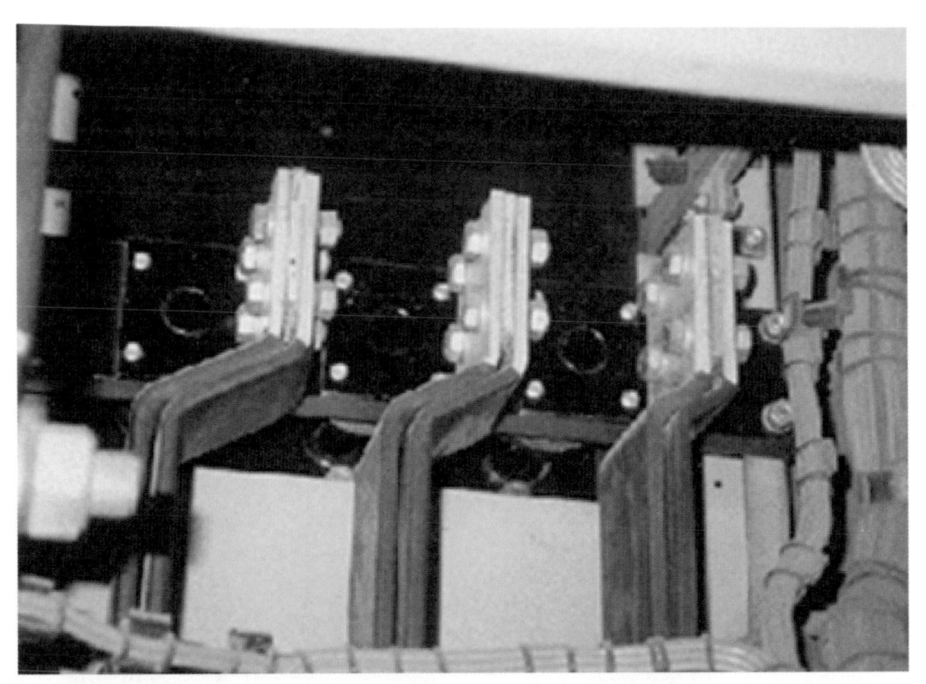

Bolted connections in a busbar

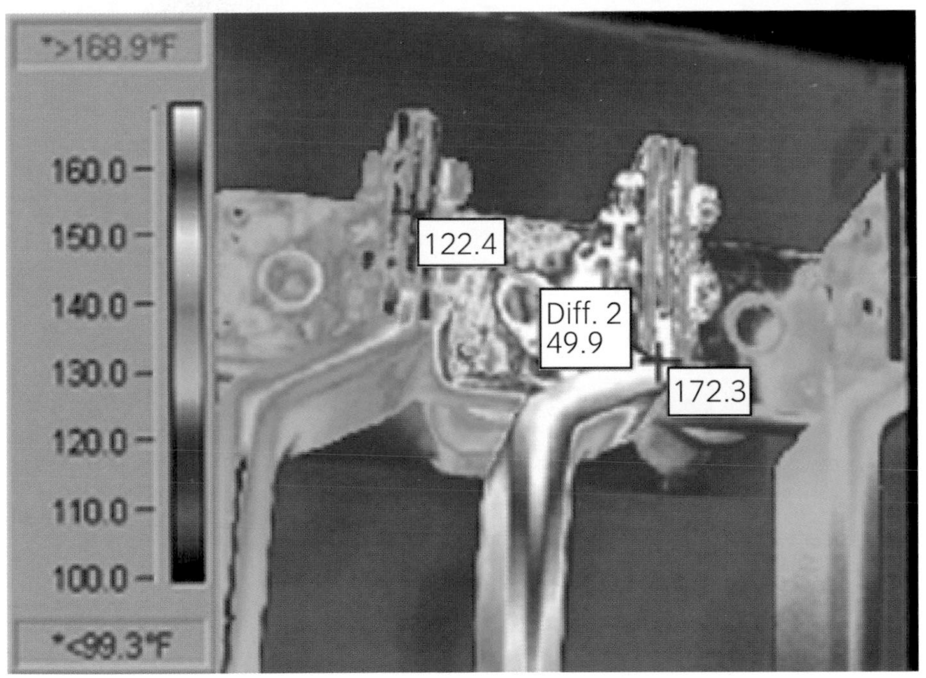


Bolted connections in a busbar, viewed using thermal imaging (centre contact running hot)

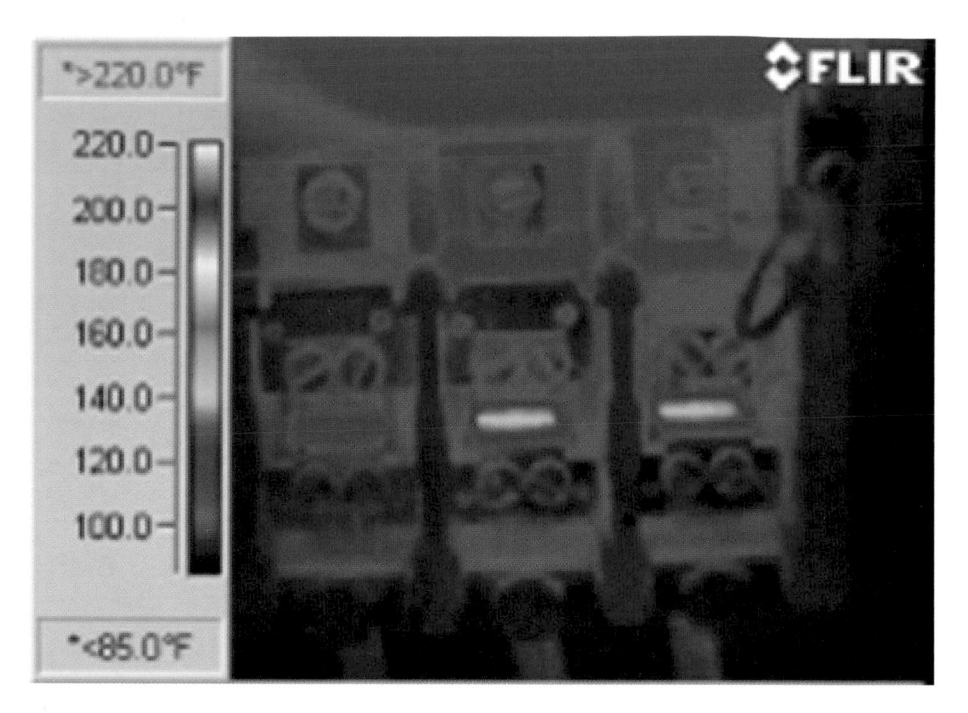


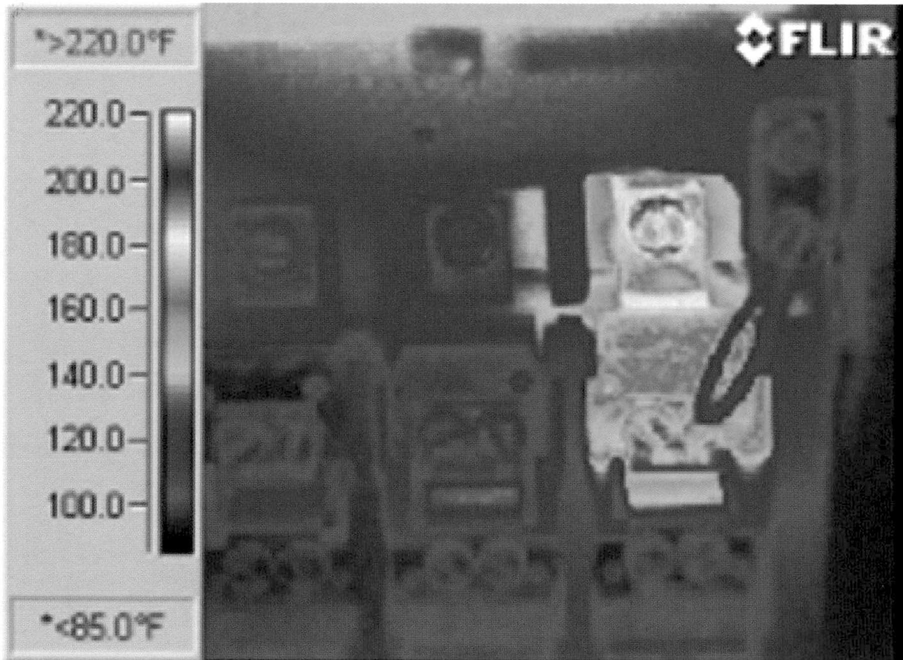


Thermal image of contactor (the right-hand termination is too hot in the bottom picture)

## Consequence of poor terminations

Poor terminations are the biggest causes of fires in electrical installations. If the resistance of a joint or termination increases, the termination or joint will increase in temperature and could ignite the accessory.

If a termination of joint on an earthing system becomes loose or of high resistance due to electrolysis, this may be undetected in a healthy circuit. This is why it is very important for all installations to be checked regularly by inspection and testing as poor joint or termination on an earthing system, if undetected, will cause the earth fault loop impedance to increase. This in turn will lead to longer disconnection times and, potentially a fatal electric shock. As earthing systems often connect to extraneous parts, these may be subject to greater external influences meaning they may require more regular checking and inspection.

Earthing systems are similar to airbags in a car. The airbag may be faulty but the car will work all the same. It is only when an accident occurs that you realise whether the airbags are working properly or not – and you want to see them working properly.

### ACTIVITY

Connections can be the weakest part of the installation, if not carried out correctly. List four possible connection faults.

OUTCOME 4

# Understand how to plan work schedules for electrical systems installation

This outcome introduces the practical aspects of working on site: organising the work, managing staff, working to a main contractor's programme and cooperating with other trades.

Assessment criteria

4.1 Describe effective working relationships between trades

## OTHER TRADES

All the trades on a construction project are dependent upon one another. No trade can work in isolation from the others. It is important that the electrical work does not hold up other trades and that other trades do not hold up the electrical work. Due to the nature of electrical installation work, an electrical installer will need to liaise with nearly every trade on a site. These are described below:

- **Bricklayer** – A trade contractor building brick and blockwork walls and finishes.
- **Joiner** – A trade contractor carrying out various tasks, including studded wall construction (usually timber internal walls), door and frame installation, skirting boards, cupboards and other fixtures to be fitted.
- **Plasterer** – A trade contractor applying wet plaster and finishes or, in some installations, providing dry-lined walls (walls without wet plaster) with specialist taped joints or minimal skimmed finish (a small amount of finishing plaster applied to the board).
- **Tiler** – A trade contractor who is responsible for tiled finishes; this is usually a finishing trade.
- **H&V fitter/engineer** – A trade contractor who carries out plumbing and heating installation works. These tradespeople normally carry out installations in steel, whereas plumbers normally carry out installations in copper.
- **Air-conditioning and refrigeration engineers** – These trades are aligned closely to the electrical trades, although the work involves an amount of pipework installation. The refrigeration engineer has to be aware of refrigeration legislation, as certain gases are banned in the UK.

- **Gas fitter** – A specialist trade contractor who is qualified to work on gas and combustion equipment installations, such as boilers. These individuals are registered with Gas Safe, which is a competent person scheme with the remit to ensure that registered tradespeople are qualified and competent to work on gas systems and boilers within the legal requirements.
- **Decorator** – A trade contractor who is responsible for decorated finishes to all surfaces and woodwork; this is usually a finishing trade.
- **Ground worker** – A manual worker associated with levelling of ground or excavations for substructure and foundation works and providing trenches for pipes or incoming services. Usually the ground worker is supported by a wide variety of earth-moving equipment and excavation tools, from a simple pick and shovel to complex hydraulic machines including the standard Backhoe (usually termed JCB, although it is a brand name) as well as other earth-moving equipment.

## PROGRAMMES OF WORK

Assessment criteria

4.2 Explain programme of work for electrical system installations

The main contractor will prepare a work schedule. All the sub-contractors, including the electrical sub-contractors, need to study the schedule.

The contractors must ascertain when they have to carry out the work and what they have to do to complete it. They can then determine:

- the materials required
- the type of work and the skills needed
- how many people are required, and when.

**Lead time** on the materials must be determined as soon as the contract is placed, so that materials are ordered at the appropriate time. It is important not to order materials too early, as you will not get paid for them until you install them. In practice, this is not simple.

**Lead time**

Lead time is the delay between the initiation and execution of a process. For example, the lead time in ordering, for instance, busbar trunking and delivery to site may be six weeks.

Once the number of people required for the work has been estimated, the contractors will have to look at their own programme of works to ensue that they can fit in with the main contractor's schedules.

If an organisation is responsible for holding up the work, they will be penalised financially. The cost of holding up other trades may far exceed the value of the electrical works. It is also absolutely essential for an organisation to programme their own jobs with a similar schedule so that customers can be told when the job will start and finish. When an organisation provides estimates for a job, it will need to include an indication of when the work can start and finish. Organisations can often run into trouble for being too busy and finding themselves unable to cope with the workload. Any profits from work can be lost through penalties, so it is essential that organisations carefully assess their workloads and plan accordingly.

Assessment criteria

4.3 Describe how to overcome difficulties that may arise when supervising electrical system installations

## OVERCOMING DIFFICULTIES WHEN SUPERVISING INSTALLATIONS

Problems will inevitably arise when carrying out work on a site. The important steps a supervisor can take to tackle problems arising are listed below:

- Keep up to date on the progress of their company's work.
- Ensure materials are ordered in advance and check they have been received *before* they are required.
- Ensure a manager knows well in advance of any need to increase or decrease staff on the job.
- Meet the main contractor regularly to receive information on progress of the job and monitor progress on the site.
- Obtain written orders, from persons with authority to issue them, before any extra work is undertaken (variation orders, or VOs).
- Consult with supervisors of other trades.
- Do what you can to assist other trades and ask for their help – in other words, cooperate.
- Make sure the site is secure: vandalism will cause a job to make a financial loss and it will delay completion. Discuss security with the client/main contractor.

Assessment criteria

4.4 Describe handover procedures

## HANDOVER

You will find more information about regulations in respect of handover procedures in BS 7671, Chapter 62.

### Inspection, testing, certification and reporting

Inspection, testing, certification and reporting must be carried out during and on completion of the work, as appropriate. This includes issuing an installation certificate and inspection and test result schedules to the person ordering the work.

### Operating instructions and manuals

Operating instructions must be provided for all equipment, including all useful information provided by suppliers and manufacturers. For large installations this will include detailed plans of the electricity distribution.

Where there are complex control systems that need to be monitored, these should be formally handed over and forms signed, or at least a letter sent advising the client that control has been handed over.

The Health and Safety at Work etc Act 1974 (HSW Act), Section 6, and the Construction (Design and Management) Regulations 2007 (CDM 2007) are concerned with the provision of information. Guidance on the preparation of technical manuals is given in the BS 4884 series (Technical manuals) and the BS 4940 series (Technical information on construction products and services). The size and complexity of the installation will dictate the nature and extent of the manual.

## Advice to customer

It will help all customers if the responsible person or the client is shown the installation and the particular features and locations of equipment noted.

For installations with residual current devices (RCDs), their function should be explained, particularly the need to trip test as per the labels and show how to reset them *before* calling out the contractor. Although most parts of an electrical installation are reasonably easy to operate, clients will need to be shown how to operate more complex items, such as home automation products and remote dimming units. It is always worth investing time in showing the client how to operate these items, to avoid wasting time later, when you are busy on other projects, visiting a client who is unable to operate something correctly.

**ACTIVITY**

Give details of two common domestic installation items which you may need to explain to the customer.

# WORK SCHEDULES

**Assessment criteria**

**4.5** Specify methods of producing and illustrating work programmes

## Job scheduling

Particular projects (jobs) can be programmed on a schedule. They can be:

- simple paper schedules, as shown below
- spreadsheets, such as those that are commercially available (see next page)
- computer programs.

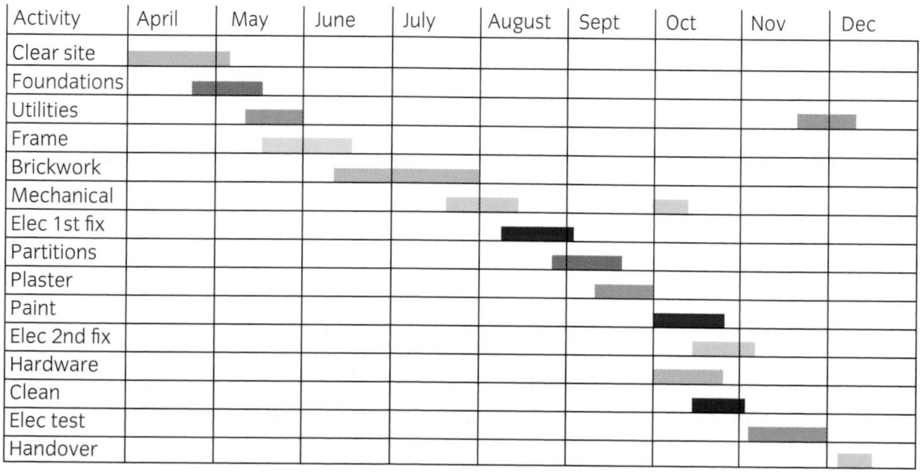

| Activity | April | May | June | July | August | Sept | Oct | Nov | Dec |
|---|---|---|---|---|---|---|---|---|---|
| Clear site | | | | | | | | | |
| Foundations | | | | | | | | | |
| Utilities | | | | | | | | | |
| Frame | | | | | | | | | |
| Brickwork | | | | | | | | | |
| Mechanical | | | | | | | | | |
| Elec 1st fix | | | | | | | | | |
| Partitions | | | | | | | | | |
| Plaster | | | | | | | | | |
| Paint | | | | | | | | | |
| Elec 2nd fix | | | | | | | | | |
| Hardware | | | | | | | | | |
| Clean | | | | | | | | | |
| Elec test | | | | | | | | | |
| Handover | | | | | | | | | |

**House construction schedule: simple paper version**

| | A | B | C | D | E | F | G | H | I | J | K | L | M | N | O | P | Q | R | S | T | U | V | W | X | Y |
|---|---|---|---|---|---|---|---|---|---|---|---|---|---|---|---|---|---|---|---|---|---|---|---|---|---|
| 1 | | | | | November 3–9 | | | | | | | November 10–16 | | | | | | | November 17–23 | | | | | | |
| 2 | **Task name** | **Duration** | **Start** | **Finish** | **M** | **T** | **W** | **T** | **F** | **S** | **S** | **M** | **T** | **W** | **T** | **F** | **S** | **S** | **M** | **T** | **W** | **T** | **F** | **S** | **S** |
| 3 | **Electrical** | | | | | | | | | | | | | | | | | | | | | | | | |
| 4 | drilling fabric | 1 | 06-Nov | 06-Nov | | | | | | | | | | | | | | | | | | | | | |
| 5 | wall chasing | 2 | 07-Nov | 08-Nov | | | | | | | | | | | | | | | | | | | | | |
| 6 | install cables | 2 | 10-Nov | 11-Nov | | | | | | | | | | | | | | | | | | | | | |
| 7 | install capping | 1 | 11-Nov | 11-Nov | | | | | | | | | | | | | | | | | | | | | |
| 8 | install boxes | 1 | 11-Nov | 11-Nov | | | | | | | | | | | | | | | | | | | | | |
| 9 | connect accessories | 2 | 21-Nov | 21-Nov | | | | | | | | | | | | | | | | | | | | | |
| 10 | test 1 | 1 | 22-Nov | 22-Nov | | | | | | | | | | | | | | | | | | | | | |
| 11 | commission | 1 | 22-Nov | 22-Nov | | | | | | | | | | | | | | | | | | | | | |
| 12 | | | | | | | | | | | | | | | | | | | | | | | | | |
| 13 | **Internal building** | | | | | | | | | | | | | | | | | | | | | | | | |
| 14 | block wall | 5 | 03-Nov | 07-Nov | | | | | | | | | | | | | | | | | | | | | |
| 15 | plaster 1 | 2 | 12-Nov | 13-Nov | | | | | | | | | | | | | | | | | | | | | |
| 16 | plaster finish | 2 | 17-Nov | 18-Nov | | | | | | | | | | | | | | | | | | | | | |
| 17 | ceiling | 1 | 11-Nov | 11-Nov | | | | | | | | | | | | | | | | | | | | | |
| 18 | ceiling finish | 1 | 18-Nov | 18-Nov | | | | | | | | | | | | | | | | | | | | | |
| 19 | paint walls 1 | 1 | 19-Nov | 19-Nov | | | | | | | | | | | | | | | | | | | | | |
| 20 | paint ceiling 1 | 1 | 19-Nov | 19-Nov | | | | | | | | | | | | | | | | | | | | | |
| 21 | paint walls finish | 1 | 20-Nov | 20-Nov | | | | | | | | | | | | | | | | | | | | | |
| 22 | paint ceiling finish | 1 | 20-Nov | 20-Nov | | | | | | | | | | | | | | | | | | | | | |
| 23 | | | | | | | | | | | | | | | | | | | | | | | | | |
| 24 | **Exterior building** | | | | | | | | | | | | | | | | | | | | | | | | |
| 25 | roofing | 6 | 28-Oct | 05-Nov | ← | | | | | | | | | | | | | | | | | | | | |
| 26 | soffit | 2 | 11-Nov | 12-Nov | | | | | | | | | | | | | | | | | | | | | |
| 27 | facia | 2 | 12-Nov | 13-Nov | | | | | | | | | | | | | | | | | | | | | |
| 28 | landscape | 12 | 05-Nov | 21-Nov | | | | | | | | | | | | | | | | | | | | | |
| 29 | | | | | | | | | | | | | | | | | | | | | | | | | |
| 30 | | | | | | | | | | | | | | | | | | | | | | | | | |
| 31 | | = | Material Delivery (against task) | | | | | | | | | | | | | | | | | | | | | | |
| 32 | | | | | | | | | | | | | | | | | | | | | | | | | |

House construction schedule: spreadsheet version

## Critical path

On larger construction sites, a work scheduling technique called the 'critical path' method (CPM) or critical path analysis is adopted. The purpose of this method is to determine the shortest time in which the project can be completed by sequencing the various construction activities. The outcome is the critical path that will lead to the minimum construction time.

The basic critical path technique is to:

- list all the activities necessary to complete the project
- add the time that each activity will take
- determine the dependency between the activities; for example, activity C cannot start until activity B is completed.

The analysis will identify those activities that are essential – that generally take the longest time to complete and form the critical path around which the other activities are to be organised.

In its simplest form, the two versions of the house construction schedule on pages 351 and 352 describe the critical path below.

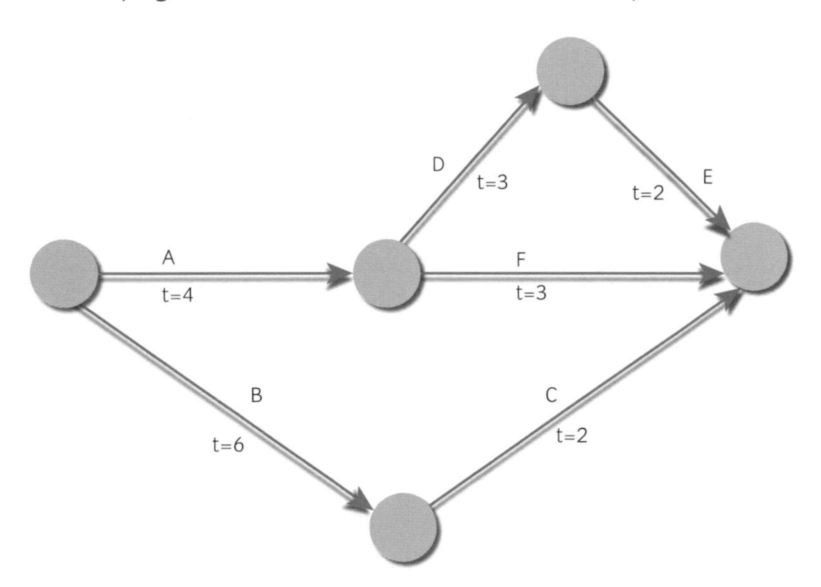


The job has 4 activities
Activity A takes 4 units of time and is not dependant on any other activity
Activity B takes 6 units of time and is not dependant on any other activity
Activity C takes 2 units of time and is dependant on completion of B
Activity D takes 3 units of time and is dependant on completion of A
Activity E takes 2 units of time and is dependant on completion of D
Activity F takes 3 units of time and is dependant on completion of A

The critical path of activity A through D and E and gives a minimum time of 9 units
Activity F is sub critical and has a float of 2 units
Path B through C is sub critical with a float of 1 unit

**Critical path method**

# ASSESSMENT CHECKLIST

## WHAT YOU NOW KNOW/CAN DO

| Learning outcome | Assessment criteria | Page number |
|---|---|---|
| 1 Understand how to interpret design information | *The learner can:* | |
| | 1 State the criteria used when selecting electrical systems | 250 |
| | 2 Explain positioning requirements when designing electrical systems | 256 |
| | 3 Describe the importance of sustainable design | 257 |
| | 4 Identify information required for electrical systems design | 258 |
| | 5 State additional considerations for special locations | 265 |
| | 6 Interpret measurements from design plans | 266 |
| | 7 Research cost of equipment used in electrical systems using different sources | 267 |
| | 8 Calculate requirements for electrical systems for a domestic property | 270 |
| | 9 Explain implications for tender content. | 272 |
| 2 Understand principles for designing electrical systems | *The learner can:* | |
| | 1 Explain how regulatory requirements impact upon the installation of electrical systems | 276 |
| | 2 Describe factors affecting selection of electrical systems | 278 |
| | 3 Explain how a circuit's maximum demand is established after diversity factors are applied | 282 |
| | 4 Assess suitability of types of wiring systems for different environments | 290 |
| | 5 Determine minimum current-carrying capacity of live conductors for given installation conditions | 295 |
| | 6 Verify circuit disconnection time is achieved | 310 |
| | 7 Verify discrimination of protective devices | 319 |
| | 8 Verify breaking capacities of protective devices | 322 |
| | 9 Verify thermal constraints. | 326 |

| Learning outcome | Assessment criteria | Page number |
|---|---|---|
| 3 Understand procedures for connecting complex electrical systems | *The learner can:* | |
| | 1 Interpret sources of information for the connection of complex electrical systems | 335 |
| | 2 Assess connection methods for given situations | 341 |
| | 3 Describe procedures for proving that terminations and connections are sound | 341 |
| | 4 Explain consequences of unsound termination. | 341 |
| 4 Understand how to plan work schedules for electrical systems installation | *The learner can:* | |
| | 1 Describe effective working relationships between trades | 348 |
| | 2 Explain programme of work for electrical systems installation | 349 |
| | 3 Describe how to overcome difficulties that may arise when supervising electrical system installations | 350 |
| | 4 Describe handover procedures | 350 |
| | 5 Specify methods of producing and illustrating work programmes. | 351 |

## ASSESSMENT GUIDANCE

The assessment for this unit is split into three written assessments.

Task A is a project with 12 questions based on plans supplied to you.

Task B is a project with seven questions based on the plans supplied to you.

Tasks A and B are open book and you may bring research materials in with you, and use the internet.

You will need writing and drawing materials and a scientific non-programmable calculator.

You may use BS 7671 and the IET On-Site Guide.

Task C has 10 questions and is an open book written paper.

This is done under exam conditions.

You are permitted to take BS 7671 and the IET On-Site Guide into the examination with you.

Make sure you arrive on time for any assessment you take.

If you are unsure of anything, ask – be clear what you are doing.

## KNOWLEDGE CHECK

These questions are typical of but not copies of questions found in the Unit 305 open book assessment. You will need to look at a copy of BS 7671 to help you answer these questions.

1 State the purpose of the following items within an electrical installation.
   - a) DNO 100 A double pole switch.
   - b) 30 mA RCD.
   - c) 100 A consumer unit DP switch.
   - d) Individual circuit breaker in consumer control unit.

2 Explain why a neutral to earth fault on a circuit protected by an RCBO may cause the RCBO to trip.

3 State why, in an electrical installation, the
   - a) value of $R_1 + R_2$ increases as the circuit length increases
   - b) value of $Z_s$ increases as the circuit length increases
   - c) value of short circuit decreases as the circuit length increase
   - d) conductor resistance increases as the temperature increases.

4 State what the following symbols represent.
   - a) $I_z$
   - b) $I_t$
   - c) $I_{\Delta n}$
   - d) $C_a$
   - e) $C_g$

5 Protective multiple earthing systems (PME) require the use of additional electrodes in the supply system. State a reason for these.

6 A direct-on-line motor starter contains two types of protection provided by the different components inside. State each component and the means of protection it provides.

7 State the difference between an emergency switch and a functional switch, giving an example of each.

8 The adiabatic equation is given as $S = \frac{\sqrt{I^2t}}{k}$

Identify what each symbol within the formula represents.

9 Complete the following table, indicating the application of each device.

Indicate Yes / No.

| Device | Standard | Isolation | Emergency | Functional |
|---|---|---|---|---|
| Device with semi-conductor | BS EN 60669-2-1 | | | |
| Plug and socket | BS 1363 | | | |
| Switched fused connection unit | BS 1363 | | | |
| Fuse | BS 1362 | | | |
| Cooker control unit | BS4177 | | | |

10 Give a simple explanation of zone 0 as applied to a bathroom and a swimming pool.

# UNIT 304
# Electrical installations: inspection, testing and commissioning

Inspection, testing, commissioning and certification are very important stages of the safety process. Electricity can be very dangerous if the guidelines in the current edition of the Institution of Engineering and Technology (IET) Wiring Regulations (BS 7671) are not followed.

Electrical inspection and testing can involve some degree of risk but if the guidelines are followed, the risk can be minimised. Every year, the Health and Safety Executive investigates many cases of injury due to bad practice and poor workmanship. Electric shock, burns and fatalities can happen; however, if a member of a workforce is killed at work, the organisation could face a corporate manslaughter case. Employers, managers and employees can be fined or sent to prison for neglecting health and safety guidelines.

This unit covers the health and safety requirements, and includes the theory and testing skills required to become competent during the process of inspection, testing and commissioning of electrical installations.

## LEARNING OUTCOMES

There are six learning outcomes to this unit. The learner will:

1. know requirements for commissioning of electrical systems
2. understand procedures for the inspection of electrical systems
3. understand procedures for completing the testing of electrical systems
4. understand requirements for documenting installing electrical systems
5. be able to inspect electrical wiring systems
6. be able to test safety of electrical systems.

In this book, the coverage of learning outcomes and assessment criteria for this unit is not in the sequence of the assessment checklist that you can find on page 492. It has been organised to provide a route through the content that best links together for overall learning.

This unit will be assessed by:

- two short-answer knowledge assignments (open book)
- a practical task to demonstrate knowledge of inspection and testing (open book)
- one short-answer knowledge test with a practical demonstration (closed book).

# OUTCOMES 1, 3, 5, 6

## Know requirements for commissioning of electrical systems
## Understand procedures for completing the testing of electrical systems
## Be able to inspect electrical wiring systems
## Be able to test safety of electrical systems

**Assessment criteria**

1.1 Specify the regulatory requirements for inspection of electrical systems

## HEALTH AND SAFETY LEGAL REQUIREMENTS

The flowchart shows the three most relevant statutory (legal) requirements that apply during electrical inspection, testing and commissioning.

If an accident occurs in the workplace, the Health and Safety at Work etc Act 1974 (HSW Act) will be addressed initially, because this is the general standard for all workplace tasks – including electrical work. It puts the duty of care on both employer and employee to ensure the safety of all persons using the work premises.

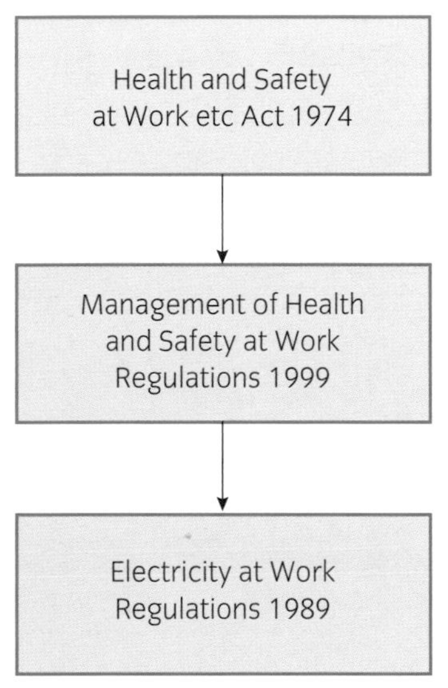
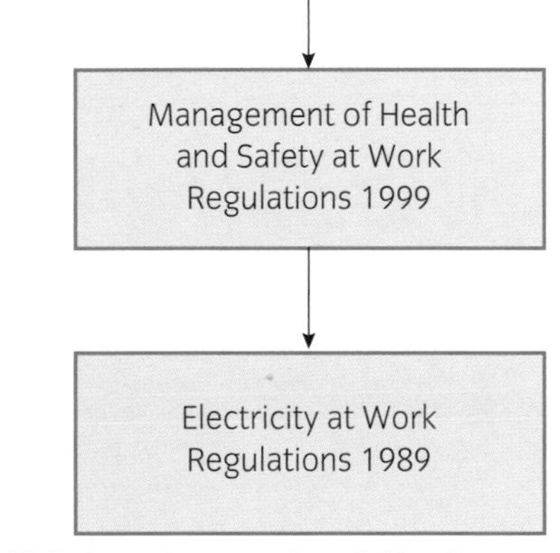


**Statutory documents relating to inspection and testing**

If the accident is electrical in nature, other statutory requirements may also be relevant:

- *The Management of Health and Safety Regulations 1999* also apply to general workplace tasks and training, and state that every employer must make a suitable and sufficient assessment of the risks to which employees are exposed. Training must be provided for employees to be able to carry out the tasks that they are expected to do in the workplace.
- *The Electricity at Work Regulations 1989* are focused on electrical work and methods of safety. They must be followed to ensure that the electrical inspector, other trades and members of the public are not put at risk during the process of inspection and testing. Property and equipment must also be guarded against misuse and damage.

### Importance of the legal requirements

It is vital to work in a safe way and that means following the legal requirements. If an accident occurs, for example, and an electrician gets blinded by arcing or by an explosion caused by a short circuit during testing, the following three questions could be asked in a court of law:

1 Did the employee take reasonable care when testing? (Regulation 7 of The Health and Safety at Work etc Act 1974)
2 Was the employee trained and capable of carrying out the task requested by the employer? (Regulation 13 of The Management of Health and Safety at Work Regulations 1999)

**SmartScreen Unit 304**

Additional resources to support this unit are available on SmartScreen.

3 Was the electrician competent and could the work have been carried out with the circuit isolated? (Regulations 16 and 14 of Electricity at Work Regulations 1989)

Contravention of any of these regulations could result in the employee, such as the electrician, and the employer being found guilty of a crime, and incurring heavy fines or being sent to prison.

## Electricity at Work Regulations

You are expected to know the Electricity at Work Regulations 1989 (EAWR) and to understand how these impose responsibilities on the **duty holder** (you), when inspecting, testing, commissioning and certificating a new circuit or installation.

Some important extracts from the EAWR 1989 are listed below.

### PART II GENERAL – Electricity at Work Regulations 1989

**Regulation 4 – Systems, work activities and protective equipment**

(1) All systems shall at all times be of such construction as to prevent, so far as is reasonably practicable, danger.

**Regulation 13 – Precautions for work on equipment made dead**

Adequate precautions shall be taken to prevent electrical equipment, which has been made dead in order to prevent danger while work is carried out on or near that equipment, from becoming electrically charged during that work if danger may thereby arise.

**Regulation 14 – Work on or near live conductors**

No person shall be engaged in any work activity on or so near any live conductor (other than one suitably covered with insulating material so as to prevent danger) that danger may arise unless –

(a) it is unreasonable in all the circumstances for it to be dead; and

(b) it is reasonable in all the circumstances for him to be at work on or near it while it is live; and

(c) suitable precautions (including where necessary the provision of suitable protective equipment) are taken to prevent injury.

**Regulation 15 – Working space, access and lighting**

For the purposes of enabling injury to be prevented, adequate working space, adequate means of access, and adequate lighting shall be provided at all electrical equipment on which or near which work is being done in circumstances which may give rise to danger.

**Regulation 16 – Persons to be competent to prevent danger and injury**

No person shall be engaged in any work activity where technical knowledge or experience is necessary to prevent danger or, where appropriate, injury, unless he possesses such knowledge or experience, or is under such degree of supervision as may be appropriate having regard to the nature of the work.

**KEY POINT**

You can find additional information with in-depth explanations in the 'Memorandum of guidance on the Electricity at Work Regulations 1989'. This can be downloaded from the Health and Safety Executive website: www.hse.gov.uk

**Duty holder**

The person responsible for actions and matters that are within their control.

**ACTIVITY**

Read Regulation 29 of the Electricity at Work Regulations and discuss its importance regarding your legal rights.

## How these regulations affect the inspection and testing process

The Electricity at Work Regulations 1989 affect the inspection and testing process as follows.

- Regulation 4(1) identifies the need to ensure that new electrical installations and circuits are subject to inspection, testing and commissioning before being put into service.
- Regulation 13 identifies the need for equipment to be 'made **dead**' during the process of initial verification when undertaking dead tests. Isolation procedures will need to be adopted (see pages 363–364 for isolation procedures).
- Regulation 14 identifies that working on or near **live** conductors will be necessary during initial verification, for example, in distribution boards and consumer units. Risk assessments, test equipment and safe working practices are covered in more detail on pages 368–370.
- Regulation 15 involves working space, **access** and lighting. Good design and installation methods will ensure that access to electrical equipment is in accordance with these requirements.
- Regulation 16 emphasises **competence** and the need to fully understand the task. There are many types of electrician – industrial, commercial and domestic, to name a few. When undertaking inspection, testing and commissioning, the duty holder must have the appropriate knowledge and experience of the system to work in a safe manner.

**KEY POINT**

The following phrase uses the highlighted terms and can act as a useful prompt for remembering Regulations 13, 14, 15 and 16 of the EAWR:

**dead** or **live, access** with **competence.**

**KEY POINT**

The terminology used in BS 7671 for inspecting, testing, commissioning and certificating a new circuit is 'initial verification'.

**ASSESSMENT GUIDANCE**

You will be assessed by means of written assignments and expected to answer questions relating to inspecting and testing electrical installations in accordance with the Electricity at Work Regulations 1989.

## THE CORRECT PROCEDURE FOR SAFE ISOLATION

**Assessment criteria**

3.3 Explain how to prepare for testing electrical systems

Isolation can be very complex due to the differing industrial, commercial and domestic working environments, some of which require experience and knowledge of the system processes.

This section deals with a basic practical procedure for the isolation and for the securing of isolation. It also looks at the reasons for safe isolation and the potential risks involved during the isolation process.

## How to undertake a basic practical procedure for isolation

### Gather together all of the equipment required for this task

You will need the following equipment:

- a voltage indicator which has been manufactured and maintained in accordance with Health and Safety Executive (HSE) Guidance Note GS38
- a proving unit compatible with the voltage indicator
- a lock and/or multi-lock system (there are many types of lock available)
- warning notices which identify the work being carried out
- relevant Personal Protective Equipment (PPE) that adheres to all site PPE rules.

**ACTIVITY**

Check your approved voltage indicator for any damage and to comply with GS38.

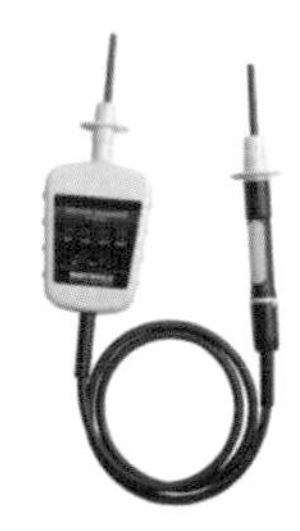

Voltage indicator    Proving unit

The equipment shown in the photographs can be used to isolate various main switches and isolators. To isolate individual circuit breakers with suitable locks and locking aids, you should consult the manufacturer's guidance.

When working on or near electrical equipment and circuits, it is important to ensure that:

- the correct point of isolation is identified
- an appropriate means of isolation is used
- the supply cannot inadvertently be reinstated while the work is in progress
- caution notices are applied at the point(s) of isolation
- conductors are proved to be dead at the point of work before they are touched
- safety barriers are erected as appropriate when working in an area that is open to other people.

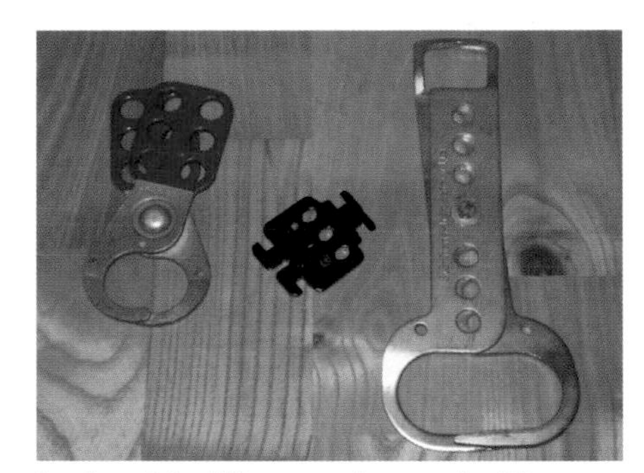

Lock-out facility – can be used with one or more locks

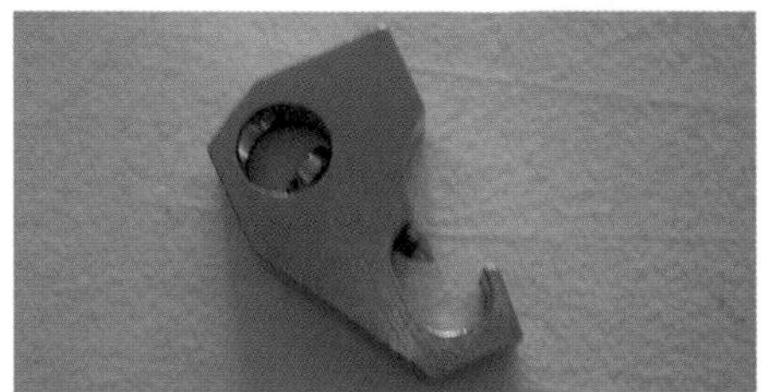

Lock-out devices for circuit breakers and RCBOs

Typical warning notices

## Carry out the practical isolation

The method of isolation is outlined below.

1 *Identify* – identify equipment or circuit to be worked on and point(s) of isolation.

2 *Isolate* – switch off, isolate and lock off (secure) equipment or circuit in an appropriate manner. Retain the key and post caution signs with details of work being carried out.

3 *Check* – check the condition of the voltage indicator leads and probes. Confirm that the voltage indicator is functioning correctly by using a proving unit.

4 *Test* – using voltage indicator, test the outgoing terminals of the isolation switch. Take precautions against adjacent live parts where necessary.

   - During single-phase isolation there are three tests to be carried out:
     L – N
     L – E
     N – E
     (L = Line, N = Neutral, E = Earth)
   - During three-phase isolation there are 10 possible tests (if the neutral is present):

| | | |
|---|---|---|
| L1 – N | L2 – N | L3 – N |
| L1 – E | L2 – E | L3 – E |
| L1 – L2 | L1 – L3 | L2 – L3 |
| N – E | | |

     (L = Line, N = Neutral, E = Earth)

5 *Prove* – using voltage indicator and proving unit, prove that the voltage indicator is still functioning correctly.

6 *Confirm* – confirm that the isolation is secure and the correct equipment has been isolated. This can be achieved by operating functional switching for the isolated circuit(s).

The relevant inspection and testing can now be carried out.

Isolator locked and tagged (secured)

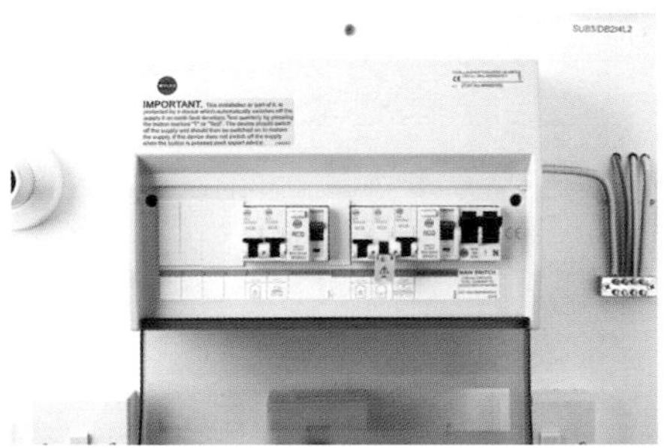

Circuit breaker locking device (not secured)

**ACTIVITY**

Why is a check between N and E necessary as part of the isolation process to ensure safety?

## Reinstate the supply

When the 'dead' electrical work is completed, you must ensure that all electrical barriers and enclosures are in place and that it is safe to switch on the isolated circuit.

1 Remove the locking device and danger/warning signs.

2 Reinstate the supply.

3 Carry out system checks to ensure that the equipment is working correctly.

**ASSESSMENT GUIDANCE**

You will be required to demonstrate to your tutor that you can safely isolate an individual circuit or full board, in compliance with industry standards.

## The requirements of the Electricity at Work Regulations 1989

When undertaking the correct procedure for isolation, you will need to abide by Regulation 13 and Regulation 14, as shown below.

### PART II GENERAL – Electricity at Work Regulations 1989

**Regulation 13 – Precautions for work on equipment made dead**

Adequate precautions shall be taken to prevent electrical equipment, which has been made dead in order to prevent danger while work is carried out on or near that equipment, from becoming electrically charged during that work if danger may thereby arise.

**Regulation 14 – Work on or near live conductors**

No person shall be engaged in any work activity on or so near any live conductor (other than one suitably covered with insulating material so as to prevent danger) that danger may arise unless–

(a) it is unreasonable in all the circumstances for it to be dead;

**KEY POINT**

You can find additional information with in-depth explanations of the EAWR 1989 Regulations in relationship to basic (and also to complex isolation) in the safe working practices booklet. This can be freely downloaded from the Health and Safety Executive website: www.hse.gov.uk

## How these regulations affect the inspection and testing process

The Electricity at Work Regulations 1989 affect the inspection and testing process as follows.

- Regulation 13 identifies the need for equipment to be 'made dead' during the process of initial verification when undertaking dead tests. Isolation procedures will be needed.
- Regulation 14 acknowledges that during isolation, an electrician may be working on or near live conductors.

**ACTIVITY**

Assemble the equipment required to isolate a circuit, and simulate isolation on various isolators and circuit breakers that are not connected to a supply. This can be done safely on a workbench or desk.

**ASSESSMENT GUIDANCE**

You will be assessed by means of written or practical assignments and expected to answer questions relating to methods of isolation, testing and securing isolation.

# SAFE ISOLATION AND IMPLICATIONS

When you isolate an electricity supply, there will be disruption. So, careful planning should precede isolation of circuits.

Consider isolating a section of a nursing home where elderly residents live. You will need to consult the nursing home staff, to consider all the possible consequences of isolation and to prepare a procedure.

The following questions are useful:

1 How will the isolation affect the staff and other personnel?
For example, think about loss of power to lifts, heating and other essential systems.

2 How could the isolation affect the residents and clients?
For example, some residents may rely on oxygen, medical drips and ripple beds to aid circulation. These critical systems usually have battery back-up facilities for short durations.

3 How could the isolation affect the members of the public?
For example, fire alarms, nurse call systems, emergency lighting and other systems may stop working.

4 How can an isolation affect systems?
For example, IT programs and data systems could be affected; timing devices could be disrupted.

In this scenario, you must make the employers, employees, clients, residents and members of the public aware of the planned isolation. Alternative electrical back-up supplies may be required in the form of generators or uninterruptable power supply systems.

## ISOLATION RISKS AND IMPLICATIONS

Before any isolation is carried out, you must assess the risks involved. This section deals with the practical implications and the risks involved during the isolation procedure, if risk assessment and method statements are not followed.

### Who is at risk and why?

If isolation is *not* carried out safely, what are the possible risks when performing inspection and testing tasks?

#### Risks to you

*Risks* to you might include:

- shock – touching a line conductor, eg if isolation is not secure
- burns – resulting from touching a line conductor and earth, or arcing
- arcing – due to a short circuit between live conductors, or an earth fault between a line conductor and earth
- explosion – arcing in certain environmental conditions, eg in the presence of airborne dust particles or gases, may cause an explosion.

**ACTIVITY**

Why is the N or E probe connected before the Line when carrying out this test?

Things that can *cause risk* to you include:

- inadequate information to enable safe or effective inspection and testing, i.e. no diagrams, legends or charts
- poor knowledge of the system you are working on (and so not meeting the competence requirements of Regulation 16 of EAWR)
- insufficient risk assessment
- inadequate test instruments (not manufactured or maintained to the standards of GS38).

## Risks to other tradespersons, customers and clients

Risks to other tradespersons, customers and clients might include:

- switching off electrical circuits – for example, switching off a heating system might cause hypothermia; if lifts stop, people may be trapped.
- applying potentially dangerous test voltages and currents
- access to open distribution boards and consumer units
- loss of service or equipment, for example:
  - loss of essential supplies
  - loss of lights for access
  - loss of production.

## Risks to members of the public

Risks to members of the public might include prolonged loss of essential power supply, causing problems, for example, with safety and evacuation systems, such as:

- fire alarms
- emergency exit and corridor lighting.

Note that, although safety services usually have back-up supplies such as batteries, these may only last for a few hours. Other safety or standby systems may have generator back-up, but this will also require isolation, leaving the building without any safety systems.

## Risks to buildings and systems within buildings

Risks to buildings and systems within buildings might involve applying excessive voltages to sensitive electronic equipment, for example:

- computers and associated IT equipment
- residual current devices (RCDs) and residual current operated circuit breakers with integral overcurrent protection (RCBOs)
- heating controls
- surge protection devices.

There might also be risk of loss of data and communications systems.

**ASSESSMENT GUIDANCE**

You will be assessed by means of a written assignment and expected to answer questions relating to risks involved during the isolation, inspection and testing of buildings (closed-book questions).

**ACTIVITY**

1. Write down a list of the risks associated with isolation and the effects isolation can have on people, livestock, systems and buildings.
2. Who is at risk if inspection and testing is not carried out correctly?
3. What might happen if you need to switch off a socket outlet circuit in a hospital?
4. What must you do if you encounter a computer server that requires a permanent supply and you need to switch off the main supply to enable safe testing procedures?

**Assessment criteria**

**5.1** Implement safe system of work for inspection of electrical systems

**6.1** Implement safe system of work for testing electrical systems

# RISK ASSESSMENTS, SAFE WORKING PROCEDURES AND EQUIPMENT

You should already be familiar with risk assessments, permits to work and method statements. This section focuses on how to apply these documents to the tasks involved in inspection and testing.

It is essential to have approved documentation, protective equipment, tools and test equipment. All of these must be supplied, maintained and updated (or replaced) as deemed necessary.

## Risk assessments for inspection and testing of electrical installations

Initial verification requires inspection and testing to be carried out correctly, and in a safe and competent manner.

The table gives the five stages of risk assessment. Use a copy of the table in the activity.

### ACTIVITY

Discuss with your class or colleagues, the dangers associated with working on or near live conductors.

**Example of risk assessment during initial verification (partially completed)**

| 1 What are the hazards? | 2 Who might be affected? Why? How? | 3 a) What are you already doing to reduce the risk of danger and injury? | 3 b) What further action is required? | 4 How will the risk assessment be implemented? | 5 When will the risk assessment be reviewed? |
|---|---|---|---|---|---|
| Working on or near live conductors when carrying out live tests in accordance with BS 7671. | Inspectors and testers of the electrical systems and … | Providing and maintaining correct and appropriate equipment, and … | Training on equipment likely to be encountered during the tasks (to ensure competence) and … | Practical assessment for various procedures, such as … | If new systems are encountered, new test equipment is supplied, new hazards are identified or … |
| Isolation to ensure safe inspection and testing of electrical installations during initial verification. | | | | | |

## ACTIVITY

1 Look around the room or workplace you are in and identify one possible risk to yourself or others.

2 Create your own table using the five stages of risk assessment shown in the table on page 368. Consider your own workplace when filling in the table.

## Reporting of unsafe situations

In order to prevent accidents, you should always report unsafe situations. Causes of unsafe situations could be:

- unsuitable or inappropriate PPE
- inadequate test equipment
- lack of training
- poor awareness of dangers
- carelessness
- incompetence.

## Safe use of tools, test equipment and personal protective equipment

You have already covered the need for PPE and the use of correct tools. Now you are going to look at a brief list of equipment and to consider the correct use and maintenance of this equipment during the initial verification process.

The tools and equipment listed below are typical of what may be needed during isolation, inspection and testing:

- locks and keys appropriate to the systems to be worked on (permits to work may be required)
- warning signs, notices and labelling
- approved voltage indicator – see GS38 information listed on page 370
- correct proving unit for the specific voltage indicator
- safety barriers to prevent access by other trades, clients and members of the public
- approved test equipment designed to perform tasks in accordance with BS 7671 – see GS38 information listed on page 370
- PPE appropriate to the work to be undertaken
- hand tools, suitably insulated and maintained.

The GS38 information listed below relates to the design and maintenance of approved electrical test equipment for use by electricians.

- Probes should have:
  - finger barriers
  - an exposed metal tip not exceeding 4 mm; however, it is strongly recommended that this is reduced to 2 mm or less
  - fuse, or fuses, with a low current rating (usually not exceeding 500 mA), or a current-limiting resistor and a fuse.
- Leads should be:
  - adequately insulated
  - colour coded
  - flexible and of sufficient capacity
  - protected against mechanical damage
  - long enough
  - sealed into the body of the voltage detector and should not have accessible exposed conductors, other than the probe tips.

Individual types of test equipment specifications are addressed on pages 403–408.

**KEY POINT**

You can find GS38 as a 10-page free download on the Health and Safety Executive website: www.hse.gov.uk

HSE Health and Safety Executive

**Electrical test equipment for use by electricians**

Guidance Note GS38

The guidance in this document is aimed at electrically competent people including electricians, electrical contractors, test supervisors, technicians, managers and/or appliance retailers.

The Electricity at Work Regulations 1989 require those in control of all or part of an electrical system to ensure it is safe to use and maintained. This document provides advice and guidance on how to achieve this.

ISBN 978 0 7176 0845 4
Price: £5.00

Contents



HSE Books

OUTCOME 1

# Know requirements for commissioning of electrical systems

## WHY YOU NEED TO INSPECT AND TEST ELECTRICAL INSTALLATIONS

Inspection and testing of electrical installations is carried out for initial verification and for periodic inspection and testing:

- *Initial verification* is carried out to ensure that a new circuit (or installation) has been designed, installed, inspected and tested in accordance with BS 7671. This confirms that the installation is in a safe and suitable condition for use.
- *Periodic inspecting and testing* is carried out to assess the on-going safety of the existing electrical installation. This periodic report is *not* a certificate, since a certificate must have all inspection and test criteria in accordance with BS 7671 with no faults or unsatisfactory comments. The periodic report allows for limitations and is used to give an overview of the on-going condition of the installation and whether it is suitable for continued use.

**Assessment criteria**

1.2 Specify the regulatory requirements for testing electrical systems

**ACTIVITY**

Guidance Note 3 gives recommended periods of time between tests for different installations. How often should domestic dwellings be tested?

## STATUTORY AND NON-STATUTORY DOCUMENTATION, AND REQUIREMENTS

Listed below are some of the documents (with explanations) that affect the processes of initial verification and periodic inspection and testing.

**Assessment criteria**

1.3 Specify the regulatory requirements for commissioning of electrical systems

### Statutory documents

There are five statutory documents that you need to abide by:

1. The Health and Safety at Work etc Act 1974 (HSW Act). This imposes general duties on employers to ensure the health, safety and welfare at work of all employees. These duties apply to virtually everything in the workplace, including electrical systems and installations.
2. The Electricity at Work Regulations 1989 (EAWR). These are concerned specifically with electrical safety.

**ASSESSMENT GUIDANCE**

You will be required to quote both statutory and non-statutory documents that relate to the process of inspection, testing and commissioning. This section will give you the information required to answer written assessments.

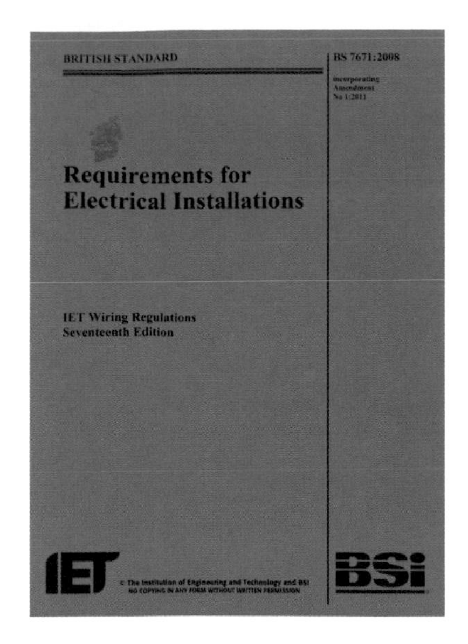

3 The Management of Health and Safety at Work Regulations 1999. Regulation 3 requires every employer to carry out an assessment of the risks to workers and any others who may be affected.

4 The Provision and Use of Work Equipment Regulations 1998. These apply to any machinery, appliance, apparatus or tool used for carrying out any work activity, whether or not electricity is involved. Regulation 5 requires such equipment to be maintained in an efficient state, in working order and in good repair. Also, any maintenance logs for machinery must be kept up to date.

5 Electricity Safety, Quality and Continuity Regulations 2002.These apply to public and consumer safety with regard to electrical distribution and supply authorities.

## Non-statutory documents

The following non-statutory documents are important and are referred to throughout this book:

1 The Institution of Engineering and Technology (IET) Wiring Regulations (BS 7671) are based on internationally agreed documents – International Electrotechnical Commission (IEC) harmonised documents – and are international safety rules for electrical installations. A British Standards Institute (BSI) IET-led committee, known as JPEL 64, compiles a UK edition of these rules, selecting regulations specifically for the UK in agreement with a European committee known as CENELEC.

2 The IET On-Site Guide (OSG) to BS 7671 is a simple guide to the requirements from the practical approach of designing, installing, inspecting and testing electrical installations. It can be used as a quick reference guide, but electrical installers should always consult BS 7671 to satisfy themselves of compliance. It is expected that people carrying out work in accordance with this guide are competent to do so.

3 The IET Guidance Note 3: Inspection and Testing to BS 7671 is one of a number of publications prepared by the IET, giving guidance to BS 7671. GN3 is a descriptive guide to the requirements of BS 7671 and provides specific guidance on inspection and testing. Electrical installers should always consult BS 7671 to satisfy themselves of compliance. It is expected that people carrying out work in accordance with this guide are competent to do so.

4 The Guidance Note GS38 – Electrical test equipment for use by electricians (published by the Health and Safety Executive, HSE) was written as a guideline to good practice when using test equipment. It is intended to be followed in order to reduce the risk of danger and injury when performing electrical tests.

## Requirements for initial verification and periodic inspection and testing

Within the non-statutory documentation, there are forms that must be completed for both initial verification and periodic inspections.

### Initial verification forms

Initial verification forms must be completed for all new installations and new circuits.

The forms shown here are completed samples. The information given in the forms is not too relevant at this stage, but you do need to know *which* forms are required for satisfactory initial verification:

1 Electrical Installation Certificate
2 Schedule(s) of Inspections
3 Generic Schedule(s) of Test Results

All documents must be completed and authenticated by a competent person(s) before certification is handed over.

The sample documents are based on a new electrical installation in a small office. The office has its own supply and metering system.

Always check you are using the latest forms, as found on the IET website: http://electrical.theiet.org

**ACTIVITY**

List who is responsible for signing an Electrical Installation Certificate.

**ASSESSMENT GUIDANCE**

You will be expected to correctly fill in all the forms for reporting and certification, as required.

Form 2 Form No: 555513....../2

ELECTRICAL INSTALLATION CERTIFICATE
(REQUIREMENTS FOR ELECTRICAL INSTALLATIONS - BS 7671 [IET WIRING REGULATIONS])

**DETAILS OF THE CLIENT** Mr D Roberts
23 Acacia Avenue Post Code: SL0 0LT
Sometown, Berks

**INSTALLATION ADDRESS** Unit 3 The Quadrant
Sometown Business Park
Sometown, Berks Post Code: SL1 0ZZ

**DESCRIPTION AND EXTENT OF THE INSTALLATION** Tick boxes as appropriate

Description of installation: Commercial office

Extent of installation covered by this Certificate: Full new installation

(Use continuation sheet if necessary) see continuation sheet No: ........

New installation ☑

Addition to an existing installation ☐

Alteration to an existing installation ☐

**FOR DESIGN**
I/We being the person(s) responsible for the design of the electrical installation (as indicated by my/our signatures below), particulars of which are described above, having exercised reasonable skill and care when carrying out the design hereby CERTIFY that the design work for which I/we have been responsible is to the best of my/our knowledge and belief in accordance with BS 7671:2008, amended to ..2011 ........... (date) except for the departures, if any, detailed as follows:

Details of departures from BS 7671 (Regulations 120.3 and 133.5):
None N/A

The extent of liability of the signatory or the signatories is limited to the work described above as the subject of this Certificate.

For the DESIGN of the installation: **(Where there is mutual responsibility for the design)

Signature: D.Jones Date: 15/08/2013 Name (IN BLOCK LETTERS): D.JONES Designer No 1

Signature: N/A Date: .............. Name (IN BLOCK LETTERS): N/A Designer No 2**

**FOR CONSTRUCTION**
I/We being the person(s) responsible for the construction of the electrical installation (as indicated by my/our signatures below), particulars of which are described above, having exercised reasonable skill and care when carrying out the construction hereby CERTIFY that the construction work for which I/we have been responsible is to the best of my/our knowledge and belief in accordance with BS 7671:2008, amended to .....2011..........(date) except for the departures, if any, detailed as follows:

Details of departures from BS 7671 (Regulations 120.3 and 133.5):
None N/A

The extent of liability of the signatory is limited to the work described above as the subject of this Certificate.

For CONSTRUCTION of the installation:

Signature: T.Smith Date: 15/08/2013 Name (IN BLOCK LETTERS): T.SMITH

**FOR INSPECTION & TESTING**
I/We being the person(s) responsible for the inspection & testing of the electrical installation (as indicated by my/our signatures below), particulars of which are described above, having exercised reasonable skill and care when carrying out the inspection & testing hereby CERTIFY that the work for which I/we have been responsible is to the best of my/our knowledge and belief in accordance with BS 7671:2008, amended to ......2011.........(date) except for the departures, if any, detailed as follows:

Details of departures from BS 7671 (Regulations 120.3 and 133.5):
None N/A

The extent of liability of the signatory is limited to the work described above as the subject of this Certificate.

For INSPECTION AND TESTING of the installation:

Signature: G.Wilson Date: 15/08/2013 Name (IN BLOCK LETTERS): G.WILSON

**NEXT INSPECTION**
I/We the designer(s), recommend that this installation is further inspected and tested after an interval of not more than ......5....... years/~~months~~.

Page 1 of ...4

Electrical Installation Certificate sample 1 (always check you are using the latest forms, as found on the IET website: http://electrical.theiet.org)

Form 2 Form No: ...SSSS13............../2

**PARTICULARS OF SIGNATORIES TO THE ELECTRICAL INSTALLATION CERTIFICATE**

**Designer (No 1)**
Name: D.Jones Company: The Electrical Design Partnership
Address: 23 High Street
Sometown, Berks Postcode: SL10 0YY Tel No: 01000 999999

**Designer (No 2)** (if applicable)
Name: ........ Company: ........
Address: ........
........ Postcode: ........ Tel No: ........

**Constructor**
Name: T.Smith Company: T.Smith Electrical Installations
Address: Unit 8a Sometown Ind.Estate
Sometown, Berks Postcode: SL3 0XX Tel No: 01000 888888

**Inspector**
Name: G.Wilson Company: Wilson and Sons
Address: 11 Crabtree Row
Sometown, Berks Postcode: SL2 0WW Tel No: 01000 777777

**SUPPLY CHARACTERISTICS AND EARTHING ARRANGEMENTS** Tick boxes and enter details, as appropriate

| Earthing arrangements | Number and Type of Live Conductors | Nature of Supply Parameters | Supply Protective Device Characteristics |
|---|---|---|---|
| TN-C ☐<br>TN-S ☐<br>TN-C-S ☑<br>TT ☐<br>IT ☐ | a.c. ☑ d.c. ☐<br>1-phase, 2-wire ☑ 2-wire ☐<br>1-phase, 3-wire ☐ 3-wire ☐<br>2-phase, 3-wire ☐ other ☐<br>3-phase, 3-wire ☐<br>3-phase, 4-wire ☐ | Nominal voltage, ~~U~~/$U_0$ (1) ....230........ V<br>Nominal frequency, f (1) .....50........ Hz<br>Prospective fault current, $I_{pf}$ (2) 1.41.. kA<br>External loop impedance, $Z_e$ (2) ......0.34Ω | Type: ...BS 88-3.....<br>Rated current..80.. A |
| Other sources of supply (to be detailed on attached schedules) ☐ | Confirmation of supply polarity ☑ | *(Note: (1) by enquiry, (2) by enquiry or by measurement)* | |

**PARTICULARS OF INSTALLATION REFERRED TO IN THE CERTIFICATE** Tick boxes and enter details, as appropriate

**Means of Earthing**
Distributor's facility ☑
Installation earth electrode ☐

**Maximum Demand**
Maximum demand (load) ...........68......... ~~kVA~~ / Amps Delete as appropriate

**Details of Installation Earth Electrode** *(where applicable)*

| Type (e.g. rod(s), tape etc) | Location | Electrode resistance to Earth |
|---|---|---|
| N/A | N/A | N/A Ω |

**Main Protective Conductors**

Earthing conductor: material ...Copper.............. csa .........16.........$mm^2$ Continuity and connection verified ☑

Main protective bonding conductors material ....Copper............. csa .........16.........$mm^2$ Continuity and connection verified ☑

To incoming water and/or gas service ☑ To other elements: ...............N/A........

**Main Switch or Circuit-breaker**

BS, Type and No. of poles ......BS EN 60497-3 (2-pole)................ Current rating .......100.....A Voltage rating .....400 ........V

Location ...Office Suite Consumer Unit.................. Fuse rating or setting......N/A............ A

Rated residual operating current $I_{\Delta n}$ = ....N/A... mA, and operating time of ....N/A.... ms (at $I_{\Delta n}$) (applicable only where an RCD is suitable and is used as a main circuit breaker)

**COMMENTS ON EXISTING INSTALLATION** (in the case of an addition or alteration see Section 633):
.........N/A........

**SCHEDULES**
The attached Schedules are part of this document and this Certificate is valid only when they are attached to it.
......1...... Schedules of Inspections and ......1...... Schedules of Test Results are attached.
(Enter quantities of schedules attached)

Page .2..of .4.

Electrical Installation Certificate sample 2 (always check you are using the latest forms, as found on the IET website: http://electrical.theiet.org)

Form 3 Form No: ...SSSS13...../3

SCHEDULE OF INSPECTIONS (for new installation work only)

**Methods of protection against electric shock**

**Both basic and fault protection:**

- ✓ (i) SELV (note 1)
- N/A (ii) PELV
- N/A (iii) Double insulation
- N/A (iv) Reinforced insulation

**Basic protection:** (note 2)

- ✓ (i) Insulation of live parts
- ✓ (ii) Barriers or enclosures
- N/A (iii) Obstacles (note 3)
- N/A (iv) Placing out of reach (note 4)

**Fault protection:**

**(i) Automatic disconnection of supply:**

- ✓ Presence of earthing conductor
- ✓ Presence of circuit protective conductors
- ✓ Presence of protective bonding conductors
- ✓ Presence of supplementary bonding conductors
- N/A Presence of earthing arrangements for combined protective and functional purposes
- N/A Presence of adequate arrangements for other sources, where applicable
- N/A FELV
- ✓ Choice and setting of protective and monitoring devices (for fault and/or overcurrent protection)

**(ii) Non-conducting location:** (note 5)

- N/A Absence of protective conductors

**(iii) Earth-free local equipotential bonding:** (note 6)

- N/A Presence of earth-free local equipotential bonding

**(iv) Electrical separation:** (note 7)

- N/A Provided for **one item** of current-using equipment
- N/A Provided for **more than one item** of current-using equipment

**Additional protection:**

- ✓ Presence of residual current devices(s)
- ✓ Presence of supplementary bonding conductors

**Prevention of mutual detrimental influence**

- ✓ (a) Proximity to non-electrical services and other influences
- ✓ (b) Segregation of Band I and Band II circuits or use of Band II insulation
- N/A (c) Segregation of safety circuits

**Identification**

- ✓ (a) Presence of diagrams, instructions, circuit charts and similar information
- ✓ (b) Presence of danger notices and other warning notices
- ✓ (c) Labelling of protective devices, switches and terminals
- ✓ (d) Identification of conductors

**Cables and conductors**

- ✓ Selection of conductors for current-carrying capacity and voltage drop
- ✓ Erection methods
- ✓ Routing of cables in prescribed zones
- N/A Cables incorporating earthed armour or sheath, or run within an earthed wiring system, or otherwise adequately protected against nails, screws and the like
- N/A Additional protection provided by 30 mA RCD for cables concealed in walls (where required in premises not under the supervision of a skilled or instructed person)
- ✓ Connection of conductors
- ✓ Presence of fire barriers, suitable seals and protection against thermal effects

**General**

- ✓ Presence and correct location of appropriate devices for isolation and switching
- ✓ Adequacy of access to switchgear and other equipment
- N/A Particular protective measures for special installations and locations
- ✓ Connection of single-pole devices for protection or switching in line conductors only
- ✓ Correct connection of accessories and equipment
- N/A Presence of undervoltage protective devices
- ✓ Selection of equipment and protective measures appropriate to external influences
- ✓ Selection of appropriate functional switching devices

Inspected by ............G.Wilson........................................ Date .....15/08/2013............................................

**NOTES:**

✓ to indicate an inspection has been carried out and the result is satisfactory
**N/A** to indicate that the inspection is not applicable to a particular item
An entry must be made in every box.

1. SELV An extra-low voltage system which is electrically separated from Earth and from other systems. The particular requirements of the Regulations must be checked (see Section 414)
2. Method of basic protection - will include measurement of distances where appropriate
3. Obstacles - only adopted in special circumstances (see Regulations 416.2 and 417.2)
4. Placing out of reach - only adopted in special circumstances (see Regulation 417.3)
5. Non-conducting locations - not applicable in domestic premises and requiring special precautions (see Regulation 418.1)
6. Earth-free local equipotential bonding - not applicable in domestic premises, only used in special circumstances (see Regulation 418.2)
7. Electrical separation (see Section 413 and Regulation 418.3)

Page 3... of .4...

Schedule of Inspections sample (always check you are using the latest forms, as found on the IET website: http://electrical.theiet.org)

**ASSESSMENT GUIDANCE**

Remember that you must complete all the boxes on the Schedule of Inspections. Blank boxes or an 'X' are not allowed.

Form 4

Form No: ......1235........./4

## GENERIC SCHEDULE OF TEST RESULTS

| | | |
|---|---|---|
| DB reference no ......Commercial office......<br>Location ..Unit 3, The Quadrant, SL1 0ZZ......<br>Zs at DB (Ω) .........0.34.........<br>$I_{pf}$ at DB (kA) .........1.41.........<br>Correct supply polarity confirmed ☑<br>Phase sequence confirmed (where appropriate) N/A | Details of circuits and/or installed equipment vulnerable to damage when testing ..........Downlighter spots – electronic SELV transformers.......... | Details of test instruments used (state serial and/or asset numbers)<br>Continuity..........Megin multi-function 10563..........<br>Insulation resistance ..........-"-..........<br>Earth fault loop impedance ..........-"-..........<br>RCD..........-"-..........<br>Earth electrode resistance ..........N/A.......... |

Tested by:
Name (Capitals) ..........G WILSON..........
Signature ..........G. Wilson.......... Date ..........15/08/2013..........

| Circuit details | | | | | | | | | Test results | | | | | | | | | | | | |
|---|---|---|---|---|---|---|---|---|---|---|---|---|---|---|---|---|---|---|---|---|---|
| | | Overcurrent device | | | | Conductor details | | | Ring final circuit continuity (Ω) | | | Continuity (Ω) ($R_1$ + $R_2$) or $R_2$ | | Insulation Resistance (MΩ) | | Polarity | $Z_s$ (Ω) | RCD (ms) | | | Remarks (continue on a separate sheet if necessary) |
| Circuit number | Circuit Description | BS (EN) | type | rating (A) | breaking capacity (kA) | Reference Method | Live (mm²) | cpc (mm²) | $r_1$ (line) | $r_n$ (neutral) | $r_2$ (cpc) | ($R_1$ + $R_2$) * | $R_2$ | Live-Live | Live-E | | | @ $I_{\Delta n}$ | @ 5 $I_{\Delta n}$ | Test button / functionality | |
| 1 | 2 | 3 | 4 | 5 | 6 | 7 | 8 | 9 | 10 | 11 | 12 | 13 | 14 | 15 | 16 | 17 | 18 | 19 | 20 | 21 | 22 |
| 1 | Socket outlets - dado | 60898 | B | 20 | 6 | B | 2.5 | 1.5 | N/A | N/A | N/A | 0.38 | N/A | >200 | >200 | ✓ | 0.75 | N/A | N/A | N/A | ✓ Checked for compliance |
| 2 | Socket outlets - wall | 61009 | B | 32 | 6 | B | 2x2.5 | 2x1.5 | 0.62 | 0.62 | 1.02 | 0.41 | N/A | >200 | >200 | ✓ | 0.71 | 85 | 16 | ✓ | ✓ -"- |
| 3 | Down lighter spots | 60898 | B | 6 | 6 | B | 1.5 | 1.0 | N/A | N/A | N/A | 0.56 | N/A | >200 | >200 | ✓ | 0.83 | N/A | N/A | N/A | ✓ -"- |
| 4 | General lighting | 60898 | B | 10 | 6 | B | 1.5 | 1.0 | N/A | N/A | N/A | 0.48 | N/A | >200 | >200 | ✓ | 0.81 | N/A | N/A | N/A | ✓ -"- |
| 5 | Water heater | 60898 | B | 16 | 6 | B | 2.5 | 1.5 | N/A | N/A | N/A | 0.11 | N/A | >200 | >200 | ✓ | 0.45 | N/A | N/A | N/A | ✓ -"- |
| | | | | | | | | | | | | | | | | | | | | | |
| | | | | | | | | | | | | | | | | | | | | | |
| | | | | | | | | | | | | | | | | | | | | | |
| | | | | | | | | | | | | | | | | | | | | | |
| | | | | | | | | | | | | | | | | | | | | | |
| | | | | | | | | | | | | | | | | | | | | | |
| | | | | | | | | | | | | | | | | | | | | | |

* Where there are no spurs connected to a ring final circuit this value is also the ($R_1$ + $R_2$) of the circuit.

Page 4 ... of ...4

Generic Schedule of Test Results sample (always check you are using the latest forms, as found on the IET website: http://electrical.theiet.org)

## Periodic inspection and testing forms

Periodic inspection and testing is carried out to assess the on-going safety of existing electrical installations.

The following forms are blank samples. These forms are *not* classed as certification; they are reports relating to the condition of the electrical installation. The **extent** and **limitations** of the report must be agreed with the person ordering the work. The reasons for these limitations must be validated.

All documents must be completed and authenticated by a competent person(s) before a full report is handed over.

Documentation to be completed for periodic inspection and testing can be found in Appendix 6 of BS 7671:

1 Electrical Installation Condition Report
2 Condition Report Inspection Schedule(s)
3 Generic Schedule(s) of Test Results

**Extent**

The amount of inspection and testing. For example, in a third floor flat with a single distribution board and eight circuits, the inspection will be visual only without removing covers and testing, and will involve sample tests at the final point of each circuit.

**Limitation**

A part of the inspection and test process that cannot be done for operational reasons. For example, the main protective bonding connection to the water system, located in the basement, could not be inspected as a key for the room was not available.

### ACTIVITY

Using statutory documents as a guide, decide who would be a competent person to carry out testing.

## ELECTRICAL INSTALLATION CONDITION REPORT

**SECTION A. DETAILS OF THE CLIENT / PERSON ORDERING THE REPORT**
Name ......
Address ......
......

**SECTION B. REASON FOR PRODUCING THIS REPORT** ......
......
Date(s) on which inspection and testing was carried out ......

**SECTION C. DETAILS OF THE INSTALLATION WHICH IS THE SUBJECT OF THIS REPORT**
Occupier ......
Address ......
......
Description of premises (tick as appropriate)
Domestic ☐ Commercial ☐ Industrial ☐ Other (include brief description) ☐ ......
Estimated age of wiring system ......years
Evidence of additions / alterations Yes ☐ No ☐ Not apparent ☐ If yes, estimate age ......years
Installation records available? (Regulation 621.1) Yes ☐ No ☐ Date of last inspection ...... (date)

**SECTION D. EXTENT AND LIMITATIONS OF INSPECTION AND TESTING**
Extent of the electrical installation covered by this report
......
......
Agreed limitations including the reasons (see Regulation 634.2) ......
......
Agreed with: ......
Operational limitations including the reasons (see page no......) ......
......
The inspection and testing detailed in this report and accompanying schedules have been carried out in accordance with BS 7671: 2008 (IET Wiring Regulations) as amended to ......
It should be noted that cables concealed within trunking and conduits, under floors, in roof spaces, and generally within the fabric of the building or underground, have **not** been inspected unless specifically agreed between the client and inspector prior to the inspection.

**SECTION E. SUMMARY OF THE CONDITION OF THE INSTALLATION**
General condition of the installation (in terms of electrical safety) ......
......
......
Overall assessment of the installation in terms of its suitability for continued use
SATISFACTORY / UNSATISFACTORY* (Delete as appropriate)
*An unsatisfactory assessment indicates that dangerous (code C1) and/or potentially dangerous (code C2) conditions have been identified.

**SECTION F. RECOMMENDATIONS**
Where the overall assessment of the suitability of the installation for continued use above is stated as UNSATISFACTORY, I / we recommend that any observations classified as *'Danger present'* (code C1) or *'Potentially dangerous'* (code C2) are acted upon as a matter of urgency.
Investigation without delay is recommended for observations identified as *'further investigation required'*.
Observations classified as *'Improvement recommended'* (code C3) should be given due consideration.

Subject to the necessary remedial action being taken, I / we recommend that the installation is further inspected and tested by ......(date)

**SECTION G. DECLARATION**
**I/We, being the person(s) responsible for the inspection and testing of the electrical installation (as indicated by my/our signatures below), particulars of which are described above, having exercised reasonable skill and care when carrying out the inspection and testing, hereby declare that the information in this report, including the observations and the attached schedules, provides an accurate assessment of the condition of the electrical installation taking into account the stated extent and limitations in section D of this report.**

| **Inspected and tested by:** | **Report authorised for issue by:** |
|---|---|
| Name (Capitals) ...... | Name (Capitals) ...... |
| Signature ...... | Signature ...... |
| For/on behalf of ...... | For/on behalf of ...... |
| Position ...... | Position ...... |
| Address ...... | Address ...... |
| Date ...... | Date ...... |

**SECTION H. SCHEDULE(S)**
......schedule(s) of inspection and ......schedule(s) of test results are attached.
The attached schedule(s) are part of this document and this report is valid only when they are attached to it.

**SECTION I. SUPPLY CHARACTERISTICS AND EARTHING ARRANGEMENTS**

| Earthing arrangements | Number and Type of Live Conductors | | Nature of Supply Parameters | Supply Protective Device |
|---|---|---|---|---|
| TN-C ☐ | a.c. ☐ | d.c. ☐ | Nominal voltage, U / $U_0$[1] ......V | BS (EN) ...... |
| TN-S ☐ | 1-phase, 2-wire ☐ | 2-wire ☐ | Nominal frequency, f[1] ......Hz | Type ...... |
| TN-C-S ☐ | 2 phase, 3-wire ☐ | 3-wire ☐ | Prospective fault current, $I_{pf}$[2] ......kA | Rated current ......A |
| TT ☐ | 3 phase, 3-wire ☐ | | External loop impedance, Ze[2] ......Ω | |
| IT ☐ | 3 phase, 4-wire ☐ | | Note: (1) by enquiry | |
| | Confirmation of supply polarity ☐ | | (2) by enquiry or by measurement | |

Other sources of supply (as detailed on attached schedule) ☐

**SECTION J. PARTICULARS OF INSTALLATION REFERRED TO IN THE REPORT**

| **Means of Earthing** | **Details of Installation Earth Electrode** *(where applicable)* |
|---|---|
| Distributor's facility ☐ | Type ...... |
| Installation earth electrode ☐ | Location ...... |
| | Resistance to Earth ...... Ω |

**Main Protective Conductors**

| | | | |
|---|---|---|---|
| Earthing conductor | Material ...... | Csa ......mm² | Connection / continuity verified ☐ |
| Main protective bonding conductors | Material ...... | Csa ......mm² | Connection / continuity verified ☐ |
| To incoming water service ☐ | To incoming gas service ☐ | To incoming oil service ☐ | To structural steel ☐ |
| To lightning protection ☐ | To other incoming service(s) ☐ Specify ...... | | |

**Main Switch / Switch-Fuse / Circuit-Breaker / RCD**

| | | **If RCD main switch** |
|---|---|---|
| Location ...... | Current rating ......A | Rated residual operating current ($I_{\Delta n}$) ......mA |
| ...... | Fuse / device rating or setting ......A | Rated time delay ......ms |
| BS(EN) ...... | Voltage rating ......V | Measured operating time(at $I_{\Delta n}$) ......ms |
| No of poles ...... | | |

**SECTION K. OBSERVATIONS**
Referring to the attached schedules of inspection and test results, and subject to the limitations specified at the *Extent and limitations of inspection and testing* section
No remedial action is required ☐ The following observations are made ☐ (see below):

| OBSERVATION(S) | CLASSIFICATION CODE | FURTHER INVESTIGATION REQUIRED (YES / NO) |
|---|---|---|
| ...... | ...... | ...... |
| ...... | ...... | ...... |
| ...... | ...... | ...... |
| ...... | ...... | ...... |
| ...... | ...... | ...... |
| ...... | ...... | ...... |
| ...... | ...... | ...... |
| ...... | ...... | ...... |
| ...... | ...... | ...... |
| ...... | ...... | ...... |
| ...... | ...... | ...... |
| ...... | ...... | ...... |
| ...... | ...... | ...... |
| ...... | ...... | ...... |
| ...... | ...... | ...... |
| ...... | ...... | ...... |
| ...... | ...... | ...... |
| ...... | ...... | ...... |
| ...... | ...... | ...... |
| ...... | ...... | ...... |
| ...... | ...... | ...... |

One of the following codes, as appropriate, has been allocated to each of the observations made above to indicate to the person(s) responsible for the installation the degree of urgency for remedial action.
C1 – Danger present. Risk of injury. Immediate remedial action required
C2 – Potentially dangerous - urgent remedial action required
C3 – Improvement recommended

Electrical Installation Condition Report (always check you are using the latest forms, as found on the IET website: http://electrical.theiet.org)

**CONDITION REPORT INSPECTION SCHEDULE FOR DOMESTIC AND SIMILAR PREMISES WITH UP TO 100 A SUPPLY**

*Note: This form is suitable for many types of smaller installation not exclusively domestic.*

| OUTCOMES | Acceptable condition ✓ | Unacceptable condition State C1 or C2 | Improvement recommended State C3 | Not verified N/V | Limitation LIM | Not applicable N/A |
|---|---|---|---|---|---|---|

| ITEM NO | DESCRIPTION | OUTCOME (Use codes above. Provide additional comment where appropriate. C1, C2 and C3 coded items to be recorded in Section K of the Condition Report) | Further investigation required? (Y or N) |
|---|---|---|---|
| **1.0** | **DISTRIBUTOR'S / SUPPLY INTAKE EQUIPMENT** | | |
| 1.1 | Service cable condition | | |
| 1.2 | Condition of service head | | |
| 1.3 | Condition of tails - Distributor | | |
| 1.4 | Condition of tails - Consumer | | |
| 1.5 | Condition of metering equipment | | |
| 1.6 | Condition of isolator (where present) | | |
| **2.0** | **PRESENCE OF ADEQUATE ARRANGEMENTS FOR OTHER SOURCES SUCH AS MICROGENERATORS (551.6; 551.7)** | | |
| **3.0** | **EARTHING / BONDING ARRANGEMENTS (411.3; Chap 54)** | | |
| 3.1 | Presence and condition of distributor's earthing arrangement (542.1.2.1; 542.1.2.2) | | |
| 3.2 | Presence and condition of earth electrode connection where applicable (542.1.2.3) | | |
| 3.3 | Provision of earthing / bonding labels at all appropriate locations (514.11) | | |
| 3.4 | Confirmation of earthing conductor size (542.3; 543.1.1) | | |
| 3.5 | Accessibility and condition of earthing conductor at MET (543.3.2) | | |
| 3.6 | Confirmation of main protective bonding conductor sizes (544.1) | | |
| 3.7 | Condition and accessibility of main protective bonding conductor connections (543.3.2; 544.1.2) | | |
| 3.8 | Accessibility and condition of all protective bonding connections (543.3.2) | | |
| **4.0** | **CONSUMER UNIT(S) / DISTRIBUTION BOARD(S)** | | |
| 4.1 | Adequacy of working space / accessibility to consumer unit / distribution board (132.12; 513.1) | | |
| 4.2 | Security of fixing (134.1.1) | | |
| 4.3 | Condition of enclosure(s) in terms of IP rating etc (416.2) | | |
| 4.4 | Condition of enclosure(s) in terms of fire rating etc (526.5) | | |
| 4.5 | Enclosure not damaged/deteriorated so as to impair safety (621.2(iii)) | | |
| 4.6 | Presence of main linked switch (as required by 537.1.4) | | |
| 4.7 | Operation of main switch (functional check) (612.13.2) | | |
| 4.8 | Manual operation of circuit-breakers and RCDs to prove disconnection (612.13.2) | | |
| 4.9 | Correct identification of circuit details and protective devices (514.8.1; 514.9.1) | | |
| 4.10 | Presence of RCD quarterly test notice at or near consumer unit / distribution board (514.12.2) | | |
| 4.11 | Presence of non-standard (mixed) cable colour warning notice at or near consumer unit / distribution board (514.14) | | |
| 4.12 | Presence of alternative supply warning notice at or near consumer unit / distribution board (514.15) | | |
| 4.13 | Presence of other required labelling (please specify) (Section 514) | | |
| 4.14 | Examination of protective device(s) and base(s); correct type and rating (no signs of unacceptable thermal damage, arcing or overheating) (421.1.3) | | |
| 4.15 | Single-pole protective devices in line conductor only (132.14.1; 530.3.2) | | |
| 4.16 | Protection against mechanical damage where cables enter consumer unit / distribution board (522.8.1; 522.8.11) | | |
| 4.17 | Protection against electromagnetic effects where cables enter consumer unit / distribution board / enclosures (521.5.1) | | |
| 4.18 | RCD(s) provided for fault protection – includes RCBOs (411.4.9; 411.5.2; 531.2) | | |
| 4.19 | RCD(s) provided for additional protection - includes RCBOs (411.3.3; 415.1) | | |

| OUTCOMES | Acceptable condition ✓ | Unacceptable condition State C1 or C2 | Improvement recommended State C3 | Not verified N/V | Limitation LIM | Not applicable N/A |
|---|---|---|---|---|---|---|

| ITEM NO | DESCRIPTION | OUTCOME (Use codes above. Provide additional comment where appropriate. C1, C2 and C3 coded items to be recorded in Section K of the Condition Report) | Further investigation required? (Y or N) |
|---|---|---|---|
| **5.0** | **FINAL CIRCUITS** | | |
| 5.1 | Identification of conductors (514.3.1) | | |
| 5.2 | Cables correctly supported throughout their run (522.8.5) | | |
| 5.3 | Condition of insulation of live parts (416.1) | | |
| 5.4 | Non-sheathed cables protected by enclosure in conduit, ducting or trunking (521.10.1) | | |
| | ▪ To include the integrity of conduit and trunking systems (metallic and plastic) | | |
| 5.5 | Adequacy of cables for current-carrying capacity with regard for the type and nature of installation (Section 523) | | |
| 5.6 | Coordination between conductors and overload protective devices (433.1; 533.2.1) | | |
| 5.7 | Adequacy of protective devices: type and rated current for fault protection (411.3) | | |
| 5.8 | Presence and adequacy of circuit protective conductors (411.3.1.1; 543.1) | | |
| 5.9 | Wiring system(s) appropriate for the type and nature of the installation and external influences (Section 522) | | |
| 5.10 | Concealed cables installed in prescribed zones (see Section D. *Extent and limitations*) (522.6.101) | | |
| 5.11 | Concealed cables incorporating earthed armour or sheath, or run within earthed wiring system, or otherwise protected against mechanical damage from nails, screws and the like (see Section D. *Extent and limitations*) (522.6.101; 522.6.103) | | |
| 5.12 | Provision of additional protection by RCD not exceeding 30 mA: | | |
| | ▪ for all socket-outlets of rating 20 A or less provided for use by ordinary persons unless an exception is permitted (411.3.3) | | |
| | ▪ for supply to mobile equipment not exceeding 32 A rating for use outdoors (411.3.3) | | |
| | ▪ for cables concealed in walls or partitions (522.6.102; 522.6.103) | | |
| 5.13 | Provision of fire barriers, sealing arrangements and protection against thermal effects (Section 527) | | |
| 5.14 | Band II cables segregated / separated from Band I cables (528.1) | | |
| 5.15 | Cables segregated / separated from communications cabling (528.2) | | |
| 5.16 | Cables segregated / separated from non-electrical services (528.3) | | |
| 5.17 | Termination of cables at enclosures – indicate extent of sampling in Section D of the report (Section 526) | | |
| | ▪ Connections soundly made and under no undue strain (526.6) | | |
| | ▪ No basic insulation of a conductor visible outside enclosure (526.98) | | |
| | ▪ Connections of live conductors adequately enclosed (526.5) | | |
| | ▪ Adequately connected at point of entry to enclosure (glands, bushes etc.) (522.8.5) | | |
| 5.18 | Condition of accessories including socket-outlets, switches and joint boxes (621.2 (iii)) | | |
| 5.19 | Suitability of accessories for external influences (512.2) | | |
| **6.0** | **LOCATION(S) CONTAINING A BATH OR SHOWER** | | |
| 6.1 | Additional protection for all low voltage (LV) circuits by RCD not exceeding 30 mA (701.411.3.3) | | |
| 6.2 | Where used as a protective measure, requirements for SELV or PELV met (701.414.4.5) | | |
| 6.3 | Shaver sockets comply with BS EN 61558-2-5 formally BS 3535 (701.512.3) | | |
| 6.4 | Presence of supplementary bonding conductors, unless not required by BS 7671:2008 (701.415.2) | | |
| 6.5 | Low voltage (e.g. 230 volt) socket-outlets sited at least 3 m from zone 1 (701.512.3) | | |
| 6.6 | Suitability of equipment for external influences for installed location in terms of IP rating (701.512.2) | | |
| 6.7 | Suitability of equipment for installation in a particular zone (701.512.3) | | |
| 6.8 | Suitability of current-using equipment for particular position within the location (701.55) | | |
| **7.0** | **OTHER PART 7 SPECIAL INSTALLATIONS OR LOCATIONS** | | |
| 7.1 | List all other special installations or locations present, if any. (Record separately the results of particular inspections applied.) | | |

**Inspected by:**
**Name (Capitals)** ........................................ **Signature** ........................................ **Date** ....................

Condition Report Inspection Schedule(s) (always check you are using the latest forms, as found on the IET website: http://electrical.theiet.org)

Form 4

Form No: ................/4

## GENERIC SCHEDULE OF TEST RESULTS

**DB reference no** ..........................................................
**Location** ..........................................................
**Zs at DB (Ω)** ..........................................................
**$I_{pf}$ at DB (kA)** ..........................................................
**Correct supply polarity confirmed** ☐
**Phase sequence confirmed (where appropriate)** ☐

**Details of circuits and/or installed equipment vulnerable to damage when testing** ..........................................................
..........................................................
..........................................................
..........................................................
..........................................................

**Details of test instruments used (state serial and/or asset numbers)**
Continuity ..........................................................
Insulation resistance ..........................................................
Earth fault loop impedance ..........................................................
RCD ..........................................................
Earth electrode resistance ..........................................................

**Tested by:**
**Name (Capitals)** ..........................................................
**Signature** .......................................... **Date** ..............................

| Circuit details | | | | | | | | | Test results | | | | | | | | | | | | |
|---|---|---|---|---|---|---|---|---|---|---|---|---|---|---|---|---|---|---|---|---|---|
| | | Overcurrent device | | | | Conductor details | | | Ring final circuit continuity (Ω) | | | Continuity (Ω) ($R_1 + R_2$) or $R_2$ | | Insulation Resistance (MΩ) | | Polarity | $Z_s$ (Ω) | RCD (ms) | | | Remarks (continue on a separate sheet if necessary) |
| Circuit number 1 | Circuit Description 2 | BS (EN) 3 | type 4 | rating (A) 5 | breaking capacity (kA) 6 | Reference Method 7 | Live ($mm^2$) 8 | cpc ($mm^2$) 9 | $r_1$ (line) 10 | $r_n$ (neutral) 11 | $r_2$ (cpc) 12 | ($R_1 + R_2$) * 13 | $R_2$ 14 | Live-Live 15 | Live-E 16 | 17 | 18 | @ $I_{\Delta n}$ 19 | @ $5I_{\Delta n}$ 20 | Test button operation 21 | 22 |
| | | | | | | | | | | | | | | | | | | | | | |
| | | | | | | | | | | | | | | | | | | | | | |
| | | | | | | | | | | | | | | | | | | | | | |
| | | | | | | | | | | | | | | | | | | | | | |
| | | | | | | | | | | | | | | | | | | | | | |
| | | | | | | | | | | | | | | | | | | | | | |
| | | | | | | | | | | | | | | | | | | | | | |
| | | | | | | | | | | | | | | | | | | | | | |
| | | | | | | | | | | | | | | | | | | | | | |
| | | | | | | | | | | | | | | | | | | | | | |
| | | | | | | | | | | | | | | | | | | | | | |
| | | | | | | | | | | | | | | | | | | | | | |
| | | | | | | | | | | | | | | | | | | | | | |
| | | | | | | | | | | | | | | | | | | | | | |
| | | | | | | | | | | | | | | | | | | | | | |

* Where there are no spurs connected to a ring final circuit this value is also the ($R_1 + R_2$) of the circuit.

Generic Schedule(s) of Test Results (always check you are using the latest forms, as found on the IET website: http://electrical.theiet.org)

## ACTIVITY

1 List:
a) **three** statutory documents relating to initial verification and periodic testing
b) **three** non-statutory documents relating to initial verification and periodic testing.

2 State the BS 7671documents that need to be completed for:
a) an initial verification
b) periodic inspection and testing.

3 Which one of the following categories of person can authenticate an Electrical Installation Certificate (as stated on the form)?
a) ordinary
b) instructed
c) skilled
d) competent

## ASSESSMENT GUIDANCE

Remember to double check all forms. All parts must be completed.

**Assessment criteria**

1.1 Specify the regulatory requirements for inspection and testing of electrical systems

1.2 Specify the regulatory requirements for testing electrical systems

1.3 Specify the regulatory requirements for commissioning of electrical systems

## ASSESSMENT GUIDANCE

Diversity takes into account the fact that not all the loads in a building will be on at the same time. How many lights, out of the maximum, are on in your house at any one time in the evening?

# INFORMATION REQUIRED TO CARRY OUT INITIAL VERIFICATION OF AN ELECTRICAL INSTALLATION

Before carrying out an initial verification of an electrical installation, you need the following information, in accordance with the IET Wiring Regulations and IET Guidance Note 3:

1 maximum demand and diversity (Regulation 311)
2 conductor and system earthing arrangements (TN or TT System) (Regulation 312)
3 nominal voltage, current, frequency, prospective short-circuit current and external earth fault loop impedance ($Z_e$) (Regulation 312)
4 compatibility of characteristics (listed in Regulation 313)
5 diagrams, documents, plans and design criteria for the building (Regulation 514.9.1).

Note that the above regulation numbers are from BS 7671.

## Maximum demand and diversity (Regulation 311)

The maximum demand of a new installation must be assessed. Appendix A of the IET OSG provides details of the methods used to calculate the maximum demand and diversity of small installations. The assessed demand after diversity must be inserted on the electrical installation certificate.

## Conductor and system earthing arrangements (TN or TT system) (Section 312)

Understanding the system earthing arrangements is one of the most important parts of the inspection and testing process. This is covered in greater detail on pages 433–439.

## Nominal voltage, current, frequency, prospective fault current and external earth fault loop impedance ($Z_e$) (Section 312)

The following values are for low voltage public electricity supply systems:

- the nominal voltage to earth ($V_0$) is 230 Volts
- the current is alternating current (a.c.)
- the frequency is 50 Hz
- the prospective fault current (PFC) will vary with every installation.
- the external earth fault loop impedance will vary with every system, supply arrangements and, therefore, installation.

### ACTIVITY

Why is alternating current used for public supplies, rather than direct current?

### KEY POINT

The nominal voltage is 230 V. It could, in fact, be between 216 and 253 volts and still be acceptable.

## Compatibility of characteristics (listed in Chapter 33)

An assessment shall be made of any characteristics of equipment likely to have harmful effects on other electrical equipment or other services, or likely to impair the supply.

A list of design considerations can be found at Section 331 of BS 7671.

The list is very technical, including overvoltages, starting currents, harmonic currents, power factor and many more compatibility design considerations.

## Diagrams, documents, plans and design criteria for the building (Regulation 514.9.1)

The inspection and certification process requires documentation to be available for the person carrying out the work.

Regulation 514.9.1 lists the following:

A legible diagram, chart or table or equivalent form of information shall be provided indicating:

(i) the type and composition of each circuit (points of utilisation served, number and size of conductors, type of wiring)

(ii) the method used for compliance with Regulation 410.3.2, for basic protective measures and independent fault protection. Additional protection may need to be considered under certain conditions

(iii) the information necessary for the identification of each device performing the functions of protection, isolation and switching, and its location

(iv) any circuit or equipment vulnerable to a typical test.

For simple installations the foregoing information may be given in a schedule. A durable copy of the schedule relating to a distribution board shall be provided within or adjacent to each distribution board.

## ASSESSMENT GUIDANCE

You may be asked to name the information required before inspection and testing can commence.

## ACTIVITY

List five items of information that should be made available to the person who is inspecting and testing.

Below is a typical example of a schedule that can be used to highlight the distribution board circuits and information.

Form 4 — Form No: ......1235........../4

**GENERIC SCHEDULE OF TEST RESULTS**

DB reference no ......Commercial office......
Location ..Unit 3, The Quadrant, SL1 0ZZ......
Zs at DB (Ω) ......0.34......
$I_{pf}$ at DB (kA) ......1.41......
Correct supply polarity confirmed ☑
Phase sequence confirmed (where appropriate) N/A

Details of circuits and/or installed equipment vulnerable to damage when testing ......Downlighter spots – electronic SELV transformers......

Details of test instruments used (state serial and/or asset numbers)
Continuity ......Megin multi-function 10563......
Insulation resistance ......– " –......
Earth fault loop impedance ......– " –......
RCD ......– " –......
Earth electrode resistance ......N/A......

Tested by:
Name (Capitals) ......G WILSON......
Signature ......G. Wilson...... Date ......15/08/2013......

| Circuit details | | Overcurrent device | | | | Conductor details | | | Test results: Ring final circuit continuity (Ω) | | | Continuity (Ω) ($R_1 + R_2$) or $R_2$ | | Insulation Resistance (MΩ) | | Polarity | $Z_s$ (Ω) | RCD (ms) | | | Remarks (continue on a separate sheet if necessary) |
|---|---|---|---|---|---|---|---|---|---|---|---|---|---|---|---|---|---|---|---|---|---|
| Circuit number 1 | Circuit Description 2 | BS (EN) 3 | type 4 | rating (A) 5 | breaking capacity (kA) 6 | Reference Method 7 | Live (mm²) 8 | cpc (mm²) 9 | $r_1$ (line) 10 | $r_n$ (neutral) 11 | $r_2$ (cpc) 12 | ($R_1 + R_2$)* 13 | $R_2$ 14 | Live-Live 15 | Live-E 16 | 17 | 18 | @ $I_{\Delta n}$ 19 | @ 5 $I_{\Delta n}$ 20 | Test button / functionality 21 | 22 |
| 1 | Socket outlets - dado | 60898 | B | 20 | 6 | B | 2.5 | 1.5 | N/A | N/A | N/A | 0.38 | N/A | >200 | >200 | ✓ | 0.75 | N/A | N/A | N/A | ✓ Checked for compliance |
| 2 | Socket outlets - wall | 61009 | B | 32 | 6 | B | 2×2.5 | 2×1.5 | 0.62 | 0.62 | 1.02 | 0.41 | N/A | >200 | >200 | ✓ | 0.71 | 85 | 16 | ✓ | ✓ - " - |
| 3 | Down lighter spots | 60898 | B | 6 | 6 | B | 1.5 | 1.0 | N/A | N/A | N/A | 0.56 | N/A | >200 | >200 | ✓ | 0.83 | N/A | N/A | N/A | ✓ - " - |
| 4 | General lighting | 60898 | B | 10 | 6 | B | 1.5 | 1.0 | N/A | N/A | N/A | 0.48 | N/A | >200 | >200 | ✓ | 0.81 | N/A | N/A | N/A | ✓ - " - |
| 5 | Water heater | 60898 | B | 16 | 6 | B | 2.5 | 1.5 | N/A | N/A | N/A | 0.11 | N/A | >200 | >200 | ✓ | 0.45 | N/A | N/A | N/A | ✓ - " - |
| | | | | | | | | | | | | | | | | | | | | | |
| | | | | | | | | | | | | | | | | | | | | | |
| | | | | | | | | | | | | | | | | | | | | | |
| | | | | | | | | | | | | | | | | | | | | | |
| | | | | | | | | | | | | | | | | | | | | | |
| | | | | | | | | | | | | | | | | | | | | | |
| | | | | | | | | | | | | | | | | | | | | | |

* Where there are no spurs connected to a ring final circuit this value is also the ($R_1 + R_2$) of the circuit.

Page 4 ... of ...4

Generic Schedule of Test Results (always check you are using the latest forms, as found on the IET website: http://electrical.theiet.org)

OUTCOMES 2, 5

# Understand procedures for the inspection of electrical systems
# Be able to inspect electrical wiring systems

## HUMAN SENSES AND INSPECTION

Human senses are vital to the inspection process. You need to know which senses are involved and be able to explain how to use them.

**Assessment criteria**

2.2 Explain how human senses can be used during the inspection process

### Using senses during inspection of electrical installations

Inspection is normally done with that part of the installation under inspection disconnected from the supply, in accordance with Regulation 611.1. The examples below show how to apply the human senses during the stages of initial verification. Those stages highlighted in red usually occur when the installation is switched on.

#### Sight

Sight is the most extensively used sense. It is used when:

1 connecting conductors
2 identifying circuits
3 routing cables
4 labelling circuits
5 correctly connecting accessories
6 burning occurs, due to bad connection.

#### Touch

Touch is used in:

1 connection of conductors (using a screwdriver to check the connections)
2 correct connection to equipment (for example, in the physical connection to a water pipe)
3 erection methods (for example, fixings for distribution boards (DBs)/conduits/trunking)
4 checking of barriers and enclosures (check for IPXXB and IPXXD in accordance with Regulation 416.2)

**KEY POINT**

There are five senses. Taste is not used in the inspection and testing process, but hearing, touch, sight and smell are used.

5 checking for careless work methods (sharp edges in conduit)
6 detecting overheating of equipment.

### Sound

Listening can detect:

1 arcing caused by loading, for example when switching on and off fluorescent fittings
2 equipment noise, such as a motor bearing problem that causes a loud grinding sound
3 arcing caused by insecure connections, for example at an accessory, junction box or distribution board.

### Smell

Smell is used to detect:

1 equipment that is overheating
2 loose connections under load conditions
3 burning of adjacent building materials (for example, recessed lights without fire protection).

**ASSESSMENT GUIDANCE**

During your assessments, you may be asked which human senses you use during the process of inspection and testing.

**Assessment criteria**

2.3 Justify choice of applicable items on an inspection checklist that apply in given situations

5.2 Record inspection of electrical systems

## INSPECTION ITEMS EXPLAINED

Some of the important items of an electrical installation that should be inspected are briefly explained below. Other items are explained on pages 481–489.

**ACTIVITY**

Using the symbols in BS 7671, show the relationship between the design current, the protective device rating and the cable rating (current-carrying capacity).

### Items you should inspect

#### Connection of conductors

Every connection between a conductor and equipment (or another conductor) should provide durable electrical continuity and adequate mechanical strength.

#### Identification of cables and conductors

Check that each core or bare conductor is identified as necessary.

#### Routing of cables

Cables and their cable management systems should be designed and installed to take into account the mechanical stresses that users will place on the installation.

#### Cable selection

Where practicable, the cable size should be assessed against the protective arrangement.

## Accessories and equipment

Correct connection (suitability and polarity) and environmental conditions, such as the presence of dust or moisture, must be checked.

### ACTIVITY

How can polarity be checked without the use of a test instrument?

## Selection and erection to minimise the spread of fire

Fire barriers, suitable seals and/or protection against thermal effects should be provided. These checks should be carried during the installation process.

## Basic protection

Basic protection is most commonly provided by insulation of live parts and/or barriers and enclosures:

- Confirmation of the insulation of live parts should be carried out to ensure no damage has occurred during the installation process.
- Barriers and enclosures need to be checked to ensure a protection level of at least IPXXB or IP2X. For horizontal top surfaces of readily accessible enclosures, you must ensure at least IPXXD or IP4X.

### ASSESSMENT GUIDANCE

You will be surprised at what gets into electrical equipment. Make sure all entries are correctly sealed or comply with the relevant IP codes.

---

# IP codes designated numbering and lettering system

The illustrations and tables that follow explain the IP codes designated numbering and lettering system.

## IP code for ingress protection

Where a characteristic numeral does not have to be specified, it can be replaced by the letter 'X' ('XX' if both numbers are omitted).

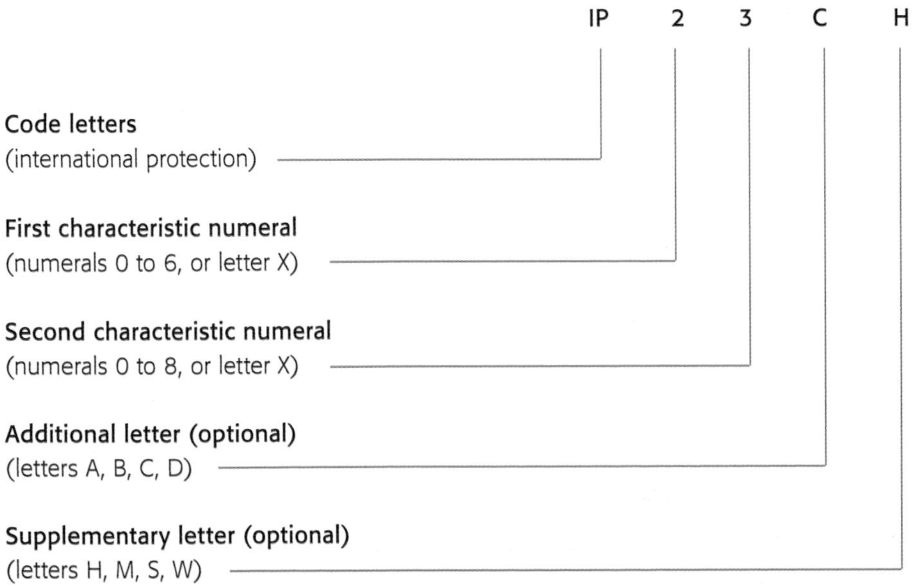

IP 2 3 C H

**Code letters**
(international protection) — IP

**First characteristic numeral**
(numerals 0 to 6, or letter X) — 2

**Second characteristic numeral**
(numerals 0 to 8, or letter X) — 3

**Additional letter (optional)**
(letters A, B, C, D) — C

**Supplementary letter (optional)**
(letters H, M, S, W) — H

**IP codes designated numbering and lettering system**

## IP characteristic numerals

The IP characteristic numerals are shown in the table.

▼ **Table B1** IP code characteristic numerals

| First characteristic numeral | Second characteristic numeral |
|---|---|
| (a) Protection of persons against access to hazardous parts inside enclosures<br>(b) Protection of equipment against ingress of solid foreign objects | Protection of equipment against ingress of water |

| No. | Degree of protection | No. | Degree of protection |
|---|---|---|---|
| 0 | (a) Not protected<br>(b) Not protected | 0 | Not protected |
| 1 | (a) Protection against access to hazardous parts with the back of the hand<br>(b) Protection against foreign solid objects of 50 mm diameter and greater | 1 | Protection against vertically falling water drops |
| 2 | (a) Protection against access to hazardous parts with a finger<br>(b) Protection against solid foreign objects of 12.5 mm diameter and greater | 2 | Protected against vertically falling water drops when enclosure tilted up to 15°. Vertically falling water drops shall have no harmful effect when the enclosure is tilted at any angle up to 15° from the vertical |
| 3 | (a) Protection against contact by tools, wires or such like more than 2.5 mm thick<br>(b) Protection against solid foreign objects of 2.5 mm diameter and greater | 3 | Protected against water spraying at an angle up to 60° on either side of the vertical |
| 4 | (a) As 3 above but against contact with a wire or strips more than 1.0 mm thick<br>(b) Protection against solid foreign objects of 1.0 mm diameter and greater | 4 | Protected against water splashing from any direction |

**IP characteristic numerals (from Guidance Note 1: Selection and Erection of Equipment, IET)**

### ASSESSMENT GUIDANCE

Only use the appropriate test equipment. DO NOT use your finger.

For a code such as IPXXD, you can ignore the numerical values in the table above and simply refer to the additional letter table illustrated here.

▼ **Table B2** IP code additional letters

| Additional letter | Brief description of protection |
|---|---|
| A | Protected against access with the back of the hand<br>(minimum 50 mm diameter sphere)<br>(adequate clearance from live parts) |
| B | Protected against access with a finger<br>(minimum 12 mm diameter test finger, 80 mm long)<br>(adequate clearance from live parts) |
| C | Protected against access with a tool<br>(minimum 2.5 mm diameter tool, 100 mm long)<br>(adequate clearance from live parts) |
| D | Protected against access with a wire<br>(minimum 1 mm diameter wire, 100 mm long)<br>(adequate clearance from live parts) |

**IP letter codes (from Guidance Note 1: Selection and Erection of Equipment, IET)**

### ACTIVITY

Which IP code relates to the BS Finger: IP4X or IPXXB?

## British Standard finger (IPXXB)

Protection against access with a finger can be assessed using the model shown in the picture.

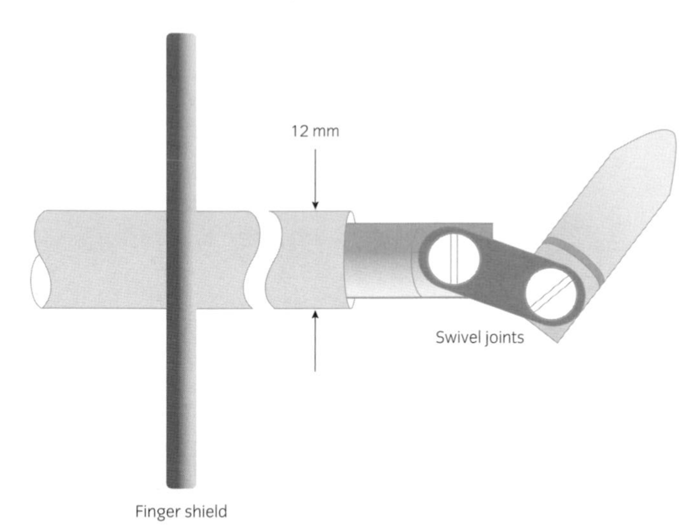


**British Standard finger (IPXXB)**

Remember that the IP codes are used in conjunction with basic protection, as a guide to protection against contact with live parts and ingress of foreign bodies.

Note also that providing obstacles and placing installations out of reach are methods of protection against contact, but they can only be used as a method of protection in a controlled and supervised environment.

## Fault protection

Fault protection inspection involves ensuring that earthing arrangements are in place. Part of this process includes automatic disconnection of supply, selection of protective devices and disconnection times, which will undergo further inspection and verification as you progress through the tests.

## Specialised systems

Specialised systems are rarely used. They can be found in Section 418 of BS 7671 and include:

- non-conducting location
- earth-free local equipotential bonding
- electrical separation for the supply to more than one item of current-using equipment.

These systems are normally not applicable (use 'n/a') when filling in the Inspection Schedule.

Electrical separation for the supply to one item of equipment is more common. This may include the shaver point in a bathroom.

Drip-proof  IPX2

Rain proof  IPX3

Splashproof  IPX4

Jet proof  IPX5

Protected against immersion in water   IPX7

**Codes and signs used on equipment where water is likely to be present**

## Prevention of mutual detrimental influence

Regulations 132.11 and 515.1 require that the electrical installation and its equipment shall not cause detriment to other electrical and non-electrical installations. Equally, other non-electrical services shall not have a detrimental impact on electrical installations. The inspector is advised to step back and think about other systems when carrying out the inspection – for example, cables under floorboards that touch central heating pipes.

## Protective devices, labelling, warning notices and adequate access

These can be observed at the distribution board or consumer unit.

## Selection of equipment and protective measures appropriate to external influences

Checks must be made to ensure that equipment has been manufactured to withstand the environmental conditions.

If water is likely to be present, signs and codes such as those shown in the diagram should be present on the equipment.

## Erection methods

Chapter 52 of BS 7671 contains detailed requirements on selection and erection. Fixings of switchgear, cables, conduit and fittings must be adequate for the environment and a detailed visual inspection is required during the erection stages, as well as at completion.

**ASSESSMENT GUIDANCE**

The IP codes for access to live parts protected by barriers and enclosures feature in questions in nearly all assessments.

**ACTIVITY**

State the IP codes which relate to the top surface of an enclosure.

**Assessment criteria**

2.1 Explain procedures for inspection

# INSPECTION PROCESS FOR INITIAL VERIFICATION

All relevant parts of an installation must be inspected during initial verification. In all types of inspection, you must be able to identify which parts of the Schedule of Inspections (see page 394) are relevant and complete the boxes with a tick (✔) or not applicable (n/a).

No boxes on the Schedule of Inspections should be left blank or marked as unsatisfactory (✗).

## Assessing the plan of an electrical system to be installed

The location plan of a small industrial unit and the accompanying information shown here and on a larger scale in Appendix 1 should allow you to practise making decisions about completing a Schedule of Inspections. Use this information as you complete the activity on page 392.

**ASSESSMENT GUIDANCE**

As part of your written assessment, you could be asked to assess the plan of an electrical system to be installed in a domestic location, an office, a garage, small industrial unit or even a special installation or location, such as a caravan park or a room containing a bath or a shower.

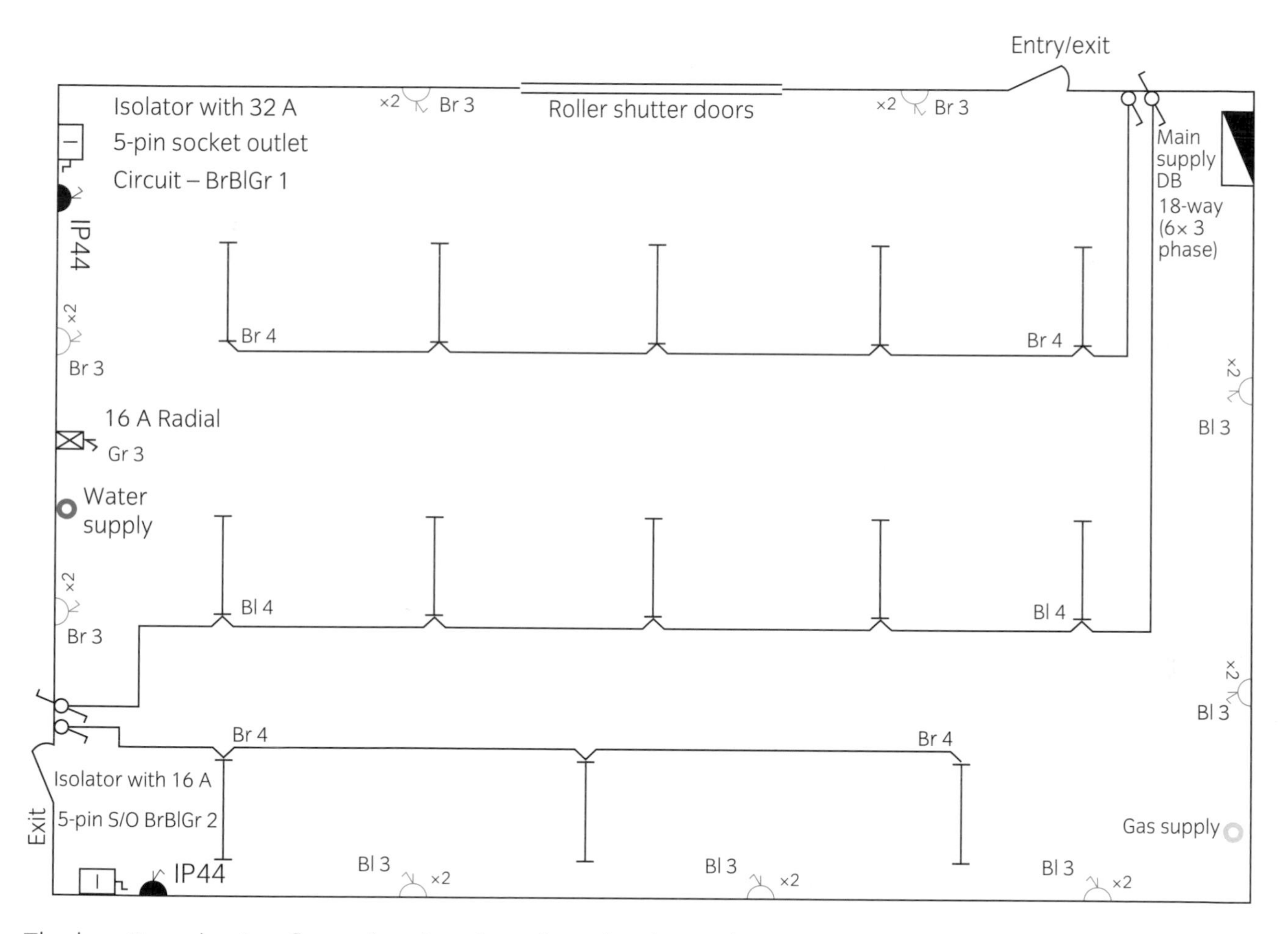


The location plan is a floor plan drawing of small industrial unit. It shows a basic outline of the electrical wiring design. The unit is one of 10 similar units on a small industrial estate.

The supply is a TN-C-S/(**PME**) system. The designer has specified that each unit has a maximum demand allowance of 100 amps per phase.

The wiring is in single-core 70 °C thermoplastic insulated cables, which are installed in a combination of metal and PVC conduit and trunking. There are no **SELV/PELV** circuits to be considered at the time of the inspection.

**SELV**

Separated extra-low voltage circuit.

**PELV**

Protective extra-low voltage circuit.

**KEY POINT**

Always analyse the information on the plans and drawings before completing the Schedule of Inspections. It is essential that you do not tick any boxes that are classed as n/a. No 'X' is allowed for initial verification.

## ACTIVITY

Use the information on pages 391–394, Appendix 1 (page 613) and Appendix 2 (page 614) to complete this activity, which is an exercise in inspection of a small industrial unit.

- You are an electrical inspector, responsible for inspecting the initial verification of the small industrial unit. During the installation process, you have visited the site and inspected the unit at relevant times in accordance with Regulation 610.1 of BS 7671.
- The appropriate design criteria, electrical plans and information that are relevant to the unit have been made available in accordance with Regulations 311, 312, 313 and 514.9.1 (see Regulation 610.2).
- You are a competent person, as required by Regulation 610.5.

Carry out the inspection process of the small industrial unit:

- The Inspection Schedule checklist below is taken from Regulation 611.3 of BS 7671. It has been put in a sequence that matches the Schedule of Inspections, as shown on page 394.
- Use the Inspection Schedule checklist to fill in a copy of the Schedule of Inspections on page 394, for the small industrial unit. The Schedule of Inspections is taken from Appendix 6 of BS 7671.
- When you have completed the Schedule of Inspections, compare your results with those on page 395.

### INSPECTION SCHEDULE CHECKLIST

**Methods of protection against electric shock**

**Both basic and fault protection:**
(i) SELV (ii) PELV (iii) Double insulation (iv) Reinforced insulation

**Basic protection:**
(i) Insulation of live parts (ii) Barriers or enclosures (iii) Obstacles
(iv) Placing out of reach

**Fault protection:**

**(i) Automatic disconnection of supply:**
Presence of earthing conductor
Presence of circuit protective conductors
Presence of protective bonding conductors
Presence of supplementary bonding conductors
Presence of earthing arrangements for combined protective and functional purposes
Presence of adequate arrangements for other sources, where applicable
**FELV**
Choice and setting of protective and monitoring devices (for fault and/or overcurrent protection)

**(ii) Non-conducting location:**
Absence of protective conductors

**(iii) Earth-free local equipotential bonding:**
Presence of earth-free local equipotential bonding

**(iv) Electrical separation:**
Provided for one item of current-using equipment
Provided for more than one item of current-using equipment

**FELV**

Functional extra-low voltage circuit (requirements can be found in Section 411.7 of BS 7671).

**Additional protection:**
Presence of residual current device(s)
Presence of supplementary bonding conductors

**Prevention of mutual detrimental influence:**
(a) Proximity to non-electrical services and other influences
(b) Segregation of Band I and Band II circuits or use of Band II insulation
(c) Segregation of safety circuits

**Identification**
(a) Presence of diagrams, instructions, circuit charts and similar information
(b) Presence of danger notices and other warning notices
(c) Labelling of protective devices, switches and terminals
(d) Identification of conductors

**Cables and conductors**
Selection of conductors for current-carrying capacity and voltage drop
Erection methods
Routing of cables in prescribed zones
Cables incorporating earthed armour or sheath, or run within an earthed wiring system, or otherwise adequately protected against nails, screws and the like
Additional protection provided by 30 mA RCD for cables concealed in walls (where required in premises not under the supervision of a skilled or instructed person)
Connection of conductors
Presence of fire barriers, suitable seals and protection against thermal effects

**General**
Presence and correct location of appropriate devices for isolation and switching
Adequacy of access to switchgear and other equipment
Particular protective measures for special installations and locations
Connection of single-pole devices for protection or switching in line conductors only
Correct connection of accessories and equipment
Presence of undervoltage protective devices
Selection of equipment and protective measures appropriate to external influences
Selection of appropriate functional switching devices

**ASSESSMENT GUIDANCE**

Circuit charts are essential but often poorly done, especially in domestic installations. Make sure yours are clear and precise.

## The Schedule of Inspections

As part of the inspection you will be required to complete a Schedule of Inspections (for new installation work only) as shown on page 394 and in Appendix 6 of BS 7671 in relationship to Regulation 610.6.

Always check you are using the latest forms, as found on the IET website: http://electrical.theiet.org

## SCHEDULE OF INSPECTIONS (for new installation work only)

### Methods of protection against electric shock

**Both basic and fault protection:**

- ☐ (i) SELV
- ☐ (ii) PELV
- ☐ (iii) Double insulation
- ☐ (iv) Reinforced insulation

**Basic protection:**

- ☐ (i) Insulation of live parts
- ☐ (ii) Barriers or enclosures
- ☐ (iii) Obstacles
- ☐ (iv) Placing out of reach

**Fault protection:**

**(i) Automatic disconnection of supply:**

- ☐ Presence of earthing conductor
- ☐ Presence of circuit protective conductors
- ☐ Presence of protective bonding conductors
- ☐ Presence of supplementary bonding conductors
- ☐ Presence of earthing arrangements for combined protective and functional purposes
- ☐ Presence of adequate arrangements for other sources, where applicable
- ☐ FELV
- ☐ Choice and setting of protective and monitoring devices (for fault and/or overcurrent protection)

**(ii) Non-conducting location:**

- ☐ Absence of protective conductors

**(iii) Earth-free local equipotential bonding:**

- ☐ Presence of earth-free local equipotential bonding

**(iv) Electrical separation:**

- ☐ Provided for **one item** of current-using equipment
- ☐ Provided for **more than one item** of current-using equipment

**Additional protection:**

- ☐ Presence of residual current device(s)
- ☐ Presence of supplementary bonding conductors

### Prevention of mutual detrimental influence

- ☐ (a) Proximity to non-electrical services and other influences
- ☐ (b) Segregation of Band I and Band II circuits or use of Band II insulation
- ☐ (c) Segregation of safety circuits

### Identification

- ☐ (a) Presence of diagrams, instructions, circuit charts and similar information
- ☐ (b) Presence of danger notices and other warning notices
- ☐ (c) Labelling of protective devices, switches and terminals
- ☐ (d) Identification of conductors

### Cables and conductors

- ☐ Selection of conductors for current-carrying capacity and voltage drop
- ☐ Erection methods
- ☐ Routing of cables in prescribed zones
- ☐ Cables incorporating earthed armour or sheath, or run within an earthed wiring system, or otherwise adequately protected against nails, screws and the like
- ☐ Additional protection provided by 30 mA RCD for cables concealed in walls (where required in premises not under the supervision of a skilled or instructed person)
- ☐ Connection of conductors
- ☐ Presence of fire barriers, suitable seals and protection against thermal effects

### General

- ☐ Presence and correct location of appropriate devices for isolation and switching
- ☐ Adequacy of access to switchgear and other equipment
- ☐ Particular protective measures for special installations and locations
- ☐ Connection of single-pole devices for protection or switching in line conductors only
- ☐ Correct connection of accessories and equipment
- ☐ Presence of undervoltage protective devices
- ☐ Selection of equipment and protective measures appropriate to external influences
- ☐ Selection of appropriate functional switching devices

Inspected by ..............................................................

Date ..........................................................................

**NOTES:**

✓ to indicate an inspection has been carried out and the result is satisfactory

**N/A** to indicate that the inspection is not applicable to a particular item

## Checklist of inspection items relevant to the small industrial unit

In the activity on page 392 you were asked to consider a floor plan for a small industrial unit and design criteria listed in an Inspection Schedule checklist, taken from Regulation 611.3. The areas from the Inspection Schedule checklist that need to be inspected and should be completed on the Schedule of Inspections are listed below. Other criteria are not applicable (n/a). Compare your results with those listed here.

**Basic protection:**
(i) Insulation of live parts (ii) Barriers or enclosures

(i) **Automatic disconnection of supply:**
Presence of earthing conductor
Presence of circuit protective conductors
Presence of protective bonding conductors
Presence of supplementary bonding conductors
Choice and setting of protective and monitoring devices (for fault and/or overcurrent protection)

**Additional protection:**
Presence of residual current device(s)
Presence of supplementary bonding conductors

**Prevention of mutual detrimental influence;**
(a) Proximity to non-electrical services and other influences

**Cables and conductors**
Selection of conductors for current-carrying capacity and voltage drop
Erection methods
Routing of cables in prescribed zones
Cables incorporating earthed armour or sheath, or run within an earthed wiring system, or otherwise adequately protected against nails, screws and the like
Connection of conductors

**General**
Presence and correct location of appropriate devices for isolation and switching
Adequacy of access to switchgear and other equipment
Connection of single-pole devices for protection or switching in line conductors only
Correct connection of accessories and equipment
Selection of equipment and protective measures appropriate to external influences
Selection of appropriate functional switching devices

**Assessment criteria**

**2.3** Justify choice of applicable items on an inspection checklist that apply in given situations

# INSPECTION REQUIREMENTS

By referring to this illustration, you can assess some of the inspection requirements below.

Remember that this relates to an inspection. All that you need to do is verify the correct connection and size (cross-sectional area) of the conductors. For testing the continuity of these conductors, please see pages 410–423.

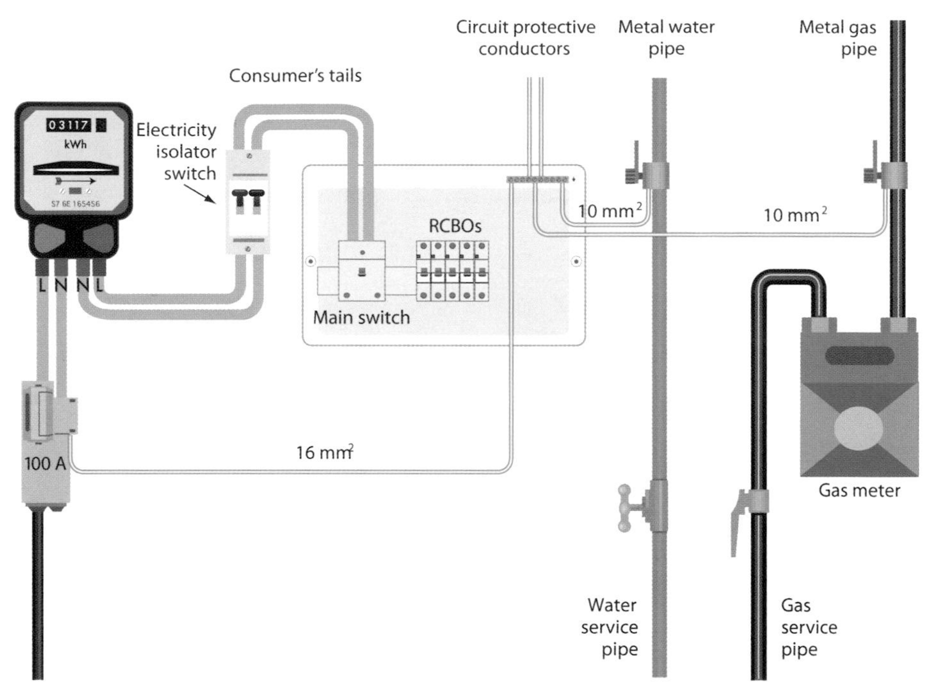


Typical single-phase domestic supply

## ASSESSMENT GUIDANCE

The electricity isolating switch allows the consumer unit to be totally isolated. It also allows the CU to be changed without contacting the meter owner.

## ACTIVITY

Assuming that a 100 A fuse is fitted to the service head, what size consumer's tails would normally be installed in this installation?

## Visual inspection

Visual checks must be carried out for inspection of the items listed below.

### Presence of earthing conductor (shown as a 16 mm² cable)

- Check that the cable is continuous from the origin of the supply to the Main Earthing Terminal (MET), consumer unit or distribution board.
- Ensure that the connections are secure.
- Ensure that the cable fixings are adequate.
- Verify the cross-sectional area of the conductor and compare with calculation or selection in accordance with Regulation 543.1.1.

- Ensure that the supply authority has supplied an earth connection at the origin (for TN-S and TN-C-S systems).
- Ensure that the earth electrode is installed correctly and that the cables are installed in accordance with Table 54.1 of BS 7671 for minimum cross-sectional area (TT system).

## Presence of main protective bonding conductors (shown as a 10 mm² cable)

- Check that the cable is continuous from the MET to the water, gas structural metalwork and any other extraneous conductive parts in accordance with Regulation 411.3.1.2. (Where the cable is hidden for part of its length, this must be verified by test.)
- Ensure that the connections are secure.
- Ensure that the cable fixings are adequate.
- Verify the cross-sectional area of the conductor and compare with Table 54.8 of BS 7671 for TN-C-S systems.
- Verify the cross-sectional area of the conductor and compare with Regulation 544.1.1 for TN-S and TT systems.

## Presence of supplementary protective bonding conductors

- Ensure the conductors are appropriately installed.
- Ensure that the connections are secure.
- Ensure that the cable fixings are adequate.
- Verify that the conductors are sized in accordance with Regulation section 544.2.3: 'A supplementary bonding conductor connecting two extraneous conductive parts shall have a cross-sectional area not less than 2.5 mm² if sheathed or otherwise provided with mechanical protection or 4 mm² if mechanical protection is not provided.'

## Presence of circuit protective conductors (cpc)

- Ensure the conductors are appropriately installed (if required).
- Ensure that the connections are secure.
- Ensure that the cable fixings are adequate.
- Verify that the conductors are selected in accordance with Table 54.7 of BS 7671 or calculated by using Regulation 543.1.3.

$$S = \frac{\sqrt{I^2 t}}{k}$$

### ACTIVITY

The use of an electricity isolating switch is a relatively new innovation. Most installations do not have them. What is the advantage of having such a switch?

### ASSESSMENT GUIDANCE

It is essential that you use the correct terminology for the various bonding and protective conductors. They are not all earth wires.

## ACTIVITY

1 Select the minimum cross-sectional area (csa) of a copper earthing conductor from the table below (Table 54.7 of BS 7671), if the supply cables (tails) for a three-phase +N distribution board are 50 mm² copper conductors (TN-C-S system earth).

**TABLE 54.7 –**
**Minimum cross-sectional area of protective conductor in relation to the cross-sectional area of associated line conductor**

| Cross-sectional area of line conductor S | Minimum cross-sectional area of the corresponding protective conductor | |
|---|---|---|
| | If the protective conductor is of the same material as the line conductor | If the protective conductor is not of the same material as the line conductor |
| (mm²) | (mm²) | (mm²) |
| $S \leq 16$ | S | $\frac{k_1}{k_2} \times S$ |
| $16 < S \leq 35$ | 16 | $\frac{k_1}{k_2} \times 16$ |
| $S > 35$ | $\frac{S}{2}$ | $\frac{k_1}{k_2} \times \frac{S}{2}$ |

**Table 54.7 of BS 7671:2008 (2011)**

2 Select the minimum csa of the corresponding main protective bonding conductors for the PME system by using the table below (Table 54.8 of BS 7671).

**TABLE 54.8 –**
**Minimum cross-sectional area of the main protective bonding conductor in relation to the neutral of the supply**

**NOTE:** Local distributor's network conditions may require a larger conductor.

| Copper equivalent cross-sectional area of the supply neutral conductor | Minimum copper equivalent* cross-sectional area of the main protective bonding conductor |
|---|---|
| 35 mm² or less | 10 mm² |
| over 35 mm² up to 50 mm² | 16 mm² |
| over 50 mm² up to 95 mm² | 25 mm² |
| over 95 mm² up to 150 mm² | 35 mm² |
| over 150 mm² | 50 mm² |

* The minimum copper equivalent cross-sectional area is given by a copper bonding conductor of the tabulated cross-sectional area or a bonding conductor of another metal affording equivalent conductance.

**Table 54.8 of BS 7671:2008 (2011)**

3 What is the minimum csa required for a supplementary protective bonding conductor from Regulation 544.2.3, if not mechanically protected?

4 Show the equation used to obtain the minimum csa of a reduced size cpc, which does not comply with the tabulated values in Table 54.7 of BS 7671.

## Isolation and isolation devices

Means of isolation should be provided as follows.

### At the origin of the installation

A main-linked switch or circuit breaker should be provided as a means of isolation and of interrupting the supply on load. For single-phase household and similar supplies that may be operated by unskilled persons, a double-pole device must be used for both TT and TN systems.

For a three-phase supply to an installation forming part of a TT system, an isolator must interrupt the line and neutral conductors. In a TN-S or

TN-C-S system only the line conductors need be interrupted as the installation has a reliable connection to earth.

### For every circuit

Other than at the origin of the installation, every circuit or group of circuits that may have to be isolated without interrupting the supply to other circuits should be provided with its own isolating device.

### For every item of equipment

All fixed electrical equipment should be provided with a means of switching which can be used for the safety and maintenance of systems, from industrial production equipment to individual items, such as immersion heaters, hand dryers and showers.

### For every motor

Every fixed electric motor should be provided with a readily accessible and easily operated device to switch off the motor and all associated equipment, including any automatic circuit breaker.

### For every supply

All isolators must be labelled or identified, if it is not obvious which circuits they control.

**ACTIVITY**

Not all switches are suitable as isolators. State one condition that would qualify a switch as an isolator.

**KEY POINT**

An isolator is a mechanical switching device which, in the open position, complies with the requirements specified for the isolating function. Most isolators are manufactured to switch on/off the supply during no-load conditions. The switching and control circuits of equipment must be used de-energise the load before the isolator switching mechanism is opened.

**ASSESSMENT GUIDANCE**

You must be able to select the correct type of overload device for the load. Failure to do so may result in frequent nuisance tripping or the circuit not being protected.

## Overcurrent protective devices

The designer of an installation will coordinate the loads, protective devices and sizing of cables. It is the inspector's job to verify that the correct protective devices have been installed as designed.

The following table explains the application of circuit breakers.

- The BS EN 60898 standard circuit breakers are Type B, C and D.
- The old versions of circuit breakers were the BS 3871 Type1, 2, 3 and 4. (These are now discontinued.)

| Circuit-breaker type | Trip current (0.1 s to 5 s) | Application |
|---|---|---|
| 1<br>B | 2.7 to 4 $I_n$<br>3 to 5 $I_n$ | Domestic and commercial installations having little or no switching surge |
| 2<br>C<br>3 | 4 to 7 $I_n$<br>5 to 10 $I_n$<br>7 to 10 $I_n$ | General use in commercial/industrial installations where the use of fluorescent lighting, small motors, etc., can produce switching surges that would operate a Type 1 or B circuit-breaker. Type C or 3 may be necessary in highly inductive circuits such as banks of fluorescent lighting |
| 4<br>D | 10 to 50 $I_n$<br>10 to 20 $I_n$ | Not suitable for general use<br>Suitable for transformers, X-ray machines, industrial welding equipment, etc., where high inrush currents may occur |

**NOTE**: $I_n$ is the nominal rating of the circuit-breaker.

**Classification of circuit breakers (from the On-Site Guide, IET)**

The cable current-carrying capacity must be checked. The cables connected to the protective devices, whether fuses or circuit breakers, must be checked for coordination.

Voltage drop calculation examples can be found on pages 468–471. Further coverage of coordination can be found on pages 422, 445, 454 and 487.

OUTCOMES 3, 6

# Understand procedures for completing the testing of electrical systems
# Be able to test safety of electrical systems

## APPROPRIATE TESTS AND SEQUENCE FOR ELECTRICAL INSTALLATIONS

The tests which need to be carried out during initial verification are listed below, in the correct sequence in accordance with BS 7671. Some of the tests will not be applicable to all electrical installations; however, the applicable tests must be carried out in order.

**Assessment criteria**

3.1 Explain why regulatory tests are undertaken

---

### Regulation section 612 – prescribed tests

#### Dead tests

| | |
|---|---|
| 612.2.1 | Continuity of protective conductors; including cpc, main and supplementary bonding conductors |
| 612.2.2 | Continuity of ring final circuit conductors |
| 612.3 | Insulation resistance |
| 612.4.1 | Protection by **SELV** |
| 612.4.2 | Protection by **PELV** |
| 612.4.3 | Protection by electrical separation |
| 612.4.4 | Functional extra-low voltage circuits |
| 612.4.5 | Basic protection by a barrier or enclosure provided during erection |
| 612.5 | Insulation resistance/impedance of floors and walls |
| 612.6 | Polarity |
| 612.7 | Earth electrode resistance (if applicable) |

#### Live tests

| | |
|---|---|
| 612.7 | Earth electrode resistance (if applicable) |
| 612.8 | Protection by automatic disconnection (verification in accordance with Chapter 41 of BS 7671) |
| 612.9 | Earth fault loop impedance |
| 612.10 | Additional protection |
| 612.11 | Prospective fault current |
| 612.12 | Check of phase sequence |
| 612.13 | Functional testing |
| 612.14 | Verification of volt drop |

**SELV**

Separated extra-low voltage circuit.

**PELV**

Protective extra-low voltage circuit.

**ACTIVITY**

What would be the first four tests to be carried out on a standard domestic ring final circuit?

## Regulation section 612 – prescribed tests for small industrial unit

If you consider the location plan of a small industrial unit, shown below and in Appendix 1 (page 613), you can see that some of the aforementioned tests will be necessary to prove that the electrical installation is safe to be put into use.

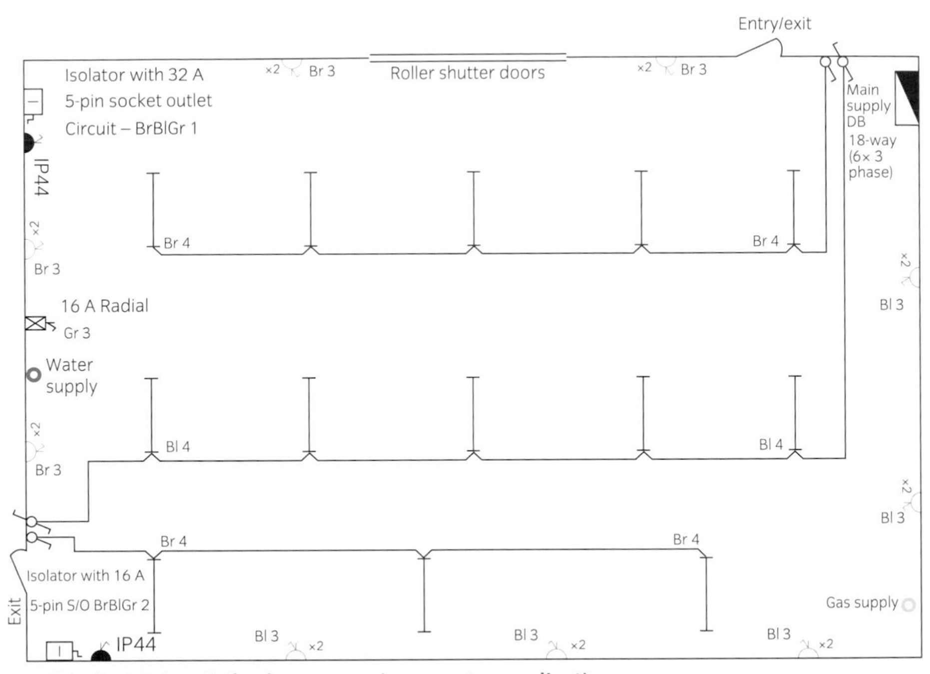


**Small industrial unit (for larger version see Appendix 1)**

### ASSESSMENT GUIDANCE

Remember that not all the tests apply to all installations. If there is no ring final circuit then there is no ring final circuit test to do.

### ACTIVITY

Why are the first tests carried out on an electrical installation as part of the initial verification, called 'dead' tests?

The applicable tests for this small industrial unit are listed below.

### Dead tests

612.2.1 Continuity of protective conductors; including main and supplementary equipotential bonding
612.2.2 Continuity of ring final circuit conductors
612.3 Insulation resistance
612.6 Polarity

### Live tests

612.8 Protection by automatic disconnection (verification in accordance with Chapter 41 of BS 7671)
612.9 Earth fault loop impedance
612.10 Additional protection
612.11 Prospective fault current
612.12 Check of phase sequence
612.13 Functional testing
612.14 Verification of volt drop

As you progress through this unit:

- the tests given in the lists on page 402, relating to the location plan of the small industrial unit, will be calculated and test methods will be explained
- the tests not selected in the lists will also be explained with the aid of illustrations, test methods, calculations and table references.

## Regulation section 612 – logging the test result data

The Generic Schedule of Test Results allows for the test results to be logged in a format that can be understood by all competent testers.

If you observe the progression through the Schedule of Test Results shown on page 404, all of the dead tests are followed by the live tests.

**ASSESSMENT GUIDANCE**

During your assignment, you will need to choose the appropriate test sequence for an installation based on the plans, diagrams and information given.

**ACTIVITY**

Without looking back at the test sequence in this section, write down 'in order' the sequence of tests required for the lighting circuit supplied by the type B, 6-amp circuit breaker (circuit Bl4) in the location plan of the industrial unit in Appendix 1.

## IDENTIFYING THE CORRECT INSTRUMENT FOR TESTS

Testing should be carried out in such a manner that no danger to person, livestock or property arises.

The multi-function test meter in the photograph shows that there are many different settings available for the various tests required by BS 7671. The meter may look complicated, but if you concentrate on one test at a time, you will understand how to use it. This section will simply outline the settings required for different tests and will give some sample test results.

**Assessment criteria**

**3.7** Specify the requirements for the safe and correct use of instruments to be used for testing

### Before you test

Before any tests are performed, you need to check the leads and probes to ensure that they are in good order. Methods of checking leads will be explained as you are shown each test in order.

The leads must be located correctly for testing. The individual meter settings for each test are briefly introduced on pages 405–406.

**ASSESSMENT GUIDANCE**

It is good practice to check the instrument for damage before and after use. Show this clearly, or describe your actions when carrying out the tests.

Form 4

Form No: ................./4

## GENERIC SCHEDULE OF TEST RESULTS

| DB reference no ..................... | Details of circuits and/or installed equipment vulnerable to damage when testing ..................... | Details of test instruments used (state serial and/or asset numbers) |
|---|---|---|
| Location ..................... | ..................... | Continuity ..................... |
| Zs at DB (Ω) ..................... | ..................... | Insulation resistance ..................... |
| $I_{pf}$ at DB (kA) ..................... | ..................... | Earth fault loop impedance ..................... |
| Correct supply polarity confirmed ☐ | ..................... | RCD ..................... |
| Phase sequence confirmed (where appropriate) ☐ | ..................... | Earth electrode resistance ..................... |

Tested by:
Name (Capitals) .....................
Signature ..................... Date .....................

| Circuit details | | | | | | | | | Test results | | | | | | | | | | | | |
|---|---|---|---|---|---|---|---|---|---|---|---|---|---|---|---|---|---|---|---|---|---|
| | | Overcurrent device | | | | Conductor details | | | Ring final circuit continuity (Ω) | | | Continuity (Ω) ($R_1 + R_2$) or $R_2$ | | Insulation Resistance (MΩ) | | Polarity | $Z_s$ (Ω) | RCD (ms) | | | Remarks (continue on a separate sheet if necessary) |
| Circuit number 1 | Circuit Description 2 | BS (EN) 3 | type 4 | rating (A) 5 | breaking capacity (kA) 6 | Reference Method 7 | Live (mm²) 8 | cpc (mm²) 9 | $r_1$ (line) 10 | $r_n$ (neutral) 11 | $r_2$ (cpc) 12 | ($R_1 + R_2$) * 13 | $R_2$ 14 | Live-Live 15 | Live-E 16 | 17 | 18 | @ $I_{\Delta n}$ 19 | @ $5I_{\Delta n}$ 20 | Test button operation 21 | 22 |
| | | | | | | | | | | | | | | | | | | | | | |
| | | | | | | | | | | | | | | | | | | | | | |
| | | | | | | | | | | | | | | | | | | | | | |
| | | | | | | | | | | | | | | | | | | | | | |
| | | | | | | | | | | | | | | | | | | | | | |
| | | | | | | | | | | | | | | | | | | | | | |
| | | | | | | | | | | | | | | | | | | | | | |
| | | | | | | | | | | | | | | | | | | | | | |
| | | | | | | | | | | | | | | | | | | | | | |
| | | | | | | | | | | | | | | | | | | | | | |
| | | | | | | | | | | | | | | | | | | | | | |
| | | | | | | | | | | | | | | | | | | | | | |
| | | | | | | | | | | | | | | | | | | | | | |
| | | | | | | | | | | | | | | | | | | | | | |
| | | | | | | | | | | | | | | | | | | | | | |

* Where there are no spurs connected to a ring final circuit this value is also the ($R_1 + R_2$) of the circuit.

**In this Schedule of Test Results, note that all of the dead tests are completed before the live tests (always check you are using the latest forms, as found on the IET website: http://electrical.theiet.org)**

## Setting up your test meter

### Low-resistance ohmmeter

The tests carried out with this meter are:

- Continuity of protective conductors, including main and supplementary equipotential bonding.
- Continuity of ring final circuit conductors.
- Polarity.

The meter setting is in ohms (Ω), as shown in the photograph.

The test meter must be capable of supplying a no-load voltage of between 4 V and 24 V (d.c. or a.c.) and a short-circuit current of not less than 200 mA.

The measuring range should cover the span 0.2 Ω to 2 Ω, with a resolution of at least 0.01 Ω for digital instruments.

Note that general purpose multi-meters are not capable of supplying these voltage and current parameters.

**ACTIVITY**

Using the photographs of the meters as guidance, set your own test instrument for:

- continuity testing
- insulation resistance testing
- earth fault loop impedance testing.

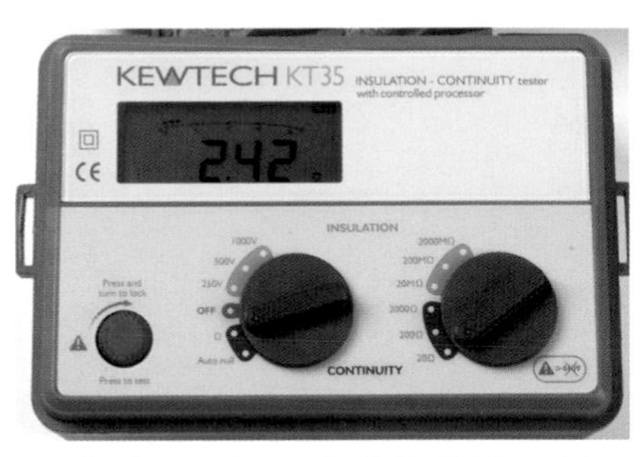


**Continuity testing – the left dial is set to the Ω scale and the right dial is set at 20 Ω**

### Insulation resistance tester

The tests carried out with this meter are:

- Insulation resistance testing.
- Separation of circuits, including
  - SELV or PELV
  - electrical separation.

The meter setting is in megohms (MΩ), as shown in the photograph.

The test meter must be capable of supplying an output test voltage of 250 V d.c., 500 V d.c. or 1000 V d.c., subject to the circuit to be tested.

The readings can range from 0.00 MΩ to to 2000 MΩ.

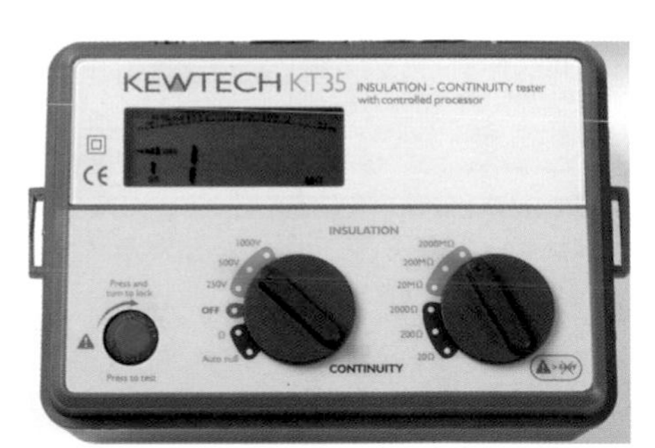


**Insulation resistance testing – the left dial is set to 500 V and the right dial is set at 2000 MΩ**

### Earth fault loop impedance tester

The tests carried out with this meter are:

- Earth fault loop impedance.
- Prospective fault current.

The photograph shows the meter set to Loop – 20 Ω scale, which is usually used for TN systems and some TT systems.

An earth fault loop impedance tester with a resolution of 0.01 Ω should be adequate for circuits rated up to 50 A. Instruments conforming to BS EN 61557-3 will fulfil the above requirements.

**Earth fault loop impedance testing – this example shows the dial set at Loop – 20 Ω. Loop 200 Ω and 2000 Ω may need to be used when testing TT systems.**

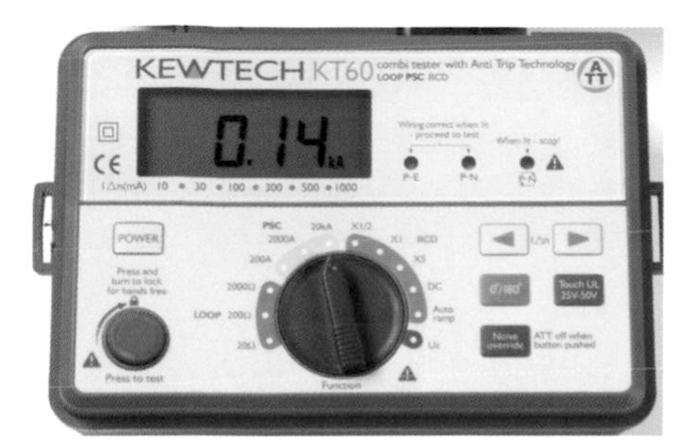


Prospective fault current testing – this example shows the dial set at PSC – 20 kA

Earth fault loop impedance instruments may also offer additional facilities for deriving prospective fault current. The basic measuring principle is generally the same as for earth fault loop impedance testers.

The photograph shows the correct settings for deriving the prospective fault current.

Prospective fault current is measured in kA and can range from 0.3 kA to 16 kA.

The readings must be compared to the rated short-circuit capacity of the protective devices.

## Residual current device (RCD) tester

The tests carried out with this meter are:

- Additional protection
- Residual current device operation.

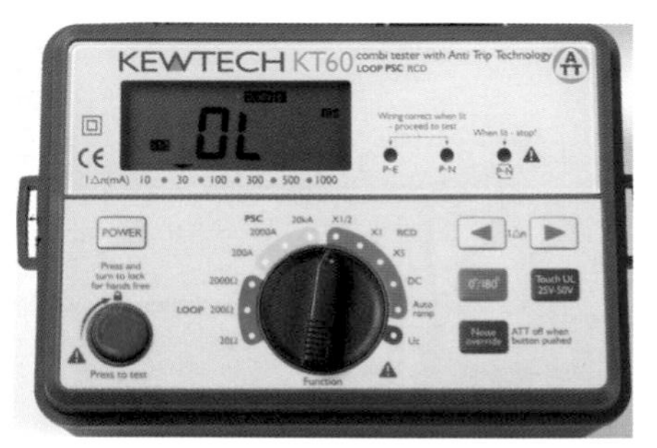


RCD testing – this example shows the dial set at RCD – x½ with the 30 mA setting displayed in the LCD

Operation of RCDs must be checked to ensure that they are operating in accordance with the manufacturer's intended time limits. These results are given in milliseconds (ms). Methods and appropriate maximum trip times can be found in Section 11 of the IET On-Site Guide.

The photograph shows just one example of a typical setting for this test.

Note that earth electrode testing, polarity testing, phase rotation testing, functional testing and verification of voltage drop will be addressed as we progress through the test sequence.

### ASSESSMENT GUIDANCE

All of these tests can be carried out by using a test meter as shown. However, when answering City & Guilds assessment questions, you will be expected to name the specific meter used in conjunction with the specific test, as shown in the activity on the right.

### ACTIVITY

State the test instrument used and the test result unit of measurement expected for each of the tests listed below. For the example of RCD tests, the answer would be 'RCDs are tested with an RCD tester and the readings are in milliseconds (ms)'.

1 Continuity of protective conductors
'Continuity testing is carried out by using a ...'

2 Insulation resistance testing
'Insulation resistance testing is ...'

## SAFE AND CORRECT USE OF INSTRUMENTS

When using test instruments, safety can be achieved by ensuring that the following points are followed.

### Steps to follow when using instruments

#### 1 Checking that the instruments being used conform to the appropriate British Standard safety specifications

- The basic instrument standard is BS EN 61557 'Electrical safety in low voltage distribution systems up to 1000 V a.c. and 1500 V d.c. Equipment for testing, measuring or monitoring of protective measures.'
- This standard includes performance requirements and requires compliance with BS EN 61010.

#### 2 Instrument accuracy

- Instrument accuracy of 5% is usually adequate for testing.
- The expected accuracy for analogue meters is 2% of full-scale deflection.

#### 3 Calibration

- Calibration must be carried out on each piece of test equipment in accordance with the manufacturer's recommendations.
- Regular checking may be carried out using known references.
- Usually, annual calibration suffices, unless the instrument is subject to excessive mechanical stresses.

#### 4 Understanding the equipment

- You should understand the equipment to be used and its ranges.
- It is important, also, to understand the characteristics of the installation where the test instrument will be used.

#### 5 Selecting and using the appropriate scales and settings

- It is essential that the correct scale and setting are selected for an instrument in a particular testing situation. As a vast range of instruments exists, always read the manufacturer's instructions before using an unfamiliar instrument. Further things to consider are:
  - When a meter displays a number 1 in the left of the display, it usually means that the instrument reading is over the range selected.
  - For continuity testing, the lowest ohms scale must be selected as values are usually low.

**ACTIVITY**

What check should be carried out on a test instrument before and after use?

**ASSESSMENT GUIDANCE**

Unless the instrument is self-ranging, set it to the highest appropriate scale and then select progressively lower scales until the most accurate reading is achieved.

- Remember that the values shown while insulation resistance testing is taking place are in megohms (millions of ohms, MΩ).
- Many instruments feature a 'no-trip' function when testing earth fault loop impedance. This should only be used where an RCD is in the circuit. In all other cases the 'high' setting should be used as this is much more accurate.
- Manufacturers instructions must be followed when selecting the method of prospective fault current testing. These tests are explained on pages 458–461.

**KEY POINT**

Guidance Note GS38 gives guidelines on test equipment, leads and probes to be used when testing on or near live conductors. This can be freely viewed and downloaded from the Health and Safety Executive website: www.hse.gov.uk

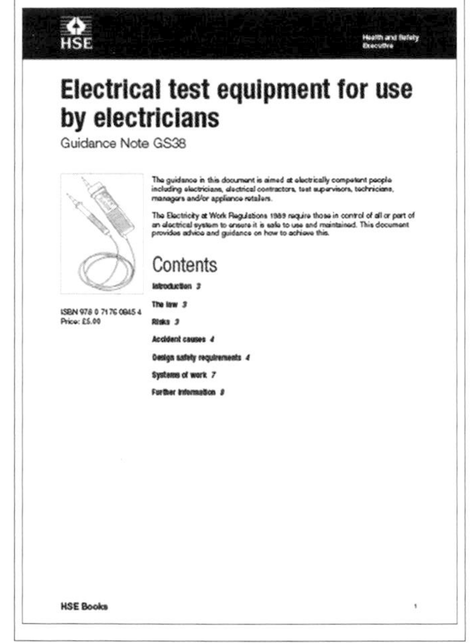

HSE — Health and Safety Executive

**Electrical test equipment for use by electricians**

Guidance Note GS38

The guidance in this document is aimed at electrically competent people including electricians, electrical contractors, test supervisors, technicians, managers and/or appliance retailers.

The Electricity at Work Regulations 1989 require those in control of all or part of an electrical system to ensure it is safe to use and maintained. This document provides advice and guidance on how to achieve this.

Contents



ISBN 978 0 7176 0645 4
Price: £5.00

HSE Books

## 6 Checking test leads

- Check that test leads, including any probes or clips, are in good order, are clean and have no cracked or broken insulation.
- Where appropriate, the requirements of the Health and Safety Executive Guidance Note GS38 should be observed for test leads.

The photograph shows a test instrument with a range of test leads, probes and clips. These leads can be used for different tests.

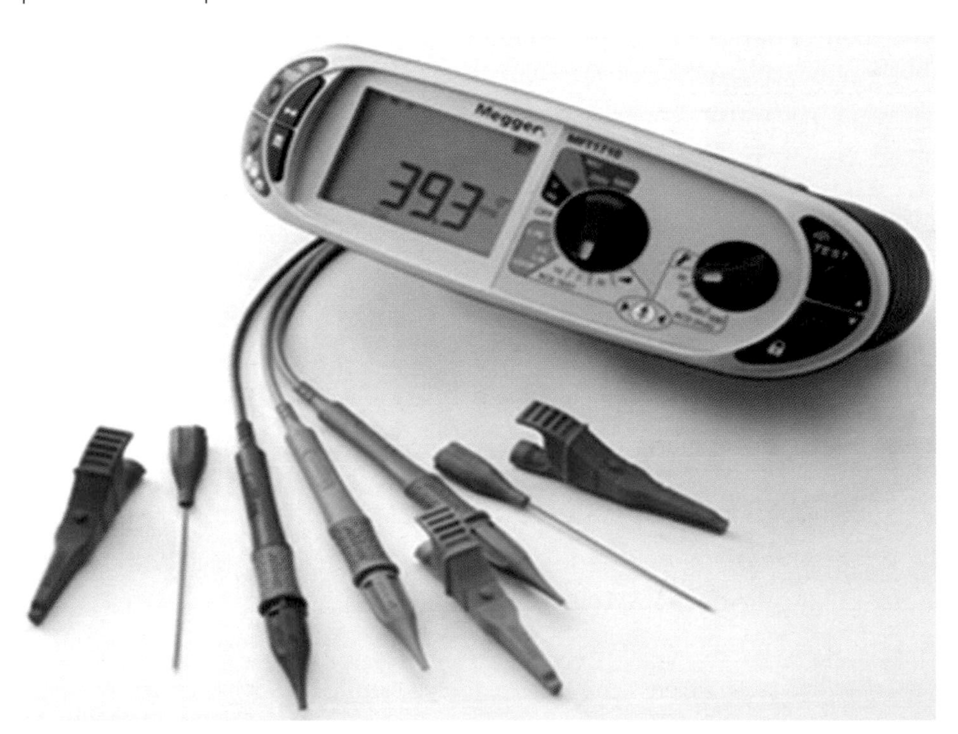


Multi-function test instrument showing a range of leads and probes (image courtesy of Megger)

## WHY TEST RESULTS NEED TO COMPLY WITH STANDARDS

Test results must be recorded on the Schedule(s) of Test Results and compared with relevant criteria. For example, in order to verify disconnection times, the relevant criteria would be the design earth fault loop impedance values provided by the designer.

Each test is carried out for a reason as outlined in the table, with a very brief description of what could happen as a consequence of the conditions revealed by the test result.

| Test | Risk(s) indicated by unsatisfactory result |
|---|---|
| Continuity of protective conductors, including main and supplementary bonding | Shock |
| Continuity of ring final circuit conductors | Overload, fire, shock |
| Insulation resistance | Fire, shock |
| Protection by SELV, PELV or by electrical separation | Shock |
| Protection by barriers or enclosures provided during erection | Shock |
| Insulation resistance of non-conducting floors and walls | Shock |
| Polarity | Shock, equipment damage |
| Earth electrode resistance | Shock |
| Protection by automatic disconnection of the supply | Shock |
| Earth fault loop impedance | Shock |
| Additional protection | Shock |
| Prospective fault current | Explosion, injury, fire damage |
| Phase sequence | Equipment damage, injury |
| Functional testing | Incorrect working of equipment, shock |
| Voltage drop | Equipment malfunction and damage |

If any of the above tests prove to be unsatisfactory during an initial verification, the fault must be rectified before any more tests (in the proper order) are undertaken.

**Assessment criteria**

**3.5** Explain implications of non-compliance of regulatory test results with regulatory values

### ACTIVITY

Which BS 7671 appendix gives disconnection times for fuses and circuit breakers?

### ASSESSMENT GUIDANCE

As you can see, shock is the greatest risk, but very few people are killed by electric shock each year. Can you think why this is?

**Assessment criteria**

**3.2** Explain why regulatory tests are carried out in the exact order as specified

## TESTING IN THE EXACT ORDER SPECIFIED

The test sequence must be carried out in the exact order prescribed in Regulations 612.2 to 612.14 of BS 7671. Testing should stop if any circuit test result is not within the prescribed limits.

Some reasons for the specified order of testing are outlined below:

1. The test sequence has been developed to reduce the risk of electric shock during testing. All of the tests from Regulation 612.2 to 612.7 should be carried out before the installation is energised (Regulation 612.1).
2. If a test result within the prescribed test sequence fails to meet its minimum or maximum value, it may be dangerous or impractical to progress. For example:
   - Impractical – if a continuity test for a lighting circuit records an open circuit, progressing to the insulation resistance test is impractical, because testing between conductors and earth is a worthless exercise.
   - Dangerous – if a very low insulation resistance value was recorded during the testing of a circuit, this would indicate possible leakage currents. If this circuit was subsequently switched on for live testing, there could be fire or electric shock risk.

**Assessment criteria**

**3.3** Explain how to prepare for testing electrical systems

**3.4** Specify procedures for regulatory tests

**6.1** Implement safe system of work for testing electrical systems

**6.2** Select the test instruments for regulatory tests

**6.3** Test electrical systems

**6.4** Record results of regulatory tests

**6.5** Verify compliance of regulatory test results

**6.6** Commission electrical systems

## CONTINUITY TESTING

### Continuity testing of the earthing conductor

Note that secure isolation must be carried out before this test is done (secure isolation is covered on pages 362–364).

#### Reason for this test

This test must be carried out to ensure that the earthing conductor is not broken; it should connect the main earthing terminal to the means of earthing.

This test is not required if the earthing conductor can be *visually* inspected throughout its entire length. If at any point, the earthing conductor is concealed, this test is recommended.

**ACTIVITY**

Except for short runs, it would be unusual for the earthing conductor to be visible throughout its length. What impact might this have on continuity testing?

#### Test meter

The test meter must be capable of supplying a no-load voltage between 4 V and 24 V d.c. or a.c. and a short-circuit current of not less than 200 mA.

The measuring range should cover the span 0.01 Ω to 2 Ω, with a resolution of at least 0.01 Ω for digital instruments.

Note that general purpose multi-meters are not capable of supplying these voltage and current parameters.

## Method

1 Select the correct meter – you will need a low-resistance ohmmeter.
2 Select the correct scale – the correct scale is ohms/continuity (see pages 407–408 for guidance on scales)
3 Check the leads for damage.
4 Insert the leads – choose the correct location on the meter (typically C and Ω, but this may differ for different instruments. Always consult the manufacturer's instructions.
5 Correctly connect the leads – the correct way to connect leads is explained in the diagram.
6 Zero or null the meter leads – the correct reading should be 0.00 Ω. If a particular test requires a longer lead or a link lead, this should be zeroed or nulled along with the meter leads.

Incorrect method of connecting leads

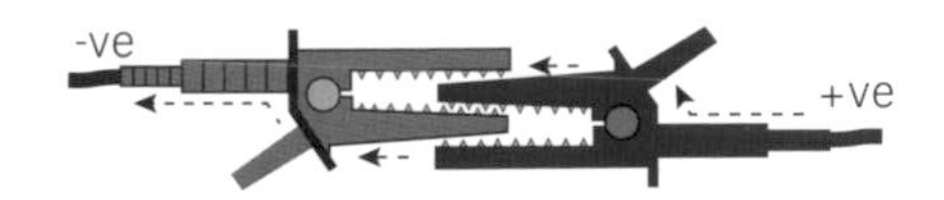


Current flow has to travel across the hinges

Correct method of connecting leads

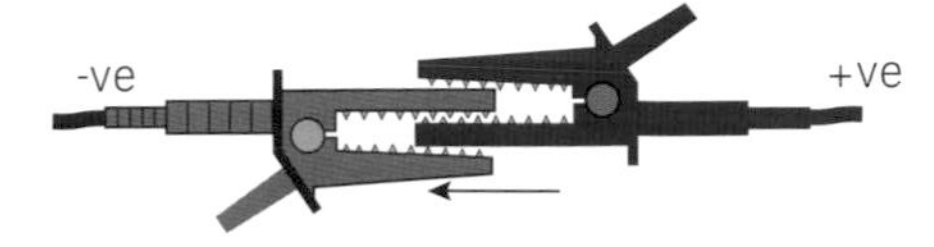


Current flow is straight from clip to clip

**Always connect leads correctly**

You are now ready to carry out the continuity testing.

## Carrying out the continuity testing of the earthing conductor

Look again at the location plan of the small industrial unit in Appendix 1 and the distribution and wiring information in Appendix 2. You need to identify the earthing conductor, the protective conductors and the circuits that require continuity testing.

The small industrial unit is fed from a TN-C-S supply. The main supply distribution board (DB) is situated 5 m away from the origin of the supply. The main three-phase (line conductors) and neutral supply tails have been calculated and designed to be 35 mm$^2$.

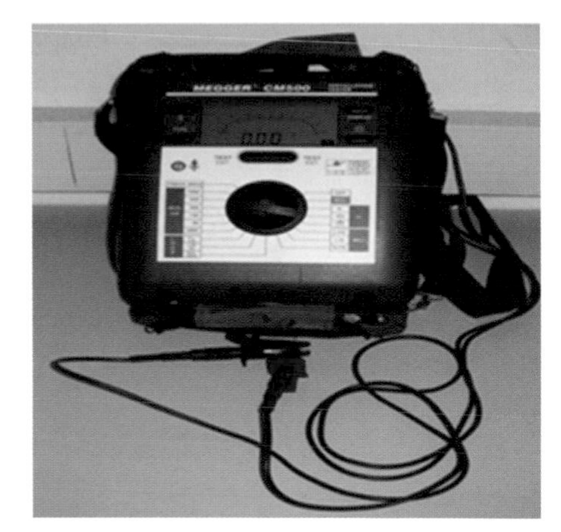

**Multi-function test instrument with leads connected together (image courtesy of Megger)**

## How to calculate the resistance

The cable cross-sectional area has been selected in accordance with Regulation 543.1.1 and Table 54.7 of BS 7671 (16 mm$^2$ earthing conductor).

The distance from the DB Main Earthing Terminal (MET) is 5 m.

Using the resistance value given in Appendix B (Table B1) of Guidance Note 3, a 16 mm$^2$ cable has a resistance of 1.15 mΩ/m.

Resistance of the cable is calculated as 5 m x 1.15 mΩ/m = 5.75 mΩ or 0.005 Ω.

| Conductor csa mm$^2$ | Copper conductor resistance mΩ/m |
|---|---|
| 4 | 4.61 |
| 10 | 1.83 |
| 16 | 1.15 |

**Resistance values for cables (information extracted from Table B1 of Guidance Note 3)**

## How to test the resistance

For this test, the meter leads are not long enough.

- You must have a flexible earth extension reel – normally about 25 m length.
- Zero the test leads by connecting them to both ends of the extension reel.
- Connect one end of the extension lead to the earthing conductor, which needs to be disconnected from the MET (isolation of the supply to the building must be carried out before disconnecting the earthing conductor).
- Connect one crocodile clip to the earthing conductor at the origin of the supply and the other crocodile clip to the extension.

A 5 m length of 16 mm$^2$ should show on the meter (to two decimal places) as 0.01 Ω.

So, test result = 0.01 Ω for the earthing conductor.

The diagram below shows the earthing conductor and the main protective bonding conductor, with the earthing conductor 16 mm$^2$.

### ASSESSMENT GUIDANCE

Remember that contact pressure can vary the resistance measurement.

### KEY POINT

The resistance of a cable is proportional to its length. If a cable is 10 m long and its resistance is 0.18 Ω, 50 m of the same cable will have a resistance of 0.90 Ω.

### ACTIVITY

1. Measure the resistance of a partly used drum or a length of cable. Zero your meter and test the resistance.
2. Compare the measured resistance value to the estimated calculation value.

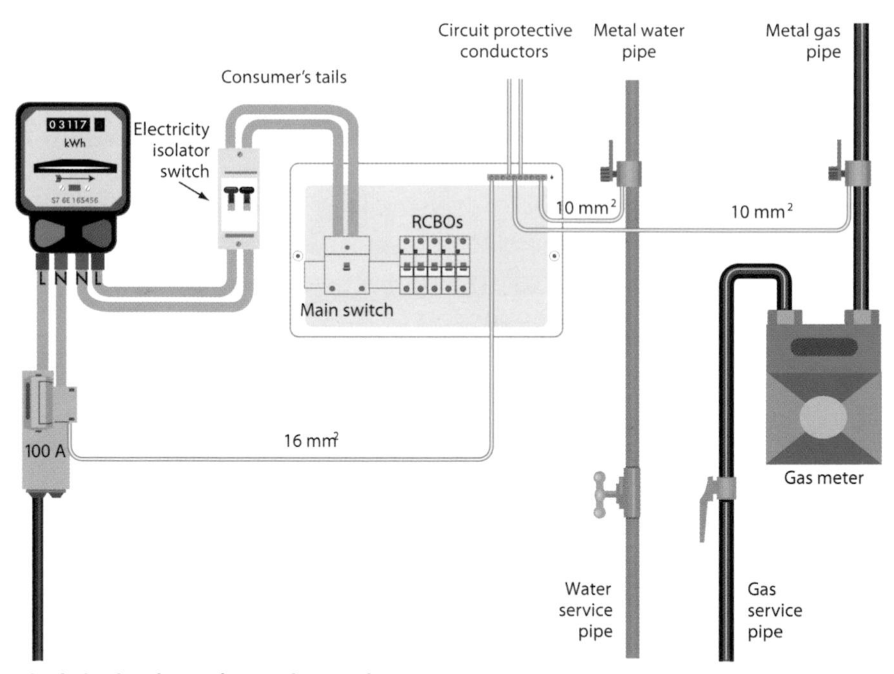


Typical single-phase domestic supply

## Continuity testing of main and supplementary protective bonding conductors

### Reason for this test

This test is carried out to ensure that all protective bonding conductors are continuous and have low resistance values. Note that secure isolation must be carried out before testing (secure isolation is covered on pages 362–364).

## Test meter

The test meter must be capable of supplying a no-load voltage between 4 V and 24 V, d.c. or a.c., and a short-circuit current of not less than 200 mA.

The measuring range should cover the span 0.2 Ω to 2 Ω, with a resolution of at least 0.01 Ω for digital instruments.

Note that general purpose multi-meters are not capable of supplying these voltage and current parameters.

### ACTIVITY

This section requires the use of Ohm's law. Make sure you can transpose the formula easily.

## Method

1 Select the correct meter – you will need a low-resistance ohmmeter.
2 Select the correct scale – the correct scale is ohms/continuity (see pages 407–408 for guidance on scales).
3 Check the leads for damage.
4 Insert leads – choose the correct location on the meter (typically C and Ω but this may differ between instruments). Always consult the manufacturer's instructions.
5 Correctly connect the leads – the correct way to connect leads is shown in the diagram.
6 Zero or null the meter leads – the correct reading should be 0.00 Ω. If a particular test requires a longer lead or a link lead, this should be zeroed or nulled along with the meter leads.

Incorrect method of connecting leads

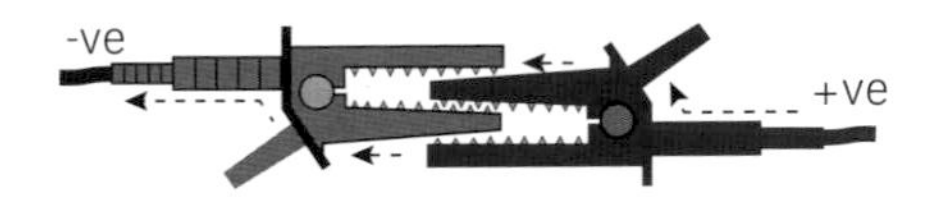


Current flow has to travel across the hinges

Correct method of connecting leads

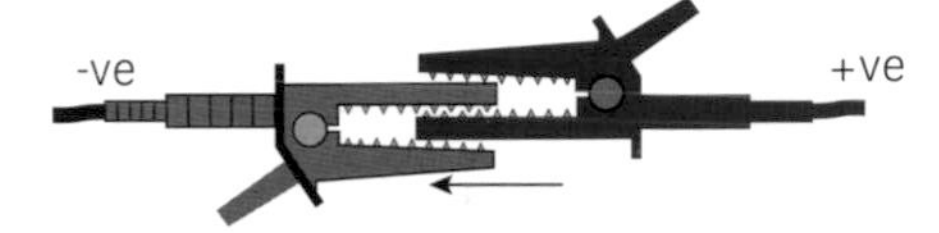


Current flow is straight from clip to clip

**Always connect leads correctly**

You are now ready to carry out the continuity testing.

## Carrying out the continuity testing of the main protective bonding conductor

If you know the cross-sectional area (csa) of the earthing conductor, you can check the minimum csa of the main protective bonding conductors. The following information will allow you to assess the resistance of the main protective bonding conductors.

Look again at the location plan of the small industrial unit in Appendix 1. The small industrial unit is fed from a TN-C-S supply. The main supply DB is situated 5 m away from the origin of the supply. The main three-phase (line conductors) and neutral supply tails have been calculated and designed to be 35 $mm^2$. The earthing conductor is 16 $mm^2$.

## How to calculate the resistance

You can calculate the value of resistance before carrying out continuity testing to give you an idea of the reading you should be obtaining.

From the location plan in Appendix 1, you can see that there are at least two main protective bonding conductors to be installed – the water utility supply and the gas utility supply.

| Conductor csa $mm^2$ | Copper conductor resistance mΩ/m |
|---|---|
| 4 | 4.61 |
| 10 | 1.83 |
| 16 | 1.15 |

**Resistance values for cables (information extracted from Table B1 of Guidance Note 3)**

The water supply main protective bonding conductor is 23 m long and selected in accordance with Regulation 544.1.1 and Table 54.8 of BS 7671 (10 $mm^2$). The neutral of the supply is 35 $mm^2$.

In order to determine the resistance of the conductor at 20 °C, the following formula is used:

$$\frac{\text{length} \times \text{m}\Omega/\text{m}}{1000}$$

Therefore, the resistance would be:

$$\frac{23 \times 1.83}{1000} = 0.04\ \Omega$$

## How to test the resistance

To test the resistance:

- Zero the test leads by connecting them to both ends of the extension earth.
- Connect one end of the extension lead to the disconnected main protective bonding conductor for the water supply pipe at the MET (isolation of the supply to the building must be carried out before disconnecting the main protective bonding conductor from the MET).
- Connect one crocodile clip to the metal water supply connection and the other crocodile clip to the remaining end of the rolled out extension lead.

The illustration shows the earthing conductor (16 $mm^2$) and the main protective bonding conductor (10 $mm^2$).

**ACTIVITY**

How would you ensure good contact between the bonding conductor and water pipe?

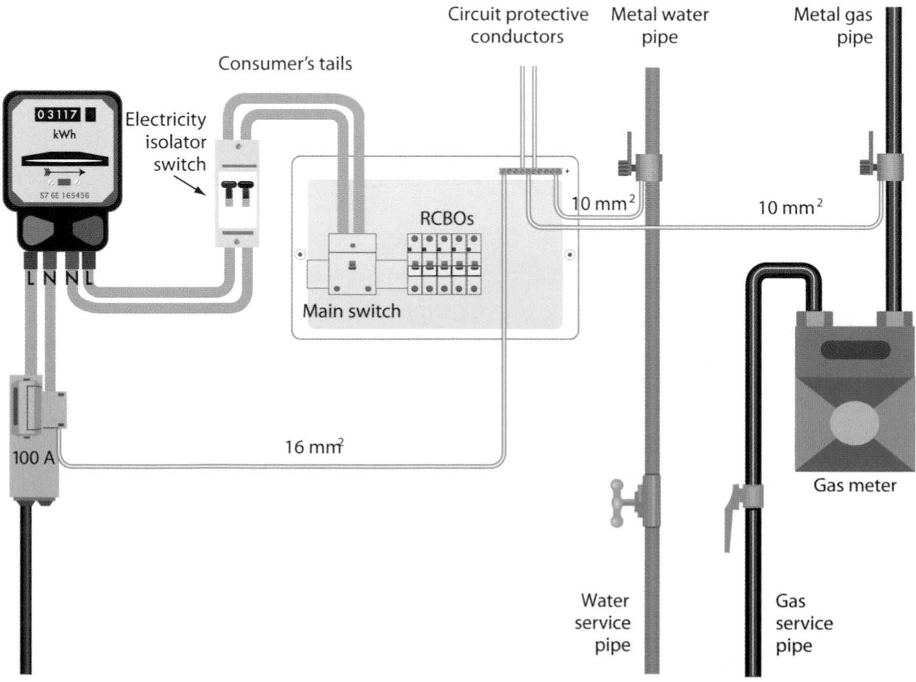


**Typical single-phase domestic supply**

A 23 m length of 10 $mm^2$ should show on the meter (to two decimal places) as 0.04 Ω, as shown in the example above.

So, test result for the water supply main protective bonding conductor is 0.04 Ω.

### Carrying out the continuity testing of supplementary protective bonding conductors

Continuity testing of supplementary protective bonding conductors should be carried out to ensure that the bonding is continuous. It is suggested that the test results between conductive parts should be in the order of 0.05 Ω or less. Further details on specific results is given in BS 7671.

In a room containing a bath or shower, tests can be carried out with a low-resistance ohmmeter to prove bonding between, for example, hot and cold water, central heating, heated towel rail, lighting circuit and shower.

**ACTIVITY**

The main protective bonding conductor to the gas utility supply is 10 $mm^2$ csa and 13 m in length. Calculate the resistance of the cable, showing the expected meter reading to two decimal places.

## Continuity of cpc for lighting circuits and radial circuits

Note that secure isolation must be carried out before this test is done (secure isolation is covered on pages 362–364).

### Reason for this test

This test is carried out to ensure that a cpc is present and effectively connected to *all* points in the circuit, including all switches, sockets, luminaires and outlets.

**ASSESSMENT GUIDANCE**

Remember, the $R_1 + R_2$ test for continuity will also cover polarity testing. Don't waste time by doing the test again.

### Test meter

The test meter must be capable of supplying a no-load voltage between 4 V and 24 V d.c. or a.c. and a short-circuit current of not less than 200 mA.

The measuring range should cover the span 0.01 Ω to 2 Ω, with a resolution of at least 0.01 Ω for digital instruments.

Note that general purpose multi-meters are not capable of supplying these voltage and current parameters.

### Method

1 Select the correct meter – you will need a low-resistance ohmmeter.
2 Select the correct scale – the correct scale is ohms/continuity (see pages 407–408 for guidance on scales).
3 Check the leads for damage.

Incorrect method of connecting leads

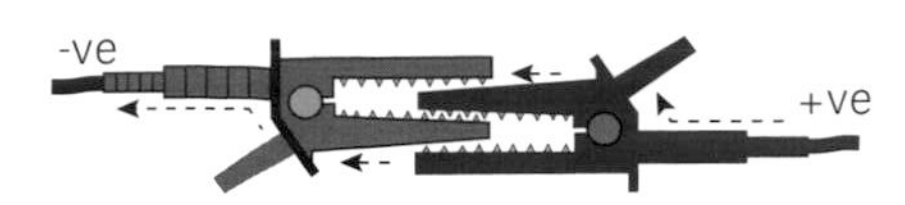


Current flow has to travel across the hinges

Correct method of connecting leads

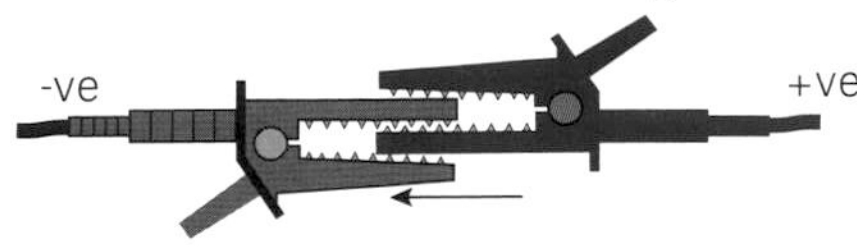


Current flow is straight from clip to clip

**Always connect leads correctly**

4 Insert leads – choose the correct location on the meter (typically C and Ω but this may differ between instruments). Always consult the manufacturer's instructions.

5 Correctly connect the leads – the correct way to connect leads is shown in the diagram.

6 Zero or null the meter leads – the correct reading should be 0.00 Ω. This test may require a link lead; this should be zeroed or nulled along with the meter leads.

You are now ready to carry out the continuity testing; in this example, for a lighting circuit and a radial circuit.

## Carrying out the continuity testing for a lighting circuit

There are two methods of testing and recording results for the continuity of a lighting circuit. Test methods 1 and 2 are illustrated below and opposite.

### Test method 1 – $R_1 + R_2$

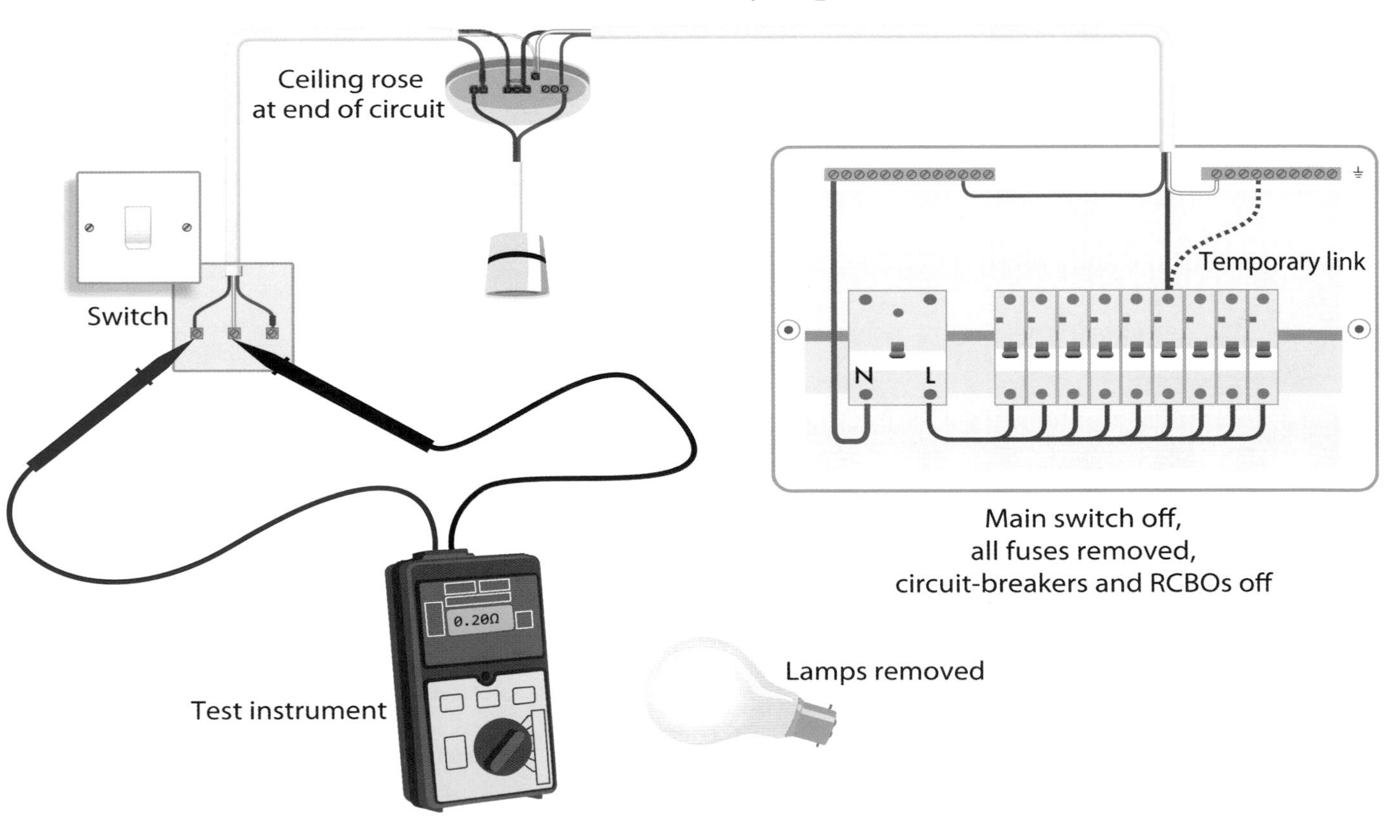


**Continuity testing at a light switch using test method 1**

Test method 1 involves temporarily linking between the line and cpc, and testing at *all* relevant points of the circuit; that is at switches and luminaires. The highest $R_1 + R_2$ value is recorded – usually at the furthest point away from the DB.

Look again at the location plan and the circuit information relating to the small industrial unit in Appendix 1.

If you use the data for the lighting circuit fed from 'Br-4', you have 37 m of 1.5 mm$^2$ line and 1.5 mm$^2$ cpc. Using the table below, the expected $R_1 + R_2$ value will be:

$$\frac{37 \times 24.2}{1000} = 0.89\ \Omega \text{ (at } 20\,°\text{C)}$$

| Line conductor csa mm$^2$ ($R_1$) | Circuit protective conductor csa mm$^2$ ($R_2$) | Copper conductor resistance mΩ/m at 20 °C |
|---|---|---|
| 1.5 | – | 12.10 |
| 1.5 | 1.5 | 24.20 |
| 2.5 | – | 7.41 |
| 2.5 | 2.5 | 14.82 |
| 4.0 | 1.5 | 16.71 |

Resistance values for cables (information extracted from Table B1 of Guidance Note 3)

## ASSESSMENT GUIDANCE

It is often difficult to know which is the end of the circuit fittings. It is best to test each outlet.

### Test method 2 (wandering lead test for $R_2$ resistance)

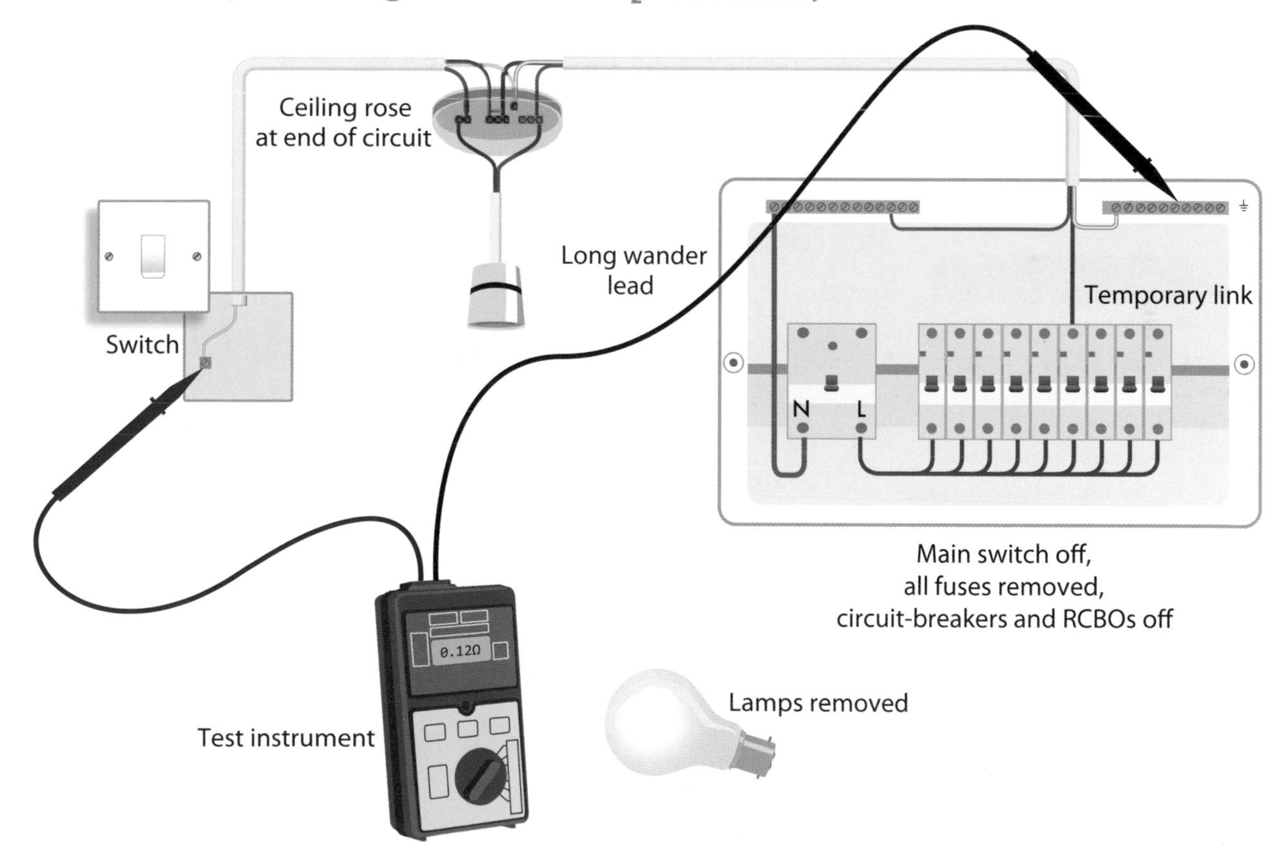


Continuity testing at a light switch using test method 2

Test method 2 involves the tester applying the long lead and zeroing the meter, as shown when testing the main protective bonding conductor on page 413.

## ACTIVITY

What are the dangers of using a long wander lead on a construction site with other workers around?

### ASSESSMENT GUIDANCE

It is often difficult to know the exact length of an installed cable. The electrician may have run the cable on a slightly different route to the planned one.

This method is rarely used to verify a radial circuit as obtaining a value of $R_1 + R_2$ is a minimum for initial verification, whereas a value of $R_2$ only proves the presence of an earth or cpc. A value of $R_1 + R_2$ will be used later when obtaining values of earth fault loop impedance.

Look again at the location plan and the circuit information relating to the small industrial unit in Appendix 1.

If you assume that the cpc is 37 m in length, the $R_2$ value will be 0.45 Ω.

## Carrying out the continuity testing for a radial circuit (eg instantaneous water heater)

Using the location plan and circuit information relating to the small industrial unit in Appendix 1, you can see that the 3 kW instantaneous water heater is fed from circuit 'Gr-3' which is protected by a B-type 16 A circuit breaker.

All conductors have a csa of 2.5 mm² and the length of run 22 m.

Before testing, you should have some idea of the expected test results. From the information given, you can calculate the approximate resistance of the line conductor ($R_1$) and the cpc ($R_2$) at an ambient temperature of 20 °C.

## Calculation of the resistance

$$\frac{22 \times 14.82}{1000} = 0.33\,\Omega$$

## Testing the resistance

### KEY POINT

A parallel earth path can occur when the cold water pipe of the instantaneous water heater connects to the main earthing terminal via main protective bonding conductor. If the cpc was connected to the earth terminal during the test, you could get a parallel earth reading – which would be substantially lower than the circuit design cpc value.

To test the resistance:

- Zero the test leads, including any temporary link lead using the ohms scale (continuity).
- Using test method 1 shown for the lighting circuit above, temporarily link between the line and cpc at the DB using a temporary link lead.
- With the instantaneous water heater cpc disconnected (to avoid parallel earth paths), test between the line conductor and the cpc.

Expected test results for $R_1+R_2$ of the instantaneous water heater = 0.33 Ω.

## Continuity testing for three-phase socket outlets and fixed equipment

1 Using test method 1 shown for the lighting circuit above, temporarily link between:
  a) line 1 and cpc at the DB
  b) line 2 and cpc at the DB
  c) line 3 and cpc at the DB.
2 At the socket outlet, test between line and cpc for each line. Record all three test results (they should all be approximately similar values). Some parts of the industry suggest that only the highest reading should be recorded as the $R_1 + R_2$ for a three-phase circuit, but it is generally best to record all three results so patterns of failure or deterioration can be detected during periodic inspections.

**KEY POINT**

Always do a quick calculation before testing. This will allow you to estimate the correct meter reading.

**ACTIVITY**

Look again at the location plan and the circuit information relating to the small industrial unit in Appendix 1.

1 Calculate the $R_1 + R_2$ of the 'Bl-3' radial circuit for sockets outlets.
2 Calculate the $R_1 + R_2$ values for the 16 A three-phase socket outlet fed from circuit 'Br/Bl/Gr-2'.

## Continuity of ring final circuit conductors

### Reason for this test

This test is carried out to ensure that the ring final circuit correctly forms a ring and that any spurs off the ring are adequately protected. This test process will also confirm polarity of the ring final circuit, as well as obtaining the $R_1 + R_2$ for the ring – confirming continuity of cpc.

**ASSESSMENT GUIDANCE**

Exam questions often ask learners to state the first three parts of the ring final circuit test. Learners often mistakenly name the resistance measurement of line, neutral and cpc conductors. This is incorrect. These comprise just the first stage.

### Test meter

Note that secure isolation must be carried out before this test is done (secure isolation is covered on pages 362–364).

The test meter must be capable of supplying a no-load voltage between 4 V and 24 V d.c. or a.c. and a short-circuit current of not less than 200 mA.

The measuring range should cover the span 0.01 Ω to 2 Ω, with a resolution of at least 0.01 Ω for digital instruments.

Note that general purpose multi-meters are not capable of supplying these voltage and current parameters.

### Method

1 Select the correct meter – you will need a low-resistance ohmmeter.
2 Select the correct scale – the correct scale is ohms/continuity (see pages 407–408 for guidance on scales).
3 Check the leads for damage.
4 Insert leads – choose the correct location on the meter (typically C and Ω but this may differ between instruments). Always consult the manufacturer's instructions.

Incorrect method of connecting leads

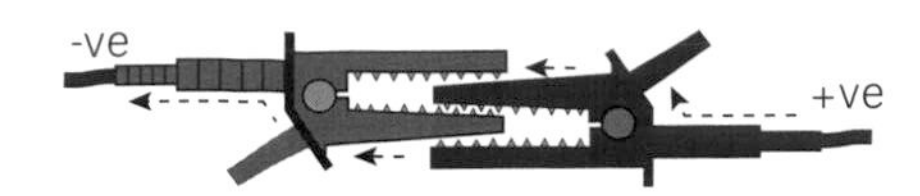


Current flow has to travel across the hinges

Correct method of connecting leads

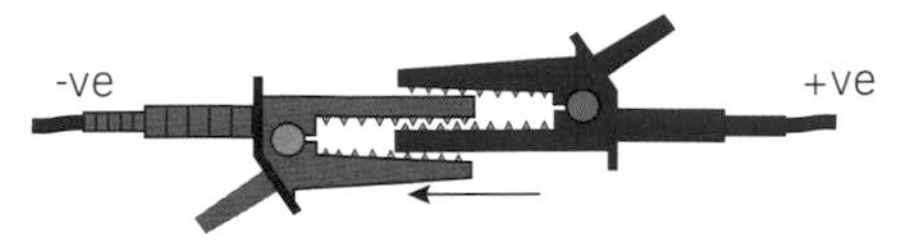


Current flow is straight from clip to clip

**Always connect leads correctly**

5 Correctly connect the leads – the correct way to connect leads is shown in the diagram.
6 Zero or null the meter leads – the correct reading should be 0.00 Ω. If a particular test requires a longer lead or a link lead, this should be zeroed or nulled along with the meter leads.

You are now ready to carry out the continuity testing.

## Carrying out continuity testing of the ring final circuit

There are three steps that you need to take to ensure that the ring final circuit has been installed correctly. These are outlined below and in the following pages.

Look again at the location plan and the circuit information relating to the small industrial unit in Appendix 1.

If you use the data for the ring final circuit, supplied from 'Br-3', you have a total ring distance of 48 m. There are no spurs off the ring final circuit.

The line and neutral have a csa of 2.5 mm$^2$ and the cpc has a csa of 1.5 mm$^2$.

### ACTIVITY

What effect would there be on the readings for the ring final circuit if a 2.5 mm$^2$ cpc were used instead of 1.5 mm$^2$ cpc?

| Line conductor csa mm$^2$ ($R_1$) | Circuit protective conductor csa mm$^2$ ($R_2$) | Copper conductor resistance mΩ/m at 20 °C |
|---|---|---|
| 2.5 | – | 7.41 |
| | 1.5 | 12.10 |
| 2.5 | 1.5 | 19.51 |
| 2.5 | 2.5 | 14.82 |

**Resistance values for cables (information extracted from Table B1 of Guidance Note 3)**

### Step 1

Calculate the 'end-to-end' resistance of the line, neutral and cpc.

$$\text{Line } (r_1) = \frac{48 \times 7.41}{1000} = 0.35\ \Omega$$

$$\text{Neutral } (r_n) = \frac{48 \times 7.41}{1000} = 0.35\ \Omega$$

$$\text{cpc } (r_2) = \frac{48 \times 12.10}{1000} = 0.58\ \Omega$$

Note the test results. You will need the test results when calculating Steps 2 and 3.

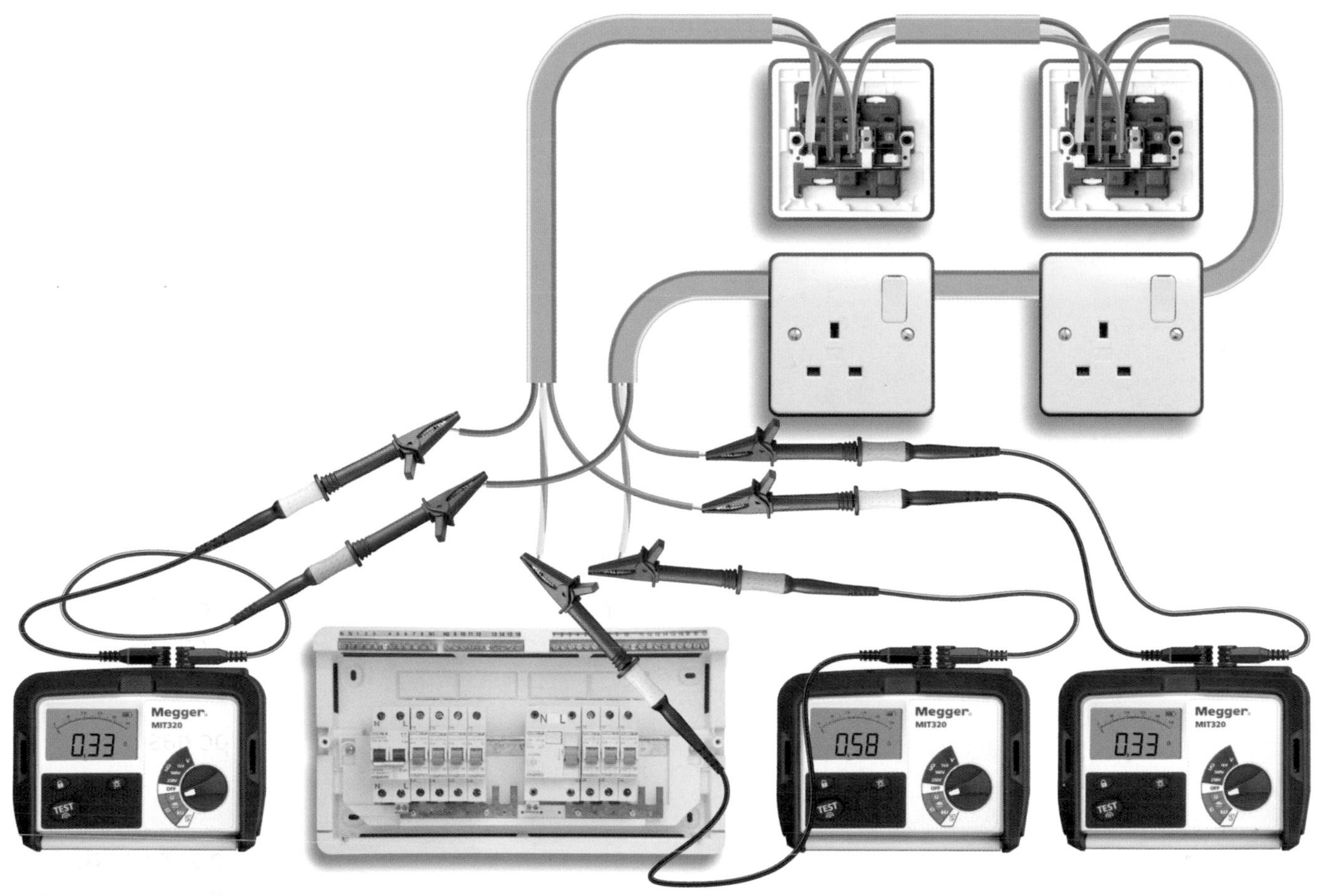


Step 1 of the ring final circuit testing

As shown in the diagram, measure the end-to-end resistance values for the line, neutral and cpc loops. These values are recorded in the Schedule of Test Results as $r_1$ (line loop), $r_n$ (neutral loop) and $r_2$ (cpc loop). These test results may vary slightly from the calculated values due to **contact resistance** or cable length estimation.

## Step 2

At the DB, cross connect the line and neutral. It is important to ensure that the line and neutral are cross connected as shown in the diagram. Incorrect cross connection will result in false readings. If readings taken at the socket outlets between line and neutral increase greatly towards the centre of the ring and then decrease again, the cross connections are incorrect.

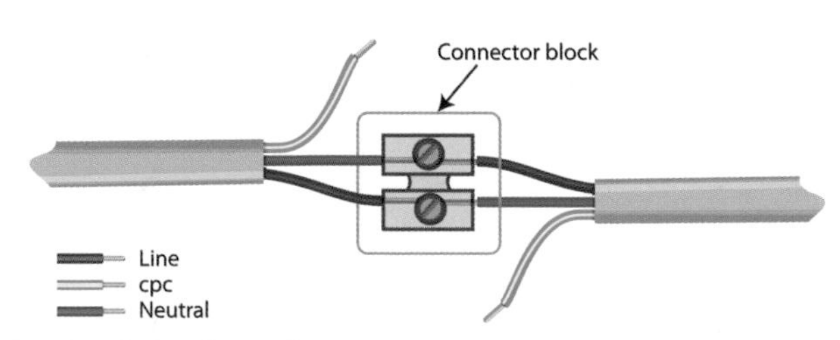


Step 2 of the ring final circuit testing

**Contact resistance**

Sometimes, a meter may give a reading in which the value does not stabilise or which is quite different from the calculated value. In this situation, reverse the test meter lead connections and re-test. This may give more accurate values.

**KEY POINT**

If the DB is crowded, this part of the test may be carried out at a convenient socket outlet. If this is done, care must be taken to ensure that the selected socket outlet is isolated.

Calculation: the easiest way to remember the calculation is to add together the resistance of each complete ring from Step 1 above and divide by 4.

$$\frac{r_1 + r_n}{4} \text{ or } \frac{0.33\,\Omega + 0.33\,\Omega}{4} = 0.16\,\Omega$$

So, the reading at each socket outlet should be approximately 0.16 Ω.

Method of testing: continuity of line and neutral connection at each of the socket outlets on the ring final circuit should be tested. All results should be the same, however slight deviations do occur due to contact resistance.

## ACTIVITY

Why is it not necessary to test the cross connected N–E loop in the ring final circuit test after the L–N and L–E test?

The use of special adaptors is recommended when testing at the socket outlets as this allows readings to be obtained from the front plate of the socket outlet, instead of accessing the connections behind.

### Typical test results

Socket outlet 1 = 0.17 Ω
Socket outlet 2 = 0.16 Ω
Socket outlet 3 = 0.18 Ω
Socket outlet 4 = 0.17 Ω

Note that these tests and results are important to ensure that there are no interconnections or extended spurs off spurs on this ring final circuit.

If a reading is substantially higher than expected, the circuit must be investigated further to ensure coordination between the current-carrying capacity of the cables and the protective devices. Before any investigation is carried out for inconsistent readings, obtain values as outlined in Step 3 – these values can give a good indication of what is causing the problem.

### Step 3

At the DB, cross connect the line and cpc. It is important to ensure that the line and cpc are cross connected as shown below.

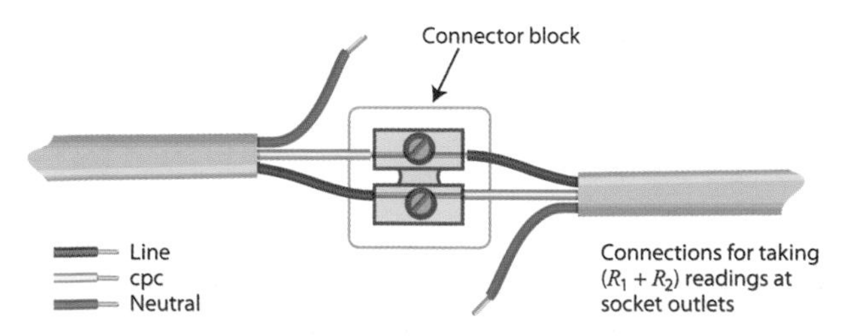


**Step 3 of the ring final circuit testing**

Calculation: again, the easiest way to remember the calculation is to add together the resistance of each complete ring from Step 1 above and divide by 4.

$$\frac{r_1 + r_2}{4} \text{ or } \frac{0.33\,\Omega + 0.58\,\Omega}{4} = 0.23\,\Omega$$

So, the reading at each socket outlet should be approximately 0.23 Ω.

Method of testing: continuity of line and cpc connection at each of the socket outlets on the ring should be tested. All results should be the same, however slight deviations can occur due to contact resistance.

**Typical test results**
Socket outlet 1 = 0.24 Ω
Socket outlet 2 = 0.23 Ω
Socket outlet 3 = 0.25 Ω
Socket outlet 4 = 0.23 Ω

Note that:

- The highest of these values will be inserted onto the Generic Schedule of Test Results as the $R_1 + R_2$ value and further evaluation will be required when assessing the earth fault loop impedance of the circuit.

$$R_1 + R_2 = 0.25\,\Omega$$

- If there is a spur off the ring final circuit, the reading at that spur could be higher than those at socket outlets on the ring. The highest reading obtained will be recorded as the $R_1 + R_2$ value.

**KEY POINT**

Always do a quick calculation before testing. This will allow you to estimate the correct meter readings for each step. Also, keep checking the instrument is zeroed or nulled, especially if special adaptors are used which may affect results.

**ACTIVITY**

Practically test a ring final circuit by using the methods explained here.

## THEORY RELATED TO INSULATION RESISTANCE TESTING

**Assessment criteria**

**3.6** Explain factors that effect insulation resistance values

The minimum acceptable values of insulation resistance of a circuit are shown in the table on page 427, taken from Table 61 of BS 7671. The purpose of insulation around a conductor is to contain the current in the conductor; the lower the insulation resistance, the greater the risk of current leaking between conductors or between conductors and earth.

### Why you need to test for insulation resistance

The purpose of this test is to verify that the insulation of conductors provides adequate insulation, is not damaged and that live conductors or protective conductors are not short-circuited or leaking overcurrent.

## What is insulation resistance?

Insulation resistance is the resistance of the insulation between conductors. The higher the resistance, the better the current is contained in the core of the cable. If insulation resistance is low, current will leak between conductors and to earth, giving a risk of electric shock or fire.

Many years ago, rubber-insulated cables were installed in buildings. Unfortunately, as some of these cables deteriorated with age, they started to crack and insulation failure occurred. Long cable runs gave low values of insulation resistance, and multiple circuits from a common source, such as a distribution board, were used.

### ACTIVITY

This calculation can also be done as product over sum. What would be the total insulation resistance of two cables in parallel having individual insulation test values of 16 MΩ and 50 MΩ?

For example, if an insulation resistance test is carried out on a 50 m length of old rubber cable (between line and neutral) and the reading is 20 MΩ, a 100 m length of the same cable would give a reading of 10 MΩ. Why? Because you doubled the length; theoretically, you have put two 50 m lengths of cable in parallel.

Two resistances in parallel: $\frac{1}{R_1} + \frac{1}{R_2} = \frac{1}{R_t}$

therefore, $\frac{1}{20} + \frac{1}{20} = \frac{1}{R_t}$

so, $\frac{1}{20} + \frac{1}{20} = 0.05 + 0.05 = 0.1$

and, finally $\frac{1}{0.1} = 10\,(\text{M}\Omega)$

### KEY POINT

The insulation resistance value decreases as the conductor length is increased.

### ASSESSMENT GUIDANCE

You could be given the insulation resistance values of three, four or even five circuits which have been individually tested from the same distribution board (DB). You may be asked to calculate the cumulative value if all of the circuits are tested together.

These calculations can be performed on a calculator using the $x^{-1}$ feature.

Try this: [20] [$x^{-1}$] [+] [20] [$x^{-1}$] [=] [$x^{-1}$] [=] and the answer displayed should be 10.

## A calculation to illustrate circuit readings accumulated in parallel

Here are five individual final circuit test results, measured between line and cpc for each circuit:

- Circuit 1 = 25 MΩ ($R_1$)
- Circuit 2 = 10 MΩ ($R_2$)
- Circuit 3 = 16 MΩ ($R_3$)

- Circuit 4 = 6 MΩ ($R_4$)
- Circuit 5 = 40 MΩ ($R_5$)

## Example questions

1 Determine the total value of these results in parallel.

2 Is the calculated value acceptable in accordance with the table taken from Table 61 of BS 7671?

## Answer to example question 1

$$\frac{1}{R_t} = \frac{1}{R_1} + \frac{1}{R_2} + \frac{1}{R_3} + \frac{1}{R_4} + \frac{1}{R_5} = \frac{1}{25} + \frac{1}{10} + \frac{1}{16} + \frac{1}{6} + \frac{1}{40}$$

Now, with a calculator, change all the fractions to decimals (to two decimal points), ie: 1 ÷ 25 = 0.04

0.04 + 0.10 + 0.06 + 0.17 + 0.03 = 0.40 total

To find the total value of these resistances in parallel, you must finally divide the total into 1.

Therefore, $\frac{1}{0.40} = 2.5\ \text{M}\Omega$

## Answer to example question 2

So, is 2.5 MΩ an acceptable value by measurement and calculation? Yes it is – the minimum value given in Table 61 of BS 7671 = 1.0 MΩ.

**KEY POINT**

The calculated value for the total resistance will always be less than the lowest resistance in the set (in this case 6 MΩ).

## Expected insulation resistance readings

The insulation resistance readings expected for a new installation are very high, for example greater than 200 MΩ for each circuit. For an older circuit that may have damage, dust or dampness in the system, readings can be considerably less than 200 MΩ.

Whatever the readings, they must not be less than the minimum values stated Table 61. If the insulation resistance value obtained for a circuit was less than 1.0 MΩ, there could be excessive leakage of current occurring within the wiring system.

If you get a reading of 0.25 MΩ at 500 V between live conductors and earth, the resultant current flowing to earth in this circuit must be:

$$\frac{500\ \text{V}}{250\,000\ \Omega} = 0.002\ \text{A (or 2 mA)}$$

This leakage current could be sufficient to cause electric shock, high touch voltages or combustion, due to tracking across between conductors.

**ASSESSMENT GUIDANCE**

There is a good reason for doing the continuity of protective conductor test before the insulation resistance test. If there is no cpc continuity, then the insulation test is worthless.

## ACTIVITY

Calculate the cumulative parallel insulation resistance value of the following four circuits:

- Circuit 1 = 25 MΩ ($R_1$)
- Circuit 2 = 10 MΩ ($R_2$)
- Circuit 3 = 16 MΩ ($R_3$)
- Circuit 4 = 6 MΩ ($R_4$)

## Practical guidance to insulation resistance testing

Note that secure isolation must be carried out before this test is done (secure isolation is covered on pages 362–364).

### Reason for this test

This test is carried out to ensure that the insulation between conductors has a high value of resistance.

### Test meter

The test meter must be capable of supplying an output test voltage of 250 V d.c., 500 V d.c. or 1000 V d.c., subject to the circuit to be tested.

See Table 61, from BS 7671, on page 427.

### Method

1. Select the correct meter – you will need an insulation resistance tester. As the voltages used by the meter are typically 250 V, 500 V or 1000 V, the requirements of Guidance Note GS38 apply.
2. Select the correct scale – the correct scale is megohms (MΩ).
3. Check the leads for damage – clips and probes can be used when carrying out these tests.
4. Insert leads – choose the correct location on the meter (typically C and Ω, but this may differ between instruments). Always consult the manufacturer's instructions.
5. Choose the voltage – the voltage must be appropriate for the circuit under test. See Table 61 from BS 7671, on page 427.
6. Check the meter and leads.
   - open circuit reading should be >200 MΩ
   - connected together reading should be 0.00 MΩ

Note that the open circuit reading may differ from that shown here and is dependent on the manufactured parameters for the test meter used. Examples range from >20 MΩ to >1000 MΩ.

### Why test insulation resistance?

The purpose of the insulation test is to verify that the insulation of conductors provides adequate insulation, is not damaged and that live conductors or protective conductors are not short-circuited or leaking overcurrent that could give rise to fire risk or electric shock.

The table below shows the minimum reading for an installation or a single circuit.

**TABLE 61 – Minimum values of insulation resistance**

| Circuit nominal voltage (V) | Test voltage d.c. (V) | Minimum insulation resistance (MΩ) |
|---|---|---|
| SELV and PELV | 250 | 0.5 |
| Up to and including 500 V with the exception of the above systems | 500 | 1.0 |
| Above 500 V | 1000 | 1.0 |

**Minimum values of insulation resistance (from BS 7671:2008 (2011))**

You are now ready to prepare and carry out the insulation resistance testing, using the methods outlined below and in the following pages.

## Before testing

1 Ensure that these items are disconnected to reduce risk of damage:
- any electronic equipment such as dimmer switches, controllers and timers
- surge protective devices (SPDs)
- residual current devices (RCDs) and residual current operated circuit breakers with overcurrent protection (RCBOs)

2 Carry out these actions to enable full testing of the wiring system and to reduce the possibility of incorrect readings:
- remove all lamps
- connect all protective conductors
- disconnect any current-using equipment
- check all fuses are in place, and circuit breakers and switches closed
- all two-way and intermediate switching is operated during testing, so that the strappers are all tested.

Note that, if current-using equipment cannot be disconnected or lamps removed, switchgear supplying these items should be in the open position. Where an initial verification is undertaken, insulation resistance testing may be carried out during and on completion of the work. This will ensure all parts of a circuit are subjected to insulation resistance tests.

### ACTIVITY

It is necessary to disconnect a dimmer before carrying out an insulation resistance test. What further action is required before the test is carried out?

### ASSESSMENT GUIDANCE

Many people think that neon indicator lights will be damaged by an IR test. This is incorrect, they just glow. This will give a false insulation resistance test reading between line and neutral.

## Single-phase testing methods

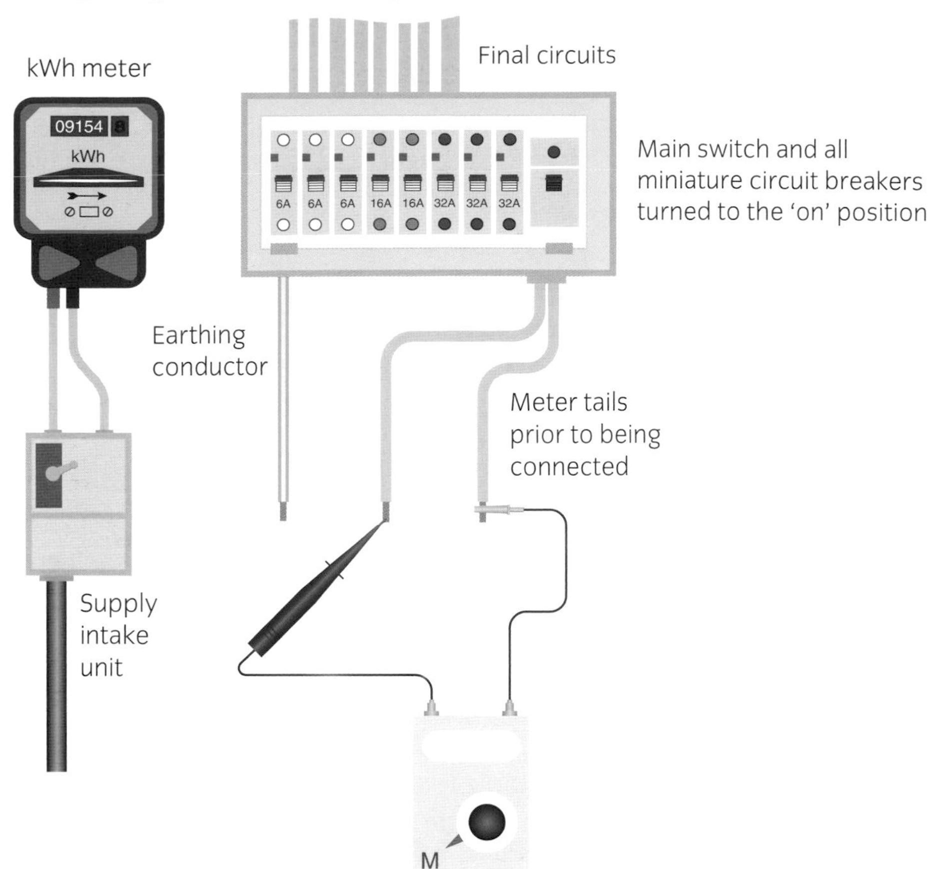


**Line to neutral test being carried out**

### ACTIVITY

The kWh meter is the property of the supply company. Where does the consumers' installation start?

Note that all earthing and bonding *must* be connected for this test.

When testing the full installation or individual circuits, three tests are used:

- Test 1 – line to neutral (live to live)
- Test 2 – line to earth (live to earth)*
- Test 3 – neutral to earth (live to earth)*.

*The lowest value of the two tests is recorded.

An alternative test is line and neutral (together) to earth. This is used for circuits or equipment that are vulnerable to the test voltage, such as capacitors in fluorescent luminaires. All earths must be connected to the main earthing terminal (MET) for this test.

| Continuity | | Insulation resistance | | |
|---|---|---|---|---|
| $(R_1 + R_2)^*$ Ω | $R_2^*$ | ring | live/live MΩ | live/earth MΩ |
| | | | > 200 | > 200 |

**Extract from the Generic Schedule of Test Results showing the typical insulation test result of a new circuit**

If the readings for the tests of each circuit are greater than 200 MΩ, that information needs to be recorded on the Generic Schedule of Test Results. Remember that most meters will only read values up to 200 MΩ. If the meter indicates an over-range value, this is recorded as >200 MΩ.

## Three-phase, neutral and earth (six-test sequence)

The six-test sequence is outlined below and on the next page. Note that Tests 5 and 6 can be carried out together if connected as shown in the diagram.

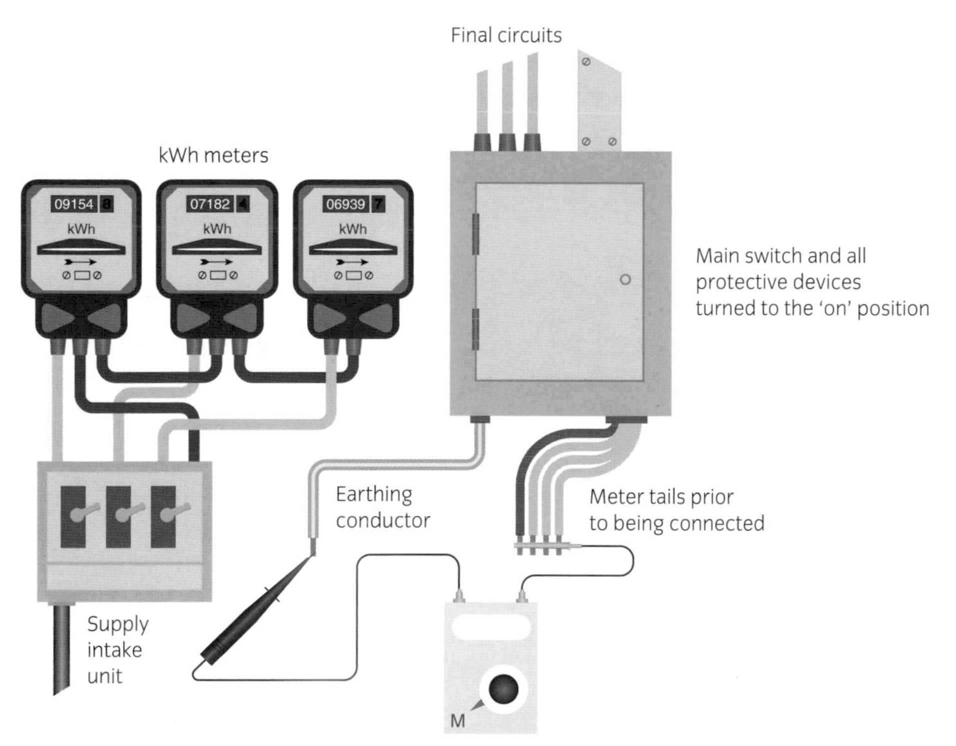


**Tests 5 and 6 being carried out together – L1, L2, L3 + neutral (connected together) to earth**

Note: all earthing and bonding *must* be connected for this test.

**KEY POINT**

Always ensure that precautions are taken to avoid damage to equipment and false readings.

**ACTIVITY**

Practice insulation resistance testing on single- and three-phase test boards at your college or training centre. Various types of boards and switchgear should be made available. Remember that these tests must only be carried out under supervision. Ensure that the test boards are fully isolated and secured before testing commences.

| | | |
|---|---|---|
| Test 1 | L1 to L2 | The lowest value of these tests is recorded as 'between live conductors' |
| Test 2 | L1 to L3 | |
| Test 3 | L2 to L3 | |
| Test 4 | L1 + L2 + L3 (connected together) to neutral | |
| Test 5 | L1 + L2 + L3 (connected together) to earth | The lowest value of these tests is recorded as 'between live conductors and earth' |
| Test 6 | neutral to earth | |

## Three-phase and earth (four-test sequence)

Tests 1, 2, 3 and 5 will apply if no neutral is present, as in the case of an armoured cable supplying a three-phase motor.

## Additional insulation resistance tests

### Protection by SELV, PELV and electrical separation

**SELV**

Separated extra-low voltage circuit.

**PELV**

Protective extra-low voltage circuit.

In certain situations within an installation, electric shock protection under fault conditions is provided by **SELV** and **PELV** or electrical separation instead of automatic disconnection of supply (ADS). Where this exists, a test for simple separation is required between the low voltage circuit supplying the separated circuit and the outgoing circuit, as well as the outgoing circuit and earth.

This is done to ensure that there is no electrical connection between the separated circuit and the rest of the installation. It is done using an insulation resistance tester. If there was an electrical connection, the protective measure would be lost.

The test is carried out with the transformer that provides separation disconnected and the test instrument set to 500 V d.c., regardless of the circuit voltage. The instrument is connected between the supply circuit live conductors and the separated circuit live conductors. The result obtained must be greater than 1 MΩ.

A further test is then carried out between the live conductors of the separated circuit and earth. This again is carried out at 500 V d.c. and the result must be greater than 1 MΩ.

## THE IMPORTANCE OF VERIFYING POLARITY

Note that verification of polarity must be carried out on all circuits.

If polarity is not correctly determined there may be a risk of electric shock during maintenance procedures. For example, if you isolate or switch the neutral of a circuit via a single-pole circuit breaker or switch, it would appear that the circuit is dead. This would not be the case, as live conductors and connections would be present at fixed equipment, sockets and switches – this would be very dangerous.

There are three recognised methods of evaluation. All three methods have their advantages and possible dangers, if they are not carried out correctly.

The methods are:

- polarity by visual inspection
- polarity by continuity testing
- live testing for polarity.

**Assessment criteria**

**3.3** Explain how to prepare for testing electrical systems

**3.4** Specify procedures for regulatory tests

**6.1** Implement safe system of work for testing electrical systems

**6.2** Select the test instruments for regulatory tests

**6.3** Test electrical systems

**6.4** Record results of regulatory tests

**6.5** Verify compliance of regulatory test results

**6.6** Commission electrical systems

### Polarity by visual inspection

By using your knowledge and sight, correct termination of cables relating to core colours can be established. It is essential that polarity is checked visually during the process of installation, especially in cases where checking by testing is impractical.

You can inspect:

- line conductors, which must be connected to circuit breakers, fuses, single-pole switches and single-pole controls
- as highlighted in both BS 7671 and Guidance Note 3, that the centre contact of all screw-type lampholders is connected to the line conductor (except for E14 and E27 lampholders which have been manufactured to BS EN 60238; these lampholders have been manufactured to reduce the risk of shock whilst inserting lamps)
- correct connection of fittings and accessories
- all fixed equipment, such as supply tails, lights, socket outlets, heaters, motors and boilers.

**ASSESSMENT GUIDANCE**

Remember that different makes of socket outlets may have their L and N terminals in different positions. Do not assume in your testing that all are the same.

**ACTIVITY**

Polarity tests are not required at a bayonet cap lampholder. Where would the nearest test point be?

## Polarity by continuity testing

You will need to use a low-resistance ohmmeter for this test.

When you continuity test radial and ring final circuits, part of the process is to test and visually inspect the polarity of fixed equipment and socket outlets.

The diagram reinforces the polarity of the switch as being supplied by the line conductor, due to the line and cpc being linked in the consumer unit.

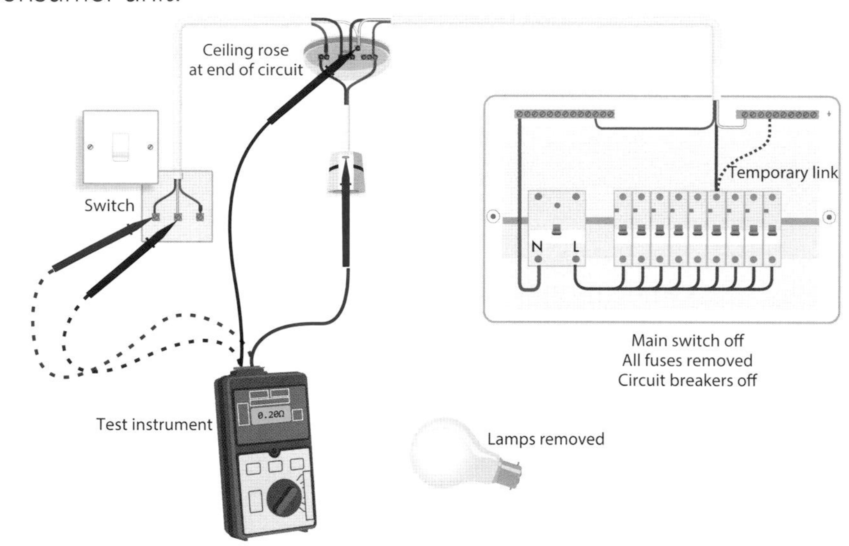


**Polarity test on lighting circuit**

Note that when testing is complete, the temporary link must be removed.

All radial circuits that are correctly tested for continuity using the $R_1 + R_2$ method, are normally visually inspected, tested and verified for polarity at the same time. All socket outlets on ring final circuits will have been tested for polarity during the three steps of testing. Three-phase circuits are tested for continuity and, therefore, polarity is also tested.

For further information and safety procedures regarding continuity testing, refer to radial and ring final circuit testing on pages 410–423.

**An ES E27 lampholder test adaptor can be used during continuity testing for polarity**

**KEY POINT**

It is important to check polarity at relevant stages of the verification process.

**ACTIVITY**

State, for **each** of the following stages, **three** items of equipment which can be checked for polarity.

1. inspecting an installation
2. continuity testing
3. live testing

## Live testing for polarity

You will need to use voltage indicators and earth fault loop impedance test meters. All test meters used for testing 'live circuits' must comply with Guidance Note GS38.

You can test these features for polarity when the system is energised:

- supply tails and earthing conductor
- distribution boards and consumer units
- socket outlets.

# METHODS OF DETERMINING EARTH FAULT LOOP PATHS

There are three types of system earthing arrangements to consider:

- **TN-S**
- **TN-C-S**
- **TT**.

**TN-S, TN-C-S, TT**

T = source earth (Terre) (first T).

N = neutral and protective conductor.

C = combined in the distribution system.

S = separated in the installation.

T = installation earth (Terre) (final T).

There are three methods of determining the external earth fault loop impedance ($Z_e$).

1. Measurement: by testing at the origin of the installation – this is the most common and accurate method used.
2. Enquiry: such as by contacting the electricity distributor, who will usually quote the maximum possible $Z_e$ value in accordance with the Electricity, Safety, Quality and Continuity Regulations 2002; you will require permission to use this system.
3. Calculation: this is the most complex method and can only be used if the designer knows all of the relevant impedances and resistances of the system to be tested.

**ASSESSMENT GUIDANCE**

This is a live test and all precautions must be taken and demonstrated to your assessor during the test. Use all the appropriate safety equipment.

## TN-S system and earth fault loop impedance

This section concentrates on the **TN-S** system. TN-S systems may be single-phase or three-phase.

**TN-S**

T = source earth (*terre*).

N = neutral and protective conductor.

S = separated in the installation.

### Three-phase supply for a TN-S system

The three-phase supply for a TN-S installation is shown in the diagram. The electricity distributor will supply the consumer with an earth, which is separated throughout the system, but connected to the neutral at centre point of the supply transformer.

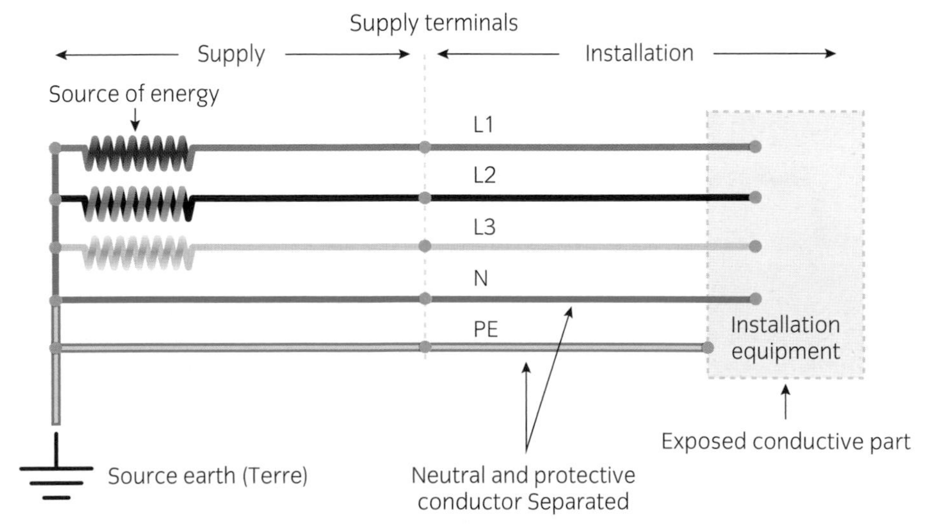


**Diagram of a three-phase supply for a TN-S installation**

## Single phase TN-S system

The diagram shows a practical arrangement for a single-phase TN-S system. It is typical of a domestic (or similar) installation.

In both single- and three-phase supplies, a TN-S system will normally use the sheath of the supplier's cable as a means of earthing the installation. From this sheath, the earthing conductor will connect to the installation's main earthing terminal (MET).

An enquiry to an electricity distributor regarding the characteristics of a domestic supply will result in the following information being provided:

- external earth fault loop impedance ($Z_e$) = 0.8 Ω, the maximum for separate earth supplies in TN-S systems
- the $Z_e$ value in practice will normally be considerably less than 0.8 Ω.

### ASSESSMENT GUIDANCE

The reference is to the metallic sheath, obviously not the PVC sheath. Make sure this is clear in any answer you may give. The metallic sheath could be a steel armour or lead in older TN-S supplies.

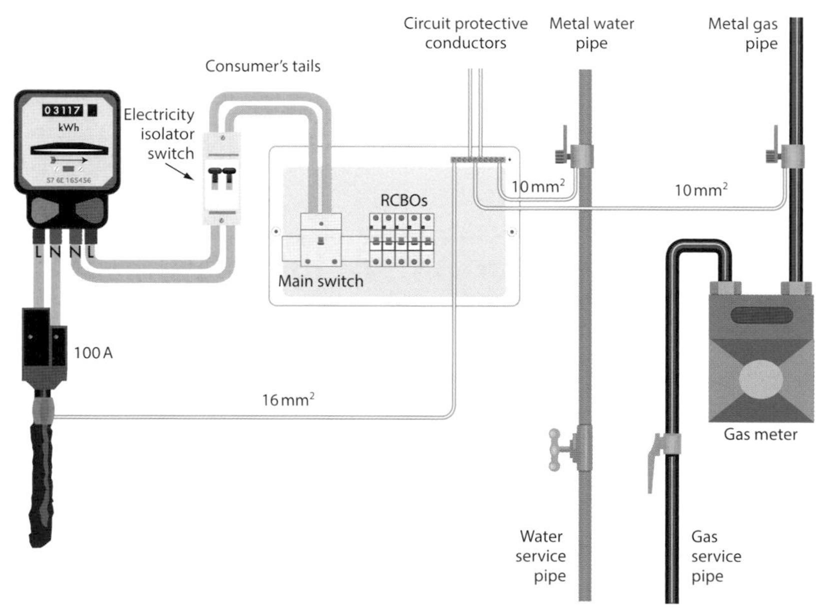


**Practical arrangement for a single-phase TN-S system**

### KEY POINT

Three methods of determining the external earth fault loop impedance are by measurement, enquiry or calculation.

The external earth fault loop impedance ($Z_e$) in ohms is a result of the impedance of the supplier's line conductor, the transformer and the separate earth added together.

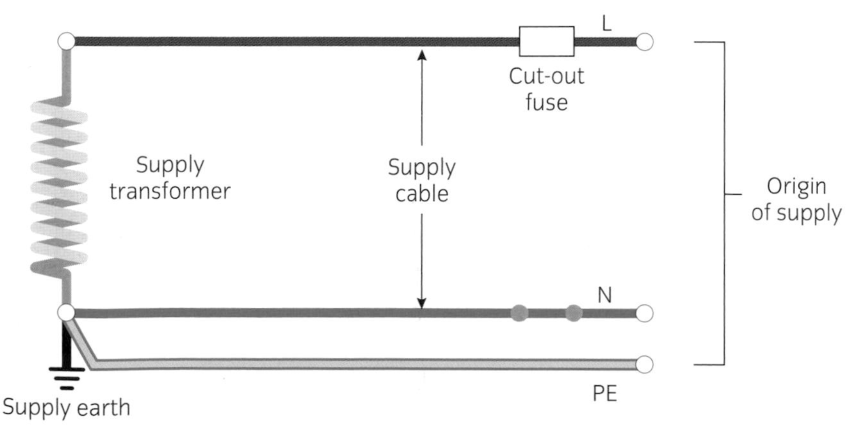


**Simple representation for a TN-S system**

### ACTIVITY

Redraw the simple TN-S system diagram and illustrate how you would attach the probes or leads of a two-lead test meter to the supply to test the earth fault loop impedance ($Z_e$) at the origin. Place as much information on the diagram as possible and show the maximum expected reading on the meter.

## TN-C-S system and earth fault loop impedance

This section concentrates on the **TN-C-S** system (**PME**).

There are three methods of determining the earth fault loop impedance ($Z_e$).

1 Measurement: by testing at the origin of the installation – this is the most common and accurate method used.
2 Enquiry: such as by contacting the electricity distributor, who will usually quote the maximum possible $Z_e$ value in accordance with the Electricity, Safety, Quality and Continuity Regulations 2002; you will require permission to use this system.
3 Calculation: this is the most complex method and can only be used if the designer knows all of the relevant impedances and resistances of the system to be tested.

**TN-C-S**

T = source earth (*terre*).

N = neutral and protective conductor.

C = combined in the distribution system.

S = separated in the installation.

**PME**

Protective multiple earthing.

### Three-phase supply for a TN-C-S system

The three-phase supply for a TN-C-S installation is shown in the diagram.

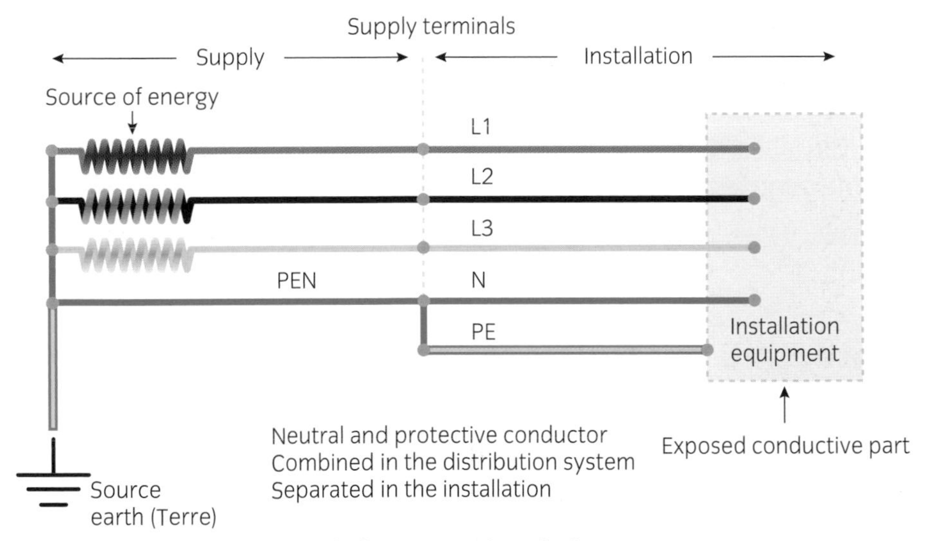


**Diagram of a three-phase supply for a TN-C-S installation**

### Single-phase TN-C-S system

The diagram on page 436 is typical of a single-phase TN-C-S system for a domestic (or similar) installation.

In both three and single phase cases the electrical distributor will supply the consumer with an earth, which is connected to the neutral at the origin of the installation. The neutral is connected to the centre point of the supply transformer. As the neutral now doubles as an earth, it is referred to as a protective earthed neutral (PEN) conductor.

In the UK the most common earthing arrangement is now TN-C-S, or PME.

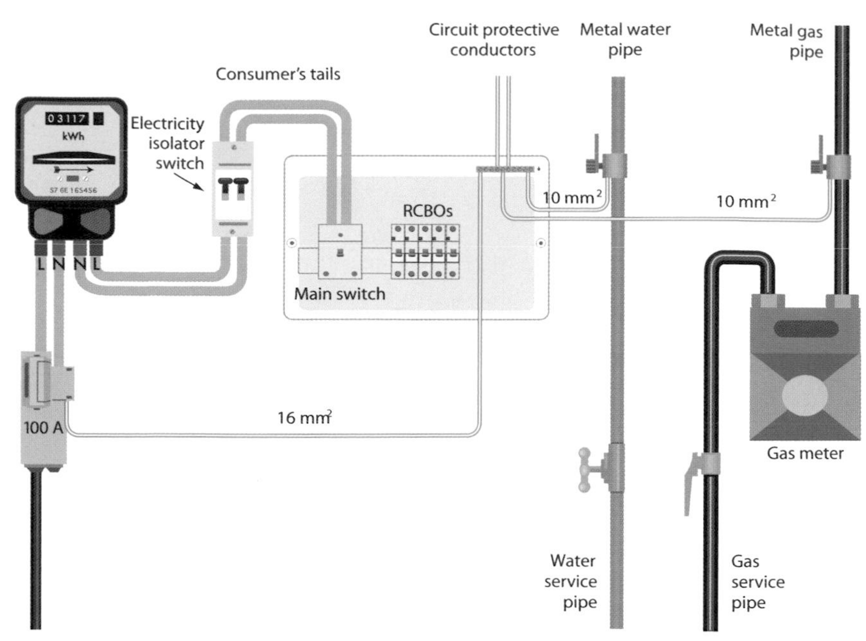


Practical arrangement for a single-phase TN-C-S system

New supplies will almost always be PME (protective multiple earthing) to provide for a TN-C-S system.

An enquiry to an electricity distributor for the characteristics of a domestic supply will result in the following information being provided:

- external earth fault loop impedance ($Z_e$) = 0.35 Ω, the maximum for PME supplies (TN-C-S systems)
- the $Z_e$ value in practice will normally be considerably less than 0.35 Ω.

## Protective multiple earthing (PME)

**KEY POINT**

Three methods of determining the external earth fault loop impedance are by measurement, enquiry or calculation.

Protective multiple earthing (PME) is a system where multiple earth electrodes are installed to provide an alternative path to the star point of a sub-station transformer. As the PEN conductor is both a live conductor and earth, if this conductor becomes part of an open circuit in the supply, earthed parts of the installation may become live as current tries to find a path back to the sub-station. Remember, if current flows into a circuit, the same current flows back on the neutral. If this neutral becomes open, that current needs an alternative path and, as the earthing of the installation is connected to the supplier's neutral, this may provide the path needed.

If the supply contains multiple earth electrodes, this risk is reduced as the electrodes should provide a more suitable return path.

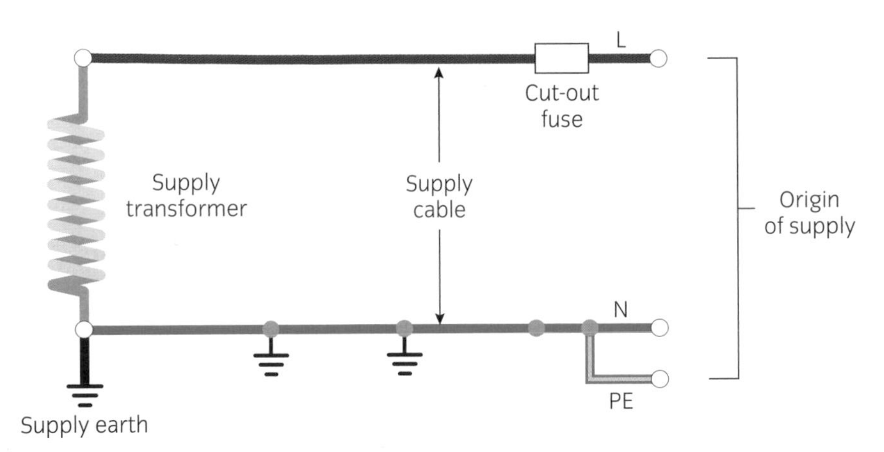


Simple representation of a TN-C-S system

**ACTIVITY**

Redraw the simple TN-C-S system diagram and illustrate how you would attach the probes or leads of a two-lead test meter to the supply to test the earth fault loop impedance ($Z_e$). Place as much information on the diagram as possible and show the maximum expected reading on the meter.

## TT system and earth fault loop impedance

This section concentrates on the **TT** system.

**TT**

T = source earth (*terre*).

T = installation earth (*terre*).

There are three methods of determining the earth fault loop impedance ($Z_e$).

1 Measurement: by testing at the origin of the installation – this is the most common and accurate method used.
2 Enquiry: such as by contacting the electricity distributor, who will usually quote the maximum possible $Z_e$ value in accordance with the Electricity, Safety, Quality and Continuity Regulations 2002.
3 Calculation: this is the most complex method and can only be used if the designer knows all of the relevant impedances and resistances of the system to be tested.

### Three-phase supply for a TT system

The three-phase supply for a TT installation is shown in the diagram.

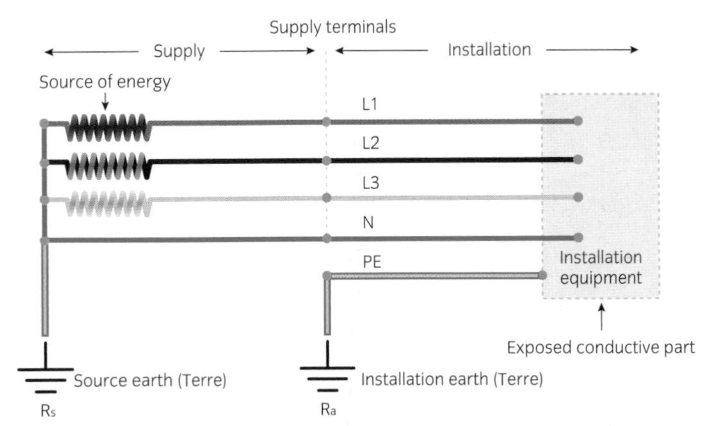


Diagram of a three-phase supply for a TT installation

A TT system is a system in which the electrical distributor does *not* supply the consumer with an earth. This may be because they can't guarantee the continuity of a protective conductor, such as in the case of a low voltage supply to a building by means of an overhead line which could deteriorate, break or become disconnected during adverse weather conditions.

TT systems may also be used on construction sites, caravan parks, marinas and many other locations. This is because the Electricity Safety Quality and Continuity Regulations will not permit the connection of PME supplies to certain locations. So, the supplier does not give an earth in these situations and it is up to the consumer to provide an earth using an earth electrode.

Generators may also use TT-type systems in order to provide an alternative, reliable earth path. If a generator is installed for standby purposes, and the power is lost, the generator starts up. This generator must have an independent earth as the supplier may be working on the supply system.

## Single-phase supply for a TT system

The diagram shows a practical arrangement for a single-phase TT system. It is typical of a domestic (or similar) installation.

An enquiry to an electricity distributor for the supply characteristics of a domestic supply will result in the following information being provided:

- external earth fault loop impedance ($Z_e$) = 21 Ω, the maximum value of the distributor's earth electrode and associated impedances of the supply transformer and line conductor (TT systems)
- the value in practice will normally be considerably less than 21 Ω.

It is very important to remember that this value *does not include* the resistance of the consumer's earth electrode.

**KEY POINT**

Three methods of determining the external earth fault loop impedance are by measurement, enquiry or calculation.

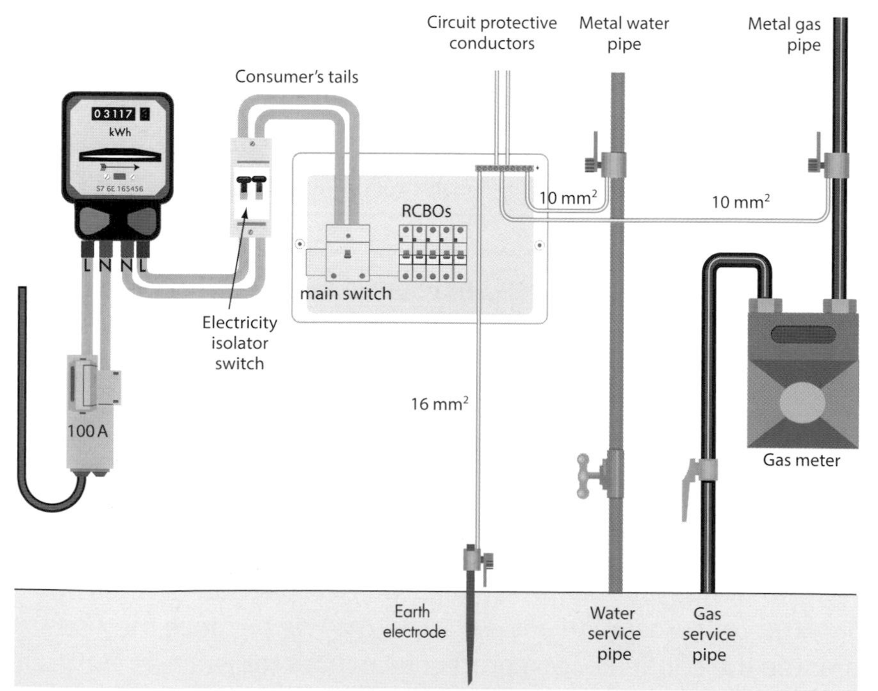


**Practical arrangement for a single-phase TT system**

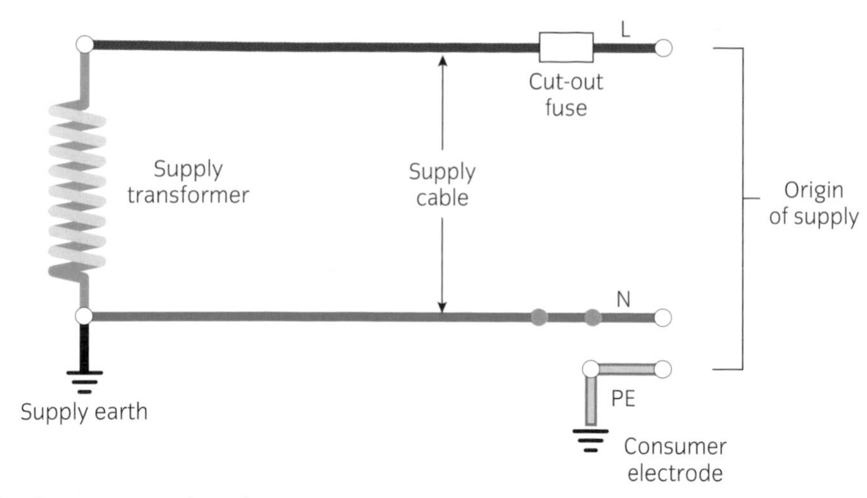


Simple representation of TT system

## ACTIVITY

Redraw the simple TT system diagram and illustrate how you would attach the probes or leads of a two-lead test meter to the supply to test the earth fault loop impedance ($Z_e$). Place as much information on the diagram as possible and show the maximum expected reading on the meter.

# EARTH ELECTRODE RESISTANCE TESTING

Earth electrode resistance testing can be carried out by using one of three test methods. This book will look at the two most common methods, using:

- a proprietary earth electrode resistance tester
- an earth fault loop impedance tester.

We will, for this book, refer to these methods as E1 and E2.

## ASSESSMENT GUIDANCE

Most electricians are unlikely to carry out this test but it is important to know the principles of the test as you never know when you might need it.

## Recognised types of earth electrode

The following types of earth electrode are recognised:

1. earth rods or pipes
2. earth tapes or wires
3. earth plates
4. underground structural metalwork, embedded in foundations
5. welded metal reinforcement of concrete, embedded in the earth
6. metal sheaths and coverings of cables (subject to Regulation 542.02 to 05 in BS 7671)
7. other suitable underground metalwork.

## Reason for earth electrode resistance testing

The purpose of this test is to establish that the resistance of the soil surrounding an earth electrode is suitable and that the electrode makes contact with the soil.

## Acceptable test values for an earth electrode

Earth electrode resistance values can differ greatly, dependent on the type of ground and environmental conditions, the material of the electrode used and area of contact with the general mass of earth.

It is recommended that the earth electrode resistance test is carried out when the ground conditions are least favourable, such as during dry weather.

Note that earth electrode resistance values above 200 Ω may not be stable, as soil conditions change due to factors such as soil drying and freezing.

## Test method E1

This method uses a proprietary earth electrode resistance tester.

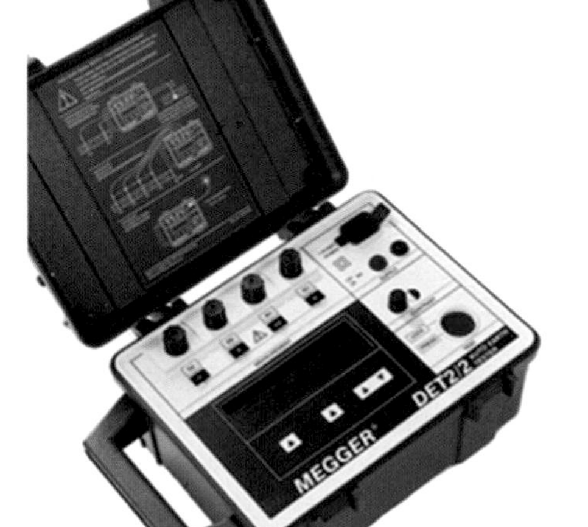

Proprietary earth electrode resistance tester (image courtesy of Megger)

### The test meter

The photograph shows a typical four-terminal digital earth electrode resistance tester. Analogue (moving coil) meters can be used.

Note that this meter can only be used if the electrode is not yet connected to the installation or if the installation relying on this earth electrode is completely isolated and the earthing conductor is disconnected from the main earthing terminal.

Test connections C1 and P1 must be connected to the electrode under test.

### Carrying out test E1

There are two possible methods used for test E1:

- Method 1 can be done by linking out C1 and P1 at the meter and supplying one lead only to the electrode, as shown in the diagram opposite.
- Method 2 is an alternative in which you can run two leads from the test meter to the electrode under test.

Method 2 may be used when ascertaining very low electrode values to eliminate the resistance of the test leads.

In most cases, method 1 will suffice as the lead resistance is generally less than 1 Ω. Always read the instrument manufacturer's instructions before use, if the instrument is not familiar to you.

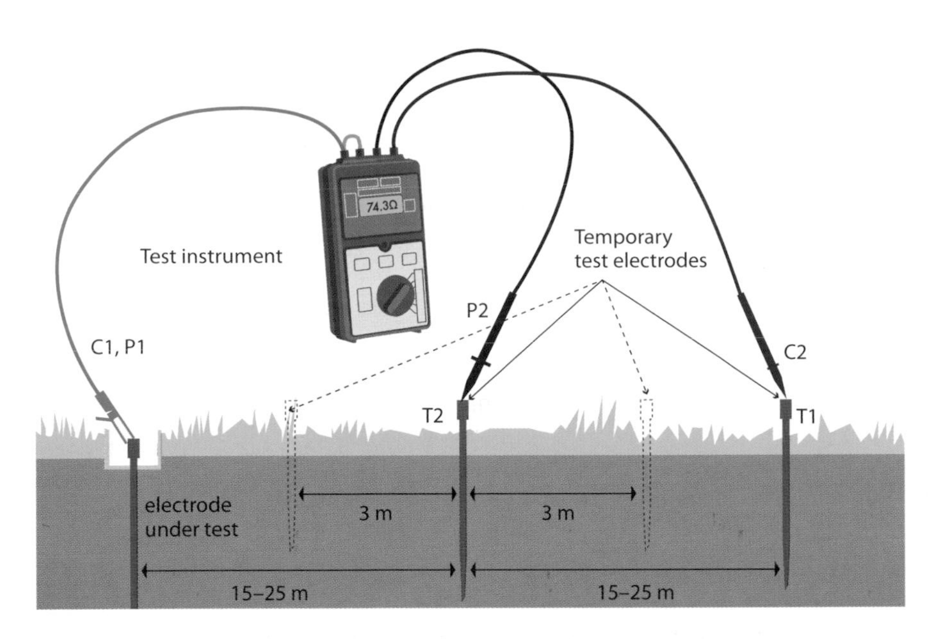


**Typical earth electrode test using a three-or-four terminal tester (E1)**

With the meter connected for method 1, there are two temporary test electrodes/spikes (T1 and T2) which must be inserted into the ground. These are normally supplied with the test instrument.

- C2 terminal on the meter is connected to T1 via a long lead, ideally 30–50 m away from the electrode under test.
- P2 terminal on the meter is connected to T2 via a long lead, and is centrally positioned between T1 and the electrode under test

**KEY POINT**

Earth electrode resistance values can differ greatly, dependent on the type of ground and environmental conditions, the material of the electrode used and the area of contact with the general mass of earth.

Ideally, the distance between the earth electrode and test spike T1 should be ten times the length of the electrode under test, but this dimension is likely to be affected by the location of the electrode and any surrounding buildings, paths or driveways, for example.

## Tests results

Three readings are taken during this test, with test spike T2 moved for each reading. The distance T2 is moved for the second and third readings depends on the distance between the electrode and spike T1. If the distance between them is 30 m then, typically, T2 will be moved 10% of that distance, which is 3 m. So, the first test is taken with spike T2 in the central position, the second test with the spike moved 10% closer to the earth electrode and the third test with the spike moved 10% from the centre, away from the earth electrode.

Here we will consider example readings for an earth electrode in good soil or clay (the earth electrode is 3 m long, so the distance between the electrode and test spike T2 is 30 m):

- with T2 central = 72 Ω
- with T2 3 m closer to the electrode under test = 70.5 Ω
- with T2 3 m closer to T1 = 73.5 Ω.

> **KEY POINT**
>
> This test can only be carried out if the surroundings allow for the installation of auxiliary test spikes.

> **ACTIVITY**
>
> Familiarise yourself with the proprietary earth electrode resistance tester and practise drawing the diagram. Indicate the designated terminal connections and lettering.

## Evaluation of test results

Once the three results have been obtained, the average of the three is found. So, using the example values given above, the average reading is:

$$\frac{72 + 70.5 + 73.5}{3} = 72\ \Omega$$

The three readings obtained should fall within a tolerance of 5% of the average, so 5% of 72 is 3.6 Ω so a tolerance of ± 5% gives 75.6 Ω and 68.4 Ω.

As the three readings all fall within this 5% tolerance, they are acceptable and the average value (72 Ω) would be recorded as the earth electrode resistance ($R_A$). If the deviation exceeds 5%, further tests must be carried out with a larger separation between the earth electrode under test and spike T1.

## Test method E2

This method uses an earth fault impedance tester and can only be used where the installation has an available supply.

### The test meter

The earth fault loop impedance tester and leads must comply with Guidance Note GS38.

Test results collected using this method will not be as accurate as those from a dedicated earth electrode tester.

### Carrying out test E2

- The main switch for the installation must be securely isolated.
- Disconnect the earthing conductor from the main earthing terminal (MET) or the earth bar, if no MET is present. At this point, for safety reasons, check all main protective bonding within the installation is securely connected together at the MET.
- Ensure that there are no parallel earth paths for the test current. The earthing conductor must connect directly to the earth electrode under test and no other parts.
- Set the meter for earth loop impedance ($Z_e$ on some meters). Check that the leads are in good condition and applicable for the tests.
- Attach the earth clip to the disconnected earthing conductor which connects to the electrode under test.
- Locate the probe to the incoming line conductor terminal on the supply side of the main switch.
- Press the test button and record the reading.

Note that some test instruments require connection of a third test probe to the neutral of the supply.

Repeat this test to ensure an accurate reading has been achieved. This reading can be accepted as your earth electrode resistance. It is recommended that this reading should be below 200 Ω to allow for soil drying and freezing.

## Test results

When using the above test method, the result will be a combination of the consumer's earth electrode resistance and the associated resistances of the line conductor, the impedance of the supply transformer, the supplier's earth electrode and the soil between the supplier's and consumer's electrodes. As this test method measures the overall loop impedance as described, we will refer to it as $Z_s$.

The diagram shows the method of testing the earth electrode using an earth fault loop impedance tester (method E2).

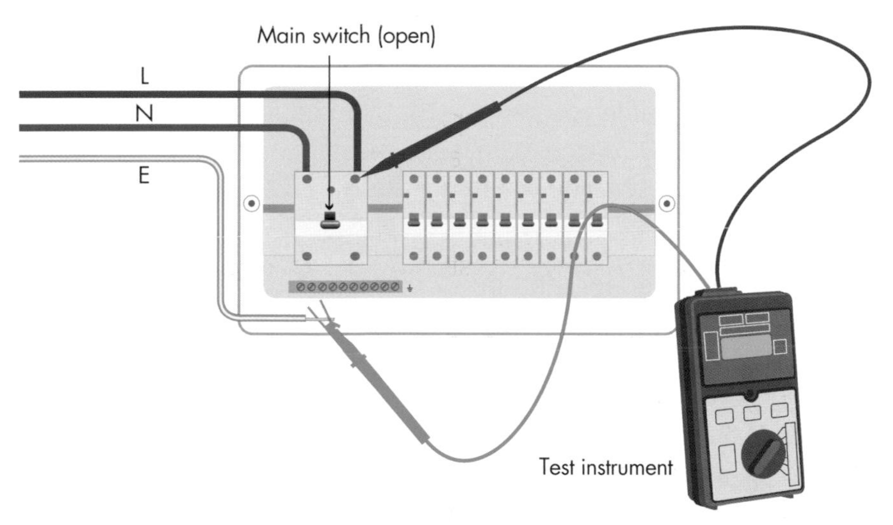


**Earth fault loop impedance tester (E2)**

## Verification of test results

For the purpose of verification of compliance, assume that a test result for an earth electrode when testing using an earth fault loop impedance tester was 115 Ω ($Z_s$).

Regulation 411.5.3 requires:

$$R_A \times I_{\Delta n} \leq 50 \text{ V}$$

or

$$Z_s \times I_{\Delta n} \leq 50 \text{ V}$$

where $R_A$ = the sum of electrode and protective conductor to exposed conductive parts
$Z_s$ = the earth fault loop impedance
$I_{\Delta n}$ = RCD rated residual operating current

Table 41.5 from BS 7671 helps us with the answers to such calculations, as long as the value recorded for $Z_s$ is below that given in the table for the particular residual current setting ($I_{\Delta n}$) of the RCD protecting the installation. (Although the table only states $Z_s$, and not $R_A$, as the formulae above are similar, either value could be used.)

If we assume, for each of our test methods that a 100 mA RCD protected each installation, we can see from the table that both results are acceptable since they are both less than 500 Ω.

| RCD rated residual operating current (mA) | Maximum value of earth fault loop impedance, $Z_s$ (Ω) |
|---|---|
| 30 | 1,667 |
| 100 | 500 |
| 300 | 167 |
| 500 | 100 |

**Table 41.5 Maximum value of earth fault loop impedance for non-delayed RCDs**

**KEY POINT**

The installation must be isolated and the earthing conductor must be disconnected from the MET to avoid parallel earth paths during this test, whichever method is used.

**ASSESSMENT GUIDANCE**

As part of assessment, you may be asked to determine the maximum $Z_s$ allowed for an RCD.

**ACTIVITY**

Calculate the maximum $Z_s$ allowed for the following RCDs, without using Table 41.5. The first one has been completed as an example.

- 30 mA RCD
  $\frac{50\text{ V}}{0.03\text{ A}} = 1667\ \Omega$
- 100 mA RCD
- 300 mA RCD
- 500 mA RCD

Equally, using this table you can see that you could use a 30 mA, 100 mA or 300 mA RCD as a main isolation switch – so the test results could be used to determine a suitably rated RCD to protect the installation.

The $Z_s$ value recorded for method E2 is too high to install a 500 mA RCD, but this RCD could be used where we recorded 72 Ω as a value of $R_A$.

## Discrimination between RCDs

Due to the design of the final circuits and their associated protection devices, if you select a 300 mA RCD as a main switch, you must ensure that there is discrimination between RCDs. This can be achieved by using a time delay S-type RCD.

If S-type RCDs are used, careful consideration must be given to ensure that requirements of Regulation 411.3 and 411.5 are met. Manufacturers' data must be verified for time delay criteria.

# AUTOMATIC DISCONNECTION OF SUPPLY

## TN systems

A TN system, in this case, can be either a TN-S or TN-C-S system. In a TN system, the electrical distributor guarantees a good earth at the origin of the supply.

It is important to note that, if an earth fault of negligible impedance develops within the fixed wiring or equipment of an installation, the protective device used to protect that circuit must operate within certain time limits given in Chapter 41 of BS 7671. This is called protection by automatic disconnection of supply.

## Earth fault loop impedance circuit design and calculation for TN systems

The earth fault loop impedance in part of a TN-C-S system is shown in the diagram.

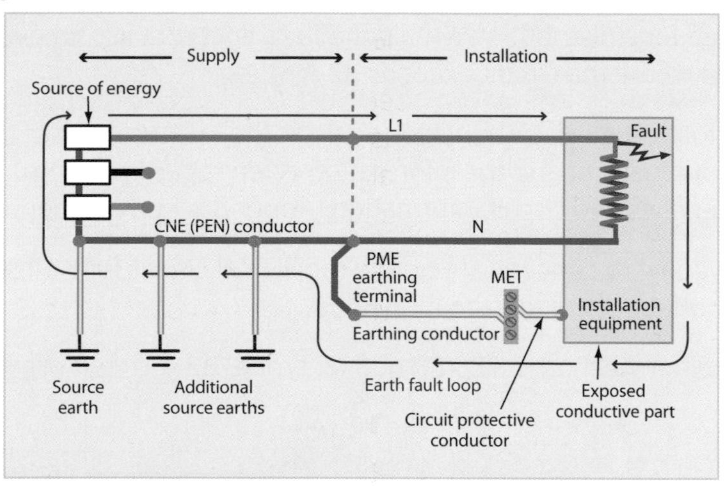


Earth fault loop impedance path of a TN-C-S system

## Protection of TN systems by automatic disconnection of supply

As part of the design process of all circuits, evaluation of the earthing and associated cables must be considered to ensure safe disconnection during earth fault conditions.

The effectiveness of measures for fault protection by automatic disconnection of supply can be verified for installations within a TN system by:

- measurement of earth fault loop impedance
- confirmation by visual inspection that overcurrent devices use suitable short-time or instantaneous tripping settings for circuit-breakers, correct current rating ($I_n$) and type for fuses and correct coordination and settings for RCDs
- testing to confirm that the disconnection times set out in Chapter 41 of BS 7671 can be achieved.

## Maximum disconnection times for a TN system

Table 41.1 from Chapter 41 of BS 7671 (shown below) gives the maximum disconnection times for final circuits not exceeding 32 A.

From the table, we can see that the maximum disconnection time for a final circuit forming part of a TN system where the nominal voltage to earth ($V_0$) is 230 V, is 0.4 seconds.

**TABLE 41.1**
**Maximum disconnection times**

| System | $50\text{ V} < V_0 \leq 120\text{ V}$ seconds | | $120\text{ V} < V_0 \leq 230\text{ V}$ seconds | | $230\text{ V} < V_0 \leq 400\text{ V}$ seconds | | $V_0 > 400\text{ V}$ seconds | |
|---|---|---|---|---|---|---|---|---|
| | a.c. | d.c. | a.c. | d.c. | a.c. | d.c. | a.c. | d.c. |
| TN | 0.8 | NOTE | 0.4 | 5 | 0.2 | 0.4 | 0.1 | 0.1 |
| TT | 0.3 | NOTE | 0.2 | 0.4 | 0.07 | 0.2 | 0.04 | 0.1 |

Maximum disconnection times for TN and TT systems

In a TN system, a disconnection time not exceeding 5 seconds is permitted for a distribution circuit and for a final circuit not covered by the table where the circuit exceeds 32 A rating.

For the following calculation, look again at the location plan of the small industrial unit and the information given regarding the supply characteristics and circuit information (Appendix 1 and Appendix 2).

You are going to concentrate on the radial final circuit for BS 1363 socket outlets, circuit designation 'Bl-3'.

The circuit protection is a BS EN 60898, type-B 32 A circuit breaker.

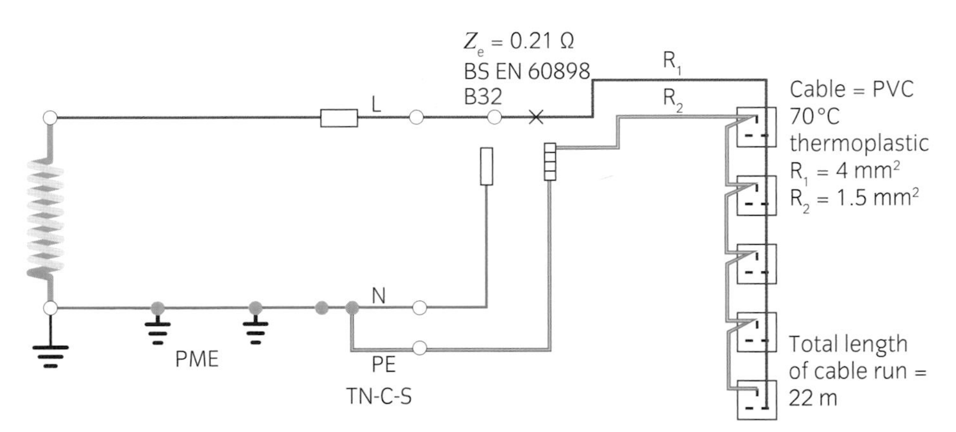


Five radial socket outlets in a circuit

The maximum disconnection time allowed for this circuit is 0.4 seconds, because the circuit is a final circuit and it does not exceed 32 A (see Table 41.1 above).

## Impedance values by design and calculation

With reference to the diagram on page 446, when you have established the cable length and cross-sectional area, you can begin to calculate the resistance of the line conductor ($R_1$) and the cpc ($R_2$).

You need to use the formula:

$$Z_s = Z_e + (R_1 + R_2)$$

$Z_e$ = 0.21 Ω (given value at the DB)

Using a BS EN 60898, type-B 32 A circuit breaker the maximum $Z_s$ allowed = 1.44 Ω (taken from Table 41.3 shown below)

**TABLE 41.3**

| **(a) Type B circuit-breakers to BS EN 60898 and the overcurrent characteristics of RCBOs to BS EN 61009-1** | | | | | | | | | | | | | | |
|---|---|---|---|---|---|---|---|---|---|---|---|---|---|---|
| **Rating (amperes)** | 3 | 6 | 10 | 16 | 20 | 25 | 32 | 40 | 50 | 63 | 80 | 100 | 125 | In |
| **$Z_s$ (ohms)** | 15.33 | 7.67 | 4.60 | 2.87 | 2.30 | 1.84 | 1.44 | 1.15 | 0.92 | 0.73 | 0.57 | 0.46 | 0.37 | 46/In |
| **(b) Type C circuit-breakers to BS EN 60898 and the overcurrent characteristics of RCBOs to BS EN 61009-1** | | | | | | | | | | | | | | |
| **Rating (amperes)** | | 6 | 10 | 16 | 20 | 25 | 32 | 40 | 50 | 63 | 80 | 100 | 125 | In |
| **$Z_s$ (ohms)** | | 3.83 | 2.30 | 1.44 | 1.15 | 0.92 | 0.72 | 0.57 | 0.46 | 0.36 | 0.29 | 0.23 | 0.18 | 23/In |
| **(c) Type D circuit-breakers to BS EN 60898 and the overcurrent characteristics of RCBOs to BS EN 61009-1** | | | | | | | | | | | | | | |
| **Rating (amperes)** | | 6 | 10 | 16 | 20 | 25 | 32 | 40 | 50 | 63 | 80 | 100 | 125 | In |
| **$Z_s$ (ohms)** | | 1.92 | 1.15 | 0.72 | 0.57 | 0.46 | 0.36 | 0.29 | 0.23 | 0.18 | 0.14 | 0.11 | 0.09 | 11.5/In |

**Maximum earth fault loop impedence ($Z_s$) for circuit breakers to comply with 0.4 s and 5 s disconnection times (from BS 7671:2008(2011))**

### ASSESSMENT GUIDANCE

Don't forget that the figures for earth loop impedance are maximums and to compare them to measured values they should be multiplied by the rule of thumb figure of 0.8 (80%).

## Establishing the resistance of $R_1 + R_2$

You need to establish the $R_1 + R_2$ of this cable by using the table below, which is a small section of Table B1 in Guidance Note 3. The cable installed for this circuit is a 4 mm²/1.5 mm². Once the $R_1 + R_2$ is calculated, you can add this value to the $Z_e$ value. We will assume an ambient temperature of 20 °C.

▼ **Table B1** Values of resistance/metre for copper and aluminium conductors and of ($R_1 + R_2$) per metre at 20 °C in milliohms/metre

| Cross-sectional area (mm²) | | Resistance/metre or ($R_1 + R_2$)/metre (mΩ/m) | |
|---|---|---|---|
| **Line conductor** | **Protective conductor** | **Copper** | **Aluminium** |
| 1 | – | 18.10 | |
| 1 | 1 | 36.20 | |
| 1.5 | – | 12.10 | |
| 1.5 | 1 | 30.20 | |
| 1.5 | 1.5 | 24.20 | |
| 2.5 | – | 7.41 | |
| 2.5 | 1 | 25.51 | |
| 2.5 | 1.5 | 19.51 | |
| 2.5 | 2.5 | 14.82 | |
| 4 | – | 4.61 | |
| 4 | 1.5 | 16.71 | |
| 4 | 2.5 | 12.02 | |
| 4 | 4 | 9.22 | |

**Resistances of copper conductors (from Guidance Note 3: Inspection and Testing, IET)**

$R_1 + R_2$ per metre = 16.71 mΩ

Total $R_1 + R_2$ = 22 m of 4 mm²/1.5 mm² cable is determined by

$$\frac{22 \times 16.71}{1000} = 0.37\ \Omega \text{ at } 20\,°\text{C}$$

If the cable is carrying its full load current, the temperature of the cable could increase to 70 °C. An increase in temperature will increase the cable's resistance.

By applying the applicable multiplier of 1.2, from the table below, we can see that the $R_1 + R_2$ value would be a maximum of

0.37 Ω x 1.20 = 0.44 Ω at normal operating temperatures.

The 1.2 factor was selected as the cables are 70 °C thermoplastic and the cpc is bunched or incorporated with the live conductors.

▼ **Table B3** Conductor temperature factor F for standard devices

**Multipliers to be applied to Table B1 for devices in Tables 41.2, 41.3, 41.4**

| Conductor installation | Conductor insulation | | |
|---|---|---|---|
| | 70°C thermoplastic (PVC) | 85°C thermosetting (note 4) | 90°C thermosetting (note 4) |
| Not incorporated in a cable and not bunched (notes 1, 3) | 1.04 | 1.04 | 1.04 |
| Incorporated in a cable or bunched (notes 2, 3) | 1.20 | 1.26 | 1.28 |

Conductor temperature factor F for standard devices – multipliers to be applied to Table B1 for devices in BS 7671 Tables 41.2, 41.3 and 41.4 (Tables B1 and B3 are from Guidance Note 3: Inspection and Testing, IET)

**KEY POINT**

It is very important to understand the theory of the maximum $Z_s$ values for circuits. You must ensure that, under fault conditions, the protective device will disconnect within the maximum duration limits for safety.

$Z_s = Z_e + (R_1+R_2)$

$= 0.21\ \Omega + 0.44\ \Omega$

$= 0.65\ \Omega$ which is less than the maximum $Z_s$ permitted by Table 41.3.

(Remember, the maximum $Z_s$ permitted = 1.44 Ω)

**ACTIVITY**

Using this method of design, calculate the $Z_s$ for circuit 'Gr-3', instantaneous water heater (see Appendix 1 and Appendix 2).

Verify that the $Z_s$ is within the maximum value permitted in Table 41.3.

The practical testing of earth fault loop impedance for this circuit is shown next.

## Verification of earth fault loop impedance test results

BS 7671 requires the inspector not only to test the installation, but also *to compare* the results with relevant design criteria (or with criteria within BS 7671). This may seem obvious, but some inspectors do pass test information back to their office without making the necessary comparisons, possibly assuming that someone else will check the results.

## Disconnection times for a TN system

Table 41.1 from Chapter 41 of BS 7671 gives the maximum disconnection times for final circuits not exceeding 32 A. From the table (opposite), we can see that the maximum disconnection time for a final circuit forming part of a TN system where the nominal voltage to earth ($V_0$) is 230 V, is 0.4 seconds.

In a TN system, a disconnection time not exceeding 5 seconds is permitted for a distribution circuit and for a final circuit not covered by the table.

**TABLE 41.1**
**Maximum disconnection times**

| System | $50\ V < V_0 \le 120\ V$ seconds | | $120\ V < V_0 \le 230\ V$ seconds | | $230\ V < V_0 \le 400\ V$ seconds | | $V_0 > 400\ V$ seconds | |
|---|---|---|---|---|---|---|---|---|
| | a.c. | d.c. | a.c. | d.c. | a.c. | d.c. | a.c. | d.c. |
| TN | 0.8 | NOTE | 0.4 | 5 | 0.2 | 0.4 | 0.1 | 0.1 |
| TT | 0.3 | NOTE | 0.2 | 0.4 | 0.07 | 0.2 | 0.04 | 0.1 |

Maximum disconnection times for TN and TT systems (from BS 7671:2008 (2011))

For the following calculation, look again at the location plan of the small industrial unit and the information given regarding the supply characteristics and circuit information, in Appendix 1.

You are going to concentrate on the radial final circuit for BS 1363 socket outlets, circuit designation (Bl-3). The circuit protection is a BS EN 60898, type-B 32 A circuit breaker.

The maximum disconnection time allowed for this circuit is 0.4 seconds because the circuit is a final circuit and it does not exceed 32 A (see Table 41.1).

## Methods of measuring earth fault loop impedance

Earth fault loop impedance values for final circuits can be obtained in two ways

- adding $Z_e$ to $R_1+R_2$
- direct measurement.

In both cases, the value of $Z_e$ at the origin of an installation needs to be established and the most accurate method is by direct measurement. We have seen how we can obtain values of $Z_e$ for TT installations on page 437.

## Practical guide to testing the $Z_e$ at the origin of a TN system

The diagram on page 450 provides a simple representation of a TN-S system that highlights the external circuit from the supply transformer to the origin of the consumer's supply.

The external earth fault loop impedance ($Z_e$) in ohms is a result of the impedance of the supplier's line conductor, the transformer and the separate earth added together.

**ASSESSMENT GUIDANCE**

Remember that the earth loop test is a live test and you must take all precautions necessary.

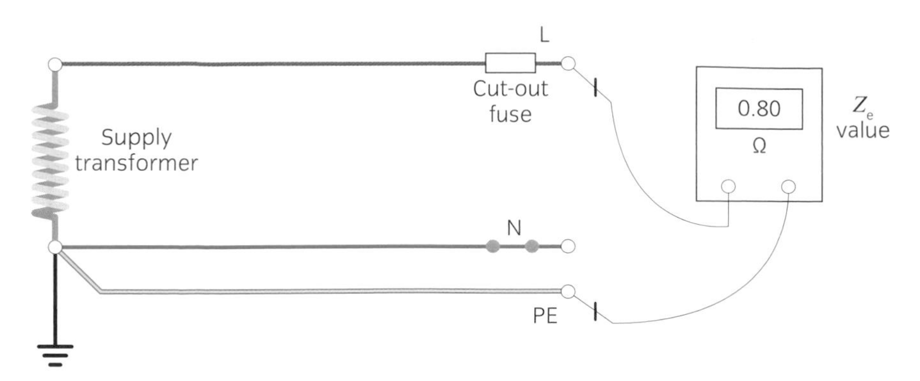


The $Z_e$ value illustrated is the maximum value of the earth fault loop impedance expected for a TN-S system.

**Simple diagrammatic representation of a TN-S system**

The test is carried out in the following way using an earth fault loop impedance tester and leads which must comply with Guidance Note GS38.

- Isolate and secure the installation main switch in the off position.
- Ensure, for reasons of safety, that all main protective bonding is connected to the main earthing terminal. If the supplier's cable is faulty, a fault current may be introduced to extraneous parts of the installation.
- Disconnect the earthing conductor from the main earthing terminal.
- Check instrument for safety and correct settings.
- Connect the earth clip to the disconnected earthing conductor, then connect the test instrument line probe to the supply line terminal of the main switch.
- Press the test button and note the result.
- For three-phase installations, repeat this test for all line conductors. For three-phase installations, the highest reading obtained is recorded as the external earth fault loop impedance.
- For a single-phase installation, record the value.
- Reconnect the earthing conductor to the MET.
- The installation supply may be switched back on if required.

Note that some test instruments may require a third lead connected to the neutral of the supply.

**KEY POINT**

Three methods of determining the external earth fault loop impedance are by measurement, enquiry or calculation.

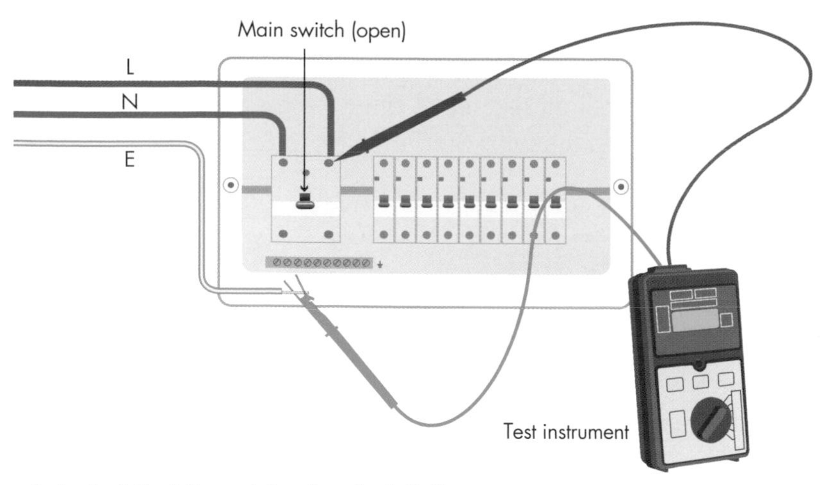


Example test of $Z_e$ at the origin of an installation

## Determining values of $Z_s$ by calculation

As we have now determined the value of $Z_e$ at the origin of the installation, we can determine the total earth fault loop impedance ($Z_s$) by adding the $R_1+R_2$ which we previously obtained through continuity testing.

Providing continuity testing is carried out thoroughly, which it should have been for any initial verification, this method will provide reliable values of $Z_s$, even if parallel earth paths change due to work carried out on mechanical services. For this reason, and for reasons of testing safely, this is the preferred method of obtaining values of $Z_s$. This method also reduces the need to revisit parts of the installation which have already been tested.

If values of $Z_s$ are obtained through direct measurement (main protective bonding must remain connected during the test, as these provide parallel earth paths), the values obtained by this method may not be a true reflection of the circuit conditions. Sample testing by direct measurement is a good exercise once $Z_s$ values are obtained by calculation; the values obtained by direct measurement are likely to be lower than those calculated.

## Method for obtaining $Z_s$ by direct measurement

1 Select the correct meter – you will need an earth fault loop impedance tester.
2 Select the correct scale – the correct scale is loop, $Z_e$ or $Z_s$ scale (depending on the instrument used).
3 Check the leads for damage – ensure that the probe tips are protruding no more than 4 mm (preferably 2 mm) in accordance with Guidance Note GS38.
4 Insert leads – choose the correct location on the meter (following the manufacturer's instructions).
5 Connect the leads correctly.

**ASSESSMENT GUIDANCE**

When testing $Z_s$ at sockets it is far safer to use a lead with a plug of the correct type and plug in directly.

You can choose either the appropriate plug-in lead for testing socket outlets or fly leads appropriate to the terminals which need to be tested. Remember that all test leads must be manufactured and maintained in accordance with Guidance Note GS38.

You are now ready to carry out the earth fault loop impedance testing.

## Earth fault loop impedance ($Z_s$) testing for radial socket outlets

Each socket outlet must be tested, with the highest value of $Z_s$ recorded on the Generic Schedule of Test Results.

The meter in the diagram is shown at the final socket outlet on the radial circuit. This will give the $Z_s$ value if all of the conductor connections are secure.

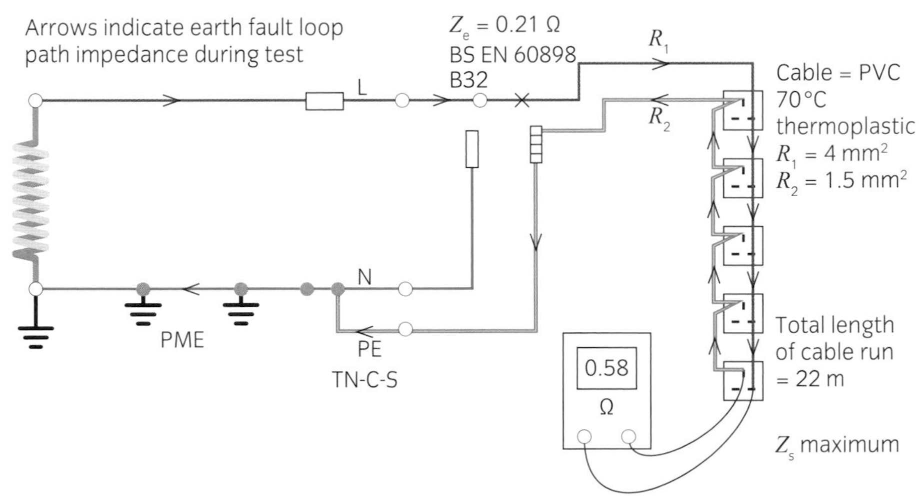


**Five radial socket outlets with 4 mm/1.5 mm cables**

## Verification of earth fault loop impedance ($Z_s$) for TN systems

Once you have established the $Z_s$, you can compare this value to the maximum value allowed, to ensure the disconnection time can be achieved.

The $Z_s$ in the circuit above = 0.58 Ω.

The maximum earth fault loop impedance allowed for this circuit can be found by using the table opposite. The table can be used to find the maximum $Z_s$ for disconnection times of 0.1 to 5 seconds for TN systems.

This table can be used for the following overcurrent protective devices: BS 3871, BS EN 60898 circuit breakers and BS EN 61009 RCBOs (residual current operated circuit breakers with integral overcurrent protection).

| Circuit-breaker type | Circuit-breaker rating (A) | | | | | | | | | | | | | | |
|---|---|---|---|---|---|---|---|---|---|---|---|---|---|---|---|
| | 5 | 6 | 10 | 15 | 16 | 20 | 25 | 30 | 32 | 40 | 45 | 50 | 63 | 100 | 125 |
| 1 | 9.27 | 7.73 | 4.64 | 3.09 | 2.90 | 2.32 | 1.85 | 1.55 | 1.45 | 1.16 | 1.03 | 0.93 | 0.74 | 0.46 | 0.37 |
| 2 | 5.3 | 4.42 | 2.65 | 1.77 | 1.66 | 1.32 | 1.06 | 0.88 | 0.83 | 0.66 | 0.59 | 0.53 | 0.42 | 0.26 | 0.21 |
| B | 7.42 | 6.18 | 3.71 | 2.47 | 2.32 | 1.85 | 1.48 | 1.24 | 1.16 | 0.93 | 0.82 | 0.74 | 0.59 | 0.37 | 0.30 |
| 3&C | 3.71 | 3.09 | 1.85 | 1.24 | 1.16 | 0.93 | 0.74 | 0.62 | 0.58 | 0.46 | 0.41 | 0.37 | 0.29 | 0.19 | 0.15 |
| D | 1.85 | 1.55 | 0.93 | 0.62 | 0.58 | 0.46 | 0.37 | 0.31 | 0.29 | 0.23 | 0.21 | 0.19 | 0.15 | 0.09 | 0.07 |

**Maximum $Z_s$ test values in ohms for circuits supplied by circuit breakers (from Guidance Note 3: Inspection and Testing, IET)**

The maximum $Z_s$ allowed for a type-B 32 A circuit breaker is 1.16 Ω.

The maximum $Z_s$ test result = 0.58 Ω.

We can conclude that the tested $Z_s$ is well within the 0.4 seconds allowed.

Note this table specifies that the ambient temperature during the testing should be no lower than 10 °C and is, therefore, temperature corrected. No further calculation is required if the $Z_s$ test result is lower than the maximum allowed in this table, unless the ambient temperature is lower than 10 °C. It is advisable to log the test temperature, as this allows future test results to be evaluated accurately. This table is also shown in the IET On-Site Guide.

If the maximum values of $Z_s$ are used, as given in BS 7671, correction for temperature will be required. As the values in BS 7671 are used for design purposes, not verification, the values must be corrected by 20% to allow for temperature changes under operating conditions.

As a rule of thumb, the measured values of $Z_s$ should not exceed 80% of the values given in Tables 41.2 to 41.4 in BS 7671.

As an example, the maximum permitted value of earth fault loop impedance for a type-B 32 A circuit breaker is 1.44 Ω; 80% of this is (1.44 × 0.8) = 1.16 Ω. Any measured value should not exceed this.

So measured $Z_s$ should be ≤ maximum $Z_s$ as BS 7671 × 0.8

If you get a higher $Z_s$ test reading than the maximum allowed, the circuit $R_1 + R_2$ must be reduced or the circuit components or the protective device may need to be redesigned or changed:

- By increasing the cross-sectional area of the cable you will reduce the $R_1 + R_2$ and therefore reduce the total $Z_s$ value.
- By changing the circuit breaker to an RCBO or by using an RCD to supply the circuit in accordance with Regulation 411.4.9, you can achieve disconnection within the 0.4 seconds.

**KEY POINT**

It is very important to understand the theory of the maximum $Z_s$ values for circuits. You must ensure that, under fault conditions, the protective device will disconnect within the maximum duration limits for safety.

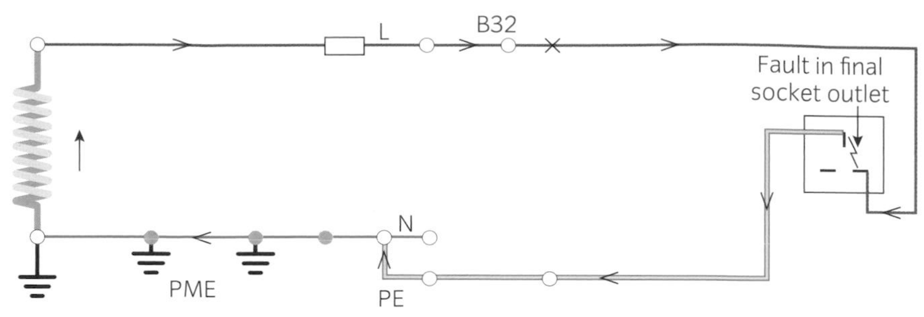


**Basic diagram of the earth fault loop impedance path at a final socket outlet to the transformer**

This diagram illustrates the earth fault loop path of the radial socket outlet circuit. You can use this method to illustrate the earth fault loop impedance of fixed equipment or lighting circuits.

## ACTIVITY

Find the maximum $Z_s$ allowed for all of the circuits on the location plan of the small industrial unit in Appendix 1.

## ASSESSMENT GUIDANCE

Comparing test results is very important, whether you are providing results for initial verification or evaluating during future periodic inspections. Making comparisons between results will indicate if the system is deteriorating.

## TT systems

### Verification of earth fault loop impedance test results in TT system

It is important to note that, if an earth fault of negligible impedance develops within the fixed wiring or equipment of an installation, the protective device must operate within the time limits given in Chapter 41 of BS 7671. This is called protection by automatic disconnection of supply.

BS 7671 requires the inspector not only to test the installation, but also *to compare* the results with relevant design criteria (or with criteria within BS 7671). This may seem obvious, but some inspectors do pass test information back to their office without making the necessary comparisons, possibly assuming that someone else will check the results.

### Protection of TT systems by automatic disconnection of supply

As part of the design process of all circuits, evaluation of the earthing and associated cables must be considered to ensure safe disconnection during earth fault conditions.

The effectiveness of measures for fault protection by automatic disconnection of supply can be verified for installations within a TT system by:

- measurement of earth fault loop impedance
- confirmation by visual inspection that overcurrent devices use suitable short-time or instantaneous tripping setting for circuit-breakers, correct current rating ($I_n$) and type for fuses and correct coordination and settings for RCDs
- testing to confirm that the disconnection times set out in Chapter 41 of BS 7671 can be achieved.

## Disconnection times for a TT system

Table 41.1 from BS 7671 gives the maximum permitted disconnection times for final circuits up to and including 32 A.

The maximum disconnection time for a final circuit forming part of a TT system, with a nominal voltage to earth ($U_0$) of 230 V with a rating not exceeding 32 A is 0.2 seconds.

**TABLE 41.1**
**Maximum disconnection times**

| System | $50\ V < V_0 \leq 120\ V$ seconds | | $120\ V < V_0 \leq 230\ V$ seconds | | $230\ V < V_0 \leq 400\ V$ seconds | | $V_0 > 400\ V$ seconds | |
|---|---|---|---|---|---|---|---|---|
| | a.c. | d.c. | a.c. | d.c. | a.c. | d.c. | a.c. | d.c. |
| TN | 0.8 | NOTE | 0.4 | 5 | 0.2 | 0.4 | 0.1 | 0.1 |
| TT | 0.3 | NOTE | 0.2 | 0.4 | 0.07 | 0.2 | 0.04 | 0.1 |

Maximum disconnection times for TN and TT Systems (from BS 7671:2008 (2011))

In a TT system, a disconnection time not exceeding 1 second is permitted for a distribution circuit and for a final circuit not covered by Table 41.1.

## Verification of test results

BS 7671 requirements for TT installations are fairly straightforward.

Regulation 411.5.3 requires:

$$R_A \times I_{\Delta n} \leq 50\ V$$

or

$$Z_s \times I_{\Delta n} \leq 50\ V$$

For the purpose of verification of compliance, assume that the test result for earth fault loop impedance including the earth electrode resistance was 113 Ω ($Z_e$).

If the lighting circuit has an $R_1 + R_2$ value of 2 Ω, we can assume the total $Z_s$ to be:

$$\begin{aligned} Z_s &= Z_e + R_1 + R_2 \\ &= 113\ \Omega + 2\ \Omega \\ &= 115\ \Omega \end{aligned}$$

Regulation 411.5.3 requires:

$$R_A \times I_{\Delta n} \leq 50\ V$$

or

$$Z_s \times I_{\Delta n} \leq 50\ V$$

$R_A$ = the sum of electrode and protective conductor to exposed conductive parts (electrode resistance added to circuit $R_1$+$R_2$)

$Z_s$ = the earth fault loop impedance

$I_{\Delta n}$ = RCD rated residual operating current

Maximum $R_A$ values can be substituted for $Z_s$ in this case.

| Non-delayed RCD rated residual operating current $I_{\Delta n}$ (mA) | Maximum value of earth fault loop impedance, $Z_s$ (Ω) |
|---|---|
| 30 | 1667 |
| 100 | 500 |
| 300 | 167 |
| 500 | 100 |

Maximum $R_A$ and $Z_S$ values for non-delayed residual current devices

This table shows that you could use a 30 mA, 100 mA or 300 mA RCD as a protective device for this circuit. Note that the $Z_s$ value is too high to install a 500 mA RCD. You would normally select a 30 mA RCBO for this circuit due to requirements for additional protection.

The guaranteed design fault current will be:

$$I_{fault} = \frac{V_0}{Z_s} = \frac{230\text{ V}}{115\ \Omega} = 2\text{ A}$$

## Fault current in relation to the disconnection time of an RCD

The fault current is calculated to be 2 A or 2000 mA.

All non-delayed 30 mA RCDs have been manufactured to operate within 40 ms (0.04 seconds) at a 150 mA fault current. Therefore, disconnection for this circuit will occur well within the maximum 0.2 seconds disconnection time shown in Table 41.1.

## Discrimination between RCDs

Due to the design of the final circuits and their associated protection devices, if you select a 300 mA RCD as a main switch, you must ensure that there is discrimination between RCDs. This can be achieved by using a time delay S-type RCD.

**KEY POINT**

It is very important to understand the theory of the maximum $Z_s$ values for circuits. You must ensure that under fault conditions, the protective device will disconnect within the maximum duration limits for safety.

**ASSESSMENT GUIDANCE**

You may be asked to calculate the maximum $Z_s$ allowed for an RCD.

The diagram opposite shows a fault on an item of Class I (earthed) equipment.

- RCD A could be a main isolator for the distribution board (300 mA $I_{\Delta n}$ S-type).
- RCD B could be a 30 mA $I_{\Delta n}$ non-delayed RCBO for the lighting circuit.

If an earth fault develops on the luminaire (Class I equipment), RCD B will trip and the supply to the other circuits will remain healthy. If S-type RCDs are used, careful consideration must be given to ensure that requirements of Regulations 411.3 and 411.5 of BS 7671 are met. Manufacturers' data must be verified for time delay criteria.

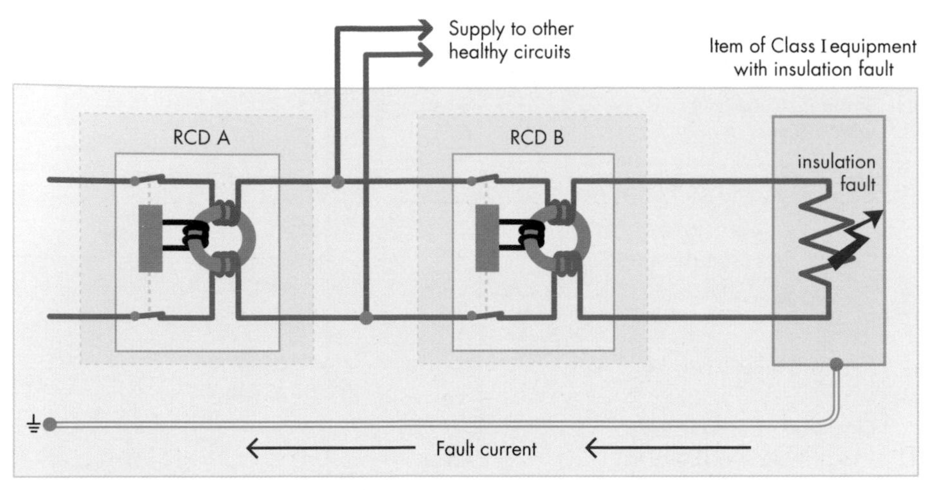


A fault on an item of Class I (earthed) equipment

## PROSPECTIVE FAULT CURRENT ($I_{PF}$)

The prospective short-circuit current (PSCC) and prospective earth fault current (PEFC) should be measured, calculated or determined by another method (such as by enquiry to the relevant electrical distributor), at the origin and at other relevant points in the installation.

### Evaluating the prospective fault current

You need to evaluate the prospective fault current to ensure that the switchgear and protective devices are capable of withstanding high fault currents without damage.

You evaluate the prospective fault current at the origin and other relevant points of the installation, i.e. main distribution boards, sub-main distribution boards and consumer units.

Regulation 434.5.1 in BS 7671 states that the **breaking capacity** rating of each protective device shall be not less than the prospective fault current at its point of installation.

The table on page 458 gives an indication of the maximum breaking capacities of our most commonly used protection devices.

#### ASSESSMENT GUIDANCE

Remember this is the prospective *maximum* fault current. It is quite likely that the fuse will blow or the circuit breaker will trip before this figure is reached. You cannot, however, reduce the rating of the device.

**Breaking capacity**

The amount of current a protective device can safely disconnect.

| Device type | Device designation | Rated short-circuit capacity (kA) | |
|---|---|---|---|
| Semi-enclosed fuse to BS 3036 with category of duty | S1A<br>S2A<br>S4A | 1<br>2<br>4 | |
| Cartridge fuse to BS 1361 type I<br>type II | | 16.5<br>33.0 | |
| General purpose fuse to BS 88-2 | | 50 at 415 V | |
| BS 88-3 type I<br>type II | | 16<br>31.5 | |
| General purpose fuse to BS 88-6 | | 16.5 at 240 V<br>80 at 415 V | |
| Circuit-breakers to BS 3871 (replaced by BS EN 60898) | M1<br>M1.5<br>M3<br>M4.5<br>M6<br>M9 | 1<br>1.5<br>3<br>4.5<br>6<br>9 | |
| Circuit-breakers to BS EN 60898* and RCBOs to BS EN 61009 | | $I_{cn}$<br>1.5<br>3.0<br>6<br>10<br>15<br>20<br>25 | $I_{cs}$<br>(1.5)<br>(3.0)<br>(6.0)<br>(7.5)<br>(7.5)<br>(10.0)<br>(12.5) |

* Two short-circuit capacities are defined in BS EN 60898 and BS EN 61009:
$I_{cn}$ – the rated short-circuit capacity (marked on the device).
$I_{cs}$ – the in-service short-circuit capacity.

Rated short-circuit capacities of protective devices (from On-Site Guide, IET)

**KEY POINT**

If a fault current greater than a breaking capacity was to flow through a device, it might explode causing injury or a fire. In the case of switches and circuit breakers, the device may weld together, which means it will not disconnect. The breaking capacity is rated in kiloamperes (kA). Most fuses and circuit breakers have a maximum kA value printed on the surround of a fuse or the casing of a circuit breaker.

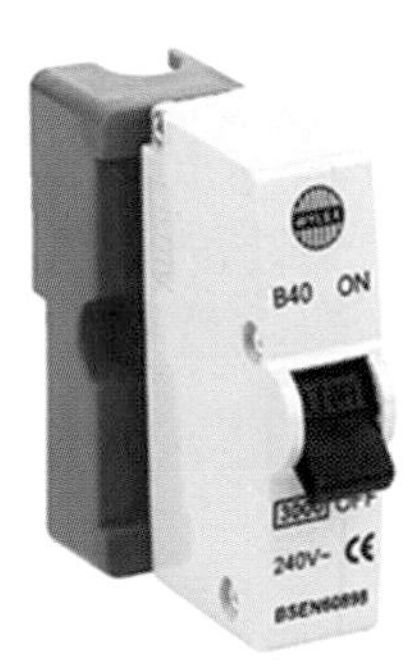


The circuit breaker above shows a 3 kA rated BS EN 60898

When evaluating the kA value of BS EN 60898 circuit breakers, you need to look for the number in the rectangular box. If you see a symbol like this:

| 6000 |
|---|

it means that the circuit breaker has been manufactured to be able to withstand 6 kA. A domestic location fault current is unlikely to exceed 6 kA.

## How to check the prospective fault current

You can check the prospective fault current by measuring it during the inspection and testing process, ensuring the recorded values of $I_{pf}$ are lower than the breaking capacities of the devices at that point or further into the installation.

## Setting up the test meter for testing the prospective fault current

Most earth fault loop impedance testers on the market are capable of evaluating the prospective fault current. Always follow the manufacturer's instructions.

## Method

1. Select the correct meter – the correct meter is a prospective fault current tester or an earth fault loop impedance tester (this will require calculation).
2. Select the correct scale – the correct scale is PSC/kA or kA, loop, or the $Z_e$ or $Z_s$ scale (depending on the individual instrument). Always read the manufacturer's instructions.
3. Check the leads for damage – ensure that the probe tips are protruding no more than 4 mm (preferably 2 mm) and all other parts are in accordance with Guidance Note GS38.
4. Insert leads – choose the correct location on the meter for the connection of the correct leads (follow the manufacturer's instructions).

**ASSESSMENT GUIDANCE**

Most instruments cover both prospective fault current and earth loop testing. When demonstrating your competence during on-site or classroom assessment, make sure the correct scale is selected.

## Carrying out the tests for prospective fault current (PFC)

### Testing the prospective earth fault current (PEFC)

- Ensure that the main protective bonding and earthing conductor are securely connected. It is important to ensure *all* possible parallel paths exist for this test, as this will produce the maximum fault current.
- Connect the earth clip of the instrument to the main earthing terminal (MET). For many three-lead instruments, the neutral clip can also be connected to the MET.
- Connect the line probe of the instrument to the incoming line connection.
- Take a measurement and note this as the prospective earth fault current (PEFC).
- For three-phase installations, repeat this test for all line conductors individually, and record the highest value.
- Remove all connections.

### Testing the prospective short circuit current (PSCC)

- Connect the earth and neutral clips to the neutral bar or connection.
- Connect the line probe of the instrument to the incoming line connection.
- Take a measurement and note this as the prospective short circuit current (PSCC).
- For three-phase installations, repeat this test for all line conductors individually, and record the highest value.
- Remove all connections.

Although some test instruments can measure fault currents between phases, do not attempt this unless you are absolutely certain of the instrument's capabilities. Remember that all test leads must be manufactured and maintained in accordance with Guidance Note GS38.

**ACTIVITY**

What might be the effect of using a single-phase prospective fault current tester on a three-phase system?

**KEY POINT**

Some instruments claim to determine the PEFC and PSCC for you. Take care as some instruments measure the earth resistance coupled with the neutral voltage. This may not provide an accurate value of fault current. The neutral and earth clips may need to be doubled up, so the test current and voltage are taken from the conductor under test only.

## Determining the prospective fault current

Once the readings have been taken, the value to be recorded as prospective fault current is determined in the following way:

- For single-phase installations – determine which value from PEFC and PSCC is the largest and record that value as the prospective fault current. If readings were obtained in ohms, divide 230 V by the lowest value (Ohm's law) to obtain the prospective fault current.
- For three-phase installations – as the voltage between phases is 400 V, but the measured value was obtained using 230 V, the PSCC has to be adjusted. It is acceptable to *double* the largest value of PSCC recorded between the line conductors and neutral. This will normally be larger than any PEFC, so would be recorded as the prospective fault current.

For example, when testing a three-phase installation for both PEFC and PSCC, the results shown in the table were obtained.

**ASSESSMENT GUIDANCE**

Virtually every device, except the lowest rated rewirable devices, will be able to handle the values shown in the table.

| | | | |
|---|---|---|---|
| L1 to E | PEFC | 1.2 kA | Largest value is 1.21 kA |
| L2 to E | PEFC | 1.21 kA | |
| L3 to E | PEFC | 1.19 kA | |
| L1 to N | PSCC | 1.42 kA | Largest value is 1.47 kA |
| L2 to N | PSCC | 1.47 kA | |
| L3 to N | PSCC | 1.39 kA | |

The value to be recorded is:

$$1.47\ \text{kA} \times 2 = 2.94\ \text{kA}$$

## Further tests for the prospective fault current

If the value of prospective fault current obtained at the origin of the installation is lower than the breaking capacity of any device within the installation at any remote distribution board or consumer unit, no further testing is required.

If, however, the prospective fault current at the origin is greater than the breaking capacity of devices at a remote distribution board, further testing, as described above, would need to be carried out at that particular distribution board. The fault current at that board is likely to be reduced due to the resistance or impedance of any distribution circuit.

Remember, the purpose of testing for prospective fault current is to ensure that protective devices can safely respond to any fault current they may experience at that point in the installation.

## Calculating the fault current value

The fault current value can be calculated during the design process. This method tends to be used on larger projects only.

A designer of a large installation will gather information about the external impedances of the supply transformer and associated conductors. Using this information, the designer will then evaluate the maximum prospective fault current value.

## Determining the fault current value by enquiry

You can determine the fault current value from the electricity distributor, by enquiry (usually by telephone). The distributor will require the address of the property, allowing them to assess the plans of the wiring system that supplies the property and enabling them to estimate the maximum prospective fault current at the origin of the supply.

In the absence of any plans, for 230 V single-phase supplies up to 100 A, the distributor will provide the consumer with an estimate of the maximum prospective short-circuit current at the distributor's cut-out, which will be based on Engineering Recommendation P25/1 and on the declared level of 16 kA at the point of connection of the service line to the LV distribution cable.

The fault level will only be this high if the installation is close to the distribution transformer. However, because changes may be made to the distribution network by the distributor over the life of an installation, the designer of the consumer's installation must specify equipment suitable for the highest fault level likely.

Remember that 16 kA will be the maximum value quoted. In most cases the $I_{pf}$ will be a lot less than this value.

**KEY POINT**

The prospective fault current must be evaluated at every relevant point of the installation.

**ACTIVITY**

Test the prospective fault current at a distribution board and check that the protective devices are capable of withstanding the maximum fault current by carrying out a visual inspection. Ensure the supply is securely isolated before any inspection is carried out. This activity must be carried out under the supervision of your tutor or assessor.

**ASSESSMENT GUIDANCE**

As part of the practical assessment, you will be required to test and verify prospective fault current.

# OPERATION AND FUNCTIONAL TESTING OF RESIDUAL CURRENT DEVICES (RCDS)

RCDs must be tested to ensure that they are working in accordance with BS and BS EN manufacturing standards.

## The theory of RCD operation

'Residual current device' or 'RCD' is the generic term for a device that operates when the residual current in the circuit reaches a predetermined value. An RCD is a protective device used to automatically disconnect the electrical supply when an imbalance is detected between the line and neutral conductors.

The diagram shows a line to earth (cpc) fault.

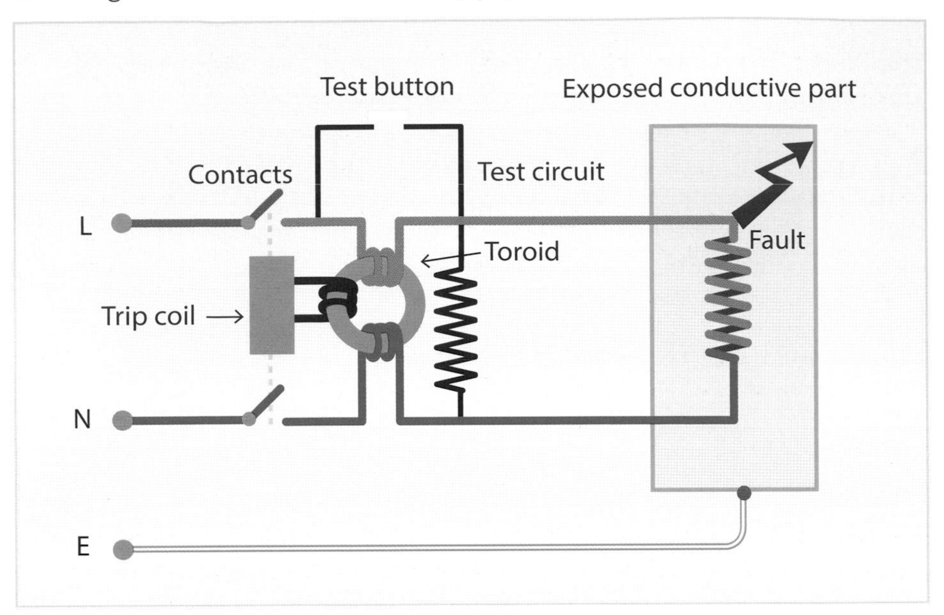


A line to earth (cpc) fault

In the presence of this fault, the RCD would detect an imbalance and the linked switch would open, as shown in the diagram.

The operating times of RCDs must be tested where they are essential for disconnection (for compliance with Chapter 41 of BS 7671) or where they are installed as additional protection as specified in Chapter 41.

## The test procedure

The table shows the maximum operating times for non-time delayed RCDs at 100% of their rated tripping current.

| | (ms) |
|---|---|
| BS 4293 | 200 |
| BS 61008 | 300 |
| BS 61009 (RCBO) | 300 |
| BS 7288 (integral socket-outlet) | 200 |

Disconnection times for RCDs (extract from Guidance Note 3: Inspection and Testing, IET)

### ACTIVITY

Look again at the location plan of the small industrial unit in Appendix 1 and use the circuit information to evaluate the correct tests and sequence for testing for the ring final circuit Br-3. The RCD is intended to be utilised for additional protection in accordance with Regulation 411.3.3.

## Setting up the test meter for testing an RCD

### Method

1 Select the correct meter – the correct meter is the RCD tester.

2 Select the correct setting for the RCD under test – this would be, for example, 30 mA, 100 mA or 300 mA.

3 Select the correct scale – the correct scale would be, where applicable, ½ rated, 1× rated or 5× rated.

### ASSESSMENT GUIDANCE

Remember, the RCD tester gives results in milliseconds.

4 Check the leads for damage – ensure that the probe tips are protruding no more than 4 mm (preferably 2 mm) and are in accordance with Guidance Note GS38, including any plug-in lead for socket outlets.
5 Insert the leads – choose the correct location on the meter (following the manufacturer's instructions).
6 Correctly connect the leads – you can choose either the correct plug-in lead for testing socket outlets or fly leads appropriate to the terminals to be tested.

Remember that all test leads must be manufactured and maintained in accordance with Guidance Note GS38.

## Carrying out the tests for residual current devices (RCDs)

For each of the tests, readings should be taken on both positive (+ve) and negative (–ve) half cycles and the longer of the operating times recorded on the Schedule of Test Results. Some instruments may show the half cycles as 0° and 180° or as two sine waves in opposite directions.

The tests that need to be carried out depend on the reason why the RCD is fitted:

- 5× rated must be used if the RCD is present for additional protection and to prove disconnection times stated in Chapter 41 of BS 7671.
- ½ rated tests are carried out to ensure that the RCD is not *too* sensitive.
- 1× rated tests are carried out to ensure that the RCD is working to the manufactured standard.

RCDs rated greater than 30 mA $I_{\Delta n}$, do not require the 5× test. The 5× test is carried out as prescribed in Regulation 415.1 of BS 7671 for additional protection against electric shock.

Where periodic testing of an RCD that provides additional protection is being carried out, do the 5× test first as this is the most important. Any test following this is likely to be fast as the mechanical parts have been exercised.

The table below illustrates the maximum durations allowed (in red) and some typical values of test results (in orange).

| BS or BS EN device | RCD rated tripping current $I_{\Delta n}$ | ½ rated $I_{\Delta n}$ | 1× rated $I_{\Delta n}$ | 5× rated $I_{\Delta n}$ |
|---|---|---|---|---|
| BS EN 61009 | 30 mA | No trip | 300 ms (max) | 40 ms (max) |
| Typical test results | | +ve – no trip<br>–ve – no trip | +ve 128 ms<br>–ve 114 ms | +ve 23 ms<br>–ve 14 ms |

## ASSESSMENT GUIDANCE

The effective operation of the RCD is reliant upon the customer pushing the test button every three months or so. The problem for many people is that the loss of supply when the RCD trips, means that many electronic devices will need to be re-set. It is important to explain to the consumer why this test should be carried out.

## RCD integral test button

The RCD integral test button is used to verify the function (and proper lubrication) of the mechanical parts, ensuring that the switching mechanism within the RCD is working correctly. Usually, this button can be seen on the front of the RCD.

Manufacturers normally recommend that the test button is pressed every three months (quarterly). This test must be done *before* all of the other tests for initial verification. However, during periodic inspection and testing, the integral test button is pressed *after* the tests.

Any device that is left in one position for a long time may stick. As RCDs rely on small amounts of energy (mA) to bring about disconnection, a sticky RCD may not trip in the time required. Regular pressing of the test button will improve the response speed of an RCD. Testing the RCD *before* pressing the test button during a periodic inspection will give a good indication of whether the RCD is being tested frequently enough!

### KEY POINT

The integral test button only tests the function of the *mechanical* parts of the RCD. Most manufacturers recommend that this button is pressed quarterly.

## Time delay RCDs

Time delay S-type RCD operating durations can be found in the table below.

There is no 5× test for S-type RCDs.

▼ **Table 2.9** Operational tripping times for various RCDs

| Device type | Non-time delayed maximum operating time at 100% rated tripping current, $I_{\Delta n}$ (ms) | With time delay operating time at 100% rated tripping current, $I_{\Delta n}$ (ms) | Notes |
|---|---|---|---|
| BS 4293 | 200 | {(0.5 to 1.0) × time delay} + 200 | |
| BS 61008 | 300 | 130 to 500 | S type |
| BS 61009 (RCBO) | 300 | 130 to 500 | S type |
| BS 7288 (integral socket-outlet) | 200 | non-applicable | |

**Disconnection times for non-time delayed and time delayed RCDs (from Guidance Note 3: Inspection and Testing, IET)**

## ACTIVITY

Some RCD test instruments have a facility known as a 'ramp test'. Using resources such as the internet, find out what a ramp test is.

# PHASE SEQUENCE TESTING

Phase sequence testing was introduced in BS 7671: 2008 (17th Edition) as Regulation 612.12. It is a requirement to verify that phase sequence is maintained for multi-phase circuits within an installation. As with many other regulations, damage or danger could result if the regulation is not followed.

## Why the correct phase sequence test is important

With three-phase rotating machines, you need to ensure that the direction of supply is correct. Starting and running a motor in the wrong direction could be disastrous. For example, consider what might happen if a unidirectional production process, pump motor or extraction system were to work in the wrong direction; production would go backwards, the pump would flood and the extractor would blow air in. Any of these could be very dangerous.

For the origin of the supply and all distribution boards, the phase rotation sequence must be correct. If these are all common, the fixed wiring within the building can be installed to maintain this phase sequence.

To keep the three-phase rotation common throughout, you must ensure that isolators, machine supplies and three-phase socket outlets are all following the same phase sequence.

### ACTIVITY

Where would be the best place to test the phase sequence for a three-phase machine?

## Equipment that requires phase sequence testing

The parts of an electrical installation that require phase sequence testing are:

- the origin of the supply
- main switches and associated distribution boards
- all sub-mains distribution boards
- three-phase and earth socket outlets (four-pin)
- three-phase, neutral and earth socket outlets (five-pin)
- three-phase isolators
- three-phase motor starters
- three-phase machine supplies and switchgear.

Do not attempt to test any equipment that may cause danger or risk of injury, such as motor terminals, or vibrating or moving machinery.

## Methods used to check the correct phase sequence

Phase sequence testing can be checked by using a rotating disc or indicator lamp type of test meter that physically shows the rotation of the system or using the electronic LCD-type display shown below.

Rotating LCD display examples

Alternatively, the meter used could simply indicate the sequence of the three phases as shown below.

Phase rotation readout examples

**KEY POINT**

If the phase sequence is not correct, equipment can be damaged and dangerous situations can arise.

**ACTIVITY**

Under the supervision of your tutor or assessor, practise testing phase sequence at a distribution board and a socket outlet.

Both types of phase sequence indicator can also be used to verify phase sequence or direction of rotation.

The test meter used should be supplied with test leads and probes in accordance with Guidance Note GS38.

The test meter will also require a plug-in facility for socket outlets. This plug-in facility must have extension adapters for different socket outlet ratings, ie 16 A, 32 A, 63 A etc. Also, four-pin and five-pin socket outlet adaptor variations may be required to test the installation socket outlets.

# THE NEED FOR FUNCTIONAL TESTING

Functional testing must be carried out on equipment to ensure that it is working correctly. All intended functional switching devices must be capable of making and breaking their intended loads in accordance with the design and/or the functional switching procedures. An example of a functional test is RCD functional testing, described on pages 461–464 and carried out by using the integral test button.

## Items that require functional testing

Equipment such as switchgear and controlgear assemblies, drives, controls and interlocks must be subjected to a functional test to show that it is properly mounted, adjusted and installed in accordance with the relevant requirements of BS 7671.

## Switchgear and isolators

Functional testing is required for all equipment that may be relied on for protection against electric shock. For example, it is essential that a firefighters' switch is functioning correctly in an emergency situation.

Isolators must be located correctly in the electrical system and be fully functional. Many isolators are not designed to make and break under load conditions. Where this is the case, such isolators should not be located in positions where they are likely to be used as switches under load conditions.

Firefighters' switch

## Control switches and interlocks

Start and stop switches, interlocks and guard switches are examples of functional switching to ensure that production processes can be stopped immediately should danger arise.

Control circuits should be designed, arranged and protected in such a way as to prevent equipment from operating inadvertently in the event of a fault. It is good practice for control systems to be as simple as possible, while being fail-safe.

Although semiconductors can't be used as a means of isolation, they may be used for functional switching purposes.

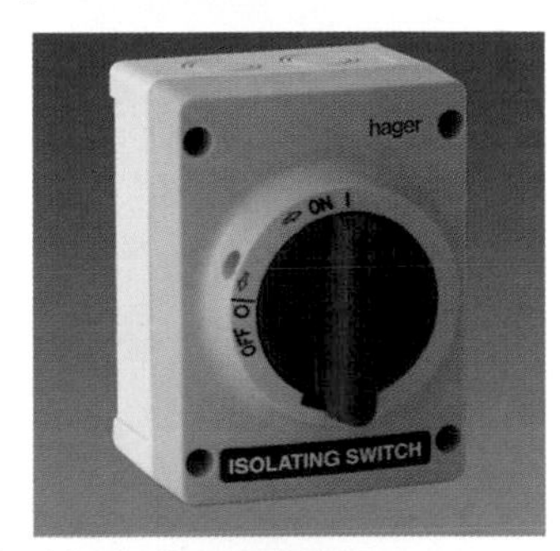


Typical isolating switch

## RCDs and RCBOs

These devices must be functionally tested by carrying out the tests recommended by BS 7671 and associated guidance notes, including by pressing the integral test button marked 'T'.

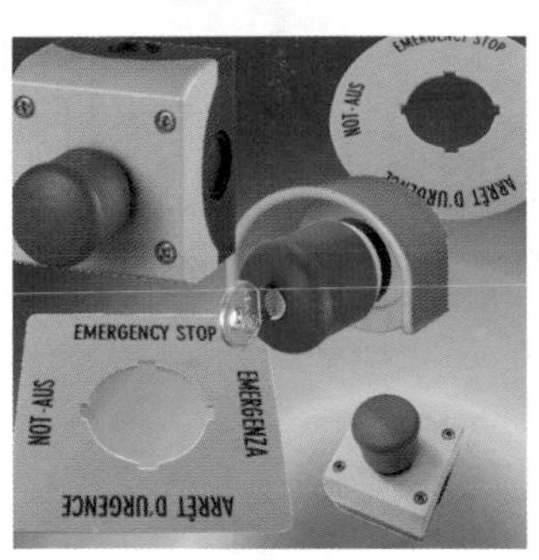


Emergency stop buttons

## Dimmer switches and speed controls

Dimmer switches and speed controls are not just simple on–off devices. These functional switches allow variable control of the luminance or the speed of a process respectively.

Residual current device

## Motor control

Some motors use reverse-current braking and, where such reversal might result in danger, measures should be taken to prevent continued reversal after the driven parts come to a standstill at the end of the braking period. Further, where safety is dependent on the motor operating in the correct direction, means should be provided to *prevent* reverse operation. Where motor control systems incorporate overload devices, these devices should be checked to ensure the settings are correct and re-sets are correctly set to manual or automatic.

## Circuit breakers

All circuit breakers should be checked for correct operation to ensure that they are opening and closing their intended circuits. Circuit protective devices are not intended for frequent load switching.

**KEY POINT**

All isolation and control switching must be checked for correct operation.

**ACTIVITY**

Make a list of all functional tests which need to be carried out, based on the location plan of a small industrial unit in Appendix 1.

# VOLTAGE DROP THEORY AND CALCULATION

Voltage drop is the product of the circuit conductor resistances and the maximum load current that the circuit conductors are expected to carry. The manufacturer of equipment normally states the minimum voltage necessary to ensure that the equipment functions correctly. Where no such limits are stated, values given in BS 7671 are considered suitable. The percentage voltage drop values given in BS7671 are shown in the table.

**TABLE 4Ab – Voltage drop**

| | Lighting | Other uses |
|---|---|---|
| (i) Low voltage installations supplied directly from a public low voltage distribution system | 3% | 5% |

**Maximum voltage drop parameters (from BS 7671:2008 (2011))**

## Voltage drop calculation

The following information is required to establish the voltage drop by calculation.

- cable length ($L$)
- design current ($I_b$)
- value of millivolts per ampere per metre (mV/A/m) as given in Appendix 4 of BS 7671, which equates to the conductor resistance at full operating temperatures.

Using the values above, voltage drop could be determined by:

$$\frac{\text{mV/A/m} \times L \times I_b}{1000} = \text{V}$$

To use an example, assume a power circuit has the following values:

- cable length = 22 m
- design current = 13 A
- mV/A/m = 18 (taken from Table 4D1B below using column 3 for a 2.5mm$^2$ cable, reference method B)

The table on page 469 (top) is an extract from Table 4D1B, BS 7671. It shows the millivolt drop per ampere per metre (mV/A/m) for given cable types and insulations. When using the tables in BS 7671, the correct table should be used, depending on the cable type and material used for insulation of the conductors.

### ACTIVITY

Cable calculations can be very time-consuming for large installations. Research the internet to find suitable IT software for the task.

VOLTAGE DROP (per ampere per metre):

| Conductor cross-sectional area | 2 cables, d.c. | 2 cables, single-phase a.c. | | |
|---|---|---|---|---|
| | | Reference Methods A & B (enclosed in conduit or trunking) | Reference Methods C & F (clipped direct, on tray or in free air) | |
| | | | Cables touching | Cables spaced* |
| 1 | 2 | 3 | 4 | 5 |
| (mm$^2$) | (mV/A/m) | (mV/A/m) | (mV/A/m) | (mV/A/m) |
| 1 | 44 | 44 | 44 | 44 |
| 1.5 | 29 | 29 | 29 | 29 |
| 2.5 | 18 | 18 | 18 | 18 |
| 4 | 11 | 11 | 11 | 11 |
| 6 | 7.3 | 7.3 | 7.3 | 7.3 |
| 10 | 4.4 | 4.4 | 4.4 | 4.4 |
| 16 | 2.8 | 2.8 | 2.8 | 2.8 |

**Voltage drop values extracted from Table 4D1B of BS 7671:2008 (2011)**

So the actual voltage drop for the example would be:

$$\frac{\text{mV/A/m} \times L \times I_b}{1000} = \text{V}$$

$$\frac{18 \times 22 \times 13}{1000} = 5.15\ \text{V}$$

Maximum volt drop allowed for a power circuit = 5% of 230 V = 11.5 V (taken from Table 4Ab below).

As the value of 5.15 V is well below 11.5 V, this is satisfactory.

**TABLE 4Ab – Voltage drop**

| | Lighting | Other uses |
|---|---|---|
| (i) Low voltage installations supplied directly from a public low voltage distribution system | 3% | 5% |

**Maximum voltage drop parameters (from BS 7671:2008 (2011))**

## ACTIVITY

What would be the maximum allowable voltage drop for a 400 V three-phase circuit?

### KEY POINT

Verification of voltage drop is not normally required during initial verification. This should be calculated during the design process, but initial verification does suggest that it should be considered.

## Voltage drop verification and testing

Voltage drop problems are quite rare but the inspector should be aware that long runs and/or high currents can sometimes cause voltage drop problems.

There are two recognised methods of testing voltage drop:

- measuring with voltage and current test equipment
- measuring resistance.

## ASSESSMENT GUIDANCE

Low voltage problems are rare these days. Low voltage might, however, be noticed in remote rural areas at the end of a long line, when a heavy load is applied. Remember that 230 V is only the nominal voltage.

## Method 1: Measuring with voltage and current test equipment

This method is *not* recommended due to the need to work on or near live parts and having exposed live terminals.

Measurement of voltage drop within an installation is not practical as this would mean simultaneously measuring the instantaneous voltage at both the origin and at the end point of the circuit. This would have to be done with the circuit operating under fully loaded conditions.

## Method 2: Measuring resistance

Practical testing is more complicated and less accurate than calculating voltage drop by analysing the design criteria and using the calculation format explained on pages 468–469.

If you do decide to test the resistance of the cable, *isolation procedures* must be followed and a risk assessment should be carried out.

## How to test a (single-phase) radial circuit

To test a (single-phase) radial circuit you will need to measure the resistance of the line and neutral conductors (in ohms). This can be done using the same procedure as is used to measure $R_1+R_2$ but, instead of measuring the cpc resistance ($R_2$), you measure the neutral resistance ($R_n$).

Once a value of $R_1+R_n$ has been obtained by measurement, it must be adjusted for temperature. If the cable has a maximum operating temperature of 70 °C, the resistance value should be adjusted from that at the temperature at the time of testing, to that at 70 °C.

As resistance changes by 2% per 5 °C change in temperature, you can determine the resistance increase using the following table.

| Ambient temperature at time of test (°C) | Temperature increase to maximum operating temperature (°C) | % increase | Factor to be applied |
|---|---|---|---|
| 10 | 60 | 24 | 1.24 |
| 15 | 55 | 22 | 1.22 |
| 20 | 50 | 20 | 1.20 |
| 25 | 45 | 18 | 1.18 |
| 30 | 40 | 16 | 1.16 |
| 35 | 35 | 14 | 1.14 |
| 40 | 30 | 12 | 1.12 |

**Multipliers for adjusting resistance of conductors such as thermoplastic 70 °C cable**

For example, the $R_1+R_n$ for a radial power circuit was 0.22 Ω at an ambient temperature of 15 °C. If the cable has an operating temperature of 70 °C, we can determine the resistance under operating conditions:

$$0.22 \times 1.22 = 0.27 \text{ at } 70\,°C$$

As voltage drop is determined using Ohm's law, we need to consider the full load current. For this example, we will assume full load current is 24 A.

So as I × R = V

24 A × 0.27 Ω = 6.48 V

**TABLE 4Ab – Voltage drop**

| | Lighting | Other uses |
|---|---|---|
| (i) Low voltage installations supplied directly from a public low voltage distribution system | 3% | 5% |

**Maximum voltage drop parameters (from BS 7671:2008 (2011))**

The calculated voltage drop is 6.48 V. As the maximum permitted voltage drop for a radial power circuit is 5% of 230 V (= 11.5 V), this value is acceptable.

**KEY POINT**

Verification of voltage drop is not normally required during initial verification. This should be calculated during the design process, but initial verification does suggest that it should be considered. If an installation has been designed correctly, and the inspector has this information available, testing for voltage drop would not be required.

**ACTIVITY**

Determine the maximum permitted voltage drop for a single-phase lighting circuit.

# APPROPRIATE PROCEDURES FOR DEALING WITH CUSTOMERS AND CLIENTS

As soon as initial verification and certification is complete, the client will want to start using the electrical equipment. It is essential that the information given to the client is correctly explained. The relevant information and requirements are outlined below and in the following pages.

**Assessment criteria**

6.7 Complete certification documentation

## Commissioning and certification

### Client awareness during handover

All installations differ in complexity, from domestic to large industrial. In every case, as part of the handover, the client or customer must be made aware of the safe and correct use of the electrical installation and all of its controls. This can be achieved by giving a tour of the installation to explain any complexities.

Any user manuals and information booklets must be given to the client. In the case of complex installations or commercial and industrial installation, this would normally form the basis of an operations and maintenance manual (O&M Manual).

**ASSESSMENT GUIDANCE**

It is important to walk round the installation with the customer and talk about the function and operation of any controls, and so on. During practical assessment you should be able to tell your assessor how components work.

### Certification presented during handover

After initial verification and commissioning is complete, the certification paperwork must be handed over to the client or the person ordering the work.

These documents are:

- Electrical Installation Certificate
- Schedule of Inspections
- Generic Schedule of Test Results.

All three documents must be complete and satisfactory before the client uses the electrical system.

Where minor additions or alterations to a single existing circuit have taken place, this would be recorded on a Minor Electrical Installation Works Certificate.

In every case, the certification cannot be completed and handed over until all work is satisfactory and complies with BS 7671.

## Identification of circuits and switches

Distribution board and consumer unit protective devices must be identified correctly and information must be available for the client in accordance with Regulation 514.9.1 of BS 7671.

A durable copy of the circuit details relating to the distribution board (DB) or consumer control unit (CCU) should be placed within or adjacent to each DB or CCU. Diagrams, plans and labelling of isolators and switchgear must be accurate.

## Why the client requires certification

There are many reasons for certification. It is required:

- in accordance with British Standards (BS 7671)
- to meet health and safety regulations in the workplace (Electricity at Work Regulations 1989)
- for domestic premises as part of Building Regulations
- for building insurance
- for proof of compliance in the event of any employee and personal insurance claims
- before letting a property (Landlord and Tenant Act 1985)
- for local authority licensing requirements, such as in the case of obtaining a public entertainments license or fire certification.

## Why certification must be maintained

After electrical installation certification is handed over, the documents must be regularly updated to record alterations and associated additional certification, and as part of the maintenance of the system.

Future periodic testing can be carried out and historical test results can be compared to ensure that there are no major changes or patterns of deterioration within the electrical system.

If the client is selling the property, the purchaser will require the certification, as well as any subsequent reports.

**KEY POINT**

The original proof of the documentation must be given to the person ordering the work. A copy of all documentation must be retained by the electrical inspector.

**ACTIVITY**

It is a worthwhile exercise to for you to put yourself in the place of a client. Consider this scenario: you are the owner of a new hotel and handover of your newly installed electrical installation is due to take place. List the documents relating to the electrical installation that should be given to you during the handover process.

# OUTCOMES 4, 6

## Understand requirements for documenting installing electrical systems

## Be able to test safety of electrical systems

## DOCUMENTATION AND CERTIFICATION FOR INITIAL VERIFICATION

This section provides guidance on completing the necessary forms and certificates associated with initial verification.

**Assessment criteria**

4.1 Explain the purpose of certification documentation

### New installations

Following an initial verification, or an addition or alteration to an existing installation, an Electrical Installation Certificate must be completed and issued, together with Inspection Schedules and Test Result Schedules.

There are two options for the Electrical Installation Certificate, as shown on Form 1 (page 474) and Form 2 (page 475).

### Form 1: Short Form of Electrical Installation Certificate

This form is used when *one* person is responsible for the design, construction, inspection and testing of an installation. This is normally used for small installations, such as domestic dwellings, where one person carries out all of the roles. Even if the person has only selected cable cross-sectional areas using tables in the IET On-Site Guide, they must take responsibility for design.

### Form 2: Electrical Installation Certificate

This form is used when the design, construction and the inspection are carried out by different groups of people (it is a three-signatory version of the form). This form is normally used for larger projects where the installation has been designed by a consultant or a member of the architects' team, has been installed by many people working for a contractor and has been inspected by a specialist inspector – who is either part of the installation organisation, or is nominated by the client.

**ASSESSMENT GUIDANCE**

Remember to complete all parts of the certificates and forms as appropriate. If any parts are left incomplete, this may mean the forms are not valid.

### Forms 3 and 4: Schedule of Inspections and Generic Schedule of Test Results

Whichever of the two Electrical Installation Certificates is used, other forms are required to accompany the Certificate: Form 3 (Schedule of Inspections) Form 4 (Schedule of Test Results).

Completed samples of Forms 3 and 4 from Guidance Note 3 are shown on pages 476 and 477.

Form 1 — Form No: 505513.../1

ELECTRICAL INSTALLATION CERTIFICATE
(REQUIREMENTS FOR ELECTRICAL INSTALLATIONS - BS 7671 [IET WIRING REGULATIONS])

**DETAILS OF THE CLIENT** Mr T Brown
32 South St
Anytown, Surrey — Post Code: TO1 1ZZ

**INSTALLATION ADDRESS**
The Coffee Bean
31 Station Road
Anytown, Surrey — Post Code: TO3 2YF

**DESCRIPTION AND EXTENT OF THE INSTALLATION** Tick boxes as appropriate

Description of installation:
Re-wire of ground floor, on change of use.

New installation ☑
Addition to an existing installation ☐
Alteration to an existing installation ☐

Extent of installation covered by this Certificate:
Complete electrical re-wire of refurbished premises, on change of use from offices to cafe/snack bar.

(Use continuation sheet if necessary) see continuation sheet No: ........

**FOR DESIGN, CONSTRUCTION, INSPECTION & TESTING**
I being the person responsible for the design, construction, inspection & testing of the electrical installation (as indicated by my signature below), particulars of which are described above, having exercised reasonable skill and care when carrying out the design, construction, inspection & testing hereby CERTIFY that the said work for which I have been responsible is to the best of my knowledge and belief in accordance with BS 7671:2008, amended to 2011 (date) except for the departures, if any, detailed as follows:

Details of departures from BS 7671 (Regulations 120.3 and 133.5):
None

The extent of liability of the signatory is limited to the work described above as the subject of this Certificate.

Signature: W Hastings Date: 21-Jan-2011 Name (IN BLOCK LETTERS): W HASTINGS

Company Address: Hastings Electrical
21 The Arches
Anytown, Surrey — Postcode: TO2 9YY Tel No: 01022 999999

**NEXT INSPECTION**
I recommend that this installation is further inspected and tested after an interval of not more than 5 years/~~months~~.

**SUPPLY CHARACTERISTICS AND EARTHING ARRANGEMENTS** Tick boxes and enter details, as appropriate

| Earthing arrangements | Number and Type of Live Conductors | Nature of Supply Parameters | Supply Protective Device Characteristics |
|---|---|---|---|
| TN-C ☐<br>TN-S ☐<br>TN-C-S ☑<br>TT ☐<br>IT ☐ | a.c. ☑ d.c. ☐<br>1-phase, 2-wire ☑ 2-wire ☐<br>1-phase, 3-wire ☐ 3-wire ☐<br>2-phase, 3-wire ☐ other ☐<br>3-phase, 3-wire ☐<br>3-phase, 4-wire ☐ | Nominal voltage, $U/U_0$ (1) 230 V<br>Nominal frequency, f (1) 50 Hz<br>Prospective fault current, $I_{pf}$ (2) 9.0 kA<br>External loop impedance, $Z_e$ (2) 0.28 Ω | Type: BS 1361 Fuse<br>Rated current 100 A |
| Other sources of supply (to be detailed on attached schedules) ☐ | Confirmation of supply polarity ☑ | (Note: (1) by enquiry, (2) by enquiry or by measurement) | |

Form 1 — Form No: 505513.../1

**PARTICULARS OF INSTALLATION REFERRED TO IN THE CERTIFICATE** Tick boxes and enter details, as appropriate

**Means of Earthing**
Distributor's facility ☑
Installation earth electrode ☐

**Maximum Demand**
Maximum demand (load) 80 ~~kVA~~/ Amps (Delete as appropriate)

**Details of Installation Earth Electrode** (*where applicable*)
Type (e.g. rod(s), tape etc): N/A — Location: N/A — Electrode resistance to Earth: N/A Ω

**Main Protective Conductors**
Earthing conductor: material Copper csa 16 mm² Continuity and connection verified ☑
Main protective bonding conductors: material Copper csa 10 mm² Continuity and connection verified ☑
To incoming water and/or gas service ☑ To other elements: N/A

**Main Switch or Circuit-breaker**
BS, Type and No. of poles BS EN 60947-3 (2- pole) Current rating 100 A Voltage rating 230 V
Location Services cupboard adjacent rear exit Fuse rating or setting N/A A
Rated residual operating current $I_{\Delta n}$ = N/A mA, and operating time of N/A ms (at $I_{\Delta n}$) (applicable only where an RCD is suitable and is used as a main circuit-breaker)

**COMMENTS ON EXISTING INSTALLATION** (in the case of an addition or alteration see Section 633):
Not Applicable

**SCHEDULES**
The attached Schedules are part of this document and this Certificate is valid only when they are attached to it.
1 Schedules of Inspections and 1 Schedules of Test Results are attached.
(Enter quantities of schedules attached).

Page 2 of 4.

Form 1: Short form of Electrical Installation Certificate (always check you are using the latest forms, as found on the IET website: http://electrical.theiet.org)

Form 2 Form No: SSSS13....../2

**ELECTRICAL INSTALLATION CERTIFICATE**
(REQUIREMENTS FOR ELECTRICAL INSTALLATIONS - BS 7671 [IET WIRING REGULATIONS])

**DETAILS OF THE CLIENT** Mr D Roberts
23 Acacia Avenue Post Code: SL0 0LT
Sometown, Berks

**INSTALLATION ADDRESS** Unit 3 The Quadrant
Sometown Business Park
Sometown, Berks Post Code: SL1 0ZZ

**DESCRIPTION AND EXTENT OF THE INSTALLATION** Tick boxes as appropriate

Description of installation: Commercial office

Extent of installation covered by this Certificate: Full new installation

(Use continuation sheet if necessary) see continuation sheet No: ........

| | |
|---|---|
| New installation | ☑ |
| Addition to an existing installation | ☐ |
| Alteration to an existing installation | ☐ |

**FOR DESIGN**
I/We being the person(s) responsible for the design of the electrical installation (as indicated by my/our signatures below), particulars of which are described above, having exercised reasonable skill and care when carrying out the design hereby CERTIFY that the design work for which I/we have been responsible is to the best of my/our knowledge and belief in accordance with BS 7671:2008, amended to 2011 (date) except for the departures, if any, detailed as follows:
Details of departures from BS 7671 (Regulations 120.3 and 133.5):
None N/A

The extent of liability of the signatory or the signatories is limited to the work described above as the subject of this Certificate.

For the DESIGN of the installation: **(Where there is mutual responsibility for the design)

Signature: D Jones Date: 15/08/2013 Name (IN BLOCK LETTERS): D JONES Designer No 1

Signature: N/A Date: ........ Name (IN BLOCK LETTERS): N/A Designer No 2**

**FOR CONSTRUCTION**
I/We being the person(s) responsible for the construction of the electrical installation (as indicated by my/our signatures below), particulars of which are described above, having exercised reasonable skill and care when carrying out the construction hereby CERTIFY that the construction work for which I/we have been responsible is to the best of my/our knowledge and belief in accordance with BS 7671:2008, amended to 2011 (date) except for the departures, if any, detailed as follows:
Details of departures from BS 7671 (Regulations 120.3 and 133.5):
None N/A

The extent of liability of the signatory is limited to the work described above as the subject of this Certificate.

For CONSTRUCTION of the installation:

Signature: T Smith Date: 15/08/2013 Name (IN BLOCK LETTERS): T SMITH

**FOR INSPECTION & TESTING**
I/We being the person(s) responsible for the inspection & testing of the electrical installation (as indicated by my/our signatures below), particulars of which are described above, having exercised reasonable skill and care when carrying out the inspection & testing hereby CERTIFY that the work for which I/we have been responsible is to the best of my/our knowledge and belief in accordance with BS 7671:2008, amended to 2011 (date) except for the departures, if any, detailed as follows:
Details of departures from BS 7671 (Regulations 120.3 and 133.5):
None N/A

The extent of liability of the signatory is limited to the work described above as the subject of this Certificate.

For INSPECTION AND TESTING of the installation:

Signature: G Wilson Date: 15/08/2013 Name (IN BLOCK LETTERS): G WILSON

**NEXT INSPECTION**
I/We the designer(s), recommend that this installation is further inspected and tested after an interval of not more than 5 years/~~months~~.

Page 1 of 4

Form 2 Form No: SSSS13........../2

**PARTICULARS OF SIGNATORIES TO THE ELECTRICAL INSTALLATION CERTIFICATE**

**Designer (No 1)**
Name: D Jones Company: The Electrical Design Partnership
Address: 23 High Street
Sometown, Berks Postcode: SL10 0YY Tel No: 01000 999999

**Designer (No 2)** (if applicable)
Name: ........ Company: ........
Address: ........
Postcode: ........ Tel No: ........

**Constructor**
Name: T Smith Company: T Smith Electrical Installations
Address: Unit 8a Sometown Ind Estate
Sometown, Berks Postcode: SL3 0XX Tel No: 01000 888888

**Inspector**
Name: G Wilson Company: Wilson and Sons
Address: 11 Crabtree Row
Sometown, Berks Postcode: SL2 0WW Tel No: 01000 777777

**SUPPLY CHARACTERISTICS AND EARTHING ARRANGEMENTS** Tick boxes and enter details, as appropriate

**Earthing arrangements**
TN-C ☐
TN-S ☐
TN-C-S ☑
TT ☐
IT ☐
Other sources of supply (to be detailed on attached schedules) ☐

**Number and Type of Live Conductors**
a.c. ☑ d.c. ☐
1-phase, 2-wire ☑ 2-wire ☐
1-phase, 3-wire ☐ 3-wire ☐
2-phase, 3-wire ☐ other ☐
3-phase, 3-wire ☐
3-phase, 4-wire ☐
Confirmation of supply polarity ☑

**Nature of Supply Parameters**
Nominal voltage, ~~U~~/$U_0$[1] 230 V
Nominal frequency, f[1] 50 Hz
Prospective fault current, $I_{pf}$[2] 1.41 kA
External loop impedance, $Z_e$[2] 0.34Ω
(Note: (1) by enquiry, (2) by enquiry or by measurement)

**Supply Protective Device Characteristics**
Type: BS 88-3
Rated current 80 A

**PARTICULARS OF INSTALLATION REFERRED TO IN THE CERTIFICATE** Tick boxes and enter details, as appropriate

**Means of Earthing**
Distributor's facility ☑
Installation earth electrode ☐

**Maximum Demand**
Maximum demand (load) 68 ~~kVA~~ / Amps Delete as appropriate

**Details of Installation Earth Electrode** (*where applicable*)
Type (e.g. rod(s), tape etc): N/A Location: N/A Electrode resistance to Earth: N/A Ω

**Main Protective Conductors**
Earthing conductor: material Copper csa 16 mm² Continuity and connection verified ☑
Main protective bonding conductors: material Copper csa 16 mm² Continuity and connection verified ☑
To incoming water and/or gas service ☑ To other elements: N/A

**Main Switch or Circuit-breaker**
BS, Type and No. of poles BS EN 60497-3 (2-pole) Current rating 100 A Voltage rating 400 V
Location Office Suite Consumer Unit Fuse rating or setting N/A A
Rated residual operating current $I_{\Delta n}$ = N/A mA, and operating time of N/A ms (at $I_{\Delta n}$) (applicable only where an RCD is suitable and is used as a main circuit breaker)

**COMMENTS ON EXISTING INSTALLATION** (in the case of an addition or alteration see Section 633):
N/A

**SCHEDULES**
The attached Schedules are part of this document and this Certificate is valid only when they are attached to it.
1 Schedules of Inspections and 1 Schedules of Test Results are attached.
(Enter quantities of schedules attached)

Page 2 of 4

Form 2: Electrical Installation Certificate (three-signatory version) (always check you are using the latest forms, as found on the IET website: http://electrical.theiet.org)

Form 3 Form No: ...SSSS13...../3

SCHEDULE OF INSPECTIONS (for new installation work only)

**Methods of protection against electric shock**

**Both basic and fault protection:**

- ✓ (i) SELV (note 1)
- N/A (ii) PELV
- N/A (iii) Double insulation
- N/A (iv) Reinforced insulation

**Basic protection:** (note 2)

- ✓ (i) Insulation of live parts
- ✓ (ii) Barriers or enclosures
- N/A (iii) Obstacles (note 3)
- N/A (iv) Placing out of reach (note 4)

**Fault protection:**

**(i) Automatic disconnection of supply:**

- ✓ Presence of earthing conductor
- ✓ Presence of circuit protective conductors
- ✓ Presence of protective bonding conductors
- ✓ Presence of supplementary bonding conductors
- N/A Presence of earthing arrangements for combined protective and functional purposes
- N/A Presence of adequate arrangements for other sources, where applicable
- N/A FELV
- ✓ Choice and setting of protective and monitoring devices (for fault and/or overcurrent protection)

**(ii) Non-conducting location:** (note 5)

- N/A Absence of protective conductors

**(iii) Earth-free local equipotential bonding:** (note 6)

- N/A Presence of earth-free local equipotential bonding

**(iv) Electrical separation:** (note 7)

- N/A Provided for **one item** of current-using equipment
- N/A Provided for **more than one item** of current-using equipment

**Additional protection:**

- ✓ Presence of residual current devices(s)
- ✓ Presence of supplementary bonding conductors

**Prevention of mutual detrimental influence**

- ✓ (a) Proximity to non-electrical services and other influences
- ✓ (b) Segregation of Band I and Band II circuits or use of Band II insulation
- N/A (c) Segregation of safety circuits

**Identification**

- ✓ (a) Presence of diagrams, instructions, circuit charts and similar information
- ✓ (b) Presence of danger notices and other warning notices
- ✓ (c) Labelling of protective devices, switches and terminals
- ✓ (d) Identification of conductors

**Cables and conductors**

- ✓ Selection of conductors for current-carrying capacity and voltage drop
- ✓ Erection methods
- ✓ Routing of cables in prescribed zones
- N/A Cables incorporating earthed armour or sheath, or run within an earthed wiring system, or otherwise adequately protected against nails, screws and the like
- N/A Additional protection provided by 30 mA RCD for cables concealed in walls (where required in premises not under the supervision of a skilled or instructed person)
- ✓ Connection of conductors
- ✓ Presence of fire barriers, suitable seals and protection against thermal effects

**General**

- ✓ Presence and correct location of appropriate devices for isolation and switching
- ✓ Adequacy of access to switchgear and other equipment
- N/A Particular protective measures for special installations and locations
- ✓ Connection of single-pole devices for protection or switching in line conductors only
- ✓ Correct connection of accessories and equipment
- N/A Presence of undervoltage protective devices
- ✓ Selection of equipment and protective measures appropriate to external influences
- ✓ Selection of appropriate functional switching devices

Inspected by ............G.Wilson.......................................... Date .....15/08/2013...............................................

**NOTES:**

✓ to indicate an inspection has been carried out and the result is satisfactory

**N/A** to indicate that the inspection is not applicable to a particular item

An entry must be made in every box.

1. SELV An extra-low voltage system which is electrically separated from Earth and from other systems. The particular requirements of the Regulations must be checked (see Section 414)
2. Method of basic protection - will include measurement of distances where appropriate
3. Obstacles - only adopted in special circumstances (see Regulations 416.2 and 417.2)
4. Placing out of reach - only adopted in special circumstances (see Regulation 417.3)
5. Non-conducting locations - not applicable in domestic premises and requiring special precautions (see Regulation 418.1)
6. Earth-free local equipotential bonding - not applicable in domestic premises, only used in special circumstances (see Regulation 418.2)
7. Electrical separation (see Section 413 and Regulation 418.3)

Page 3... of 4...

Form 3: Schedule of Inspections (always check you are using the latest forms, as found on the IET website: http://electrical.theiet.org)

Form 4

Form No: ......1235........./4

## GENERIC SCHEDULE OF TEST RESULTS

DB reference no ......Commercial office......
Location ..Unit 3, The Quadrant, SL1 0ZZ......
Zs at DB (Ω) ......0.34......
$I_{pf}$ at DB (kA) ......1.41......
Correct supply polarity confirmed ☑
Phase sequence confirmed (where appropriate) N/A

Details of circuits and/or installed equipment vulnerable to damage when testing ......Downlighter spots – electronic SELV transformers......

Details of test instruments used (state serial and/or asset numbers)
Continuity......Megin multi-function 10563......
Insulation resistance ...... - " - ......
Earth fault loop impedance ...... - " - ......
RCD ...... - " - ......
Earth electrode resistance ......N/A......

Tested by:
Name (Capitals) ......G WILSON......
Signature ......G. Wilson...... Date ......15/08/2013......

| Circuit details | | | | | | | | | Test results | | | | | | | | | | | | |
|---|---|---|---|---|---|---|---|---|---|---|---|---|---|---|---|---|---|---|---|---|---|
| | | Overcurrent device | | | | Conductor details | | | Ring final circuit continuity (Ω) | | | Continuity (Ω) ($R_1 + R_2$) or $R_2$ | | Insulation Resistance (MΩ) | | Polarity | $Z_s$ (Ω) | RCD (ms) | | | Remarks (continue on a separate sheet if necessary) |
| Circuit number | Circuit Description | BS (EN) | type | rating (A) | breaking capacity (kA) | Reference Method | Live (mm²) | cpc (mm²) | $r_1$ (line) | $r_n$ (neutral) | $r_2$ (cpc) | ($R_1 + R_2$) * | $R_2$ | Live-Live | Live-E | | | @ $I_{\Delta n}$ | @ 5 $I_{\Delta n}$ | Test button / functionality | |
| 1 | 2 | 3 | 4 | 5 | 6 | 7 | 8 | 9 | 10 | 11 | 12 | 13 | 14 | 15 | 16 | 17 | 18 | 19 | 20 | 21 | 22 |
| 1 | Socket outlets - dado | 60898 | B | 20 | 6 | B | 2.5 | 1.5 | N/A | N/A | N/A | 0.38 | N/A | >200 | >200 | ✓ | 0.75 | N/A | N/A | N/A | ✓ Checked for compliance |
| 2 | Socket outlets - wall | 61009 | B | 32 | 6 | B | 2x2.5 | 2x1.5 | 0.62 | 0.62 | 1.02 | 0.41 | N/A | >200 | >200 | ✓ | 0.71 | 85 | 16 | ✓ | ✓ - " - |
| 3 | Down lighter spots | 60898 | B | 6 | 6 | B | 1.5 | 1.0 | N/A | N/A | N/A | 0.56 | N/A | >200 | >200 | ✓ | 0.83 | N/A | N/A | N/A | ✓ - " - |
| 4 | General lighting | 60898 | B | 10 | 6 | B | 1.5 | 1.0 | N/A | N/A | N/A | 0.48 | N/A | >200 | >200 | ✓ | 0.81 | N/A | N/A | N/A | ✓ - " - |
| 5 | Water heater | 60898 | B | 16 | 6 | B | 2.5 | 1.5 | N/A | N/A | N/A | 0.11 | N/A | >200 | >200 | ✓ | 0.45 | N/A | N/A | N/A | ✓ - " - |
| | | | | | | | | | | | | | | | | | | | | | |
| | | | | | | | | | | | | | | | | | | | | | |
| | | | | | | | | | | | | | | | | | | | | | |
| | | | | | | | | | | | | | | | | | | | | | |
| | | | | | | | | | | | | | | | | | | | | | |
| | | | | | | | | | | | | | | | | | | | | | |
| | | | | | | | | | | | | | | | | | | | | | |

* Where there are no spurs connected to a ring final circuit this value is also the ($R_1 + R_2$) of the circuit.

Page 4 ... of ...4

Form 4: Generic Schedule of Test Results (always check you are using the latest forms, as found on the IET website: http://electrical.theiet.org)

## New circuits

For installation of one or more new circuits, certification and schedules are required:

- Form 1 or 2: Electrical Installation Certificate
- Form 3: Schedule of Inspections
- Form 4: Generic Schedule of Test Results.

## Additions and alterations to existing circuits

When completing the Electrical Installation Certificate for additions or alterations to an existing electrical installation, you may need to comment on the existing installation. If there are apparent problems or deficiencies, you must inform the person ordering the work about any possible risks if remedial work is not carried out.

If the additions or alterations do not include the provision of a new circuit, the Form 5 certificate needs to be issued.

### Form 5: Minor Electrical Installation Works Certificate

A sample of Form 5, taken from Guidance Note 3, is shown opposite. The Minor Electrical Installation Works Certificate can only be used if the minor electrical work does not include the provision of a new circuit. When replacing equipment and accessories, or adding to an existing final circuit, this form can be used.

**ASSESSMENT GUIDANCE**

A new socket added to an existing ring final circuit would need a Minor Electrical Installation Works Certificate. The same socket wired back to the consumer unit or distribution board would need an Electrical Installation Certificate plus Schedule of Inspections and Generic Schedule of Test Results.

**KEY POINT**

Remember that, if a new circuit is installed, you must use the Electrical Installation Certificate format.

**ACTIVITY**

What form is most likely to be used after the replacement of a faulty water heater?

Form 5 Form No: ....1234.../5

MINOR ELECTRICAL INSTALLATION WORKS CERTIFICATE
(REQU REMENTS FOR ELECTRICAL INSTALLATIONS - BS 7671 [IET W RING REGULATIONS])
**To be used only for minor electrical work which does not include the provision of a new circuit**

**PART 1 Description of minor works**

1. Description of the minor works **2 new lighting points to home office/bedroom 3 of dwelling.**
2. Location/Address **41 Larkspur Drive, Newtown. E. Sussex** Post Code **EA1 2BB**
3. Date minor works completed **27 Jan 2012**
4. Details of departures, if any, from BS 7671:2008, amended to ...2011..... (date)
**None. Electricity supplier's terminal equipment in need of attention. Cut-out fuse carrier cracked. Customer advised to contact supplier.**

**PART 2 Installation details**

1. System earthing arrangement TN-C-S ☑ TN-S ☐ TT ☐
2. Method of fault protection **ADS**
3. Protective device for the modified circuit Type **BS EN 61009 Type B** Rating ..........6. A

Comments on existing installation, including adequacy of earthing and bonding arrangements (see Regulation 132.16):
**Existing circuit not provided with additional protection by RCD. Lighting circuit MCB converted to RCBO in order adequately to provide protection against damage to cables in walls; Reg. 522.6.101 refers.**

**PART 3 Essential Tests**

Earth continuity satisfactory ☑

Insulation resistance:

Line/neutral ..............................+299.. MΩ

Line/earth ..............................+299.. MΩ

Neutral/earth..............................+299.. MΩ

Earth fault loop impedance ..................................1.2.. Ω

Polarity satisfactory ☑

RCD operation (if applicable). Rated residual operating current $I_{\Delta n}$ ...30. mA and operating time of .... 28.ms (at $I_{\Delta n}$)

**PART 4 Declaration**

I/We CERTIFY that the said works do not impair the safety of the existing installation, that the said works have been designed, constructed, inspected and tested in accordance with BS 7671:2008 ( ET Wiring Regulations), amended to **2011**........ (date) and that the said works, to the best of my/our knowledge and belief, at the time of my/our inspection, complied with BS 7671 except as detailed in Part 1 above.

Name: **G Thompson**

For and on behalf of: **T and G Electrical**

Address: **25 Whiteleaf Close**
**Newtown**
**E Sussex**
Post code **EA4 5XX**

Signature: *G Thompson*

Position: **Proprietor**

Date: **27-Jan-2012**

Page 1 of 1

Form 5: Minor Electrical Installation Works Certificate (always check you are using the latest forms, as found on the IET website: http://electrical.theiet.org)

**Assessment criteria**

**4.2** Specify the responsibilities of personnel in relation to the completion of the certification documentation

## INFORMATION CONTAINED WITHIN DOCUMENTATION AND CERTIFICATION

This section deals with the information that must be contained within the documentation and certification for initial verification.

### Information contained within an Electrical Installation Certificate (Forms 1 and 2)

An Electrical Installation Certificate contains details of the person(s) responsible for the design, construction, inspection and testing of the installation.

It also identifies the:

- details of the installation covered by the certificate
- nature of the works; e.g. new, addition, alteration
- extent of the work covered by the certificate
- supply characteristics
- nature of supply parameters
- earthing arrangements
- maximum demand
- protective conductors and their cross-sectional area
- main switch type and standard
- comments on the condition of any existing electrical systems which are not covered by the certificate but are part of the general electrical installation.

**ACTIVITY**

There is a danger that a like-for-like change (shower unit for shower unit) does not take into account a possible increase in load current. This may well require a new circuit, with consequent design considerations. What certification would be needed in such a case?

### Information contained within a Minor Electrical Installation Works Certificate (Form 5)

The minor works which have been undertaken must be clearly described on this certificate. The purpose of the certificate is to verify that the new works comply with BS 7671 and that the existing parts of the circuit and affected parts of the installation are suitable for the additional works and remain compliant.

The relevant provisions of Part 6 'Inspection and testing' of BS 7671 must be applied in full to all minor works. For example, where a socket outlet is added to an existing circuit it is necessary to:

- establish that the earthing contact of the socket outlet is connected to the main earthing terminal
- measure the insulation resistance of the circuit that has been added to, and establish that it complies with Table 61 of BS 7671

**ASSESSMENT GUIDANCE**

Don't forget to date the forms. They are not useful unless they are dated.

- measure the earth fault loop impedance to establish that the maximum permitted disconnection time is not exceeded
- check that the polarity of the socket outlet is correct
- verify the effectiveness of the RCD, if the new work is protected by an RCD.

## Information contained within a Schedule of Inspections (Form 3)

A Schedule of Inspections contains boxes which need to be filled in with a tick (for compliance) or with 'n/a' for items which are not applicable. The inspection methods and processes have been addressed on pages 386–400.

This form must accompany the Electrical Installation Certificate.

The table below shows each tick box within the Schedule of Inspections, together with the questions that the inspector must ask while inspecting the installation. The table also provides some additional information to help the inspector with the inspection. Where aspects of the inspection need more detailed information the relevant BS 7671 Regulation number or section is given.

### Schedule of Inspections (Form 3) – information and guidance

| **Methods of protection against electric shock** | |
|---|---|
| **Both basic and fault protection** | |
| SELV | Does the installation contain SELV equipment such as extra-low voltage lighting? Is the transformer compliant with Regulation 414.3?<br>SELV circuits must be separated from the source and earth. The equipment will not have a connection to earth. |
| PELV | Does the installation contain any PELV equipment?<br>This is where an item of equipment is supplied using extra-low voltage but may be earthed using the supply circuit cpc. An example may be a server system for data where the cables and equipment casings are earthed but the circuit is separated. |
| Double Insulation | Does the installation or part of an installation rely on double insulation as a method of shock protection, instead of ADS?<br>It is tempting to tick this box where the supply tails are double insulated, but this case is not relevant to this box as the tails only have double insulation for mechanical protection. See the requirements of Section 413.<br>It is rare for this box to be applicable. |

<table>
<tr><td>Reinforced insulation</td><td>Does the installation or part of the installation rely on reinforced insulation as a method of shock protection, instead of ADS?<br><br>See the requirements of Section 413.<br><br>It is rare for this box to be applicable.</td></tr>
<tr><td colspan="2">Basic protection</td></tr>
<tr><td>Insulation of live parts</td><td>Is basic shock protection provided by insulation of live parts (such as conductors within insulated cables)?<br><br>This box is relevant to all electrical installations and is addressed through inspection.<br><br>See Section 416.1.</td></tr>
<tr><td>Barriers and enclosures</td><td>Do all barriers and enclosures (such as switch or socket outlet boxes and DBs etc.) have suitable IP (international protection) ratings? Are they secured by a tool or key to stop people accidently touching live connections?<br><br>See Section 416.2.</td></tr>
<tr><td>Obstacles</td><td>Are obstacles present and sufficient?<br><br>Obstacles are rare in installations but aim to stop people touching live parts. They are not necessarily secured by a tool or key and do not have an IP rating like an enclosure.<br><br>An example could be where open switch contacts are behind a safety rail within a restricted switch room.<br><br>See Section 417.2.</td></tr>
<tr><td>Placing out of reach</td><td>Is ‘placing out of reach’ used?<br><br>Placing out of reach, such as in a bare overhead conductor or bus bar system, is rare in installations.<br><br>The requirements of Section 417.3 must be met.</td></tr>
</table>

| **Fault protection**<br>**(i) Automatic disconnection of supply** | |
|---|---|
| Presence of earthing conductor | Is the earthing conductor present, connecting the MET to the means of earthing? Is it continuous and correctly sized in accordance with Chapter 54?<br>This is applicable in most installations. |
| Presence of cpc | As required by 411.3.1.1, is a cpc present in all circuits where automatic disconnection of supply (ADS) is the protective measure?<br>Is the cpc suitably sized in accordance with Chapter 54?<br>This is applicable in most installations. |
| Presence of main protective bonding conductors | Does the installation contain any extraneous parts as listed in Regulation 411.3.1.2? If so, they must be bonded to the main earthing terminal (MET) by a conductor in accordance with Chapter 54.<br>This is applicable in most installations. |
| Presence of supplementary bonding conductors | Is this a special location, such as a bathroom, where supplementary bonding is required to link all exposed and extraneous parts?<br>Supplementary bonding is becoming less common with the use of RCD protection. However, it is recommended that supplementary bonding is installed in some situations. For example, where RCDs are not maintained in accordance with the manufacturer's recommendations (such as by pressing the test button quarterly), there is no guarantee that the RCD will function correctly under fault conditions. Therefore, in a domestic situation where ordinary persons are present and where the installation is not under effective supervision, it may be prudent to install supplementary protective bonding conductors in locations containing a bath or a shower. |
| Presence of earthing arrangements for combined protective and functional purposes | Is there a protective conductor that also acts as a zero volt reference for equipment, such as telecoms, to function?<br>This is not common in most installations. The cable should be coloured cream to identify this situation. |

| | |
|---|---|
| Presence of adequate arrangements for other sources where applicable | Is there adequate switching and protection given where additional sources of energy exist, such as PV sources or standby generators/uninterruptable power supplies (UPS)?<br><br>Further detail exists in Chapter 55. |
| FELV | Is FELV used to reduce voltage to ELV for the purpose of function, such as in machine controls?<br><br>This is not common in installations.<br><br>The requirements of ADS must apply to the source circuit.<br><br>See Section 411.7. |
| Choice and setting of protective and monitoring devices | Is ADS used as fault protection?<br><br>The inspector must check that all circuits are adequately protected using protective devices of a suitable type and rating.<br><br>Monitoring devices, such as residual current monitoring devices (RCMs) and insulation monitoring devices (IMDs), are less common. |
| **(ii) Non-conducting location** | |
| Absence of protective conductors | This is very rarely found, although it may be present in specialised installations which are designed in accordance with Section 418. |
| **(iii) Earth-free local equipotential bonding** | |
| Presence of earth-free local equipotential bonding | This is very rarely found, although it may be present in specialised installations which are designed in accordance with Section 418. |

| (iv) Electrical separation | |
|---|---|
| Provided for one item of equipment | Is separation provided for one item? Does the transformer comply with the relevant standards?<br>This section is common in many installations. An example could be a shaver point in a bathroom. |
| Provided for more than one item of equipment | Is separation provided for more than one item, such as where an isolating transformer supplies socket outlets in a laboratory or in a workshop where RCD protection is not suitable? It may also be used where regular work is required in a conducting location with restricted movement.<br>This is less common but may apply in special circumstances.<br>The requirements of Section 413 must be met. |
| **Additional protection** | |
| Presence of RCDs | Are RCDs present and do they comply with Section 415.1?<br>This is applicable in most installations where RCDs rated no more than 30 mA are used for additional protection, such as for socket outlets for general use and in mobile equipment that is used outdoors.<br>If cables are concealed in the fabric of a building, the relevant box under the section 'Cables and conductors' applies. |
| Presence of supplementary bonding | Do we need to carry out supplementary bonding?<br>Supplementary protective bonding conductors are installed to ensure that shock risk is substantially reduced under both earth fault and earth leakage conditions. It is essential to reduce the risk of touch voltage during fault or leakage conditions. See Regulations 415.2.1 and 415.2.2.<br>Also see Regulation 411.3.2.6 for information regarding disconnection times. |

| Prevention of mutual detrimental influences | |
|---|---|
| Proximity to non-electrical services | Is all electrical equipment and cables suitably distanced from any services that may affect the installation?<br>An example may be hot water pipes near cables. If close spacing is unavoidable, the electrical parts must be suited for proximity to the services. |
| Segregation of Band I and Band II circuits | Are any electrical circuits operating at ELV adequately segregated from low-voltage circuits (unless the ELV circuits are insulated to low-voltage standards)?<br>An example may be a door bell circuit using bell wire that must be segregated from lighting circuits. See Section 528. |
| Segregation of safety circuits | Are safety circuits such as fire-alarm circuits or centrally-fed emergency lighting completely segregated from all other circuits?<br>These should never share containments, such as conduits. See Chapter 56. |
| **Identification** | |
| Presence of diagrams, instructions, circuit charts | Does the electrical installation contain suitable diagrams and charts detailing specific information about the installation?<br>See Regulation 514.9.1. |
| Presence of danger notices and other warning notices | Is there a range of danger and warning notices to satisfy the requirements of Section 514.9?<br>*All* electrical installations require these. |
| Labelling of protective devices, switches and terminals | Are all terminals clearly marked? Do all switches and devices, such as fuses and circuit breakers, clearly identify their purpose?<br>See Section 514. |
| Identification of conductors | Are all conductors clearly identified by colour or marking?<br>This includes sleeving of conductors.<br>The inspector must be satisfied that all conductors are in accordance with Section 514. |

| Cables and conductors | |
|---|---|
| Selection of conductors for current-carrying capacity and voltage drop | Are all conductors suitably sized for the intended load and voltage drop constraints?<br><br>The inspector will be reliant on the designer's specification for this, but a good level of experience is also required to make an informed judgement. The inspector must ensure correct coordination exists: $I_b \leq I_n \leq I_z$. |
| Erection methods | Are all electrical equipment, cables and containment systems suitably and securely installed?<br><br>As an example, does a conduit have adequate saddles which are correctly spaced? |
| Routing of cables in prescribed zones | Are all cables that are concealed in the fabric of the building, within the 'zones of protection'?<br><br>For example, do the concealed cables run vertically above or below an accessory as detailed in the IET On-Site Guide?<br><br>Installations should be compliant with Section 522.6. This requirement works in conjunction with the two requirements that follow. |
| Cables incorporating earthed armour or sheath, or run within an earthed wiring system | Are cables present that are not in the zones of protection as detailed above? If so, they must be protected by an earthed metallic covering. Also, where cables are concealed in a wall and are not provided with additional protection by an RCD as below, they must comply with this section.<br><br>Installations under effective supervision may not be applicable. |
| Additional protection provided by a 30 mA RCD for cables | Are cables concealed in a wall where the installation is not under effective supervision? If so, the applicable circuits must be provided with additional protection by an RCD in accordance with Section 415.1.<br><br>If the requirements of the section above are satisfied and cables are suitably incorporated in an earthed metallic covering, this section is not applicable. |
| Connection of conductors | Are all connections secure and, where applicable, readily accessible for maintenance? |
| Presence of fire barriers, suitable seals and protection against thermal effects | Does all electrical equipment have suitable protection to stop the spread of fire and minimise thermal effects in accordance with Chapter 42?<br><br>It must also be verified that elements of the building's structure that are affected by the installation, such as trunking that passes through floors, are adequately sealed. |

| General | |
|---|---|
| Presence and correct location of appropriate devices for isolation and switching | Are all isolators and switches located correctly and suitable for their intended use?<br>An example of this may be a switch for mechanical maintenance adjacent to a machine. The switch must be rated for on-load switching and be located in a suitable position in close proximity to the machine, as well as being accessible.<br>See Chapter 53. |
| Adequacy of access to switchgear and other equipment | Is all switchgear and control equipment, as well as any accessories, fully accessible for use and maintainability?<br>The particular requirements of Section 729 may also need to be met. |
| Particular protective measures for special installations and locations | Is this a special installation or location, for example a bathroom, swimming pool or caravan park?<br>If the installation contains any locations or is an installation as detailed in Part 7 of BS 7671, the particular requirements must be satisfied. This is due to the additional risks associated with these special locations. |
| Connection of single-pole devices for protection or switching in line conductors only | The inspector must be satisfied that all single-pole devices control the line conductor of the circuit and not the neutral. This includes circuit breakers, fuses, switches etc. |
| Correct connection of accessories and equipment | Is the polarity of all equipment correct? |
| Presence of undervoltage protective devices | Could the loss of supply and subsequent restarting cause a danger? If so, undervoltage protection is provided.<br>An example may be a machine controlled by a contactor such that, should the supply fail and be restored, the motor wouldn't automatically restart. |

| | |
|---|---|
| Selection of equipment and protective measures for external influences | Is all electrical equipment suitably selected and erected, taking into account potential external influences?<br><br>A complete list of external influences can be seen in Appendix 5 of BS 7671. |
| Selection of appropriate functional switching devices | Does all current-using equipment have a suitably rated and located switch in order for the equipment to be used safely? This includes light switches, auxiliary circuits and motor control.<br><br>The requirements of Section 537.5 must be met. |

## Information contained within a Schedule of Test Results (Form 4)

The completed Schedule of Test Results is a vital part of initial verification. It contains all the technical details which are used to verify that the installation is safe and suitable for use.

Each new circuit must be tested and all results need to be analysed to ensure that BS 7671 has been met.

The testing methods and processes have already been addressed at length in this publication.

This form must accompany the Electrical Installation Certificate.

# THE CERTIFICATION PROCESS FOR A COMPLETED INSTALLATION

**Assessment criteria**

4.3 Explain the regulatory requirements for documenting electrical systems

All documents must be completed and signed by a competent person or persons.

## Electrical Installation Certificate

The Electrical Installation Certificate, Schedule of Inspections and Schedule of Test Results must be handed over to the person ordering the work. All three documents must be completed and relevant to the work carried out.

### Form 1: Short form of Electrical Installation Certificate

Form 1 applies when one person is responsible for the design, construction, inspection and testing of the electrical installation. All documents must be signed by the same person.

### Form 2: Electrical Installation Certificate

Form 2 applies when there is more than one person involved in the process of certification.

Only authorised signatories may sign on behalf of the companies executing the design, construction, inspection and testing respectively. A signatory who is authorised to certify more than one category of work should sign in each of the appropriate places.

The designer must sign the first page of the certificate under 'Design'.

The installer must sign the first page of the certificate for the construction of the electrical installation.

The person inspecting and testing must sign as the person responsible for certifying the electrical installation. All Schedules of Inspection must also be signed.

### Form 5: Minor Electrical Installation Works Certificate

The Minor Electrical Installation Works Certificate (or Minor Works Certificate) can *only* be used if the minor electrical work does not include the provision of a new circuit. A Minor Works Certificate indicates the responsibility for design, construction, inspection and testing of the work described on the certificate.

**Assessment criteria**

6.7 Complete certification documentation

## PROCEDURES AND REQUIREMENTS FOR DOCUMENTS AND CERTIFICATION

All electrical installation work must be designed, constructed, inspected and tested. When the inspection and testing is complete, the customer will be supplied with the relevant copies of the certification.

### Electrical Installation Certificate

This safety certificate will be issued to confirm that the electrical installation work to which it relates has been designed, constructed, inspected and tested in accordance with British Standard 7671 (the IET Wiring Regulations). As part of the certification, the Schedule of Inspections and the Schedule of Test Results must be completed with satisfactory outcomes.

The recipient should receive an 'original' certificate and the contractor should retain a duplicate. If you are the person ordering the work, but not the owner of the installation, you should pass this certificate, or a full copy of it including the schedules, immediately to the owner.

The 'original' certificate should be retained in a safe place to be shown to any person inspecting or undertaking further work on the electrical installation in the future. If the property is later vacated, this certificate will demonstrate to the new owner that the electrical installation complied with the requirements of British Standard 7671 at the time the certificate was issued. The Construction (Design and Management) Regulations require that, for a project covered by those regulations, a copy of this certificate, together with schedules, is included in the project health and safety documentation.

## Minor Electrical Installations Works Certificate

This certificate will be issued to confirm that the electrical installation work to which it relates has been designed, constructed, inspected and tested in accordance with British Standard 7671 (IET Wiring Regulations).

The recipient should receive an 'original' certificate and the contractor should retain a duplicate. If you are the person ordering the work, but not the owner of the installation, you should pass this certificate, or a copy of it, to the owner.

A separate certificate should be received for each existing circuit on which minor works have been carried out. This certificate is not appropriate if the contractor has undertaken more extensive installation work, for which an Electrical Installation Certificate is required.

The certificate should be retained in a safe place and be shown to any person inspecting or undertaking further work on the electrical installation in the future. If the property is later vacated, this certificate will demonstrate to the new owner that the minor electrical installation work carried out complied with the requirements of British Standard 7671 at the time the certificate was issued.

## Schedule of Inspections

The Schedule of Inspections is completed and supplied as part of the electrical installation certificate. This document must be completed with no faults recorded and presented as part of the Electrical Installation Certificate.

## Generic Schedule of Test Results

The Schedule of Test Results is completed and supplied as part of the Electrical Installation Certificate. This document must be completed with no faults recorded and presented as part of the Electrical Installation Certificate. Complex installations may require more than one Schedule of Test Results. One schedule must be completed for each distribution board and consumer control unit (including distribution circuits).

**KEY POINT**

The original proof of the documentation must be given to the person ordering the work.

# ASSESSMENT CHECKLIST

## WHAT YOU NOW KNOW/CAN DO

| Learning outcome | Assessment criteria | Page number |
|---|---|---|
| 1 Know requirements for commissioning of electrical systems | *The learner can:* | |
| | 1 Specify the regulatory requirements for inspection of electrical systems | 360, 382 |
| | 2 Specify the regulatory requirements for testing electrical systems | 371, 382 |
| | 3 Specify the regulatory requirements for commissioning of electrical systems. | 371, 382 |
| 2 Understand procedures for the inspection of electrical systems | *The learner can:* | |
| | 1 Explain procedures for preparing for inspection | 390 |
| | 2 Explain how human senses can be used during the inspection process | 385 |
| | 3 Justify choice of applicable items on an inspection checklist that apply in given situations. | 386, 396 |
| 3 Understand procedures for completing the testing of electrical systems | *The learner can:* | |
| | 1 Explain why regulatory tests are undertaken | 401 |
| | 2 Explain why regulatory tests are carried out in the exact order as specified | 410 |
| | 3 Explain how to prepare for testing electrical systems | 362, 410, 431 |
| | 4 Specify procedures for regulatory tests | 410, 431 |
| | 5 Explain implications of non-compliance of regulatory test results with regulatory values | 409 |
| | 6 Explain factors that effect insulation resistance values | 432 |
| | 7 Specify the requirements for the safe and correct use of instruments to be used for testing. | 403 |

| | | |
|---|---|---|
| 4 Understand requirements for documenting installing electrical systems | *The learner can:*<br>1 Explain the purpose of certification documentation<br>2 Specify the responsibilities of personnel in relation to the completion of the certification documentation<br>3 Explain the regulatory requirements for documenting electrical systems. | <br>473<br>480<br>489 |
| 5 Be able to inspect electrical wiring systems | *The learner can:*<br>1 Implement safe system of work for inspection of electrical systems<br>2 Record inspection of electrical systems. | <br>368<br>386 |
| 6 Be able to test safety of electrical systems | *The learner can:*<br>1 Implement safe system of work for testing electrical systems<br>2 Select the test instruments for regulatory tests<br>3 Test electrical systems<br>4 Record results of regulatory tests<br>5 Verify compliance of regulatory test results<br>6 Commission electrical systems<br>7 Complete certification documentation. | <br>368, 410, 431<br>410, 431<br>410, 431<br>410, 431<br>410, 431<br>410, 431<br>471, 490 |

## ASSESSMENT GUIDANCE

### For your assignments

- This unit concentrates on inspection and testing of electrical installations.
- Make sure you have the books and drawing materials you require. These would include BS 7671 IET Wiring Regulations and On-Site Guide. You will need a calculator, not a mobile phone.
- Each assignment is based on a drawing. You must read the notes on the drawing to familiarise yourself with the building detail and the details of the electricity supply and earthing arrangements.
- You should use either a black or a blue pen to complete the assessment. Use other coloured pens to draw any wiring systems or earthing arrangements. Use a straight-edge rule for neatness.
- Make sure you read each question thoroughly. Mistakes are often made as a result of not reading the question fully before answering it.
- The assessment is **not** carried out under exam conditions and you will be allowed sufficient time to complete it.
- You may discuss questions with others but the work submitted must be your own.
- The questions will be supplied beforehand to allow research.
- Make sure you have all the paperwork you need before you begin.
- Arrive on time for the assessment.
- Answer all questions as far as you can.
- When answering questions that require you to describe a process, such as a test procedure, use bullet points to help delineate each stage.
- Plan your approach to the assessment and keep a record of your progress so that you do not have to rush at the end.

## For your practical assessments

- Task B requires you to carry out electrical installation testing on a rig.
- Make sure you know the sequence of testing.
- You may have guidance material with you.
- You must demonstrate competence to your assessor.
- Ensure you check all instruments before use, to ensure they work correctly and are in a safe condition.
- Keep rechecking your instruments for accuracy when carrying out continuity tests; changing leads will necessitate re-nulling or zeroing.
- Ensure you use all necessary Personal Protection Equipment; this should be made available to you.
- Testing is not a race. Doing it right is far better than doing it quickly.
- When you complete a particular stage or test, ensure everything is reconnected or put back before moving on.
- Do not create clutter in your work area. Work tidily and things become easier.
- Ask before you switch off: ask before you switch on.
- Make sure you fill in any paperwork fully. Do not leave any gaps.

Above all else – WORK SAFELY.

## KNOWLEDGE CHECK

1 Test leads complying with GS 38 must be used when the a.c. circuit voltage exceeds what value?

2 Name the test instrument used to carry out a continuity test on a bonding conductor.

3 An insulation resistance test on an existing distribution board produces an overall result of 5 MΩ. What would be the overall value if an additional circuit with an IR value of 10 MΩ were added to the board?

4 Identify which human senses would be used when verifying:

   a) correct termination of conductors.

   b) correct circuit labelling.

   c) compliance with IP codes.

   d) rough edges on trunking.

5 A ring final circuit test produces the following loop values:

   L1–L2 = 0.8Ω  N1– N2 = 0.8Ω  cpc1–cpc2 = 1.3Ω

   a) State why the cpc loop has a higher reading than the other two.

   b) Calculate the overall value of the L–N and L–cpc figure of eight tests.

6 A cable 20 m long has a millivolt drop of 11 mv/am. Calculate the voltage drop when carrying 22 A.

7 Which document would be completed following the replacement of a faulty 6 A lighting switch?

8 To be complete the Electrical Installation Certificate must be accompanied by two other documents. Which documents are they?

9 A new installation has four distribution boards. What documentation needs to be produced?

10 Identify two methods of confirming continuity of the cpc on a lighting circuit.

# UNIT 303
# Electrical installations: fault diagnosis and rectification

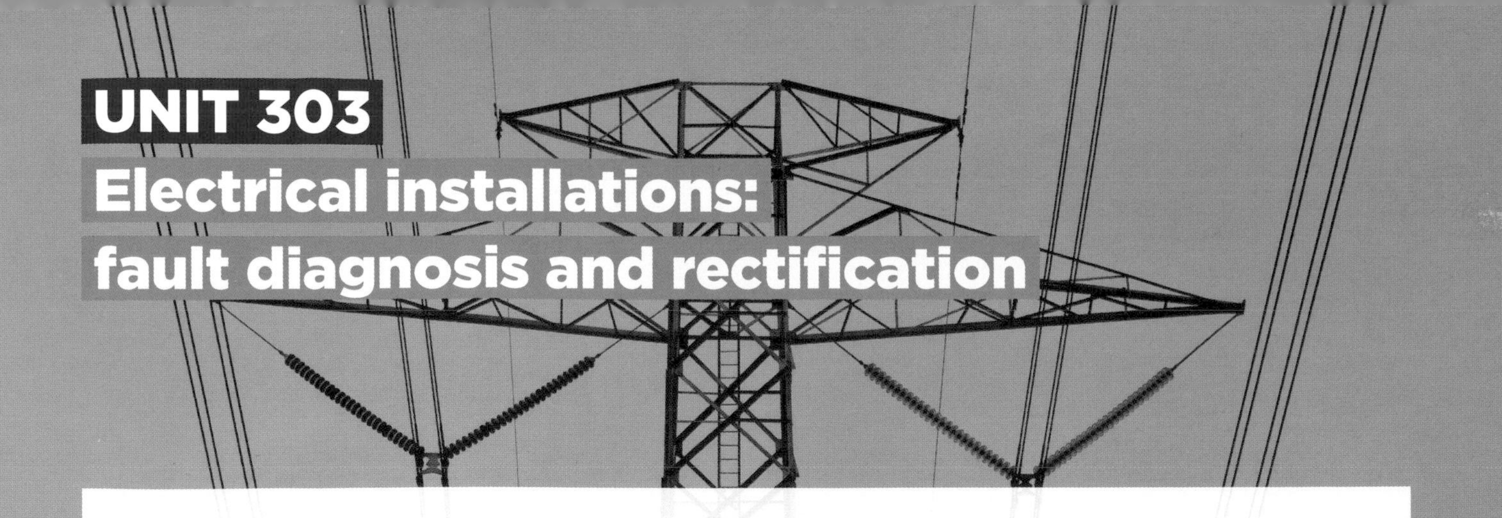

There are many types of fault that can occur in electrical circuits. These include not only the failure of physical components but also errors that may have been introduced during the original design or during the installation. The purpose of this module is to enable learners to understand how to carry out fault diagnosis and rectification of complex electrical systems safely, and in accordance with regulatory requirements, and to develop the skills to apply their understanding of fault diagnosis in simulated environments.

## LEARNING OUTCOMES

There are four learning outcomes to this unit. The learner will:

1 understand how electrical fault diagnosis is reported
2 understand how electrical faults are diagnosed
3 understand process for fault rectification
4 be able to diagnose faults on electrical systems.

This unit will be assessed by:

- practical assignment
- project-based assignments.

# OUTCOME 1

## Understand how electrical fault diagnosis is reported

**SmartScreen Unit 303**

Additional resources to support this unit are available on SmartScreen.

**Assessment criteria**

1.1 Describe procedures for recording information on electrical fault diagnosis

A fault is defined in BS 7671: 2008 Requirements for Electrical Installations (the IET Wiring Regulations) as: 'A circuit condition in which current flows through an abnormal or unintended path. This may result from an insulation failure or a bridging of insulation.'

### PROCEDURES FOR RECORDING INFORMATION ON ELECTRICAL FAULT DIAGNOSIS

Electrical faults do not occur at convenient times, and fault diagnosis and repair will often be undertaken in difficult circumstances. Whatever the circumstances, the information gathered during the fault diagnosis and after repair must be recorded and retained for future use.

The Electricity at Work Regulations 1989 (EAW Regulations) is the principal legislation relating to work on electrical systems. In particular, Regulation 4(3) requires that: 'Work on or near to an electrical system shall be carried out in such a manner as not to give rise, so far as is reasonably practicable, to danger.'

Regulation 14 places a strict prohibition on working on or near live conductors unless:

a) it is unreasonable for the equipment to be dead
b) it is reasonable for the work to take place on or near the live conductor and
c) suitable precautions have been taken to prevent injury.

In addition, employers are required under Regulation 3 of the Management of Health and Safety at Work Regulations 1999 to assess the risks to the health and safety of their employees while they are at work, in order to identify and implement the necessary precautions to ensure safety.

BS 7671, the Requirements for Electrical Installations, requires that, following the inspection and testing of all new installations, alterations and additions to existing installations or periodic inspections, an Electrical Installation Certificate (EIC), together with a Schedule of Test Results, should be given to the person ordering the work; this person is normally the client, **duty holder** or **responsible person**. Model forms for certification and reporting are contained in Appendix 6 of BS 7671.

**Duty holder**

The person in control of the danger is the duty holder. This person must be competent by formal training and experience, and with sufficient knowledge to avoid danger. The level of competence will differ for different types of work.

**Responsible person**

The person who is designated as the responsible person has delegated responsibility for certain aspects of a company's operational functions, such as fire safety, electrical operational safety or the day-to-day responsibility for controlling any identified risk such as *Legionella* bacteria.

## Using the correct forms

There are two options available for the Electrical Installation Certificate (EIC): Form 1 or Form 2.

- *Form 1:* short form EIC, is to be used when one person is responsible for the design, construction, inspection and testing of an installation. An example of this form is shown on page 500.
- *Form 2:* this EIC is to be used when more than one person is responsible for the design, construction, inspection and testing of an installation. An example of this form is shown on page 501.

Whichever EIC is used, appropriate numbers of the following forms are required to accompany the certificate:

- Schedule of Inspections
- Schedule of Test Results.

When an addition to an electrical installation does not involve the installation of a new circuit a Minor Electrical Installation Works Certificate (MEIWC) may be used. This certificate is intended for use when work such as the addition of a socket outlet or lighting point to an existing circuit or a repair or modification to a circuit is undertaken.

Electrical Installation Certificates and Minor Electrical Installation Works Certificates must be completed and signed by a competent person or persons in respect of the design, the construction and the inspection and test of the installation.

A competent person is defined in BS 7671 as: 'A person who possesses sufficient technical knowledge, relevant practical skills and experience for the nature of the electrical work undertaken and is able at all times to prevent danger and, where appropriate, injury to him/herself and others'.

Therefore, competent persons must have a sound knowledge and relevant experience of the type of work being undertaken and of the technical requirements of BS 7671. They must also have a sound knowledge of the inspection and testing procedures contained in the Regulations and must use suitable testing equipment.

EICs and MEIWCs must identify who is responsible for the design, construction, inspection and testing, whether this is new work or an alteration or addition to an existing installation.

**ASSESSMENT GUIDANCE**

Make sure you select the correct form for the purpose intended. Make sure you use the correct terminology when identifying the form. It is not an inspection schedule, it is a Schedule of Inspections; it is not a test certificate but a Schedule of Test Results etc.

**KEY POINT**

An Electrical Installation Condition Report (EICR) will be provided when an inspection and test on an electrical installation has been undertaken in order to highlight any safety shortcomings, defects or deviations from the current version of the Requirements for Electrical Installations (BS 7671).

Form 1 — Form No: 505513.../1

ELECTRICAL INSTALLATION CERTIFICATE
(REQUIREMENTS FOR ELECTRICAL INSTALLATIONS - BS 7671 [IET WIRING REGULATIONS])

**DETAILS OF THE CLIENT** Mr T Brown
32 South St
Anytown, Surrey ... Post Code: TO1 1ZZ

**INSTALLATION ADDRESS** The Coffee Bean
31 Station Road
Anytown, Surrey ... Post Code: TO3 2YF

**DESCRIPTION AND EXTENT OF THE INSTALLATION** Tick boxes as appropriate

Description of installation:
Re-wire of ground floor, on change of use.

New installation ☑
Addition to an existing installation ☐
Alteration to an existing installation ☐

Extent of installation covered by this Certificate:
Complete electrical re-wire of refurbished premises, on change of use from offices to cafe/snack bar.

(Use continuation sheet if necessary) see continuation sheet No: ........

**FOR DESIGN, CONSTRUCTION, INSPECTION & TESTING**
I being the person responsible for the design, construction, inspection & testing of the electrical installation (as indicated by my signature below), particulars of which are described above, having exercised reasonable skill and care when carrying out the design, construction, inspection & testing hereby CERTIFY that the said work for which I have been responsible is to the best of my knowledge and belief in accordance with BS 7671:2008, amended to 2011..... (date) except for the departures, if any, detailed as follows:

Details of departures from BS 7671 (Regulations 120.3 and 133.5):
None

The extent of liability of the signatory is limited to the work described above as the subject of this Certificate.

Signature: *W Hastings* Date: 21-Jan-2011 Name (IN BLOCK LETTERS): W HASTINGS

Company Address: Hastings Electrical
21 The Arches
Anytown, Surrey ... Postcode: TO2 9YY Tel No: 01022 999999

**NEXT INSPECTION**
I recommend that this installation is further inspected and tested after an interval of not more than ...5....... years/~~months~~.

**SUPPLY CHARACTERISTICS AND EARTHING ARRANGEMENTS** Tick boxes and enter details, as appropriate

| Earthing arrangements | Number and Type of Live Conductors | Nature of Supply Parameters | Supply Protective Device Characteristics |
|---|---|---|---|
| TN-C ☐<br>TN-S ☐<br>TN-C-S ☑<br>TT ☐<br>IT ☐ | a.c. ☑ d.c. ☐<br>1-phase, 2-wire ☑ 2-wire ☐<br>1-phase, 3-wire ☐ 3-wire ☐<br>2-phase, 3-wire ☐ other ☐<br>3-phase, 3-wire ☐<br>3-phase, 4-wire ☐ | Nominal voltage, $U/U_0$ (1) ....230 V<br>Nominal frequency, f (1) ....50 Hz<br>Prospective fault current, $I_{pf}$ (2) ..9.0 kA<br>External loop impedance, $Z_e$ (2) 0.28 Ω | Type: BS 1361 Fuse<br><br>Rated current....100....A |
| Other sources ☐ of supply (to be detailed on attached schedules) | Confirmation of supply polarity ☑ | (Note: (1) by enquiry, (2) by enquiry or by measurement) | |

Form 1 — Form No: 505513.../1

**PARTICULARS OF INSTALLATION REFERRED TO IN THE CERTIFICATE** Tick boxes and enter details, as appropriate

**Means of Earthing**
Distributor's facility ☑
Installation earth electrode ☐

**Maximum Demand**
Maximum demand (load) ..........80 ~~kVA~~/ Amps Delete as appropriate

**Details of Installation Earth Electrode** (*where applicable*)

| Type (e.g. rod(s), tape etc) | Location | Electrode resistance to Earth |
|---|---|---|
| N/A | N/A | N/A Ω |

**Main Protective Conductors**
Earthing conductor: material Copper ...... csa ....16....mm² Continuity and connection verified ☑
Main protective bonding conductors material Copper ...... csa ....10....mm² Continuity and connection verified ☑
To incoming water and/or gas service ☑ To other elements: N/A

**Main Switch or Circuit-breaker**
BS, Type and No. of poles BS EN 60947-3 (2- pole) Current rating ......100 A Voltage rating ......230 V
Location Services cupboard adjacent rear exit Fuse rating or setting........N/A A
Rated residual operating current $I_{\Delta n}$ = ...N/A... mA, and operating time of ..N/A.. ms (at $I_{\Delta n}$) (applicable only where an RCD is suitable and is used as a main circuit-breaker)

**COMMENTS ON EXISTING INSTALLATION** (in the case of an addition or alteration see Section 633):
Not Applicable

**SCHEDULES**
The attached Schedules are part of this document and this Certificate is valid only when they are attached to it.
.....1.... Schedules of Inspections and .....1...... Schedules of Test Results are attached.
(Enter quantities of schedules attached).

Page 2 of 4

Form 1: Short form of Electrical Installation Certificate (always check you are using the latest forms, as found on the IET website: http://electrical.theiet.org)

Form 2 Form No: SSSS13...../2

## ELECTRICAL INSTALLATION CERTIFICATE

(REQUIREMENTS FOR ELECTRICAL INSTALLATIONS - BS 7671 [IET WIRING REGULATIONS])

**DETAILS OF THE CLIENT** Mr D Roberts
23 Acacia Avenue ........ Post Code: SL0 0LT
Sometown, Berks

**INSTALLATION ADDRESS** Unit 3 The Quadrant
Sometown Business Park
Sometown, Berks ........ Post Code: SL1 0ZZ

**DESCRIPTION AND EXTENT OF THE INSTALLATION** Tick boxes as appropriate

Description of installation: Commercial office — New installation ☑

Extent of installation covered by this Certificate: Full new installation — Addition to an existing installation ☐

(Use continuation sheet if necessary) see continuation sheet No: ........ — Alteration to an existing installation ☐

**FOR DESIGN**
I/We being the person(s) responsible for the design of the electrical installation (as indicated by my/our signatures below), particulars of which are described above, having exercised reasonable skill and care when carrying out the design hereby CERTIFY that the design work for which I/we have been responsible is to the best of my/our knowledge and belief in accordance with BS 7671:2008, amended to ..2011.......... (date) except for the departures, if any, detailed as follows:

Details of departures from BS 7671 (Regulations 120.3 and 133.5):
None N/A

The extent of liability of the signatory or the signatories is limited to the work described above as the subject of this Certificate.

For the DESIGN of the installation: **(Where there is mutual responsibility for the design)

Signature: D Jones Date: 15/08/2013 Name (IN BLOCK LETTERS): D JONES Designer No 1

Signature: N/A Date: ........ Name (IN BLOCK LETTERS): N/A Designer No 2**

**FOR CONSTRUCTION**
I/We being the person(s) responsible for the construction of the electrical installation (as indicated by my/our signatures below), particulars of which are described above, having exercised reasonable skill and care when carrying out the construction hereby CERTIFY that the construction work for which I/we have been responsible is to the best of my/our knowledge and belief in accordance with BS 7671:2008, amended to .....2011.........(date) except for the departures, if any, detailed as follows:

Details of departures from BS 7671 (Regulations 120.3 and 133.5):
None N/A

The extent of liability of the signatory is limited to the work described above as the subject of this Certificate.

For CONSTRUCTION of the installation:

Signature: T Smith Date: 15/08/2013 Name (IN BLOCK LETTERS): T SMITH

**FOR INSPECTION & TESTING**
I/We being the person(s) responsible for the inspection & testing of the electrical installation (as indicated by my/our signatures below), particulars of which are described above, having exercised reasonable skill and care when carrying out the inspection & testing hereby CERTIFY that the work for which I/we have been responsible is to the best of my/our knowledge and belief in accordance with BS 7671:2008, amended to ......2011.........(date) except for the departures, if any, detailed as follows:

Details of departures from BS 7671 (Regulations 120.3 and 133.5):
None N/A

The extent of liability of the signatory is limited to the work described above as the subject of this Certificate.

For INSPECTION AND TESTING of the installation:

Signature: G Wilson Date: 15/08/2013 Name (IN BLOCK LETTERS): G WILSON

**NEXT INSPECTION**
I/We the designer(s), recommend that this installation is further inspected and tested after an interval of not more than .....5....... years/~~months~~.

Page 1 of ...4

Form 2 Form No: ...SSSS13............./2

**PARTICULARS OF SIGNATORIES TO THE ELECTRICAL INSTALLATION CERTIFICATE**

**Designer (No 1)**
Name: D Jones Company: The Electrical Design Partnership
Address: 23 High Street
Sometown, Berks Postcode: SL10 0YY Tel No: 01000 999999

**Designer (No 2)** (if applicable)
Name: ........ Company: ........
Address: ........
Postcode: ........ Tel No: ........

**Constructor**
Name: T Smith Company: T Smith Electrical Installations
Address: Unit 8a Sometown Ind Estate
Sometown, Berks Postcode: SL3 0XX Tel No: 01000 888888

**Inspector**
Name: G Wilson Company: Wilson and Sons
Address: 11 Crabtree Row
Sometown, Berks Postcode: SL2 0WW Tel No: 01000 777777

**SUPPLY CHARACTERISTICS AND EARTHING ARRANGEMENTS** Tick boxes and enter details, as appropriate

| Earthing arrangements | Number and Type of Live Conductors | Nature of Supply Parameters | Supply Protective Device Characteristics |
|---|---|---|---|
| TN-C ☐<br>TN-S ☐<br>TN-C-S ☑<br>TT ☐<br>IT ☐ | a.c. ☑ d.c. ☐<br>1-phase, 2-wire ☑ 2-wire ☐<br>1-phase, 3-wire ☐ 3-wire ☐<br>2-phase, 3-wire ☐ other ☐<br>3-phase, 3-wire ☐<br>3-phase, 4-wire ☐ | Nominal voltage, ~~U~~/$U_0$[1] ...230...... V<br>Nominal frequency, f [1] ....50..... Hz<br>Prospective fault current, $I_{pf}$[2] .1.41.. kA<br>External loop impedance, $Z_e$[2] ....0.34Ω | Type: ...BS 88-3.....<br>Rated current..80. A |
| Other sources of supply (to be detailed on attached schedules) ☐ | Confirmation of supply polarity ☑ | (Note: (1) by enquiry, (2) by enquiry or by measurement) | |

**PARTICULARS OF INSTALLATION REFERRED TO IN THE CERTIFICATE** Tick boxes and enter details, as appropriate

**Means of Earthing**
Distributor's facility ☑
Installation earth electrode ☐

**Maximum Demand**
Maximum demand (load) ........68....... ~~kVA~~ / Amps Delete as appropriate

**Details of Installation Earth Electrode** (*where applicable*)
Type (e.g. rod(s), tape etc) ....N/A.... Location ....N/A.... Electrode resistance to Earth ....N/A.... Ω

**Main Protective Conductors**
Earthing conductor: material ...Copper.... csa ......16......mm² Continuity and connection verified ☑
Main protective bonding conductors material ....Copper.... csa ......16......mm² Continuity and connection verified ☑
To incoming water and/or gas service ☑ To other elements: .....N/A.....

**Main Switch or Circuit-breaker**
BS, Type and No. of poles ....BS EN 60497-3 (2-pole)....... Current rating .....100.....A Voltage rating ....400......V
Location ...Office Suite Consumer Unit........ Fuse rating or setting.....N/A.......... A
Rated residual operating current $I_{\Delta n}$ = ....N/A... mA, and operating time of ....N/A.... ms (at $I_{\Delta n}$) (applicable only where an RCD is suitable and is used as a main circuit breaker)

**COMMENTS ON EXISTING INSTALLATION** (in the case of an addition or alteration see Section 633):
.......N/A........

**SCHEDULES**
The attached Schedules are part of this document and this Certificate is valid only when they are attached to it.
.....1...... Schedules of Inspections and ......1...... Schedules of Test Results are attached.
(Enter quantities of schedules attached)

Page .2..of ..4

## Form 1 (short form) Electrical Installation Certificate

Form 1, the short form EIC is a shortened version of Form 2. It is used where one person is responsible for the design, construction, inspection and testing of a new installation or where a major alteration or addition to an existing installation is carried out and the installation, inspection and testing of that work is the responsibility of one person.

## Form 2 Electrical Installation Certificate (three-signatory version)

Form 2, the EIC three-signatory version, is used where more than one person is responsible for the design, construction, inspection and testing of a new installation or where a major alteration or addition to an existing installation is carried out. The Form 2 certificate also has space for the name, address, telephone number and signatures of two different design organisations. This can be used where different sections of the installation are designed by different individuals or companies (for example, the electrical distribution system may be designed by one consultant and the final circuits by another). In this situation the details and signatures of both parties are required. BS 7671 provides advice notes for the completion of each type of form and also guidance notes for the person receiving the certificate.

## Electrical Installation Condition Report

The EICR should only be used for reporting on the condition of an electrical installation. It is intended primarily for the person who is ordering the work to be undertaken (the client, duty holder or responsible person for the installation) and anyone subsequently involved in additional or remedial work or additional inspections, to confirm, so far as is reasonably practicable, whether or not the electrical installation is in a satisfactory condition for continued service.

Each observation of a problem or concern relating to the safety of an installation should be given an appropriate classification code selected from standard classification codes as follows.

- *C1: danger present*. Risk of injury. Immediate remedial action required.
- *C2: potentially dangerous*. Urgent remedial action required.
- *C3: improvement recommended*.

## Minor Electrical Installation Works Certificate

The Minor Electrical Installation Works Certificate (MEIWC) is used for minor works, which are defined as an addition to an electrical installation that does not extend to the installation of a new circuit. This could cover, perhaps, the addition of a new socket outlet or of a lighting point to an existing circuit. This certificate includes space for the recording of essential test results but does not require the addition of a Schedule of Test Results.

**ACTIVITY**

Look up each certificate/report/schedule and make sure you understand how to fill them in.

Each certificate or report will have attached to it a set of guidance notes explaining the purpose of the certificate or report being issued.

The original copy of the certificate should be retained in a safe place so that it can be shown to any person inspecting the installation or carrying out further work in the future. If the property or building is later vacated, this certificate will demonstrate to the new owner that the minor works covered by the certificate met the requirements of the Regulations at the time that the report was issued.

When an electrical fault has been diagnosed, identified and then repaired the IET Wiring Regulations require that the circuit, system or individual piece of equipment be inspected and tested and that functional tests should be carried out. These inspections and tests must be carried out in accordance with Part 6 of the Regulations and the results recorded in accordance with Part 6, Chapter 63. The requirements that apply to fault diagnosis and repair are:

- 631.2 – completion of an EICR
- 631.4 – the EICR shall be compiled and signed by a competent person or persons
- 631.5 – the report can be produced in any durable medium, such as written hard copy, or by electronic means.

Depending on the extent of the fault and subsequent report, either an EIC or a MEIWC will be issued to the person requesting the work.

## USING FAULT CODES

**Assessment criteria**

**1.2** Identify codes used in electrical condition report (BS 7671) for different faults

Electrical installations degrade with time due to physical damage to switches, sockets and other fittings, together with deterioration of cables. The severity of degradation is more pronounced in installations within buildings where construction work is in progress, where adverse elements (such as corrosive chemicals or extreme temperatures) are involved or where maintenance has been poor. To ensure the safety of everyone who may come into contact with an electrical installation, such as users or people who undertake maintenance, it is vital that the installation is regularly inspected and tested to identify any faults or potential failures. Regulation 4(2) of the EAWR (1989) states: 'As may be necessary to prevent danger, all systems shall be maintained so as to prevent, so far as reasonably practicable, such danger.'

Regulation 4(2) recognises that the integrity of any electrical system can only be preserved in its 'initially as installed' condition if it is regularly maintained and repaired as necessary. This is no more or less than what one would expect with the use of any item of electrical or mechanical plant.

BS 7671 is the national standard in the United Kingdom for low-voltage electrical installations. It is non-statutory, but if wiring installation and maintenance work is undertaken in accordance with BS 7671 it is almost certain to meet the requirements of EAW Regulation 4(2). BS 7671 is largely based on the European Committee for Electrotechnical Standardisation (CENELEC) harmonised documents and is very similar to the current wiring regulations of other European countries. The Regulations deal with the design, selection, erection, inspection and testing of electrical installations operating at voltages up to 1000 V a.c.

**Person in control of the premises**

The person in control of the premises is likely to have physical possession of the premises or is the person who has responsibility for, and control over, the condition of the premises, the activities conducted on those premises and the people allowed to enter.

As BS 7671 is non-statutory there is no legislative requirement which stipulates that a duty holder, responsible person or **person in control of the premises** must have or retain any kind of electrical inspection report. However, from a liability and safety perspective, it is advisable and provides proof that the requirements of the EAWR are being met. Rented properties and certain types of public place, such as theatres, restaurants, cinemas, clubs and hotels, are generally required to have some kind of report for insurance purposes and an EICR (as recommended in BS 7671) fulfils this role. In the event of a death or serious injury, resulting from failure to maintain the electrical integrity of an installation, some form of legal action could result.

## BS 7671 EICR

When an existing installation has been subjected to a periodic inspection and test, BS 7671 clause 631.2 recommends the provision of an EICR. The report should include details of the extent of the installation, and the limitations of the inspection and testing, together with records of inspection, the results of testing and a recommendation for the interval until the next periodic inspection. The EICR must be compiled and signed by a competent person or persons and, in the context of low-voltage (LV) systems, BS 7671 contains definitions for '**competent person**', 'instructed person', 'skilled person' and 'ordinary person'.

**Competent person**

A person who possess sufficient knowledge, relevant practical skills and experience for the nature of the electrical work undertaken and is able at all times to prevent danger and, where appropriate, injury to him/herself and others.

As its title suggests, the EICR is a report, not a certificate. It relates to an assessment of the in-service condition of an electrical installation against the requirements of the issue of BS 7671 current at the time of the inspection, irrespective of the age of the installation.

The results, measurements and values taken during the inspection and testing are clearly recorded in a report and appropriate recommendations, if applicable, are made to rectify any damage, deterioration or defects, dangerous conditions and non-compliance with the requirements of the Regulations that may give rise to danger, along with any limitations to the inspection and testing, and the reasons for these.

The EICR contains 11 sections, which are identified alphabetically from A to K. Section K, 'Observations' has three columns to be completed. Observation(s) are entered in the first column, the second column requires a classification code (C1, C2 or C3) with reference to the observation(s) and the third column is used to record whether or not further investigation is required.

## Code C1 – danger present

This code indicates that there is a risk of injury and that immediate remedial action is required to remove the dangerous condition.

Code C1 allows those carrying out an inspection to report to the client or responsible person that a risk of injury exists, which could be, for example, accessible live conductors due to damage, poorly modified enclosures or removed maintenance panels. Incorrect polarity would also attract a Code C1 as it may allow conductive parts, not normally expected to be live, to become live.

A reported Code C1 warrants immediate action to be taken. This involves immediately informing the client or responsible person for the installation, both verbally and in writing, that a risk of injury exists. A detailed explanation of this risk should be recorded on the report, together with details of any verbal and written warnings. If possible, dangerous situations should be made safe or rectified before further work or inspections are carried out.

## Code C2 – potentially dangerous

This code indicates that urgent remedial action is required. The EICR should declare the nature of the problem, but not the remedial actions required.

The phrase 'potentially dangerous' indicates a risk of injury from contact with live parts following a 'sequence of events'. A sequence of events could mean that an individual would need to move, open or gain access to live parts when undertaking a daily task that would not normally be expected to give access to live parts. An example of this would be an isolator with a damaged casing, in a locked cupboard. This might leave exposed live parts. But these could only be accessed with the use of access equipment such as a specialist tool or key. An individual would need to gain access to the cupboard before coming into contact with live parts, but nevertheless the potential for risk of injury is high.

The lack of an adequate earthing arrangement for an installation, the use of utility pipes as the means of earthing or an undersized earthing conductor (in accordance with BS 7671 Regulation 543.1.3) would also warrant a Code C2 because a primary fault would be needed in order for these scenarios to become potentially dangerous.

### ACTIVITY

An installation is 30 years old and, apart from minor wear and tear, is satisfactory. It does, however, have rewireable fuses in the consumer unit. Outline what changes you may suggest following replacement of old pendants.

Which codes, if any, should you use in this situation?

## ASSESSMENT GUIDANCE

Remember that just because an installation does not comply with current regulations this does not make it unsafe. Most installations will still comply with the regulations in force when they were installed.

**Assessment criteria**

1.3 Explain implications of recorded information

## Code C3 – improvement recommended

This code implies that, while the installation may not comply with the current edition of the Regulations, it does comply with a previous edition and is deemed safe, although improvements could be made.

## ACTIONS TO BE TAKEN IN RESPONSE TO DANGER CODES

An EICR is intended primarily for the person who is ordering the work to be undertaken (the client, duty holder or responsible person for the installation) and anyone subsequently involved in additional or remedial work or additional inspections to confirm, so far as reasonably practicable, whether or not the electrical installation is in a satisfactory condition for continued service.

## Code C1

Wherever practicable, items classified as 'danger present' (C1) should be made safe on discovery. The duty holder or responsible person ordering the report should be advised of the action taken and on the necessary remedial work to be undertaken. If that is not practical you must take other appropriate action, such as switching off and isolating the affected parts of the installation to prevent danger.

### KEY POINT

Many accidents occur due to a failure to isolate all sources of supply to, or within, equipment (consider control and auxiliary supplies, uninterruptable power supply systems and parallel circuit arrangements) giving rise to back feeds.

There are two separate and distinct requirements for making a section of an electrical system safe when a Code C1 has been observed and recorded.

- Cutting/switching off the supply – depending on the equipment and the circumstances, this may be no more than carrying out normal functional switching (on/off) or emergency switching by means of a stop button or a trip switch.
- Isolation – this means the disconnection and separation of the electrical equipment from *'every source of electrical energy* in such a way that this disconnection and separation is *secure'*.

**Key term**

Security can best be achieved by locking off with a safety lock – such as a lock with a unique key. The posting of a warning notice also serves to alert others to the isolation.

**Photo of secure isolation**

As has already been stated, the purpose of a condition report is to confirm, so far as is reasonably practicable, whether or not the electrical installation is in a satisfactory condition for continued service.

## Code C2

For items classified as 'potentially dangerous' (C2), the safety of those using the installation may be at risk and it is recommended that a competent person undertakes any necessary remedial work as a matter of urgency.

If this potentially dangerous situation can be rectified by an addition or a simple alteration to the installation, the client will probably request that this is carried out. It should be remembered that this is not part of the initial agreement. If the work does not involve major system changes, such as the installation of a new circuit, then a MEIWC should be issued. Remember an EIC is to be used only for the initial certification of a new installation or for an alteration or addition to an existing installation where new circuits have been introduced.

## Code C3

For items classified as Code C3, 'improvement recommended', the inspector has found some problems that are not immediate risks but may become risks if not improved. One example may be that the inspection has revealed an apparent deficiency that could not, due to the extent or limitations of the inspection, be fully identified. A further examination of the installation will be necessary, to determine the nature and extent of the apparent deficiency. Such investigations should be carried out as soon as possible. For safety reasons, the electrical installation will need to be reinspected at appropriate intervals by a competent person. The recommended date by which the next inspection is due should be included in the report, under 'Recommendations', and on a label near to the consumer unit or distribution board.

On completion of the inspection, the client, duty holder or responsible person ordering the report should receive the original report and the inspector should retain a duplicate copy. The original report should be retained in a safe place and be made available to any person inspecting or undertaking work on the electrical installation in the future. If the property is vacated, the report will provide the new owner or occupier with details of the condition of the electrical installation at the time the report was issued.

# OUTCOME 2

## Understand how electrical faults are diagnosed

A fault can be defined as a defect that may or may not result in the failure of a particular component or the whole system. A fault that occurs irregularly is an intermittent fault; one that persists is a permanent or solid fault.

A fault is defined in BS 7671: 2008 Requirements for Electrical Installations as: 'A circuit condition in which current flows through an abnormal or unintended path. This may result from an insulation failure or a bridging of insulation.'

In fact, an electrical fault could result from factors that are covered by combination of both these definitions, complicating the job for the person who has been employed to find and repair the fault. Consider a fault that is reported as simply: 'The socket in the sitting room gets hot when we put the electric fire on.' The result of a fault diagnosis may reveal a loose connection on one of the terminals at the rear of the socket outlet, a breakdown of the insulation in the socket outlet case moulding perhaps caused by a manufacturing issue, or a combination of both of these aggravated by the increased load of the 3kW electric fire. The original fault may have developed from either of these separate faults. This simple example can be magnified many fold where complex electrical installations are involved.

**ASSESSMENT GUIDANCE**

You have a tongue; use it. Ask what has happened. Maybe someone remembers turning on a particular switch or hearing a noise when the supply went off.

**Assessment criteria**

2.1 Explain safe working procedures that should be adopted for completion of fault diagnosis

### SAFE WORKING PROCEDURES FOR FAULT DIAGNOSIS

Electricity is a safe, clean and powerful source of energy and is in use in practically every factory, office, workshop and home in the country. However, this energy source can also be very hazardous – with a risk of causing death, if it is not treated with care. Injury can occur when live electrical parts are exposed and can be touched, or when metalwork that is meant to be earthed becomes live at a dangerous voltage.

The possibility of touching live parts is increased during electrical testing and fault-finding, when conductors at dangerous voltages are often exposed. This risk can be reduced if testing is done while the equipment is isolated from any source of electrical supply. However, this is not always possible and, if this is the case, it is important to follow procedures that prevent contact with any hazardous internally produced voltages.

**ACTIVITY**

Identify four faults that could occur from the input of a 20 A double-pole switch and the element of an immersion heater.

But before fault finding and testing work commences reference must be made to the following relevant legal duties.

- *The Electricity at Work Regulations 1989* (EAWR): this is the principal legislation relating to work on electrical systems. Regulation 4(3) requires that: 'work on or near to an electrical system shall be carried out in such a manner as not to give rise, so far as is reasonably practicable, to danger'. The regulations provide for two basic types of system of work: work on de-energised conductors (Regulation 13) and work on live conductors (Regulation 14).
- *The Management of Health and Safety at Work Regulations 1999*: these require employers to assess the risks to the health and safety of their employees while they are at work, in order to identify and put in place the necessary precautions to ensure safety. Depending on the extent of the work, this requires either a formal written risk assessment or one that relies on a generic system for simple work activities. A sample risk assessment is shown below.

**KEY POINT**

Regulation 13 covers the preferred system of working, which is to remove the danger at source by making the conductors dead.

Regulation 14 requires adequate safeguards to protect the person at work from the hazards of live conductors.

| What are the hazards? | Who might be harmed and how? | What actions have been taken? | Any further actions required to manage this risk? | Action by whom? | Action by when? | Done |
|---|---|---|---|---|---|---|
| *Electric shock* | *Staff carrying out fault finding.*<br><br>*Failure to carry out correct isolation procedure prior to testing;*<br>*Working on equipment that is live;*<br>*Failure to provide suitable barriers where live equipment needs to be shrouded;*<br>*Use of faulty test equipment;*<br>*Failure to use appropriate PPE*<br>*Unsafe situations left unresolved.*<br><br>*Visitors to and occupants of the building.*<br><br>*Test area not properly guarded and notices not posted;*<br>*Making contact with exposed conductive parts during testing* | *All electricians have received training in the following:*<br><br>*Correct method of isolation;*<br>*Safe testing procedures;*<br>*Safe use of test instruments*<br>*First aid training;*<br>*Correct selection and use of PPE*<br>*Correct procedure regarding action to be taken if unsafe situations are identified*<br><br>*All electricians have received training in the following:*<br><br>*Safe work area demarcation – barriers, notices.* | *Regular audits by supervisor to confirm correct testing procedures are being observed;*<br>*Refresher training for all electricians.* | *Supervisor* | *01/02/2013* | *01/02/2013* |
| *Lone working* | *Lone worker.*<br><br>*Increased risk of injury due to being unable to summon assistance.* | *All electricians have received effective means of communication (eg landline or mobile phone). Any person working alone will notify their supervisor of their itinerary including where they will be working and what time they expect to finish.* | *No* | *Supervisor* | *01/02/2013* | *01/02/2013* |
| *Asbestos* | *Staff carrying out fault finding.*<br><br>*Failure to identify presence of asbestos and drilling or abrading asbestos board or ceiling tiles* | *All electricians have received training in the following:*<br>*Asbestos awareness*<br><br>*A risk assessment has been carried out and a safe system of work prepared.* | | | | |
| *Slips trips and falls* | *Staff and visitors.*<br><br>*May be injured if they trip over objects or slip on spillages* | *General good housekeeping. All areas are well lit including stairs. There are no trailing leads or cables. Staff to keep work areas clear, eg no boxes left in walkways.* | *Better housekeeping is needed in staff kitchen, eg on spills.* | *Site responsible person. All staff, supervisor to monitor* | *01/02/2013* | *01/02/2013* |

**Sample risk assessment**

## The method statement

Once the risk assessment has been carried out, a safe system of work, sometimes called a method statement, can be prepared to enable the work to be undertaken in a safe manner.

The safe system of work or method statement will include such items as:

- who is authorised to undertake testing and, where appropriate, how to access a test area and who should not enter the test area
- arrangements for isolating equipment and how the isolation is secured
- provision and use of personal protective equipment (PPE) where necessary
- the correct use of tools and equipment
- the correct use of additional protection measures, for example, flexible insulation that may need to be applied to the equipment under test while its covers are removed
- use of barriers and positioning of notices
- safe and correct use of measuring instruments
- safe working arrangements to be agreed with client, duty holder or responsible person
- how defects are to be reported and recorded
- instructions regarding action to be taken if unsafe situations are identified.

Refer to this publication for advice on live and dead working

Details of safe systems of work or safe working procedures for fault diagnosis, testing and fault repair activities should, wherever it is reasonably practicable to do so, be written down.

Guidance on live and dead working can be found in the HSE publication *Memorandum of guidance on the Electricity at Work Regulations 1989* (HSR 25) and this can be downloaded free of charge from the Health and Safety Executive website: www.hse.gov.uk

Safe working procedures should be reviewed regularly, to make sure that they are being followed and are still appropriate for the work that is being carried out. If any changes are made to the procedures, all people who are involved in the fault diagnosis regime should be given relevant instruction and training.

## Permit to work

A permit to work (PTW) procedure is a specialised written safe system of work that ensures potentially dangerous work, such as work on high-voltage electrical systems (above 1000 V) or complex lower-voltage electrical systems, is done safely. The PTW also serves as a means of communication between those controlling the danger (the duty holder) and those who carry out the hazardous work. Essential features of PTW systems are:

- clear identification of who may authorise particular jobs (and any limits to their authority), and who is responsible for specifying the necessary precautions
- that the permit should only be issued by a technically competent person, who is familiar with the system and equipment, and who is authorised in writing by the employer to issue such documents
- provision of training and instruction in the issue, use and closure of permits
- monitoring and auditing to ensure that the PTW system works as intended
- clear identification of the types of work considered hazardous
- clear and standardised identification of tasks, risk assessments, permitted task duration and any additional activity or control measure that occurs at the same time.

A permit to work should state clearly:

- the person the permit is addressed to, that is the person carrying out the work or the leader of the group or working party who will be present throughout the work
- the exact equipment that has been made dead and its precise location
- the points of isolation
- where the conductors are earthed (on high-voltage systems)
- where warning notices are posted and special safety locks fitted
- the nature of the work to be carried out
- the presence of any other source of hazard, with cross-reference to other relevant permits
- further precautions to be taken during the course of the work.

The effective operation of a PTW system requires involvement and cooperation from a number of people, and the procedure for issuing a PTW should be written down and adhered to.

**ASSESSMENT GUIDANCE**

Clearly write the instructions on the permit to work so there is no chance of misunderstanding what is required.

## Planning and agreeing procedures

Before fault diagnosis is carried out, the safe-working arrangements must be discussed and agreed with the client and/or the duty holder (or responsible person for the installation) in a clear, concise and courteous manner. The testing and inspection procedure must be a planned activity as it will almost certainly affect people who work or live in the premises where the installation is being tested. This ensures that everyone who is concerned with the work understands what actions need to be taken, such as:

- which areas of the installation may be subject to disconnection
- anticipated disruption times
- who might be affected by the work
- health and safety requirements for the site
- which area will have restricted access
- whether temporary supplies will be required whilst the fault diagnosis is underway
- the agreement reached on who has authority for the diagnosis and repair.

It may be that a specific person has responsibility for the safe isolation of a particular section of an installation and that person should be identified and the isolation arrangements agreed. By entering into dialogue with the client before work commences, the potential for unforeseen events will be minimised and good customer relations will be fostered.

For example, in an office block where the electrical installation is complex and provides supplies to many different tenants located on a number of floors, the safe isolation of a sub-circuit for testing purposes may require a larger portion of the installation to be turned off initially. In order to achieve this with minimal disruption, an agreement must be reached between the competent person tasked to carry out the work and the person responsible for the installations affected. This responsible person could be the office manager, the designated electrical engineer for the site or, in some cases, the landlord of the building.

Everyone involved in the work (for example, client, electrician and those in the workplace) has a responsibility for their own health and safety and that of others who may be affected by the work; communication between all parties will ensure compliance with the respective health and safety requirements.

**KEY POINT**

You should appreciate the difference between the duty holder and the responsible person.

The person in control of the danger is the *duty holder*. This person must be competent by formal training and experience and with sufficient knowledge to avoid danger. The level of competence will differ for different items of work.

The person who is designated the *responsible person* has delegated responsibility for certain aspects of a company's operational functions such as fire safety, electrical operational safety or the day-to-day responsibility for controlling any identified risk such as *Legionella* bacteria.

**ACTIVITY**

Using any source of reference material, obtain a copy of a permit to work and determine when a PTW might be used on a low voltage installation.

# HAZARDS OF FAULT DIAGNOSIS

The Electricity at Work Regulations 1989 (EAWR) require those in control of part or all of an electrical system to ensure that it is safe to use and that it is maintained in a safe condition. The process of identifying and rectifying faults will inevitably involve exposure to some hazards and these must be dealt with in accordance with the requirements of the EAWR.

Electrical injuries are the most likely hazard and the following may occur due to contact with mains 230 V electricity:

- electric shock
- electrical burns
- loss of muscle control
- fires arising from electrical causes
- arcing and explosion.

**Assessment criteria**

2.2 Describe precautions that should be taken in relation to hazards of fault diagnosis

## Electric shock

Electric shock may arise either from direct contact with a live part during the testing process or, indirectly, by contact with an exposed conductive part that has become live as a result of a fault condition such as:

- contact with live electrical parts if basic protection has been removed
- a cracked equipment case causing 'tracking' from internal live parts to the external surface
- poor installation practice exposing bare live conductors at terminations
- exposure to static electricity from industrial processes or something as simple as walking on a carpet.

**ASSESSMENT GUIDANCE**

Ensure you keep up to date with the latest treatments for electric shock. You may be asked to describe the actions you would take.

The magnitude (size) and duration of the shock current are the two most significant factors determining the severity of an electric shock. The magnitude of the shock current depends on the contact voltage and impedance (electrical resistance) of the shock path. A possible shock path always exists through ground contact (for example, through a hand-to-feet route); in this case the shock path impedance is the body impedance plus any external impedance. A more dangerous situation is a hand-to-hand shock path, when one hand is in contact with an exposed conductive part, such as an earthed metal equipment case, while the other simultaneously touches a live part. In this case, the current will be limited only by the body impedance.

As the voltage increases, so the body impedance decreases – this increases the shock current. When the voltage decreases the body impedance increases, which reduces the shock current.

**KEY POINT**

At 230 V the average person has a body impedance of approximately 1300 Ω. At mains voltage and frequency (230 V, 50 Hz), currents as low as 50 milliamps (0.05 A) can prove fatal, particularly if flowing through the body for a few seconds.

## ACTIVITY

Using any source of reference material, find out why an electric cattle fence with an operating voltage of 8000 V does not cause electrocution.

Systems where voltages are below 50 V a.c. or 120 V d.c. (extra-low voltage) reduce the risk of electric shock to a low level. If the system energy levels are low, then arcing is unlikely to cause burns.

## Electric burns

Electric burn injury may arise due to:

- the passage of shock current through the body, particularly if at high voltage
- exposure to high-frequency radiation, for example, from radio transmission antennas.

An electrical burn may not show on the skin at all or may appear minor, but the damage can extend deep into the tissues beneath the skin. Medical advice should always be sought if an electrical burn is sustained.

## Loss of muscle control

People who experience an electric shock often get painful muscle spasms that are strong enough to break bones or dislocate joints. This loss of muscle control may mean that the person cannot let go to escape the electric shock. Alternatively, the person may fall if they are working at height or be thrown into nearby machinery and structures.

## Fires arising from electrical causes

It is believed that in Britain each year there are over 30 000 fires in domestic and commercial premises that have electricity as a factor in their cause. The principal causes of fires arising from electricity is wiring with defects such as insulation failure, the overloading of conductors, lack of electrical protection, poor connections and the incorrect storage of flammable materials.

## Arcing and explosion

This frequently occurs due to short-circuit flashover accidentally caused while working on live equipment (either intentionally or unintentionally). Arcing generates UV radiation, causing severe sunburn; molten metal particles are also likely to be ejected onto exposed skin surfaces.

There are two main electrical causes of explosion; short circuit, due to an equipment fault, and ignition of flammable vapours or liquids, caused by sparks or high surface temperatures.

## Minimising hazards of fault diagnosis

### Isolation and switching of supply

The preparation of electrical equipment for fault diagnosis and repair purposes often requires effective disconnection from all live supplies and a means for securing that disconnection, for example, by locking off with a suitable lock and single key. There is an important distinction between switching and isolation.

Switching is cutting off the supply while isolation is the secure disconnection and separation from all sources of electrical energy. A variety of control devices are available for switching, isolation or a combination of these functions, some incorporating protective devices. Before starting work on a piece of isolated equipment, checks should be made, using an approved testing device, to ensure that the circuit is dead. See page 524 for details of a standard isolation procedure.

There are also secondary hazards associated with electrical testing that must be considered when risk assessments and a safe work practice document are being prepared.

**ASSESSMENT GUIDANCE**

When you carry out the fault-finding exercise you will be under direct observation. Keep up a running commentary of what you are doing.

### Minimising hazards of lone working

Fault diagnosis, which may require some form of live electrical work, is often undertaken by people working alone and without close or direct supervision. There is no specific requirement contained in the Electricity at Work Regulations for two people to be present in live-work situations. It is the responsibility of the duty holder to assess the need for a second person in terms of the nature of the danger and other precautionary measures to be adopted (such as role planning and relevant training). If a second person is *not* present, consideration must be given to special procedures for the lone worker such as:

- making sure that every site visit is recorded in a visitor log book
- keeping in contact with the employer by making regular telephone calls
- signing out when leaving the site for any reason
- making contact with the site owner on returning to site, to confirm that there have been no material changes to the system since the last visit.

It is likely that an electrician visiting a site for fault rectification purposes will be required to undergo some form of induction training which will explain the requirement to observe site-specific elements appropriate to their own work and/or site-specific activities, such as:

- site hazards and risks, for example, open excavations, presence of overhead power lines, confined spaces and hazardous areas
- fire risks and site fire procedures
- area of work that will require specific authorisation to proceed such as a permit to work
- restricted areas and the reasons for the control measures in place.

## Minimising hazards from static electricity

Static electricity is produced by the build-up of electrons in weak electrical conductors or insulating materials. These materials may be gaseous, liquid or solid and may include flammable liquids, powders, plastic films and granules.

A simple example of static electricity is a person walking across a carpeted floor, who builds up a static charge because of the friction of their shoes on the carpet. When the person touches an object that is either uncharged or that has an opposite charge, a short pulse of very high-voltage static charge is released quickly. This can cause damage to sensitive circuitry in computers, laptops and communication systems. Anti-static carpet sprays can be used to combat this problem.

Large static charges may develop on drive belts between motors and machinery, such as those involving laminate manufacturing processes or large rolls of paper for toilet tissue or kitchen roll running at high speeds. If not diverted through proper earthing, the built-up static charges may discharge suddenly and cause damage to electronic equipment such as proximity switches and programmable logic controllers. Earthed anti-static strips, which look similar to tinsel, can be strategically positioned to continuously discharge these unwanted voltages.

In areas containing explosive materials, great care must be taken to prevent static discharges, since a spark could set off a violent explosion. The use of large-diameter pipes for the transfer of liquids reduces the flow rate and, hence, the build-up of static charge. Airborne fibres, dust or flecks of paper – such as those produced in industrial processes – should be removed at source and not be allowed to accumulate as this may create a fire risk if static is likely.

The static effect is increased in environments with low humidity and in buildings with air conditioning and high levels of heating, such as computer suites, and this should be considered when undertaking electrical work in such areas.

### ACTIVITY

Static electricity can be generated when a person simply walks across a room. Clothing movement causes problems, especially when man-made fibres are involved. How could static problems be prevented in an operating theatre?

## Minimising hazards from fibre-optic cabling

As more fibre-optic cable systems are introduced into installations, the maintenance and fault-finding demands increase. Probably the greatest hazard to deal with, for approved electricians, is fibre-optic cable's benign appearance. It doesn't carry electricity, so there is no electrocution risk; it isn't a source of heat or combustion so there is no fire risk; and it's not possible to know when it is operating.

Energy is transferred along fibre-optic cables as digital pulses of laser light and, therefore, direct eye contact with the light should be avoided. However, damage to the eye caused by looking into a damaged fibre-optic cable is rare as the broken surface of the cable tends to scatter the light coming through it.

**Electricians should use appropriate PPE when dealing with fibre-optic cables**

A more serious hazard of optical-fibre work comes from the fibres themselves. Fibres are pieces of optical glass which, like any piece or sliver of glass, can cause injury. Hazards arise when cables are opened and fibres penetrate the hands or are transferred to the respiratory or alimentary system. Therefore, suitable PPE (such as gloves and face masks) should be used when working with fibre-optic cabling.

Fibre-optic cables have a similar appearance to steel wire armour cables (SWA) but are much lighter. They should be installed in the same way and given a similar level of protection as steel wire armour cables. Fibre-optic terminations should always be carried out in accordance with the manufacturer's instructions.

## Minimising hazards from electronic devices

The use of electronic circuits in all types of electrical equipment is now commonplace, with components being found in motor starting equipment, control circuits, emergency lighting, discharge lighting, intruder alarms, special-effect lighting and dimmer switches, and domestic appliances. Electrical installations usually operate at currents in excess of 1 A and up to many hundreds of amperes; all electronic components and circuits are low voltage and usually operate in mA or µA. Therefore, this fact must be considered when the choice of electrical test equipment is made.

**KEY POINT**

Fibre-optic cables are termed Band I circuits when used for data transmission and must be segregated from other mains cable. Band I covers installations where the voltage is limited for operational reasons (telecommunications, signalling, bell, control and alarm installations). Extra-low voltage (ELV) will normally fall within Band I.

The working voltage of a standard insulation resistance tester (250 V/500 V) may cause damage to any of the electronic devices described above, if not rated for this voltage. When carrying out insulation resistance tests as part of a prescribed test or after fault repair all electronic devices must be disconnected. Should resistance measurements be required on electronic circuits and components, a battery-operated ohmmeter with high impedance must be used.

## Minimising hazards from high-frequency or capacitive circuits

**ASSESSMENT GUIDANCE**

Capacitors are not only used in equipment but as stand-alone units for power factor correction. These need to be isolated and discharged to be safe.

Capacitors that are used in fluorescent fittings and single-phase motors are a potential source of electric shock arising from:

- the discharge of electrical energy retained by the capacitor unit(s) after they have been isolated
- inadequate precautions to guard against electric shock as a result of any charged conductors or associated fittings
- charged capacitors that are inadequately short circuited
- equipment retaining or regaining a charge.

Live working is to be avoided where possible, but any equipment containing a capacitor must be assumed to be live or charged until it is proven, beyond any doubt, that the circuits are dead and the capacitors are discharged and are unable to recharge.

## Minimising hazard from storage batteries

Batteries are used to store electrical energy. Their ability to provide instant output makes them a regular source of standby power for emergency lighting, emergency trip coils and uninterruptable power supplies (UPS) at computer, data storage and telecommunication facilities, for example. They can prove invaluable for maintaining supplies where a supply interruption could cause business issues. There is, however, risk attached to these larger battery systems which, if used incorrectly, can be dangerous and can cause explosions.

There are two main classes of battery.

- *Lead–acid batteries* are the most common large-capacity rechargeable batteries. They are fitted extensively in UPS of computer and communication facilities, and process and machinery control systems.
- *Alkaline rechargeable batteries*, such as nickel–cadmium, nickel–metal hydride and lithium ion, are widely used in small items such as laptop computers, but large-capacity versions of these cells are now being used in UPS applications.

**Batteries can explode if used incorrectly**

The two main hazards from batteries are described below.

- *Chemical* – lead–acid batteries are usually filled with electrolytes such as sulphuric acid or potassium hydroxide. These very corrosive chemicals can permanently damage the eyes and produce serious chemical burns to the skin. If acid gets into the eyes, they should be flushed immediately with water for 15 minutes, and prompt medical attention should be sought. If acid gets on the skin, the affected area should immediately be rinsed with large amounts of water and prompt medical attention should be sought if the chemical burn appears serious. Emergency wash stations should be located near lead–acid battery storage and charging areas. Under severe overcharge conditions, hydrogen gas may be vented from lead–acid batteries and this may form an explosive mixture with air if the concentration exceeds 3.8% by volume. Adequate ventilation and correct charging arrangements should always be observed.
- *Electrical* – Batteries contain a lot of stored energy. Under certain circumstances, this energy may be released very quickly and unexpectedly. This can happen when the terminals are short circuited, for example with an uninsulated metal spanner or screwdriver. When this happens, a large amount of electricity flows through the metal object, making it very hot, very quickly. If the battery explodes, the resulting shower of molten metal and plastic can cause serious burns and can ignite any explosive gases present around the battery. The short circuit may also produce ultraviolet (UV) light which can damage the eyes.

**ASSESSMENT GUIDANCE**

Battery voltage may be quite low but the current flow can be very high, especially under short-circuit conditions.

Most batteries produce quite low voltages, and so there is little risk of direct electric shock. However, as some large battery combinations produce more than 120 V d.c. precautions should be taken to protect users and those working in the vicinity of these installations.

**KEY POINT**

Remember, you cannot isolate the energy in a battery. You can disconnect it from the load, but the battery is still charged.

- Ensure that live conductors are effectively insulated or protected.
- Keep metal tools and jewellery away from the battery.
- Post suitable notices and labels warning of the danger.
- Control access to areas where dangerous voltages are present.
- Provide effective ventilation to stop dangerous levels of hydrogen and air or oxygen accumulating in the charging area.

If battery cables are removed for any reason, ensure that they are clearly marked ‘positive’ and ‘negative’ so that they are reconnected with the correct polarity.

## Minimising hazards from information and technology (IT) equipment

All modern offices contain computers that are operated as standalone units or networked (linked together). Most computer systems are sensitive to variations in the mains supply, which can be caused by external mains switching operations, faults on adjacent electricity distribution network circuits, which cause a voltage rise, and industrial processes such as induction furnaces and electric arc welding. These system disturbances are often well within the electricity supply companies' statutory voltage limits but may be sufficient to cause computers to crash.

To avoid the effects of this system 'noise', computer networks are normally provided with a clean supply, including a 'clean earth'. This is obtained by taking the supply to the computer network from a point as close as possible to the supply intake position of the building. The fitting of noise suppressors or noise filters provides additional protection to the computer system. If a production process disruption or data loss is a serious business risk, an uninterruptable power supply (UPS) is a sensible solution.

A UPS provides emergency power by supplying power from a separate source (normally batteries) when the supply from the local supply company is unavailable. It differs from an auxiliary or emergency power system or standby generator, as these do not provide instant protection from a temporary power interruption. A UPS can be used to provide uninterrupted power to equipment, typically for 5 to 15 minutes, until an auxiliary power supply can be turned on, mains power restored, or equipment safely shut down. UPS units come in sizes ranging from units that will back up a single computer without monitor (around 200 W) to units that will power entire data centres or buildings (several megawatts, MW).

**Many single computers and installations have uninterruptable power supplies, so that they may be live even if the mains power is off**

All of the above features are designed to keep power supplies constant for IT and similar equipment to prevent data loss, but fault diagnosis can impact on this requirement. Before any fault diagnosis is carried out, confirm whether the standby or UPS will present a danger to the electrician undertaking the work. Remember, the preferred method of work is with the system dead and, as such, it must be isolated from *all* points of supply, including auxiliary circuits, and dual or alternative supplies, such as UPS or a standby generator (that may be set for automatic start-up).

If it is necessary to isolate the computer supply network, permission must be sought from the duty holder or the responsible person for the computer systems to avoid any loss of data or damage to the computers. However, standby or UPS backup systems can also be used to avoid IT shutdowns during fault diagnosis and testing by maintaining supplies to critical areas.

**KEY POINT**

Proving dead at the point of work is the most important part of ensuring a safe system of work. If the isolation has been incorrectly applied, the act of proving dead should identify that the circuit is, in fact, still live. A proper procedure, using a suitable proving device, is therefore essential (see page 524).

## Minimising hazards from hazardous areas

Although all the hazards discussed above could be considered to give rise to hazardous areas, a 'hazardous area' is defined in the Dangerous Substances and Explosive Atmospheres Regulations 2002 (DSEAR) as 'any place in which an explosive atmosphere may occur in quantities such as to require special precautions to protect the safety of workers'.

These regulations provide a specific legal requirement to carry out a hazardous area study and document the conclusions in the form of zones. In the context of the definition, 'special precautions' relate to the construction, installation and use of apparatus, as given in BS EN 60079-10.

Hazardous areas are classified into zones based on an assessment of the frequency of the occurrence and duration of an explosive gas atmosphere, as follows:

- Zone 0: an area in which an explosive gas atmosphere is present continuously or for long periods
- Zone 1: an area in which an explosive gas atmosphere is likely to occur in normal operation
- Zone 2: an area in which an explosive gas atmosphere is not likely to occur in normal operation and, if it occurs, will only exist for a short time.

Most volatile materials (those that disperse readily in air) only form explosive mixtures between certain concentration limits. The flash point of a volatile material is the lowest temperature at which it can vaporise to form an ignitable mixture in air; flash point refers to both flammable liquids and combustible liquids.

The primary risk associated with combustible gases and vapours is the possibility of an explosion. Explosion, like fire, requires three elements: fuel, oxygen and an ignition source. Each combustible gas or vapour will ignite only within a specific range of fuel/oxygen mixtures.

**KEY POINT**

**Flash point** is a term used to refer to liquids that are **flammable**.

Gases are **ignitable** and we use LEL and UEL to define limits when a gas will ignite or explode.

**Flammability** is a measure of how easy it is for something to burn or ignite, causing a fire.

**Combustion** is a chemical reaction between fuel, oxygen and an ignition source (eg matches). The fuel can be a gas, a liquid or a solid such as wood or paper.

**KEY POINT**

An area in which an explosive mixture is present is called a **hazardous area** and precautions need to be taken when installing or working on any electrical equipment.

Too little or too much gas and the gas or vapour will not ignite. These conditions are defined by the lower explosive limit (LEL) and the upper explosive limit (UEL). Any amount of gas between the two limits is explosive.

Sources of ignition should be effectively controlled in all hazardous areas by a combination of design measures and systems of work. All electrical equipment must carry the appropriate markings if the integrity of the wiring system is to be maintained. A luminaire to IP 65 (this International Protection code means the fitting is dust-tight and water-jet protected) is an example of equipment used in hazardous areas.

**ACTIVITY**

Can you name three appliances which may come under the heading of '**Hot work**'?

**Hot work**

Work of any type which involves actual or potential sources of ignition and which is done in an area where there may be a risk of fire or explosion (for example welding, flame cutting and grinding).

When any work is to be undertaken on electrical equipment in a hazardous area the following procedures must be observed:

- a written safe work procedure (PTW system) that includes control of activities which may cause sparks, hot surfaces or naked flames should be in place
- control of the working area with signs and barriers
- use of appropriate PPE
- only authorised and competent people to be engaged in the work activity
- only approved tools and equipment to be used
- prohibition of smoking, use of matches and lighters.

If work is to be undertaken in a hazardous area, the hazards and the requirements to be observed will normally be explained during the site induction process.

**Assessment criteria**

**2.3** Explain the logical stages of fault diagnosis that should be followed

## STAGES OF FAULT DIAGNOSIS

Diagnosis of faults requires knowledge, experience and a logical and disciplined approach. Past experience or detailed knowledge will help in some situations, but fault situations are rarely, if ever, the same and each problem has to be investigated on an individual basis. An intuitive approach can be used, but this must be accompanied by a logical technique.

**ASSESSMENT GUIDANCE**

Do not rush in like a bull in a china shop. Even if it *looks* like a previous fault, that doesn't mean it is.

The fundamental steps in the logical diagnostic process for all types of equipment are:

- analysis and identification of the symptoms
- collection and analysis of data, information and maintenance records
- visual inspection
- fault location
- circuit checks and testing
- repair or replacement of damaged component
- functional tests
- restoration.

Before a fault on an electrical installation can be diagnosed, the approved electrician must have:

- an approved safe working procedure or method statement, as detailed on page 508, including a risk assessment
- a thorough knowledge and understanding of the electrical installation and electrical equipment (including access to site-specific circuit diagrams and drawings, manufacturers' information and operating instructions, design data, copies of previous Schedules of Inspection or certificates, installation specification and maintenance records)
- information relating to the fault, including any available knowledge of events leading up to the fault (such as unusual smells or observations); this can be obtained from users, the duty holder or the responsible person
- details of personnel relevant to the operation of the installation.

Once the probable cause of a fault has been predicted by analysis, it must be proven. In some situations one fault may be masked by a secondary fault and both need to be identified. This can sometimes be achieved by an inspection but, in many cases, test instruments will be used to help narrow down the faulty component or circuit.

The inspection referred to above is not simply a visual check. Other senses, such as smell and touch, must also be employed during the inspection. For example, the sense of touch is used to test conductors to see if they are loose and the sense of smell can identify overheating thermoplastic material.

If a fault has occurred, it is likely that users of the system will have reported a loss of supply or changes in the way the system operates; they will probably be able to identify the areas affected. Determining which protective device has operated (main circuit breaker, individual circuit breaker or residual current device) will narrow the search area and a flow diagram can often assist in this process.

Questions to ask users include the following.

- Which parts of the system are without supply – the whole installation or a particular item of equipment?
- Is the failure a permanent problem or does it only happen at certain times or on certain days?
- What are the events that led up to the fault occurring?

Once all of the available information has been obtained, the electrician can begin to determine the likely cause of the fault, using their own skills, expertise and competence and that of colleagues, if available. At this point, no physical work should be undertaken on the installation.

The next stage in the procedure is to analyse the information and begin to predict the probable cause of the fault. The predictions should then be tested, using a logical approach to attempt to identify the cause of the fault.

If it is not possible to diagnose the fault from the information gathered, it may be necessary to carry out some form of testing. This will require the following procedures to be adopted.

- Decide whether the testing is to be carried using dead or live working practices.
- Agree the extent of electrical installation to be tested with the responsible person or duty holder and, if necessary, obtain approval for circuits to be isolated.
- Carry out a visual inspection of the installation.
- If dead working is the chosen option, isolate circuits using a safe isolation procedure (a standard isolation procedure is given below).
- Notify all persons who are going to be affected by the loss of supply.
- Ensure test instrument has a valid calibration certificate.
- Check and test the supply and protective devices.
- Interpret test results and, using previous pre-test data, diagnose the cause of the fault.
- Correct the fault.
- Carry out the relevant safety and functional tests before restoring the supply.

A standard procedure for isolation is as follows.

1. Select an approved voltage indicator to GS 38 and confirm operation.
2. Locate correct source of supply to the section needing isolation.
3. Confirm that the device used for isolation is suitable and may be secured effectively.
4. Power down circuit loads if the isolator is not suitable for on-load switching.
5. Disconnect using the located isolator (from step 2).
6. Secure in the off position, keep key* on person and post warning signs.
7. Using voltage indicator, confirm isolation by checking ALL combinations.
8. Prove voltage indicator on known source, such as proving unit.

*If the device used for the purpose of isolation is a fuse or removable handle instead of a lockable device, keep this securely under supervision while work is undertaken.

Carrying out a safe isolation procedure will safeguard not only the person undertaking work on the installation, but also other people who may be working within the building, such as:

- occupiers and other trades people ( who may be inconvenienced by loss of supply to essential equipment/machinery)
- the customer or client for whom the work is being done (who may suffer from loss of service and downtime)
- members of the public (who are exposed to possible danger due to loss of essential services such as fire-alarm systems, emergency or escape lighting)
- those who require the continued provision of a supply for data and communication systems.

Remember that, if isolation is required to be carried out on a distribution board that is in a communal area, such as at the entrance to flats, a shop or hotel, there will be additional requirements such as barriers and notices to prevent unauthorised access to the work area.

Most electrical accidents occur because people are working on or near equipment that is:

- thought to be dead but which is live
- known to be live, but those involved do not have adequate training or appropriate equipment, or they have not taken adequate precautions.

If the correct isolation procedure is *not* undertaken, it can have implications not only for the person who is carrying out the work on the installation but to other people, such as the occupiers of the building, clients or customers, other trades or personnel working within the building and members of the public.

The implications:

- to the person carrying out the work and to members of the public are the possibility of electric shock or burns
- to the occupiers of the building include a risk of contact with electrical parts when basic protection has been removed
- to clients and customers will be risk of shock or burns and damage to equipment.

The incorrect isolation procedure may also present a risk of damage to electrical equipment and to building fabric.

**ASSESSMENT GUIDANCE**

Never take any action that may put yourself or anyone else in danger. When undertaking the unit assessment you must work safely at all times.

## ACTIVITY

Think of four groups of people who may be affected by a power loss in a shopping centre.

### KEY POINT

Remember that, although live testing may be required for fault finding, the fault repair must not be undertaken with the installation live. The individual circuit or piece of equipment must be isolated before repair work is commenced. For example, a lighting circuit which does not operate correctly may require linking at a switch and testing for continuity at the light point, then opening again to test the switch lines are operating correctly. A radial power circuit may need breaking at mid-point to determine the section of an open circuit and then, when the direction of the fault is located, that section may be broken down into halves to locate the open circuit.

If live testing is the only option for fault diagnosis, the person carrying out the live work must:

- be appropriately trained and have sufficient technical knowledge and experience to avoid danger
- use approved test equipment and test probes
- have appropriate insulated tools, equipment and protective clothing
- use insulated barriers or screens
- put in place effective control of the work area, such as restricting access (by use of appropriate barriers and warning notices).

Other systems that may be affected during testing must be identified and taken out of service. Some equipment, such as computers or other IT machinery, can be severely disrupted or damaged by a sudden loss of supply or by being subjected to a high test voltage from an insulation resistance test. Standby generators are standard equipment in many data centres and hospitals, and isolation of circuits for testing may cause the emergency systems to operate. Therefore, alternative arrangements will need to be made, and agreed in advance, when supplies are to be disrupted or fault-finding is to take place. Any resistance measurements of electronic equipment or electronic circuits must be carried out using battery-operated ohmmeters to avoid damage to the electronic circuits.

**Assessment criteria**

**2.4** Specify the requirements for the safe and correct use of instruments to be used for fault diagnosis

## USE OF INSTRUMENTS FOR FAULT DIAGNOSIS

In fault diagnosis, the use of suitable and safe voltage-indicating devices and measuring instruments is as important as the competency of the person undertaking the fault-finding activities. The possibility of touching live parts is increased during electrical testing and fault finding, when conductors at dangerous voltages are often exposed. The risks can be reduced if testing is done while the equipment or part of an installation is made dead and is isolated from any dangerous source of electrical supply. Special attention should be paid when carrying out tests with instruments capable of generating test voltages greater than 50 V or which use the supply voltage for the purpose of earth-loop testing or a residual-current device test. Refer back to pages 513–514 for more information about electric shock.

The HSE has produced Guidance Note GS 38 (Electrical test equipment for use by electricians), which is intended to provide guidance for electrically competent people, including electricians, electrical contractors, test supervisors, technicians, managers and/or appliance retailers. It offers advice in the selection and use of test probes, leads, lamps, and voltage-indicating devices and measuring equipment for circuits with rated voltages not exceeding 650 V. It recommends the use of fused test leads aimed primarily at reducing the risks associated with arcing under fault conditions. Where possible, it is recommended that tests are carried out at reduced voltages, which will usually reduce the risk of injury.

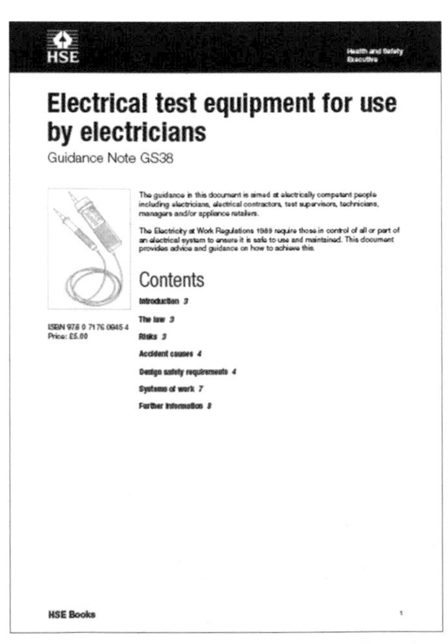
HSE Health and Safety Executive

**Electrical test equipment for use by electricians**

Guidance Note GS38

The guidance in this document is aimed at electrically competent people including electricians, electrical contractors, test supervisors, technicians, managers and/or appliance retailers.

The Electricity at Work Regulations 1989 require those in control of all or part of an electrical system to ensure it is safe to use and maintained. This document provides advice and guidance on how to achieve this.

ISBN 978 0 7176 0945 4
Price: £5.00

Contents

Introduction 3
The law 3
Risks 3
Accident causes 4
Design safety requirements 4
Systems of work 7
Further information 8

HSE Books 1

**Guidance Note GS 38, an essential tool for electricians and others**

## Probes and leads

Guidance Note GS 38 recommends that probes and leads used in conjunction with voltmeters, multi-meters and electricians' test lamps or voltage indicators should be selected to prevent danger.

Probes should:

- have finger barriers or stops, or be shaped so that the hand or fingers cannot make contact with live conductors under test
- be insulated to leave an exposed metal tip not exceeding 4 mm measured across any surface of the tip. Where practicable, it is strongly recommended that this is reduced to 2 mm or less, or that spring-loaded retractable screened probes are used.

Leads should:

- be adequately insulated and coloured so that any one lead is readily identifiable from any other
- be flexible and sufficiently robust
- be long enough for the purpose – but not too long
- not have accessible exposed conductors, even if they become detached from the probe or from the instrument
- have suitably high breaking capacity (hbc), sometimes known as hrc, fuse or fuses with a low current rating (usually not exceeding 500 mA), or a current-limiting resistor and a fuse.

GS 38 also recommends that, if a test for the presence or absence of voltage is being made, the preferred instrument to be used is a proprietary test lamp or voltage indicator. The use of multi-meters for voltage indication has often resulted in accidents due to the multi-meter being set on the incorrect range.

**ASSESSMENT GUIDANCE**

When using a two-lead testing device, always place the first lead on the earth or neutral terminal. This will ensure that the free lead is not live.

**KEY POINT**

Voltage indicators and test lamps do not fail safe and, as such, must always be tested on a known voltage source before and after use.

## Standards for test instruments

The basic instrument standard is BS EN 61557: 'Electrical safety in low voltage distribution systems up to 1000 V a.c. and 1500 V d.c. Equipment for testing, measuring or monitoring of protective measures'. This standard includes performance requirements and requires compliance with BS EN 61010: 'Safety requirements for electrical equipment for measurement control and laboratory use' which is the basic safety standard for electrical test instruments.

**ACTIVITY**

Multi-meters (as opposed to multi-function meters) are not recommended for live fault-finding. Give reasons for this.

The following table shows the test instrument along with the associated harmonised standard.

| Instrument | Standard | Instrument | Standard |
|---|---|---|---|
| insulation resistance ohmmeters | BS EN 61557-2 | earth electrode resistance testers | BS EN 61557-5 |
| earth fault loop impedance testers | BS EN 61557-3 | RCD testers | BS EN 61557-6 |
| low-resistance ohmmeters | BS EN 61557-4 | | |

## Instrument accuracy

A basic measuring accuracy of 5% is usually adequate for test instruments constructed in accordance with BS EN 61557 and for analogue instruments, a basic accuracy of 2% of full-scale deflection will provide the required accuracy measurement over the useful proportion of the scale.

## Using instruments safely

When using test instruments to carry out fault diagnosis, follow these basic precautions to achieve safe working.

- *Understanding the equipment* – make sure you are familiar with the instrument to be used and its ranges; check its suitability for the characteristics of the installation it will be used on.
- *Self test* – many organisations regularly test instruments on known values to ensure they remain accurate. These tests would be documented.
- *Calibration* – all electrical test instruments should be calibrated on a regular basis. The time between calibrations will depend on the amount of usage that the instrument receives, although this should not exceed 12 months under any circumstances. Instruments have to be calibrated under laboratory conditions, against standards that can be traced back to national standards; therefore, this usually means returning the instrument to a specialist laboratory. Once calibrated, the instrument will have a calibration label attached to it stating the date the calibration took place and the date the next calibration is due. It will also be issued with a calibration certificate, detailing the tests that have been carried out, and a reference to the equipment used. Instruments that are subject to any electrical or mechanical misuse (for example, if the instrument undergoes an electrical short circuit or is dropped) should be returned for recalibration before being used again. Electrical test instruments are relatively delicate and expensive items of equipment and should be handled with care. When not in use, they should be stored in clean, dry conditions at normal room temperature. Care should also be taken of instrument leads and probes, to prevent damage to their insulation and to maintain them in safe working condition.
- *Check test leads* – make sure that these and any associated probes or clips are in good order, are clean and have no cracked or broken insulation. Where appropriate, the requirements of the HSE Guidance Note GS38 should be observed for test leads.
- *Select appropriate scales and settings* – it is essential that the correct scale and settings are selected for an instrument. Manufacturers' instructions must be observed under all circumstances.

**KEY POINT**

The user of the instrument should always check to ensure that the instrument is within the calibration period before using it.

## Insulation resistance testers (BS EN 61557-2)

Insulation resistance test meters are used for insulation resistance testing and the separation of circuits, including SELV or PELV. Insulation tests should be made on electrically isolated circuits and, before commencing the test, it should be verified that pilot or indicator lamps and capacitors are disconnected from the circuits being tested, to avoid inaccurate test results. All voltage-sensitive electronic equipment, such as residual current circuit breakers (RCCBs), residual-current circuit breakers with overload protection (RCBOs), dimmer switches, touch switches, delay timers, power controllers, electronic starters for fluorescent lamps, emergency lighting and residual current devices (RCDs), should be disconnected so that they are not subjected to the test voltage. If these items cannot be disconnected, a measurement to protective earth with the line and neutral connected together should be made.

The operating accuracy of insulation resistance testers can be affected by 50 Hz currents induced into the cables under test and by capacitance in the test object. This capacitance may be as high as 5 μF. It is necessary for the test instrument to have a facility that discharges this capacitance safely. The test leads should not be touched while the tester is being used and, after the test, the instrument should be left connected until the capacitance within the installation has fully discharged.

Remember that the values shown when insulation resistance testing are in megohms (MΩ, millions of ohms).

### KEY POINT

As the tester will be applying at least 250 V (and most often 500 V) to the circuit being tested, care must be taken not to touch exposed terminals or other exposed metalwork during testing. The guidance contained in GS 38 for test probes and leads should be followed. If premises are occupied while the testing is underway, the occupants and any possible visitors must be advised that testing is taking place. If practical, barriers and notices should be posted to warn people that testing is taking place.

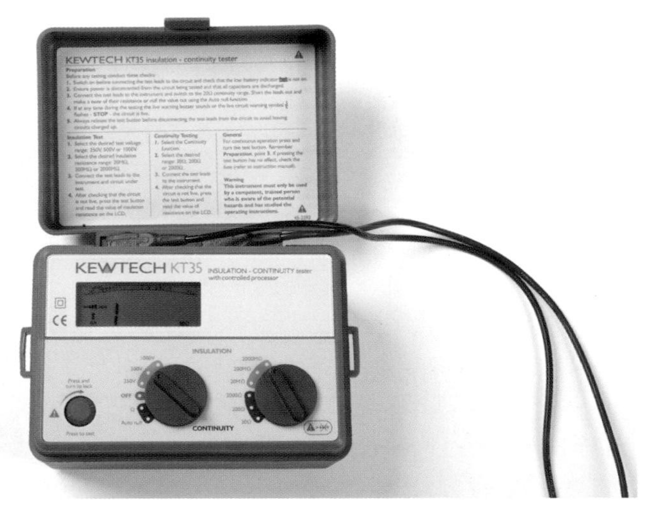

An insulation resistance tester

Finally, never measure resistance in a circuit when power is applied. If the circuit has sufficient power, the meter can explode or burst into flames. Most modern meters are fuse and diode protected, in order to prevent explosions, but would still be damaged by an overload of this magnitude.

Details of the method for completing insulation resistance tests can be found in Unit 304, pages 423–430.

### ACTIVITY

If a meter is inadvertently connected to the mains supply it is reasonable to assume that some damage may have occurred. What course of action would you take?

## Earth fault loop impedance tester (BS EN 61557-3)

These instruments operate by drawing a test current (25 A) from the supply and causing it to flow around the earth fault loop. Earth fault loop impedance is required to verify that there is an earth connection and that the value of the earth fault loop impedance is less than or equal to the value determined by the designer and which was used in the design calculations.

The earth fault current loop comprises the following elements, starting at the point of fault on the phase to earth loop:

- the protective conductor
- the main earthing terminal and earthing conductor
- for TN systems (TN-S and TN-C-S), the metallic return path or, in the case of TT systems, the earth return path
- the path through the earthed neutral point of the transformer
- the source phase winding
- the line conductor from the source to the point of fault.

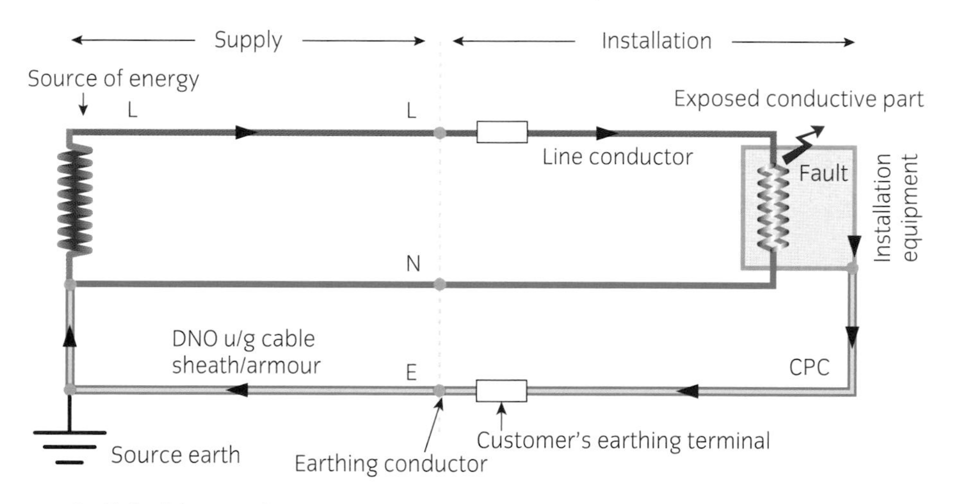


**Earth fault loop**

## ACTIVITY

Using an earth loop impedance tester to measure the earth electrode resistance involves disconnecting which conductor?

Earth fault loop impedance testing is not recommended for fault diagnosis but can be used to confirm that any repair or rectification has not altered the maximum permissible earth fault loop impedances.

Details of the method for carrying out earth fault loop impedance tests can be found in Unit 304, pages 433–444.

## Low-resistance ohmmeters (BS EN 61557-4)

These are used for checking protective conductor continuity, including main and supplementary equipotential bonding, ring final circuit continuity tests and polarity. Other applications include the checking of the integrity of welded joints, bolted lap joints on bus bars and verifying winding resistance in large motors and transformers.

The test current may be d.c. or a.c. and should be derived from a source with a no-load voltage of not less than 4 V, and no greater than 24 V and a short-circuit current of not less than 200 mA.

The measuring range should cover the span 0.2 Ω to 2 Ω with a resolution of at least 0.01 Ω for digital instruments and, for continuity testing, the lowest ohms scale must be selected – low values are expected.

Note that general-purpose multi-meters are not capable of supplying these voltage and current parameters.

Field effects contributing to in-service errors are contact resistance, test lead resistance, a.c. interference and thermocouple effects in mixed metal systems.

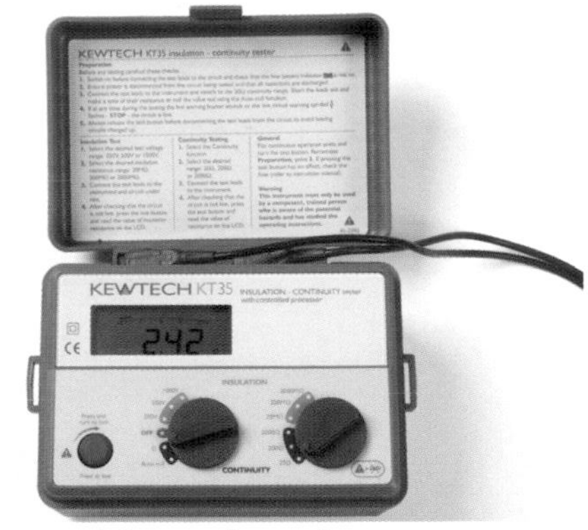

Low resistance ohmmeter

## Earth electrode testers (BS EN 61557-2)

There are three methods of measuring the resistance of an earth electrode:

- E1 uses a dedicated earth electrode tester (fall-of-potential three- or four-terminal type)
- E2 uses a dedicated earth electrode tester (stakeless or probe type)
- E3 uses an earth fault loop impedance tester.

Earth electrode resistance testing will not be described in this unit. It is covered in Unit 304, page 439.

## Residual current tester (RCD) (BS EN 61557-6)

RCDs are sensitive devices, typically operating on earth fault currents as low as 30 mA and with response times as fast as 20 to 40 ms. BS 7671 provides for wider use of RCDs, for indoor circuits as well as for socket outlets intended to supply portable outdoor electrical equipment that is outside the zone of earthed equipotential bonding.

Where an installation incorporates an RCD, Regulation 514.12.2 requires a notice to be fixed in a prominent position at or near the origin of the installation. The notice should be in indelible characters not smaller than those illustrated in the Regulation.

Note that the integral test button incorporated in RCDs only verifies the correct operation of the mechanical parts of the RCD and does not provide a means of checking the continuity of the earthing conductor, the earth electrode or the sensitivity of the device. This can only be done by use of an RCD tester designed for testing RCDs.

**ASSESSMENT GUIDANCE**

RCDs do not operate on overloads or short circuits, RCBOs do. Make sure you understand the difference between the two.

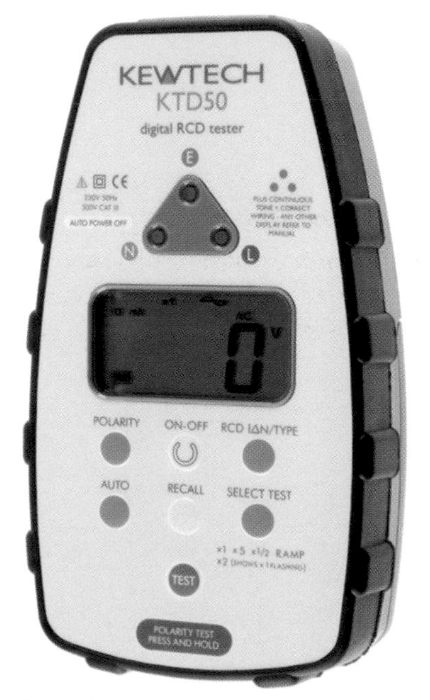

RCD tester

RCD testers should be capable of applying the full range of test current to an in-service accuracy, as given in BS EN 61557-6 (10%). This in-service reading accuracy will include the effects of voltage variations around the nominal voltage of the tester. To check the RCD operation and to minimise danger during the test, the test current should be applied for no longer than 2 s.

Details of the method for testing the correct operation of residual current devices are shown on page 461 in Unit 304.

## Measuring the prospective fault current, $I_{pf}$

The current that is expected to flow from line to neutral in the event of a fault is the prospective short-circuit current ($I_p$ or PSCC) and from line to earth it is the prospective earth-fault current ($I_f$ or PEFC).

The fault current depends on the total circuit impedance at the point of the fault. If the value of either the prospective short-circuit current or the prospective earth-fault current is higher than that which can be safely interrupted by the protective devices, catastrophic damage to the protective devices or associated equipment may occur. Typically, such fault currents are measured in kA. Regulation 612.11 requires both the prospective short-circuit current and the prospective earth-fault current to be measured, calculated or determined by another method, at the origin and at every other relevant point in the installation.

The reference to 'every other relevant point' means every point where a protective device is required to operate under fault conditions and this includes the origin of the installation.

Prospective fault levels should be measured at the distribution board, with all main bonding in place, between live conductors and between line conductors and earth.

To measure the prospective fault current, the prospective fault-current range of a suitable earth fault loop impedance tester can be used (BS EN 61557-3). Two-lead and three-lead testers are available for measuring prospective fault current. It is important that any instrument being used is set on the prospective fault current range and is connected in accordance with the manufacturer's instructions; note that instrument accuracy decreases as scale reading increases.

Remember, this is a live test and, as the tester will be using the mains supply, care must be taken not to touch exposed terminals or other exposed metalwork during testing. The guidance contained in GS38 for the test probes and leads should be followed.

Details of the method for checking prospective fault current can be found on page 457 of Unit 304.

# CAUSES OF ELECTRICAL FAULTS

A fault is defined in BS 7671: 2008 Requirements for Electrical Installations as: 'a circuit condition in which current flows through an abnormal or unintended path. This may result from an insulation failure or a bridging of insulation'. A fault is not a normal occurrence and the abnormal current flow could be caused by one of the following:

- poor system design
- designer's specification incorrectly implemented
- installation workmanship to a poor standard
- poor maintenance or neglect of equipment or installation
- misuse, abuse or deliberate ill-treatment of equipment.

**Assessment criteria**

**2.5** Explain causes of electrical faults

**ASSESSMENT GUIDANCE**

It could be that the original installation is perfectly acceptable but changes to the building layout may adversely affect the installation.

## Poor system design

The proper specification and design of electrical systems and equipment is fundamental to the safe operation of installations. The designer of a low-voltage system will need a good working knowledge of the current edition of BS 7671, associated IET Guidance Notes and codes of practice, technical and equipment standards. They should also understand the duties imposed by Regulations 4 to15 of the EAWR and the technical reasons behind them.

## Failure to interpret the design correctly

However good the design of a system, it has to be installed correctly to function as intended. Those concerned with installing the system must be competent to ensure that the designer's specification is correctly implemented and all workmanship is to a high standard.

## Installation workmanship to a poor standard

Typical areas of poor installation practice are:

- poor termination of conductors, which creates overheating due to poor electrical contact
- loose bushings and couplings, which may lead to poor earth continuity and risk of shock
- use of incorrectly sized conductors, leading to increased current flow and overheating
- cable damage that occurs when drawing in cables
- heavily populated trunking or conduit, leading to overheating
- incorrect connections at components, which may lead to crossed polarity (so that a circuit remains live although the protective device has operated).

Those concerned with installation work need to know not only *how* to do a job, but *why* it is important to do it in the specified way.

## Misuse of equipment

EAW Regulation 4(3) requires that: 'every work activity, including operation, use and maintenance of a system and work near a system, shall be carried out in such a manner as not to give rise, so far as is reasonably practicable, to danger and users (the duty holder) have a responsibility to use equipment in the manner for which it was intended.' Deliberate or unintentional misuse, such as overloading a radial circuit supplying socket outlets with too many electric heaters, could be seen as misuse.

### Abuse or deliberate ill-treatment of equipment

Abuse or deliberate ill-treatment of equipment, such as removal of barriers or covers, may result in contact between live conductors and a short circuit.

## Common faults

**ACTIVITY**

Some items listed as faults are not so much faults but the result of a fault further down the line. One such is a loss of phase (line). What could cause a loss of phase?

An electrical fault can reveal itself in a number of ways or in a combination of those ways, for example:

- complete loss of power throughout the whole installation
- partial or localised loss of power
- failure of an individual component.

### Failure of an individual component

Faults in components such as a switches, contactors, time switches and photo electric cells may result in particular circuits not operating as was intended. For example, heaters that are programmed to operate via a time switch can only be turned off by operation of the protective device.

### Total failure of a piece of equipment

An example might be the failure of a heating element in a fire or shower, resulting in an earth fault or short circuit.

### Low insulation resistance

An insulation fault can allow current leakage. As insulation ages, its insulating properties and performance deteriorates; this is accelerated in harsh environments, such as those with temperature extremes and/or chemical contamination. This deterioration can result in dangerous conditions for people and poor reliability in installations. It is important to recognise this deterioration so that corrective steps can be taken.

## Overload

If excess current is drawn by a circuit, protective devices will operate frequently. This could be caused by too many items of equipment being connected at the same time.

## Short circuit

Short circuits can occur between conductors in a circuit or component parts. These may be caused by a loose connection at a termination or joint.

## Open circuit

An open circuit may result from loose connections or an overheated joint; this may be identified by high resistance readings during continuity testing.

## Transient overvoltage

This is caused by heavy current switching and distribution network operator (DNO) system faults. Occasionally, large internal loads may cause such transient overvoltages.

The designer of a system must ensure that all system components are selected according to their intended use and are appropriately rated. Two main factors must be considered:

- switchgear, fuse gear and cables must be of adequate rating to carry both normal and any likely fault current safely
- the equipment should meet appropriate British, European or International Standards or CENELEC Harmonisation Documents (HDs) in order to withstand prospective fault conditions in the system.

However, design work does not cater for abnormal transient voltages caused by distribution network operator (DNO) faults on the supply network, heavy current switching that causes voltage drops, lightning strikes on overhead line conductors, electronic equipment and earth fault voltages. In some installations, such as IT networks, where transient voltages can cause problems, additional measures are included. Filters added to IT networks can suppress transient voltages and provide stabilised voltage levels.

## Undervoltage

This is often identified when electromagnetic relays fail to latch. Protection against undervoltage is normally required for motor drive machines. The Regulations require that, where unexpected restarting of a motor may cause danger, the provision of a motor starter that is designed to prevent automatic restarting must be provided. If motors do not start it may be due a motor fault or an undervoltage issue.

### ASSESSMENT GUIDANCE

Ring final circuits are designed to supply low levels of current to many outlets. In theory, 10 13 A sockets could supply a total load of 130 A but there is only a 32 A fuse at the consumer unit. Most of the installed load is only a fraction of an ampere.

**Assessment criteria**

**2.6** Specify types of electrical faults found in different locations

## WHERE ARE ELECTRICAL FAULTS FOUND?

Electrical faults can occur in a range of different areas and components. The following section is not exhaustive but highlights the major areas where faults may occur.

### Wiring systems

Wiring faults do occur in low-voltage systems but cable damage is not a common occurrence if the installation has been constructed in accordance with Chapter 52 of BS 7671.

However, if the installation has not been constructed to a high standard, damage may occur and a latent fault may exist, causing a failure at a later date. PVC insulated and sheathed wiring is extensively used for wiring in domestic and light industrial installations, and the very act of installing cable runs can result in mechanical damage to the cable due to impact, abrasion, compression or penetration (which must be minimised at all times). Where cables have to be turned or bent, the radius should be such that it does not damage the conductors. Where cables are installed in walls behind plaster or in a stud wall, mechanical protection is only required if the depth of the cable is less than 50 mm, or if they are not placed within accepted zones that are 150 mm wide around wall edges and ceiling–wall joints, or if the cables are not run horizontally or vertically to an accessory point. If it is not possible to provide such protection and the installation is used by people with no technical background, additional protection must be provided by the installation of a residual current device.

**ACTIVITY**

Many years ago there was a tendency to run cables down inside the cavity walls of domestic premises. This lead to compression of the cable at the top and possible insulation failure. Is this practice still allowed?

When designers calculate the conductor sizes for installations, based on load factors, they also specify how the cables should be installed; any deviation from this specification could have an adverse effect – cable current-carrying capacities will be reduced by such factors as grouping, surrounding the cables with thermal insulation and siting the cables in a hot environment that will cause overheating. Bear in mind that this could take place after the installation was originally commissioned. Poor installation practice could also result in cables being subjected to compression, pinching or abrasive damage.

Surface cabling in industrial workplaces, or where the movement of materials or goods may cause damage, must be protected from mechanical damage. This may be achieved by the use of cables incorporating mechanical protection (such as steel wire armoured or mineral insulated cable) or by the installation of cables in steel conduit or trunking, or by locating the cables where they will not be liable to damage.

Wiring faults often occur where cables are jointed or terminated. A termination is a connection between a conductor and other equipment (for example distribution boards or fixed appliances), or an accessory. Connections between conductors are usually termed 'joints'.

A poorly constructed joint or termination may give rise to arcing with the passage of load current, resulting in localised heating (and an increased fire risk) and progressive deterioration of the joint or termination, with evidence of pitting, arcing, carbon build-up or corrosion that could lead to eventual failure and an open circuit. Loose connections developing into a short circuit are often accompanied by audible sounds such as 'crackling' or 'fizzing' and by referred problems in adjacent parts of the installation.

When a fault occurs elsewhere on an associated part of an electrical system, a poor joint or termination may constitute a hazard by limiting the flow of fault current, thereby causing delayed operating of the circuit protective device. In an extreme case, the passage of high fault current may cause the defective joint to burn into an open-circuit fault before the circuit protective device has time to operate, with the result that the original fault remains uncleared. This is particularly hazardous in the case of an earth fault, since exposed conductive parts may become and remain live at the system voltage, giving an increased risk of electric shock.

## Equipment and accessories

Faults can also occur in equipment and accessories such as switches, sockets, control equipment, motor contactors and electric appliances or at the point of connection with electronic equipment. Contacts that make and break a circuit are a potential source of failure, depending on the amount of operation they are subjected to. Socket outlets and isolating switches that control fans in kitchens or bathrooms, or showers in bathrooms, that are subjected to regular use with capacity loads may fail due to overheating and loose connections.

Failure of an individual component, such as an electric shower or immersion heater element, may lead to an open-circuit fault. Photo-electric cells and time switches may fail in the 'closed position' and then do not operate as intended. Faulty RCDs may cause nuisance tripping. The correct operation of a suspected faulty RCD would need to be determined by the use of an approved RCD tester (Unit 304, page 461).

Modern RCD testers include a facility called a ramp test. The ramp test slowly increases the tripping current until the RCD trips. The display is in mA and will identify if an RCD is susceptible to nuisance tripping or is out of specification. It can also be used to indicate which circuits have high leakage currents and can be used when there is evidence of frequent trips on a particular RCD.

**ACTIVITY**

Remember that all equipment has some earth leakage, however small. It may be the sum of many leakages that causes the RCD to trip. What would you do to resolve this problem?

Under certain circumstances these tests can result in potentially dangerous voltages appearing on exposed and extraneous conductive parts within the installation. Therefore, suitable precautions must be taken to prevent contact by persons or livestock with any such part. Other users of the building should be made aware of the tests being carried out and warning notices should be posted as necessary.

Cables that are terminated at an accessory contained in, or passing through, a metal box should be bushed with a rubber grommet to prevent abrasion of the cable.

## Instrument and metering panels

Faults can occur in instrument and metering panels as a result of a faulty component, such as a burnt-out current transformer (CT) or voltage transformer (VT), or as a result of faulty test probes being applied to the instrument.

## Protective devices

These may be subject to loose connections. If an incorrect device was selected for the circuit and does not offer the correct protection or discrimination, a fault may result.

**ASSESSMENT GUIDANCE**

An old choke nearing the end of its life may become noisy as the laminations loosen.

## Luminaires

Light fittings often fault because the lamp has expired. Fluorescent fittings (discharge lighting) grow dimmer with age, and may even begin to flicker or to flash on and off. These are warning signals and the necessary repairs should be made as soon as there is any change in the lamp's normal performance. A dim tube usually requires replacement, and failure to replace it can strain other parts of the fixture. Likewise, repeated flickering or flashing will wear out the starter, causing the insulation at the starter to deteriorate.

## Flexible cords

Many faults occur when flexible cords are terminated because they are:

- of too small cross sectional area for the load
- not adequately anchored to reduce mechanical stresses
- not suitable for the ambient temperature at the point of termination.

It should be remembered that components are less likely than poor installation practices to cause a fault on an installation. Therefore, adherence to EAW Regulation 16 (Competence), the specific technical requirements of EAW Regulations 4 to 15 (concerned with the design, construction and maintenance of an electrical system) and the relevant IET Regulations is vital to ensuring the installation remains fault free.

## DIAGNOSING FAULTS BY TESTING

**Assessment criteria**

**2.7** Explain how faults are diagnosed by tests

Diagnosis of faults requires knowledge, experience and a logical and disciplined approach. Past experience or detailed knowledge will help in some situations, but fault situations are rarely, if ever, the same and each problem has to be investigated on an individual basis. An intuitive approach can be used, but this must be accompanied by a logical technique.

The fundamental steps in the logical diagnostic process for all types of equipment are:

- analysis and identification of the symptoms
- collection and analysis of data, information and maintenance records
- visual inspection
- fault location
- circuit checks and testing
- repair or replace damaged component
- functional tests
- restoration.

### Preparing for fault diagnosis by testing

The identification of the following data, information and documents will be of great assistance in the fault diagnosis work:

- copy of BS 7671 and associated Guidance Notes
- manufacturer's information and operating instructions
- copies of previous Schedule of Inspections or certificates
- installation specifications
- drawings and diagrams
- maintenance records
- reports from personnel relevant to the operation of the installation.

Once the probable cause of a fault has been predicted by analysis, it must be proven. In some situations one fault may be masked by a secondary fault and both need to be identified. This can sometimes be achieved by an inspection but, in many cases, test instruments will be used to help pinpoint the faulty component or circuit.

**ASSESSMENT GUIDANCE**

The majority of tests required for fault finding can be carried out by using multi-function test meters. However, when answering City & Guilds assessment questions, you are expected to give the name of the specific meter used in conjunction with a specific test.

For example: when carrying out continuity testing, a low-resistance ohmmeter should be used and the readings should be in ohms (Ω).

The inspection referred to above is not simply a visual check. Other senses, such as smell and touch, must also be employed during the inspection. For example, the sense of touch is used to test conductors to see if they are loose and the sense of smell can identify overheating thermoplastic material.

Remember these points.

- Before making a test, it is advisable to be aware of the anticipated reading if the circuit is operating normally.
- If the reading is anything other than the predicted value, it is likely that the part of the circuit under test is being affected by the fault.
- The next stage is to decide whether the testing is to be carried out using dead or live working practices.
- If dead working is the chosen option, use a safe isolation procedure (see page 524) to isolate circuits. Next, agree the extent of the electrical installation to be tested with the duty holder or responsible person and, if necessary, obtain approval for circuits to be isolated.
- Ensure the test instrument has a valid calibration certificate.
- Carry out the test.

### Sectionalising

In some cases, even though an analysis has taken place, the problem area may represent a very large portion of the installation and it would not be feasible to test all of the circuits, equipment and fittings in the installation. The best method of dealing with this situation to test the installation a section at a time; this technique is commonly called sectionalising and it reduces the size of the area under test, narrowing down the search area until the fault is identified.

Faults can generally be categorised into open circuits, short circuits, earth faults, high resistance, overloads and crossed polarity.

**ACTIVITY**

Testing of circuits can often be done using a 'half' test. What is meant by this term?

## Tests for different faults

### Open circuits

An open circuit generally occurs when there is a break in the circuit. This could be the result of a broken conductor, a loose connection caused by excessive mechanical stress or a component that has failed due to age, damage, overloading, suitability for continued use and overheating.

If the installation can be energised safely, a voltmeter is probably the most suitable instrument for finding an open circuit. Once the problem has been identified as an open circuit and the general area of the fault has been located, confirm that the instrument is operating correctly by connecting the negative lead to a known reference. The neutral or earth on an a.c. installation is preferable.

Test through the affected circuit with the other lead, making sure all switches are in the closed position. The conductor or device between the last test position where a full voltage was recorded and the first point where a less than full voltage is recorded is the point at which the open circuit is located.

Continuity testing may be used to locate an open-circuit conductor. To locate the conductor, the circuit may need to be divided into sections. For example, to test a lighting circuit that does not operate correctly, link at a switch and test for continuity at the light point, then open again to test the switch lines.

A radial power circuit may be broken at mid-point to determine the section of an open circuit and, when the direction of the fault is located, that section is broken into halves to locate the open circuit.

**ASSESSMENT GUIDANCE**

Ring final circuits may also be split in this way. One ring equals two radials when split in the middle.

## Short circuits

These occur when two or more components, which would normally be separated by insulation or barriers, come in contact with each other. For example, in an overload condition the circuit conductors are carrying more current than the manufacturer's design specification allows and, if this is allowed to continue indefinitely, the conductors will become very hot, causing deterioration of the electrical insulation surrounding the live conductors. This may eventually lead to breakdown of the insulation and a short circuit. In this overload situation, it is likely that the protective device, if specified correctly, will operate before a short-circuit condition is reached. Continual operation of a protective device should be investigated immediately, by isolating the circuit or carrying out an insulation resistance test, or by using a clamp meter or clip-on meter to check load readings.

If it is suspected that insulation is beginning to fail, indicated by the frequent operation of a protective device or the operation of a RCD, the fault can be located by undertaking an insulation resistance test on the whole installation (if it is a relatively small system) or (if it is a complex installation) on the various component parts. This will need to be carried out with the circuit isolated and the fitting disconnected (if it contains electronic equipment).

Alternatively, the isolated line and neutral can be joined together and tested to earth with the fitting(s) connected.

If there appears to be no low insulation-resistance reading or fault, the correct functioning of a 30 mA RCD can be checked by injecting ½ × rated current and 1 × rated current. If the RCD trips out at, say, 15 mA or it was felt necessary to test the sensitivity of the RCD, a ramp test could be employed.

This facility, which most modern RCD testers have incorporated, slowly increases the tripping current until the RCD trips. The display is in milliamps and will identify if an RCD is susceptible to nuisance tripping or is out of specification. It can also be used to indicate which circuits have high leakage currents and can be used when there is evidence of frequent trips on a particular RCD.

Under certain circumstances these tests can result in potentially dangerous voltages appearing on exposed and extraneous conductive parts within the installation. Therefore, suitable precautions must be taken to prevent contact with any such part by persons or livestock. Other building users should be made aware of the tests being carried out and warning notices should be posted as necessary.

## Earth faults

These occur when the fault current flows to earth. In practice, an earth fault usually occurs inside equipment containing exposed conductive parts. An exposed conductive part is a conductive part of equipment that can be touched and which is not a live part but which may become live under fault conditions; an example would be a metal equipment case. This type of fault happens when a live conductor makes contact with the conductive case, which then becomes charged at the same voltage. This may arise by, for example, a breakdown of electrical insulation or a conductor coming adrift from a terminal connection. The exposed conductive parts are likely to remain live since, although some leakage current to earth may arise, this will be insufficient to operate the circuit protective device. Thus, an indirect shock hazard is created and, if the live exposed conductive parts are touched, the body will provide a path for fault current to flow to earth. Earth faults may also be caused when a line conductor makes contact with earth or the heating element in a domestic iron, space heater or shower breaks down.

## High-resistance faults

These faults usually occur in a circuit where a cable or conductor is poorly joined in an accessory such as a socket outlet, switch, light fitting, or junction box. The most common cause is simply a loose screw terminal connection, because the connection was not made sufficiently well in the first place, debris is present in the connection (for example pieces of stripped insulation), or it might be that the connection has become loose over time. Connections can work loose through the normal thermal cycling of a circuit. If a circuit routinely carries a significant proportion of its maximum design current, the cable will be subject to repeated heating and cooling cycles. This can result in expansion and contraction, which can loosen the terminal screws. Vibration can also affect joints and connections, and that is why connections for generators need to be crimped or brazed.

## Faults in polarity

The polarity of all circuits must be verified before a connection is made to the supply, to ensure all single-pole devices (fuses, switches and circuit breakers) are connected in the line conductor only (BS 7671, Regulation 612.6). The test creates a circuit, using the line conductor and the single-pole device in question. Breaking the circuit when operating the device means that the reading on the instrument will change, thus confirming that the device must be connected in the line conductor. This test is carried out with either an ohmmeter or the continuity range of an insulation and continuity tester. A polarity check will also identify incorrect connections at components, which may lead to crossed polarity and a circuit remaining live when a protective device has operated.

There are three recognised methods of evaluation. All three methods have their advantages and possible dangers, if they are not carried out correctly. The methods are:

- polarity by visual inspection
- polarity by continuity testing
- live testing for polarity.

## Voltage testing

Within commercial, industrial and, possibly, some domestic installations different supply voltages will be encountered. It is important that suitable methods are used to verify these voltages. Instruments used solely for voltage verification fall into two categories:

- test lamps, which rely on an illuminated lamp to show if a voltage is present
- voltage indicating devices, which give a visual reading (analogue or digital) of the voltage.

### ASSESSMENT GUIDANCE

GS 38 requires test meter leads to be colour coded but some approved voltage indicator (AVI) leads are not as they do not rely on correct polarity for safety or function. Always check to see if the leads require correct polarity when used.

### ACTIVITY

Why do the test leads of the multimeter shown fail to comply with GS 38? What can be done to ensure that they comply?

Test lamp

Voltage indicator

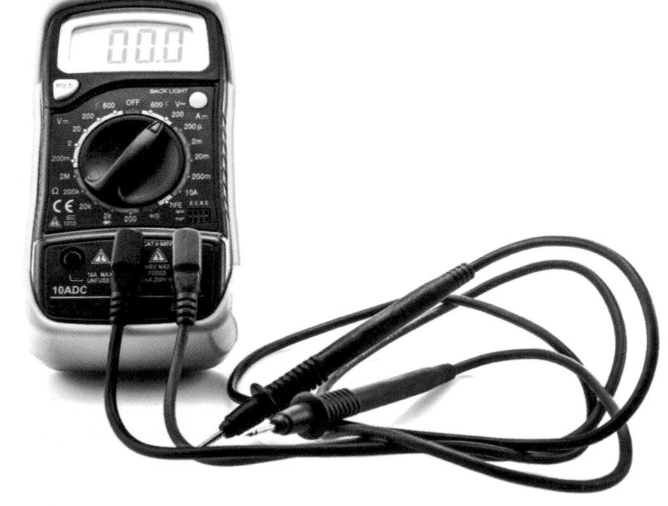


Multimeter

Test lamps are fitted with a 15 W bulb (that may be protected by a guard) and should not give rise to danger if the lamp is broken. This type of detector requires overcurrent protection, which may be provided by a suitable high breaking capacity (hbc or hrc) fuse or fuses with a low current rating (usually not exceeding 500 mA), or by means of a current-limiting resistor and a fuse. These protective devices are housed in the probes themselves. The test lead or leads are held captive and sealed into the body of the voltage detector.

Voltage indicators/detectors usually incorporate two or more independent indicating systems – visual and audible – and limit energy input to the detector by the internal circuitry. They are provided with in-built test features to check the functioning of the detector before and after use. If the detector does not have an in-built test feature it must be tested before and after application to the test area.

It is recommended that test lamps and voltage indicators are clearly marked with the maximum voltage that may be tested by the device and any short-time rating for the device, if applicable. This rating is the recommended maximum current that should pass through the device for a few seconds, as these devices are generally not designed to be connected for more than a few seconds. Knowledge of the expected voltage will assist in making sure that the correct instrument is chosen and its maximum voltage is not exceeded.

## ACTIVITY

Clamp meters only work when the clamp is round one conductor at a time. Can you explain why this is?

## Current testing

Clamp-meters, clip-on ammeters and 'tong-testers' are hand-held devices that can be used to determine the current in a current-carrying conductor, that is a circuit on-load.

**Clamp-on ammeter**

These devices can be used to identify whether a circuit is drawing too much current, due to overloading by too many items of electrical equipment. Their operation relies on a current transformer (CT) that transfers a high-rated current to a low-rating meter and gives the reading on a calibrated scale. When using clamp-meters or clip-on ammeters it is important to ensure that the CT clamp is fully closed without any air gap, as erroneous readings can be recorded if the clamp is not correctly closed.

Where current measurements are to be made, using instruments other than insulated tong-testers, the connections should be made with the apparatus dead, and should be made secure before the power is switched on. Any such temporary connections need to be adequately rated both for current and voltage.

Voltage and current checks can be used to confirm the operation of newly installed equipment. If a new piece of equipment fails to function after installation it should be confirmed that equipment is connected in accordance with the manufacturer's instructions, a voltage is present at the supply terminals and the equipment is drawing the correct current.

There are some simple rules that should be observed when using either a voltmeter or an ammeter.

- Familiarise yourself with the meter before use.
- Ensure it is safe to use – with no obvious damage to the meter or the meter leads.
- Ensure the test leads are in the correct sockets and the rotary switch is in the correct position for the desired measurement.
- Set the meter to the highest range available and reduce the range to obtain more accurate readings.
- If the reading is lower than the next available lower range on the meter, it is acceptable to set the meter to a lower range while the circuit is switched on.
- Keep fingers behind the finger guards on the test probes when making measurements.
- Always replace the battery as soon as the low-battery sign appears.

## Phase sequence testing

For multiphase circuits BS 7671 Regulation 612.12 requires verification that the phase sequence in multiphase (that is three-phase) circuits is maintained.

Correct phase sequence means that the phase sequence of the incoming line conductors (Brown–L1, Black–L2, Grey–L3) is the same throughout the installation and phase sequence testing confirms that this is the case.

It is not uncommon, in large industrial installations, which have a number of sub-distribution boards, to find that the incoming line conductors at each distribution board are connected in a different sequence to that at the incoming three-phase supply to the building.

You should understand that a phase-rotation test instrument will not verify the phase sequence of a multiphase circuit. Phase-rotation verification confirms the phase rotation at a particular item of equipment but it does not verify that the phase rotation at that item of equipment is the same as the phase rotation at the incoming supply. In order to confirm that the phase sequence is correct, the installation must be isolated from the supply and continuity testing of the individual conductors must be carried out from the origin of the installation to each distribution board, and from there to the furthest point of each circuit. This testing may be carried out in sections.

Once the supply polarity, including phase rotation, has been confirmed for each section, the power can be restored and the next section tested until the phase sequence has been confirmed throughout the installation. Phase sequence testing can then be carried out at each item of equipment.

As each item of equipment is energised, it can be tested to confirm that its phase sequence is correct, using a phase sequence/rotation instrument.

With three-phase circuits, problems often arise during commissioning stages or after fault diagnosis and rectification – items of equipment, such as motors, may operate in the wrong direction. Motors that start and run in the wrong direction will, in most circumstances, cause problems, especially in manufacturing businesses where motors are used for extraction, drive assemblies and pumping duties. The correct method of rectifying this problem is to alter the equipment's internal wiring at the terminal block. Unfortunately, the method often employed to correct this fault is to change the connections at the local isolator or at the distribution board supplying the equipment. This creates a situation where the phase sequence of the multiphase circuit is not maintained and, therefore, is in contravention of Regulation 612.12.

Although Regulation 612.12 refers to multiphase circuits, it is important, in a multiphase installation that supplies a number of single-phase circuits, that the phases L1, L2, & L3 are confirmed as part of the verification procedure. This confirms that any single-phase circuits of the installation that are in close proximity to each other, or in common enclosures, are identified correctly.

Remember: if the phase sequence is not correct, equipment could be damaged and people in the vicinity of a machine or motors could be put in danger. In addition, production could be lost.

Details of the method for checking the correct phase sequence can be found on page 465.

**ASSESSMENT GUIDANCE**

Phase sequence testing is carried out to ensure that what is identified at Board A as Brown line is the same as at Board B. Remember, sequence-wise Brown, Grey, Black is the same as Grey, Black, Brown to a motor but not to a distribution board.

## WHAT TO DO NEXT?

Assessment criteria

2.8 Explain responsibilities where unsatisfactory results are obtained

Following an inspection and test on an electrical installation, an Electrical Installation Condition Report will be provided to highlight any safety shortcomings, defects or deviations from the current version of BS 7671.

An Electrical Installation Condition Report is intended primarily for the person who is ordering the work to be undertaken (the client, duty holder or responsible person for the installation) and anyone subsequently involved in additional or remedial work or additional inspections, to confirm, so far as reasonably practicable, whether the electrical installation is in a satisfactory condition for continued service.

Each observation relating to a problem or safety concern should be given an appropriate classification code selected from standard classification codes.

### Code C1: danger present

This code signifies a risk of injury and that immediate remedial action is required.

Where practicable, items classified as C1 should be made safe on discovery by the person undertaking the inspection. The duty holder or responsible person ordering the report should be advised of the action taken and asked for permission to allow remedial work to be undertaken straight away. If that is not practical, other appropriate action should be taken, such as switching off and isolating the affected parts of the installation to prevent danger.

### Code C2: potentially dangerous

This code signifies that urgent remedial action is required.

For items classified as C2, the safety of those using the installation may be at risk and it is recommended that a competent person undertakes any necessary remedial work as a matter of urgency.

### Code C3: improvement recommended

For items classified as C3, the inspection has revealed an apparent deficiency which could not, due to the extent or limitations of the inspection, be fully identified.

It should be reported to the person requesting the inspection that further investigation is required, as soon as possible ,to determine the nature and extent of the apparent deficiency. For safety reasons, the electrical installation should be reinspected at appropriate intervals by a competent person.

# ELECTRICAL INSTALLATION CONDITION REPORT

**SECTION A. DETAILS OF THE CLIENT / PERSON ORDERING THE REPORT**

Name ..................................................

Address ..................................................

..................................................

**SECTION B. REASON FOR PRODUCING THIS REPORT** ..................................................

..................................................

Date(s) on which inspection and testing was carried out ..................................................

**SECTION C. DETAILS OF THE INSTALLATION WHICH IS THE SUBJECT OF THIS REPORT**

Occupier ..................................................

Address ..................................................

..................................................

Description of premises (tick as appropriate)

Domestic ☐ Commercial ☐ Industrial ☐ Other (include brief description) ☐ ..................................................

Estimated age of wiring system ................years

Evidence of additions / alterations Yes ☐ No ☐ Not apparent ☐ If yes, estimate age ................years

Installation records available? (Regulation 621.1) Yes ☐ No ☐ Date of last inspection .............................. (date)

**SECTION D. EXTENT AND LIMITATIONS OF INSPECTION AND TESTING**

Extent of the electrical installation covered by this report

..................................................

..................................................

Agreed limitations including the reasons (see Regulation 634.2) ..................................................

..................................................

Agreed with: ..................................................

Operational limitations including the reasons (see page no.............) ..................................................

..................................................

The inspection and testing detailed in this report and accompanying schedules have been carried out in accordance with BS 7671: 2008 (IET Wiring Regulations) as amended to ........................................

It should be noted that cables concealed within trunking and conduits, under floors, in roof spaces, and generally within the fabric of the building or underground, have **not** been inspected unless specifically agreed between the client and inspector prior to the inspection.

**SECTION E. SUMMARY OF THE CONDITION OF THE INSTALLATION**

General condition of the installation (in terms of electrical safety) ..................................................

..................................................

..................................................

Overall assessment of the installation in terms of its suitability for continued use

SATISFACTORY / UNSATISFACTORY* (Delete as appropriate)

*An unsatisfactory assessment indicates that dangerous (code C1) and/or potentially dangerous (code C2) conditions have been identified.

**SECTION F. RECOMMENDATIONS**

Where the overall assessment of the suitability of the installation for continued use above is stated as UNSATISFACTORY, I / we recommend that any observations classified as *'Danger present'* (code C1) or *'Potentially dangerous'* (code C2) are acted upon as a matter of urgency.

Investigation without delay is recommended for observations identified as *'further investigation required'*.

Observations classified as *'Improvement recommended'* (code C3) should be given due consideration.

Subject to the necessary remedial action being taken, I / we recommend that the installation is further inspected and tested by ................(date)

**SECTION G. DECLARATION**

**I/We, being the person(s) responsible for the inspection and testing of the electrical installation (as indicated by my/our signatures below), particulars of which are described above, having exercised reasonable skill and care when carrying out the inspection and testing, hereby declare that the information in this report, including the observations and the attached schedules, provides an accurate assessment of the condition of the electrical installation taking into account the stated extent and limitations in section D of this report.**

| **Inspected and tested by:** | **Report authorised for issue by:** |
|---|---|
| Name (Capitals) .................... | Name (Capitals) .................... |
| Signature .................... | Signature .................... |
| For/on behalf of .................... | For/on behalf of .................... |
| Position .................... | Position .................... |
| Address .................... | Address .................... |
| Date .................... | Date .................... |

**SECTION H. SCHEDULE(S)**

...........schedule(s) of inspection and .........schedule(s) of test results are attached.

The attached schedule(s) are part of this document and this report is valid only when they are attached to it.

**SECTION I. SUPPLY CHARACTERISTICS AND EARTHING ARRANGEMENTS**

| Earthing arrangements | Number and Type of Live Conductors | | Nature of Supply Parameters | Supply Protective Device |
|---|---|---|---|---|
| TN-C ☐ | a.c. ☐ | d.c. ☐ | Nominal voltage, U / $U_0$[1] ................V | BS (EN) .................... |
| TN-S ☐ | 1-phase, 2-wire ☐ | 2-wire ☐ | Nominal frequency, f[1] ................Hz | Type .................... |
| TN-C-S ☐ | 2 phase, 3-wire ☐ | 3-wire ☐ | Prospective fault current, $I_{pf}$[2] ................kA | |
| TT ☐ | 3 phase, 3-wire ☐ | | External loop impedance, Ze[2] ................ Ω | Rated current ................A |
| IT ☐ | 3 phase, 4-wire ☐ | | Note: (1) by enquiry | |
| | Confirmation of supply polarity ☐ | | (2) by enquiry or by measurement | |

Other sources of supply (as detailed on attached schedule) ☐

**SECTION J. PARTICULARS OF INSTALLATION REFERRED TO IN THE REPORT**

| **Means of Earthing** | **Details of Installation Earth Electrode** *(where applicable)* |
|---|---|
| Distributor's facility ☐ | Type .................... |
| Installation earth electrode ☐ | Location .................... |
| | Resistance to Earth ................ Ω |

**Main Protective Conductors**

| | | | |
|---|---|---|---|
| Earthing conductor | Material .................... | Csa ................$mm^2$ | Connection / continuity verified ☐ |
| Main protective bonding conductors | Material .................... | Csa ................$mm^2$ | Connection / continuity verified ☐ |
| To incoming water service ☐ | To incoming gas service ☐ | To incoming oil service ☐ | To structural steel ☐ |
| To lightning protection ☐ | To other incoming service(s) ☐ Specify .................... | | |

**Main Switch / Switch-Fuse / Circuit-Breaker / RCD**

| | | **If RCD main switch** |
|---|---|---|
| Location .................... | Current rating ................A | Rated residual operating current ($I_{\Delta n}$) ................mA |
| .................... | Fuse / device rating or setting ................A | Rated time delay ................ms |
| BS(EN) .................... | Voltage rating ................V | Measured operating time(at $I_{\Delta n}$) ................ms |
| No of poles .................... | | |

**SECTION K. OBSERVATIONS**

Referring to the attached schedules of inspection and test results, and subject to the limitations specified at the *Extent and limitations of inspection and testing* section

No remedial action is required ☐ The following observations are made ☐ (see below):

| OBSERVATION(S) | CLASSIFICATION CODE | FURTHER INVESTIGATION REQUIRED (YES / NO) |
|---|---|---|
| .................... | .................... | .................... |
| .................... | .................... | .................... |
| .................... | .................... | .................... |
| .................... | .................... | .................... |
| .................... | .................... | .................... |
| .................... | .................... | .................... |
| .................... | .................... | .................... |
| .................... | .................... | .................... |
| .................... | .................... | .................... |
| .................... | .................... | .................... |
| .................... | .................... | .................... |
| .................... | .................... | .................... |
| .................... | .................... | .................... |
| .................... | .................... | .................... |
| .................... | .................... | .................... |
| .................... | .................... | .................... |
| .................... | .................... | .................... |
| .................... | .................... | .................... |
| .................... | .................... | .................... |
| .................... | .................... | .................... |

One of the following codes, as appropriate, has been allocated to each of the observations made above to indicate to the person(s) responsible for the installation the degree of urgency for remedial action.

C1 – Danger present. Risk of injury. Immediate remedial action required

C2 – Potentially dangerous - urgent remedial action required

C3 – Improvement recommended

The recommended date by which the next inspection is due should be included in the report, under 'Recommendations' and on a label near to the consumer unit or distribution board.

On completion of the inspection, the client, duty holder or responsible person ordering the report should receive the original and the inspector should retain a duplicate copy. The original report should be retained in a safe place and be made available to any person inspecting or undertaking work on the electrical installation in the future. If the property is vacated, the report will provide the new owner or occupier with details of the condition of the electrical installation at the time the report was issued.

Failure to take action when a deviation from BS 7671 has been identified and recorded would be seen as a failure under the provisions of Health and Safety at Work etc Act (1974) and the Electricity at Work Regulations 1989, and directors and managers of any company that employs more than five employees can be held personally responsible for failure to control health and safety.

## IMPLICATIONS OF FAULT DIAGNOSIS

Depending on the type of fault diagnosed and the method of rectification required, there may be a number of implications for those with responsibility for the electrical installation and users of the system.

There may be a requirement to disturb the fabric or structure of the building and, if this is the case, it is very important for all aspects of the rectification to be discussed with the client. Agreement must be obtained for the work to be undertaken, for the extent of the repair necessary (to brick, block, plaster, concrete, screed, plasterboard and decorations, for example) and for the contractual arrangements (who is paying for the repair).

Once agreement has been reached on the contractual issues, the repairs should be carried out in a logical manner, while taking into account the effect that circuit disconnections will have on users of the system. When only a small area or limited amount of equipment is affected by a fault, it is preferable not to request a total shutdown of the whole installation. By analysing circuit and schematic diagrams, it should be possible to isolate individual circuits to avoid disruption. If it is necessary to isolate individual circuits, especially those that supply IT equipment, then permission must be sought from the duty holder or the responsible person for the equipment to avoid any loss or damage to such items as computers.

There may, however, be circumstances when minor disruption or even a 'risk of trip' are not acceptable and alternative arrangements, such as 'out of hours working', will need to be considered.

**Assessment criteria**

**2.9** Explain implications of fault diagnosed

### ACTIVITY

Like any other installation work, maintenance work will cause disturbance to the work area and create a mess on the floor, carpets etc. Make a list of items required to keep the area clean and tidy.

### ASSESSMENT GUIDANCE

Any temporary repairs made while awaiting delivery of replacement parts must be carried out to the same regulatory requirements as permanent installations.

# OUTCOME 3

# Understand process for fault rectification

**Assessment criteria**

**3.1** Explain safe working procedures that should be adopted for completion of fault rectification

## SAFE WORKING PROCEDURES FOR FAULT RECTIFICATION

The possibility of touching live parts is increased during fault finding and repair, when conductors at dangerous voltages are often exposed. This risk can be reduced if work is done while the equipment is isolated from any dangerous source of electrical supply. However, this may not always be possible and, if this is the case, procedures should be followed to prevent contact with any hazardous internally produced voltages.

Before fault finding and repair work commences, anyone carrying out the work must be aware of the following relevant legal requirements:

- the Health and Safety at Work etc Act 1974 in relation to safe systems of work
- the Electricity at Work Regulations 1989 regarding work on or near to an electrical system
- the Management of Health and Safety at Work Regulations 1999 which covers risk assessment.

### The method statement

Once the risk assessment in accordance with the Management of Health and Safety at Work Regulations has been carried out, a safe system of work, sometimes called a method statement, can be prepared to enable the work to be undertaken in a safe manner.

The safe system of work or method statement will include such items as:

- who is authorised to undertake testing and, where appropriate, how to access a test area and who should not enter the test area
- arrangements for isolating equipment and how the isolation is secured
- provision and use of personal protective equipment (PPE) where necessary
- the correct use of tools and equipment

- the correct use of additional protection measures, for example, flexible insulation that may need to be applied to the equipment under test while its covers are removed
- use of barriers and positioning of notices
- safe and correct use of measuring instruments
- safe working arrangements to be agreed with client, duty holder or responsible person
- how defects are to be reported and recorded
- instructions regarding action to be taken if unsafe situations are identified.

Guidance on live and dead working can be found in the HSE publication *Memorandum of guidance on the Electricity at Work Regulations 1989* (HSR 25) and can be downloaded free of charge from the Health and Safety Executive website: www.hse.gov.uk,

Safe working procedures should be reviewed regularly, to make sure that they are being followed and are still appropriate for the work that is being carried out. If any changes are made to the procedures, all people who are involved in the fault diagnosis regime should be given relevant instruction and training.

A permit to work (PTW) procedure, as described on page 511 of Outcome 2, may be required if the installation is complex in nature.

**ACTIVITY**

A lathe is driven by a three-phase motor. It is one of a number of machines in a factory making automobile components. The motor is believed to have developed an internal fault. Complete a method statement for the fault-finding operation on the machine.

## Planning and agreeing procedures

Before fault rectification is carried out, the safe working arrangements must be discussed and agreed with the client and/or the duty holder (or responsible person for the installation) in a clear, concise and courteous manner. The procedure must be a planned activity as it will almost certainly affect people who work or live in the premises where the installation is being tested. This ensures that everyone who is concerned with the work understands what actions need to be taken, such as:

- which areas of the installation may be subject to disconnection
- anticipated disruption times
- who might be affected by the work
- health and safety requirements for the site
- which area will have restricted access
- whether temporary supplies will be required whilst the fault repair is underway
- the agreement reached on who has authority for the repair.

It may be that a specific person has responsibility for the safe isolation of a particular section of an installation, in which case that person should be identified and the isolation arrangements agreed. Entering into dialogue with the client, before work commences, will minimise the potential for unforeseen events and foster good customer relations.

For example, in an office block where the electrical installation is complex and provides supplies to many different tenants located on a number of floors, the safe isolation of a sub-circuit for testing purposes may require a larger portion of the installation to be turned off initially. To achieve this with minimal disruption, an agreement must be reached between the competent person tasked to carry out the work and the person responsible for the installations affected. This responsible person may be the office manager, the designated electrical engineer for the site or, in some cases, the landlord of the building.

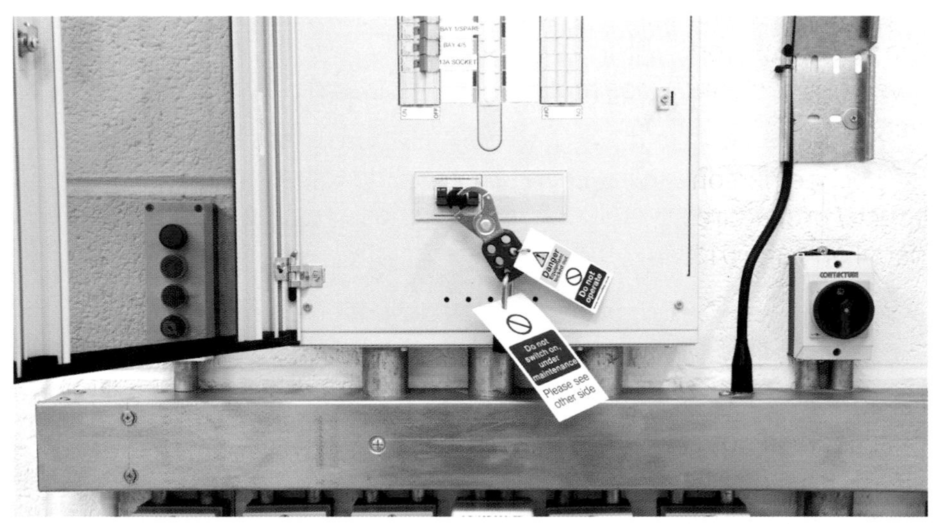

**Isolation of an installation should be part of a planned and agreed procedure**

Everyone involved in the work (for example, client, electrician and those in the workplace) has a responsibility for their own health and safety and that of others who may be affected by the work; communication between all parties will ensure compliance with the respective health and safety requirements.

**Assessment criteria**

**3.2** Describe precautions that should be taken in relation to hazards of fault rectification

# HAZARDS OF FAULT RECTIFICATION

## Electrical injuries

Electrical injuries are the most likely hazard and the following may occur due to contact with mains 230 V electricity:

- electric shock
- electrical burns
- loss of muscle control
- fires arising from electrical causes
- arcing and explosion.

## Lone working

The task of fault diagnosis and repair, which sometimes requires some form of live electrical work, is often undertaken by people working alone and without close or direct supervision.

There is no specific requirement contained in the Electricity at Work Regulations 1989 (EAW) for two people to be present in live-work situations. It is the responsibility of the duty holder to assess the need for a second person in terms of the nature of the danger and other precautionary measures to be adopted (such as role planning and relevant training).

Where a second person is present, they must clearly understand their role. The accompanying person must *not* be distracting, such as by making idle conversation, but should make a positive contribution to the system of work. This means that careful consideration must be given to the competence and training requirements for accompanying persons.

If a second person is *not* employed, consideration must be given to special procedures for the lone worker. This was covered in more detail on page 515 in Outcome 2.

**ASSESSMENT GUIDANCE**

Two-person working is preferable but not always possible. Apart from the safety aspects, the advantage of two-person working is that the shared experience of fault finding helps enormously.

## Static electricity

Large static charges may develop on drive belts between motors and machinery; if not diverted through proper earthing, the built-up static charges may discharge suddenly and cause damage to equipment. In areas containing explosive materials, great care must be taken to prevent static discharges, since a spark could set off a violent explosion. The static effect is increased in environments with low humidity and in buildings with air conditioning and high levels of heating, such as computer suites, and this should be considered when undertaking electrical work in such areas.

On a smaller scale, the sudden flow of static electricity can interfere with the operation of electronic circuits. The static discharge can induce very high voltages in sensitive circuitry and care must be taken to discharge any static that may have built up before touching or handling a static-sensitive piece of equipment.

## Fibre-optic cabling

Energy is transferred along fibre-optic cables as digital pulses of laser light. Direct eye contact with this light should be avoided. However, damage to the eye caused by looking into a damaged fibre-optic cable is rare as the broken surface of the cable tends to scatter the light coming through it.

A more serious hazard of optical-fibre work is the fibres themselves. Fibres are pieces of optical glass which, like any piece or sliver of glass, can cause injury. Hazards arise when cables are opened and fibres penetrate the hands or are transferred to the respiratory or alimentary system.

Fibre-optic cables look like steel wire armour (SWA) cables but are much lighter. They should be installed in the same way and given a similar level of protection as steel wire armour cables.

## Electronic devices

The use of electronic circuits in all types of electrical equipment is now commonplace, with components being found in motor-starting equipment, control circuits, emergency lighting, discharge lighting, intruder alarms, special-effect lighting and dimmer switches, and domestic appliances. Electrical installations usually operate at currents in excess of 1 A and up to many hundreds of amperes; all electronic components and circuits are low voltage and usually operate in mA or µA. Therefore, this fact must be considered when the choice of electrical test equipment is made.

The working voltage of a standard insulation resistance tester (250 V/500 V) will cause damage to any of the electronic devices described above. When carrying out insulation resistance tests as part of a prescribed test or after fault repair all electronic devices must be disconnected. Should resistance measurements be required on electronic circuits and components, a battery operated ohmmeter with high impedance must be used.

## High-frequency or capacitive circuits

Capacitors that are used in fluorescent fittings and single-phase motors are a potential source of electric shock arising from:

- the discharge of electrical energy retained by the capacitor unit(s) after they have been isolated
- inadequate precautions to guard against electric shock as a result of any charged conductors or associated fittings
- charged capacitors that are inadequately short circuited
- equipment retaining or regaining a charge.

Live working is to be avoided where possible, but any equipment containing a capacitor must be assumed to be live or charged until it is proven, beyond any doubt, that the circuits are dead, the capacitors are discharged and are unable to recharge.

## Storage batteries

Batteries are used to store electrical energy and their ability to provide instant output makes them a regular source of standby power for emergency lighting, emergency trip coils and uninterruptable power supplies (UPS) at computer, data storage and telecommunication facilities, for example. They can prove invaluable for maintaining supplies where a supply interruption could cause business issues. There is, however, risk attached to these larger battery systems which, if used incorrectly, can be dangerous and can cause explosions.

There are two main classes of battery.

- *Lead–acid batteries* are the most common large-capacity rechargeable batteries. They are fitted extensively in UPS of computer and communication facilities, and process and machinery control systems.
- *Alkaline rechargeable batteries*, such as nickel–cadmium, nickel–metal hydride and lithium ion, are widely used in small items such as laptop computers, but large-capacity versions of these cells are now being used in UPS applications.

The two main hazards from batteries are described below.

- *Chemical* – lead–acid batteries are usually filled with electrolytes such as sulphuric acid or potassium hydroxide. These very corrosive chemicals can permanently damage the eyes and produce serious chemical burns to the skin. If acid gets into the eyes, they should be flushed immediately with water for 15 minutes, and prompt medical attention should be sought. If acid gets on the skin, the affected area should immediately be rinsed with large amounts of water and prompt medical attention should be sought if the chemical burn appears serious. Emergency wash stations should be located near lead–acid battery storage and charging areas. Under severe overcharge conditions, hydrogen gas may be vented from lead–acid batteries and this may form an explosive mixture with air if the concentration exceeds 3.8% by volume. Adequate ventilation and correct charging arrangements should always be observed.
- *Electrical* – Batteries contain a lot of stored energy. Under certain circumstances, this energy may be released very quickly and unexpectedly. This can happen when the terminals are short circuited, for example with an uninsulated metal spanner or screwdriver. When this happens, a large amount of electricity flows through the metal object, making it very hot, very quickly. If the battery explodes, the resulting shower of molten metal and plastic can cause serious burns and can ignite any explosive gases present around the battery. The short circuit may also produce ultraviolet (UV) light which can damage the eyes.

**ASSESSMENT GUIDANCE**

Never leave a working area unguarded. You never know who may come wandering along.

**ACTIVITY**

Name three safety precautions to be taken when working on or near lead–acid batteries.

Most batteries produce quite low voltages, and so there is little risk of direct electric shock. However, as some large battery combinations produce more than 120 V d.c. precautions should be taken to protect users and those working in the vicinity of these installations.

- Ensure that live conductors are effectively insulated or protected.
- Keep metal tools and jewellery away from the battery.
- Post suitable notices and labels warning of the danger.
- Control access to areas where dangerous voltages are present.
- Provide effective ventilation to stop dangerous levels of hydrogen and air or oxygen accumulating in the charging area.

If battery cables are removed for any reason, ensure that they are clearly marked 'positive' and 'negative' so that they are reconnected with the correct polarity.

## Information technology (IT) equipment

All modern offices contain computers that are operated as standalone units or networked (linked together). Most computer systems are sensitive to variations in the mains supply that can be caused by external mains switching operations, faults on adjacent electricity distribution network circuits that cause a voltage rise and industrial processes such as induction furnaces and electric arc welding. These system disturbances are often well within the electricity supply companies' statutory voltage limits but may be sufficient to cause computers to crash.

To avoid the effects of this system 'noise', computer networks are normally provided with a clean supply, including a 'clean earth'. This is obtained by taking the supply to the computer network from a point as close as possible to the supply intake position of the building. The fitting of noise suppressors or noise filters provides additional protection to the computer system. If a production process disruption or data loss is a serious business risk, an uninterruptable power supply (UPS) is a sensible solution.

A UPS provides emergency power by supplying power from a separate source (normally batteries) when the supply from the local supply company is unavailable. It differs from an auxiliary or emergency power system or standby generator, as these do not provide instant protection from a temporary power interruption. A UPS can be used to provide uninterrupted power to equipment, typically for 5 to 15 minutes, until an auxiliary power supply can be turned on, mains power restored, or equipment safely shut down. UPS units come in sizes ranging from units that will back up a single computer without monitor (around 200 W) to units that will power entire data centres or buildings (several megawatts, MW).

All of the above features are designed to keep power supplies constant for IT and similar equipment to prevent data loss, but fault rectification can impact on this requirement.

Before any fault repair is carried out, confirm whether the standby or UPS will present a danger to the electrician undertaking the work. Remember, the preferred method of work is with the system dead and, as such, it must be isolated from *all* points of supply, including auxiliary circuits, and dual or alternative supplies, such as UPS or a standby generator (that may be set for automatic start-up).

**KEY POINT**

Proving dead at the point of work is the most important part of ensuring a safe system of work. If the isolation has been incorrectly applied, the act of proving dead should identify that the circuit is, in fact, still live. A proper procedure using a suitable proving device is therefore essential (see page 522).

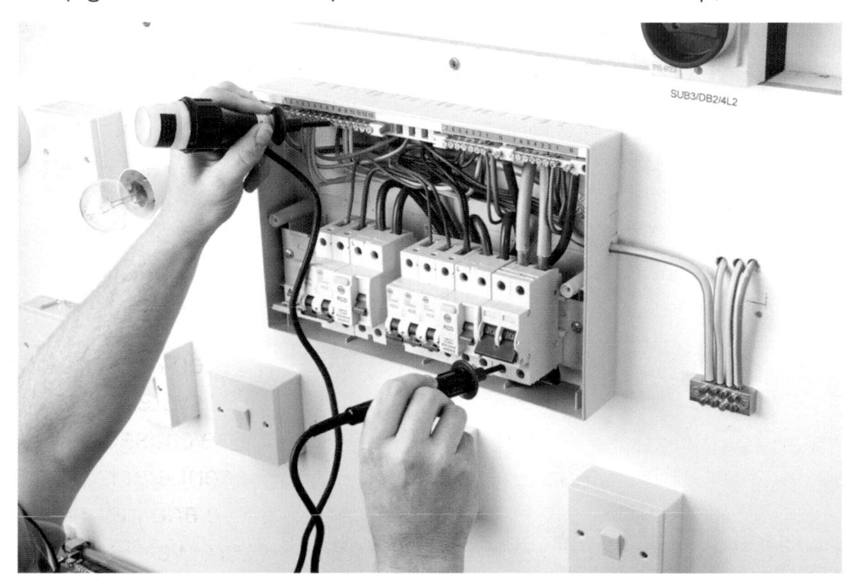


**Using test lamps to prove dead at point of work**

If it is necessary to isolate the computer supply network, permission must be sought from the duty holder or the responsible person for the computer systems to avoid any loss of data or damage to the computers.

However, standby or UPS backup systems can also be used to avoid IT shutdowns during fault rectification by maintaining supplies to critical areas.

## Hazardous areas

Although all the hazards discussed above could be considered to give rise to hazardous areas, a 'hazardous area' is defined in the Dangerous Substances and Explosive Atmospheres Regulations 2002 (DSEAR) as: 'any place in which an explosive atmosphere may occur in quantities such as to require special precautions to protect the safety of workers'.

These regulations provide a specific legal requirement to carry out a hazardous area study and document the conclusions in the form of zones. In the context of the definition, 'special precautions' relate to the construction, installation and use of apparatus, as given in BS EN 60079-10.

Most volatile materials (those that disperse readily in air) only form explosive mixtures between certain concentration limits. The flash point of a volatile material is the lowest temperature at which it can vaporise to form an ignitable mixture in air; flash point refers to both flammable liquids and combustible liquids (page 521).

The primary risk associated with combustible gases and vapours is the possibility of an explosion. Explosion, like fire, requires three elements: fuel, oxygen and an ignition source. Each combustible gas or vapour will ignite only within a specific range of fuel/oxygen mixtures. Too little or too much gas and the gas or vapour will not ignite. These conditions are defined by the lower explosive limit (LEL) and the upper explosive limit (UEL). Any amount of gas between the two limits is explosive.

**ACTIVITY**

Explain how a $CO_2$ or foam fire extinguisher removes the source of oxygen from a fire.

Sources of ignition should be effectively controlled in all hazardous areas by a combination of design measures and systems of work. All electrical equipment must carry the appropriate markings if the integrity of the wiring system is to be maintained. A luminaire to IP 65 (this International Protection code means the fitting is dust tight and water jet protected) is an example of equipment used in hazardous areas.

When any work is to be undertaken on electrical equipment in a hazardous area the following procedures must be observed:

- a written safe work procedure (PTW system) that includes control of activities which may cause sparks, hot surfaces or naked flames should be in place
- control of the working area with signs and barriers
- use of appropriate PPE
- only authorised and competent people to be engaged in the work activity
- only approved tools and equipment to be used
- prohibition of smoking, use of matches and lighters.

If work is to be undertaken in a hazardous area, the hazards and the requirements to be observed will normally be explained during the site induction process.

**Assessment criteria**

**3.3** Explain process of fault rectification

## THE PROCESS OF FAULT RECTIFICATION

Electrical faults do not occur at convenient times, and fault diagnosis and repair is often undertaken in difficult circumstances. For the duty holder or person responsible for the electrical installation it can be difficult to determine the most suitable arrangements for fault rectification, due to the potential for danger to people and equipment, time delay and lost production.

The consequences of a fault on an electrical installation can be considerable: lost production due to the unavailability of a machine, lost business where goods, products or services are not being made available to a customer and data loss if, for example, the backup UPS fails to function. Therefore, a suitable process needs to be established in order to keep inconvenience and costs to an acceptable level, without compromising the safety of users of the system and technicians undertaking the fault repair.

Having identified the fault, the most suitable rectification process should be identified. This could as simple as correcting a loose connection at an accessory terminal or undertaking a repair by replacing a faulty conductor.

## Agreeing the process of fault rectification

Often the solution will be fairly straightforward but, in some circumstances, options must be discussed and agreed with the duty holder or responsible person. Some or all of the following issues may be open to consideration.

- What is the cost of replacement?
- What is the availability of suitably trained staff?
- Is it possible to replace a lesser number of components?
- What is the cost comparison of replacing or repairing a component?
- What is the potential disruption to the manufacturing process or computer availability?
- What is the availability of replacement parts, if required?
- Can other parts of the system be energised whilst the repair is carried out?
- What is availability of emergency or standby supplies?
- What is the anticipated 'down time' while a repair is carried out?
- Is there a requirement for a continuous supply?
- Will out-of-hours working be required?
- Are there any legal or personnel issues to be considered (warranties, contracts)?

Once the decision on how the fault rectification is to proceed has been made and agreed with the duty holder or responsible person, a work activity plan should be prepared. This, once again, must be agreed with the duty holder. The plan should include a safe system of work and requirements for circuit outages.

Unless the fault diagnosis and rectification are being undertaken by an in-house electrician, site access must be granted by the duty holder or responsible person. Site access passes may be issued and there may be a logging-in procedure (visitors' book) for each visit.

The preceding arrangements are all concerned with a fault or supply failure on the client's system. However, the initial fault diagnosis may determine that the problem is connected with the system of the distribution network operator (DNO). If this is the case, the responsibility for the fault rectification rests with the network operator. Under these circumstances, it is good business practice to advise the client on how to deal with the DNO, requesting the provision of temporary standby supplies in the form of a backup generator if required.

The work activity should be managed with appropriate technical support and resource allocations to ensure that an effective repair is completed within the agreed timescales.

**ASSESSMENT GUIDANCE**

Both overhead and underground cables are liable to damage. Overhead lines are susceptible to lightning strikes and inadvertent contact (by fishing poles, masts on dinghys, ladders etc), while underground cables can be damaged during ground excavation works or by sharp tools used in street works and similar operations.

There may be a requirement to disturb the fabric or structure of the building and, if this is the case, it is very important for all aspects of the rectification to be discussed with the client. Agreement must be obtained for the work to be undertaken, for the extent of the repair necessary (to brick, block, plaster, concrete, screed, plasterboard and decorations, for example) and for the contractual arrangements (who is paying for the repair). The fabric and structure of the building must always be left in a condition that does not compromise either fire safety or the building's structure performance.

Minor cosmetic repair works such as patch plastering, disturbance to stud walls or decoration are often within the capability of an experienced electrical technician, but you must always recognise your own limitations. Expert advice, such as from a specialist contractor, should be sought if any structural modifications are required.

## Testing the circuit

When the repair work is complete, the circuit, system or individual piece of equipment must be inspected, tested and functional checks carried out in accordance with BS 7671 Chapter 61. This inspection and function testing will confirm the electrical integrity of the system before it is energised.

The tests recommended in Regulation 612, Testing of BS 7671, are divided into those tests that are conducted with the system dead and those that are done with the system live.

Tests conducted before the supply is reconnected include:

- continuity of protective conductors, including main and supplementary bonding
- continuity of ring final circuit conductors, including protective conductors
- insulation resistance.

Tests conducted with the supply connected are:

- to check the polarity of the supply, using an approved voltage tester
- earth electrode resistance, using a loop impedance tester
- earth fault loop impedance
- prospective fault current measurement
- functional testing, including RCDs and switchgear.

## Waste disposal

Another important part of the fault repair process is the safe disposal of waste. This ensures both good customer relations and compliance with the relevant legislation, such as the Waste Electrical and Electronic Equipment (WEEE) Regulations 2006, the Waste (England and Wales) Regulations 2011 and the Control of Asbestos at Work Regulations 2012.

The UK has implemented an EU Directive through the WEEE Regulations 2006, which came into force on 2 January 2007. The regulations apply to all electrical and electronic equipment placed on the market in the UK in any of following 10 product categories:

- large household appliances
- small household appliances
- IT and telecoms equipment
- consumer equipment
- lighting equipment
- electrical and electronic tools
- toys, leisure and sports equipment
- medical devices (except implants and infected products)
- monitoring and control equipment
- automatic dispensers.

The regulations require any 'producer' of such equipment, that is a manufacturer, rebrander or importer of electrical and electronic equipment, to finance the costs of collection and treatment of waste electrical and electronic equipment that arises over a calendar year, in proportion to the amount by weight placed on the market. Producers meet their obligations by registering with an approved *producer compliance scheme*. Through this scheme, producers fund reuse, recovery and recycling of electrical goods at an approved authorised treatment facility (AATF) or approved exporter (AE).

In 2009 there were several amendments made to the UK WEEE Regulations that mainly affect producer compliance schemes, approved authorised treatment facilities and approved exporters.

The UK has also implemented an EU Directive (The Waste Framework Directive), which is the primary European legislation for the management of waste, through a series of regulations dealing with waste. The directive has been revised and these revisions have been implemented in England and Wales through the Waste (England and Wales) Regulations 2011 and ancillary legislation in Wales.

The on-site disposal of waste materials following electrical installation work will be dealt with in a number of ways.

- The packaging material from the electrical fittings and accessories (mainly cardboard) is normally stored and arrangements made for collection, transportation and recycling.
- Small amounts of non-recyclable material can be disposed of in the electrical contractor's skip or in the client's skip, if agreement has been reached for that to take place.
- Off-cuts of cable, conduit, trunking, cable tray and general ferrous and non-ferrous materials are often collected for disposal at a metal recycling plant.
- Useable off-cuts of cable, conduit, trunking and cable tray should be returned to stock for future use.

## Asbestos

The Control of Asbestos at Work Regulations 2012 affect anyone who owns, occupies, manages or otherwise has responsibilities for the maintenance and repair of buildings that may contain asbestos.

Asbestos materials may be encountered by electricians during the course of their work. Asbestos materials in good condition are usually safe. However, if asbestos fibres become airborne they are very dangerous; this may happen when materials are damaged, due to demolition or remedial works on or in the vicinity of ceiling tiles, asbestos cement roofs, wall sheets, sprayed asbestos coating on steel structures, and lagging.

If asbestos is discovered during electrical installation or remedial work, work must be stopped immediately. Specialist contractors must be engaged to ascertain the condition of the asbestos and to determine any actions necessary for its removal, treatment or retention. In particular, the disposal of asbestos should only be undertaken by specialist contractors.

## Fluorescent tubes

Fluorescent tubes generally contain 94% glass, 4% ferrous and non-ferrous metals and 2% phosphor powder, which itself contains mercury. Fluorescent tubes are classified as hazardous waste in England and Wales and as special waste in Scotland; preferably, they should be recycled or, if absolutely necessary, taken to specialist disposal sites. They must not be disposed of as general waste.

### ACTIVITY

Describe the action to be taken in disposal of the following items:

**a)** part drums of cable
**b)** polystyrene packaging
**c)** cardboard packaging
**d)** fluorescent tubes.

## Leaving the installation safe

Remember that Section 3 of the Health and Safety at Work etc Act and the EAW Regulations require that installations are left in a safe condition. People who have been working on an installation must not leave it in an unsafe condition which could affect contractors, visitors or the general public. For example, where there are accessible live parts due to blanks missing from a consumer unit, suitable temporary barriers should be provided to prevent direct contact with those live parts.

If there is a risk due to a classification C1 defect (as described on page 506 in Outcome 1, it is *not* sufficient just to make the duty holder aware of the danger when submitting the report. The installation should be made safe, on discovery of the defect, by the person undertaking the inspection and testing. The duty holder or responsible person ordering the report should be advised immediately of the action taken. You should seek agreement for any necessary remedial work to be undertaken straight away or, if that is not practical, you must take other appropriate action, such as switching off and isolating the affected parts of the installation to prevent danger.

# FACTORS THAT AFFECT FAULT RECTIFICATION

**Assessment criteria**

**3.4** Explain factors which can affect fault rectification

Some faults are minor and these can be dealt with quickly and right after the fault has been diagnosed, with little impact on materials and manpower. However, in the case of major faults in large industrial or commercial enterprises, the repair time could be weeks or months, depending on the extent of the failure. If this is the case, careful planning of the work process is required and consideration should be made of whether the likely value of the repair work will necessitate a tendering process.

## Agreeing the scope of the work through a tendering process

The next stage following a tender is the preparation of a contract document between the successful tenderer and the client, to ensure that all parties are aware of their obligations. Failure to prepare a contract could result in future conflict over costs, exact work requirements and the procedure for dealing with variations from the agreed job specification.

A contract of this nature usually includes:

- scope of the work to be done
- location of the work
- the provision of information, such as drawings, plans, certificates

**ASSESSMENT GUIDANCE**

It is not always possible to provide an exact quotation for the cost — an estimate is more likely. The customer must be kept aware of any developments.

- details of the payment schedule – a fixed price, or time and materials arrangement
- the completion dates and schedules
- details of responsibilities that may impact on the schedules, including those of any third parties
- responsibility for additional fees or charges that may arise over the course of the project
- responsibilities and requirement for safety of the contractor's staff and others who may be affected by the work
- responsibility for losses, material damage or personal injuries that may occur during the project
- agreed variation procedure
- guarantee or warranty period for the completed work.

Downtime and costs are probably the two issues of greatest concern that must be agreed and then adhered to. Most clients prefer to fix a price and will not countenance additional costs, unless it can be proven that these result from a departure from the original specification, that could not have been foreseen.

When a quote for work is accepted, it is essential that the quoted work is undertaken; any deviation from the quote, such as not using specified components or using different-size cabling, could be construed as a breach of contract.

## Procurement of a backup supply

Depending on the nature of the business affected by the fault, the client may request out-of-hours fault repair work. If this is not an option, the provision of temporary standby supplies (in the form of a backup generator) may be requested. The procurement and provision of suitable standby supplies could have an effect on the timescale for the fault rectification. Again, the supply and management of such equipment must be agreed before it is connected to the installation.

## Other factors that affect timescale

Even with all agreements and arrangements in place, remember that, before the repair work can take place, an approved safe working procedure including a risk assessment must be prepared.

Any delay in providing information, such as past inspection and tests results, schematics or wiring diagrams, may have a bearing on how quickly the repair can be actioned.

Other factors may also influence the timescale. For example, if the fault has affected a functioning oven or furnace, some considerable time may need to elapse to allow the oven or furnace to cool sufficiently to allow personnel to enter the area.

OUTCOME 4

# Be able to diagnose faults on electrical systems

If possible, fault finding should be planned to avoid inconvenience to other workers and to avoid disruption of the normal working routine. However, an equipment fault or a fault in an installation is not normally a planned event and these often occur at the most inconvenient time. The diagnosis and rectification of a fault may, therefore, be carried out in very stressful circumstances. A safe system of work is necessary to ensure that fault finding and repair are carried out in a manner that will not endanger the technician or other people who may be affected by the fault-finding activity.

## A SAFE SYSTEM OF WORK FOR FAULT DIAGNOSIS

**Assessment criteria**

**4.1** Implement safe system of work for fault diagnosis of electrical systems

The possibility of touching live parts is increased during electrical testing and fault finding, when conductors at dangerous voltages are often exposed. This risk can be reduced if testing is done while the equipment is isolated from any dangerous source of electrical supply. However, this may not always be possible and, if it is necessary to undertake the work whilst the system is live, procedures must be in place to prevent contact with any hazardous internally produced voltages.

**ACTIVITY**

Within complex installations, individual circuits may be isolated but they will most likely be installed with circuits which are still live. What device could be used to give an indication of which circuits are live?

A safe system of work for fault diagnosis and rectification was explained in some detail in Outcome 2 (page 565). The components of a safe work procedure include:

- the names of people authorised to undertake testing
- arrangements for accessing a test area
- arrangements for isolating equipment and details of how the isolation will be secured
- provision and use of personal protective equipment, where necessary
- the correct use of tools and equipment
- the correct use of additional protection measures
- use of barriers and positioning of notices
- safe and correct use of measuring instruments
- how defects are to be reported and recorded

- safe working arrangements that have been agreed with the client and/or duty holder/responsible person
- instructions regarding action to be taken if unsafe situations are identified.

Everyone involved in the work (for example, client, electrician and those in the workplace) has a responsibility for their own health and safety and that of others who may be affected by the work; communication between all parties will ensure compliance with the respective health and safety requirements.

**Assessment criteria**

**4.2** Use logical approach for locating faults on electrical systems

## A LOGICAL APPROACH FOR LOCATING ELECTRICAL FAULTS

Before a fault on an electrical installation can be diagnosed, the approved electrician must have:

- an approved safe working procedure or method statement as detailed on page 508, including a risk assessment
- a thorough knowledge and understanding of the electrical installation and electrical equipment (including access to site-specific circuit diagrams and drawings, manufacturers' information and operating instructions, design data, copies of previous Schedules of Inspection or certificates, installation specification and maintenance records)
- information relating to the fault, including any available knowledge of events leading up to the fault (such as unusual smells or observations); this can be obtained from users, the duty holder or the responsible person
- details of personnel relevant to the operation of the installation.

If a fault has occurred, it is likely that users of the system will have reported a loss of supply or changes in the way the system operates; they will probably be able to identify the areas affected. Determining which protective device has operated (main circuit breaker, individual circuit breaker or residual current device) will narrow the search area and a flow diagram can often assist in this process.

Questions to ask users include the following:

- Which parts of the system are without supply – the whole installation or a particular item of equipment?
- Is the failure a permanent problem or does it only happen at certain times or on certain days?
- What are the events that led up to the fault occurring?

## Checks that an approved electrician can undertake

### Visual inspection

A visual inspection is a quick way of identifying defects, damage or deterioration of the electrical installation. Although this check is commonly called a visual inspection, other senses such as smell, touch and hearing are also used. Touch can be used to test conductors to see if they are loose at joints and connections and smell can be used to identify overheating of thermoplastic material caused by a high resistance fault. Hearing can be used to detect arcing at a loose termination.

Here is an initial checklist for fault finding.

- Joints and connections – are conductors loose or is a connection missing?
- Conductors – do they show signs of overheating?
- Flexible cables and cords – are there signs of exposed insulation or conductors?
- Switching devices – any signs of overheating?
- Protective devices – have they operated?
- Enclosures and mechanical protection – any signs of overheating or physical damage?

### Check the supply voltage

Ask these questions.

- Is the complete system, together with all circuits, dead?
- Is a particular section of the installation dead?
- Has an individual piece of equipment failed?

Use an approved voltage tester to check for a supply voltage. If there is no incoming supply from the distribution network operator (DNO), this may be due to a main fuse failure, circuit breaker operation or a fault on the DNO network. It will be necessary to contact the distribution company to effect a repair.

**ASSESSMENT GUIDANCE**

There will be a visual inspection in your assessment. Make sure you carry it out efficiently and effectively.

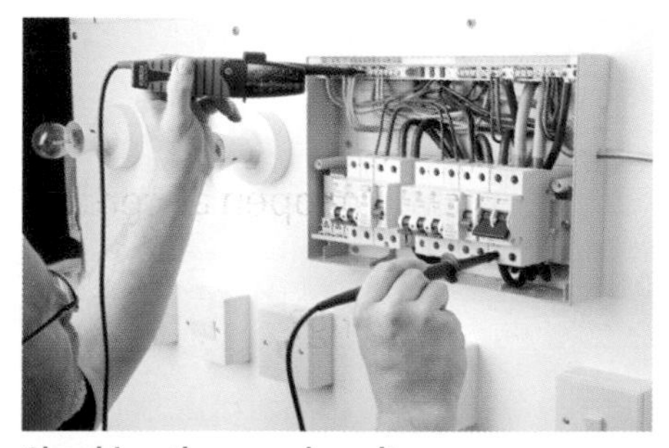

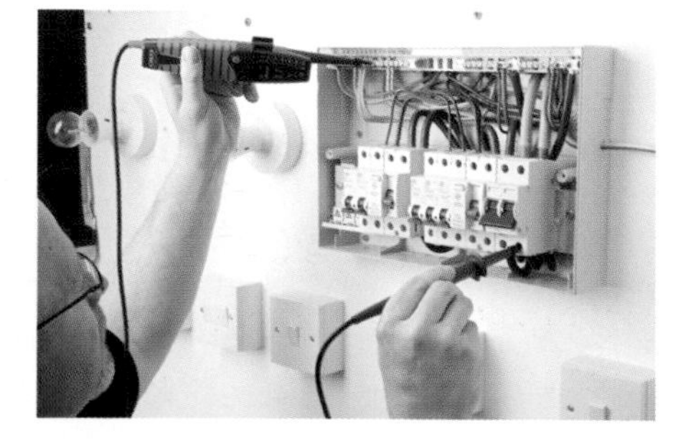

**Checking the supply voltage**

### Check the protective devices

Ask these questions.

- Has the main circuit breaker tripped?
- Is the system overloaded; have unauthorised items of equipment been connected to the system?
- Is the wiring in the distribution board damaged in some way, or is there a short circuit somewhere in the system between live and neutral conductors?

### Check individual items of equipment

Check whether the fault is restricted to one item of equipment or to the radial circuit feeding a particular piece of equipment.

For example, if the protective device operates every time a large item of equipment is switched on, it is probably the equipment that is faulty rather than the supply cable. A cable fault would normally cause the protective device to operate even when the equipment is switched off. Carry out a visual inspection of the equipment, looking for signs of damage or burning. If no physical signs of damage are apparent, undertake an insulation resistance test on the item of equipment.

Remember, before testing and fault diagnosis is carried out, the safe working arrangements must be discussed and agreed with the client, duty holder or responsible person for the installation. Everyone concerned should understand what the work entails, which areas of the installation may be subject to disconnection and who might be affected by the testing.

**Assessment criteria**

**4.3** Use testing instruments for completing fault diagnosis work

## USING TESTING INSTRUMENTS FOR FAULT DIAGNOSIS

### Standards for test instruments

The basic instrument standard is BS EN 61557: 'Electrical safety in low voltage distribution systems up to 1000 V a.c. and 1500 V d.c. Equipment for testing, measuring or monitoring of protective measures'. This standard includes performance requirements and requires compliance with BS EN 61010: 'Safety requirements for electrical equipment for measurement control and laboratory use' which is the basic safety standard for electrical test instruments.

The requirements for the safe and correct use of instruments to be used for testing and commissioning has been dealt with in detail in Outcome 2, page 565. However, the following sections give guidance on how test instruments may be used for completing fault diagnosis work.

## Insulation resistance testers (BS EN 61557-2)

Insulation resistance testers are used to verify that there is adequate insulation resistance between live conductors, and also between live conductors and the protective conductor connected to earth. The test will verify that insulation is satisfactory to withstand the supply voltage and that live conductors or protective conductors are not short-circuited. A low insulation resistance reading could indicate damaged insulation, for example, a nail in a cable, dampness in a fitting, a faulty accessory or a faulty fitting.

Before use, the condition of the instrument and leads should be checked and confirmation made that it operates correctly (in open circuit and short circuit).

In an overload condition, the circuit conductors are carrying more current that the manufacturer's design specification. If this is allowed to continue indefinitely the conductors will become very hot, causing deterioration of the electrical insulation surrounding the live conductors. This may eventually lead to breakdown of the insulation and a short circuit. It is likely, however, that the protective device, if specified correctly, will operate before a short circuit condition is reached and so prevent an insulation failure occurring. Continual operation of a protective device should be investigated immediately.

If it is suspected that insulation is beginning to fail, indicated by the frequent operation of the protective device, or even the operation of a RCD, the fault can be can be located by undertaking an insulation resistance test on the whole installation (if it is a relatively small system) or, if it is a complex installation, on the various component parts. If it is established that a fault condition exists when testing at the tails to the distribution board, the next step is to check each circuit in turn. Where the fault involves the line conductor(s), testing must be done with only one fuse inserted at a time or with one circuit breaker closed at a time. For a neutral fault, neutral conductors must be removed from the neutral bar in order to identify the fault circuit.

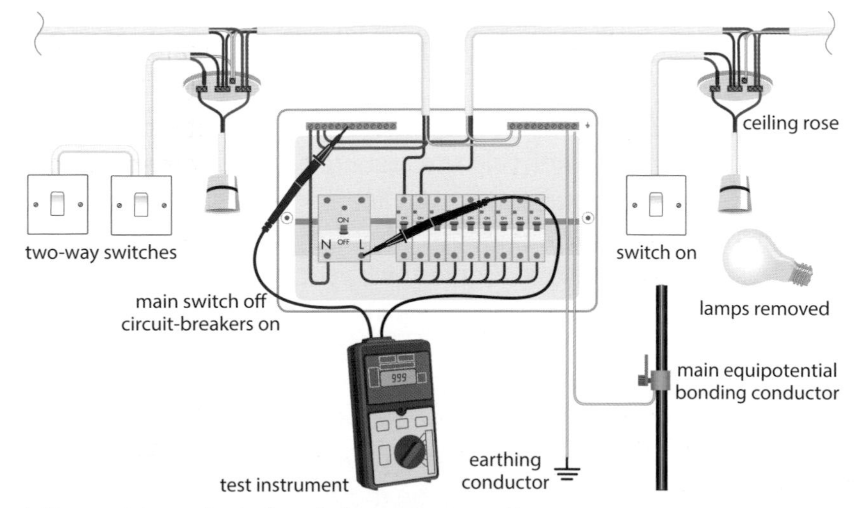


**Insulation resistance test of a whole consumer unit**

### ACTIVITY

What would be the typical output current of an insulation resistance tester with an output of 500 V d.c.?

The test is carried out between the live (line and neutral) conductors and the circuit protective conductors at the distribution board. (All equipment should be removed before the test is carried out.) If a low reading is recorded, a socket at the centre of the ring should be selected and the conductors at the socket disconnected. Carry out an insulation test in both directions. It is likely that one side of the ring will indicate a fault. If so, subdivide the section being worked on – test and subdivide until the faulty section of cable or fitting is identified. When trying to identify a fault, all circuit wiring should be included in the test – all switches must be closed and all current-using equipment, such as lamps and fixed loads, must be disconnected.

Remember, if functional switching is likely to exclude part of the circuit from the test, the switches must be operated during the test. For example, two-way switching in a lighting circuit should be operated one switch at a time to ensure that both strappers and the switch have all been tested.

**ASSESSMENT GUIDANCE**

The same rule applies to the two-way 'converted' arrangement.

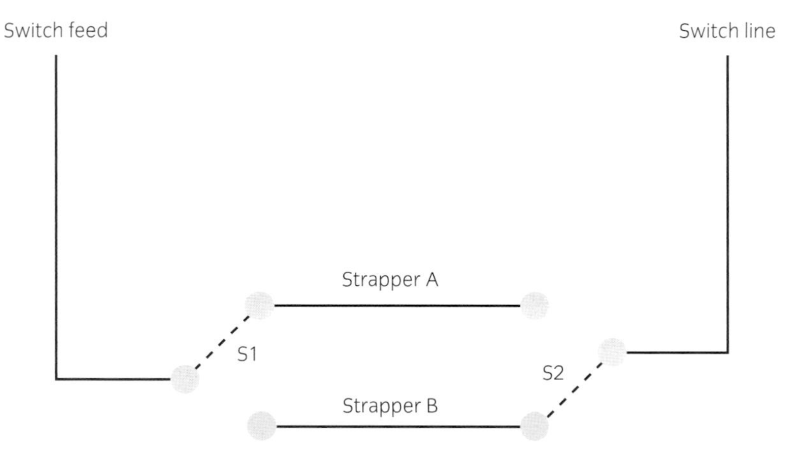


If the test is carried out with switches as shown, Strapper B and switch line will not be tested.
Operating switch S2 and retesting will test switch line but not Strapper B.
To include Strapper B, switch S1 must be operated as well.

**Two-way switching**

Low-voltage installations operating at 230/400 V must be tested at 500 V d.c. Where surge protection devices may influence the test results and it is not practicable to disconnect them, the test may be conducted at 250 V but a minimum 1.0 MΩ is required.

Insulation resistance tests should be carried out on electrically isolated circuits and, before commencing the test, it should be verified that pilot or indicator lamps and capacitors are disconnected from the circuits being tested (to avoid inaccurate test values being obtained). All voltage-sensitive electronic equipment, such as RCCBs, RCBOs, dimmer switches, touch switches, delay times, power controllers, electronic starters for fluorescent lamps, emergency lighting and residual current devices (RCDs), should be disconnected so that they are not subjected to the test voltage. If these items cannot be disconnected, a measurement to protective earth, with the line and neutral connected together, should be made.

For three-phase installations, similar tests should be conducted, with a test being carried out between all 10 pairs of conductors.

Remember that the values shown when insulation resistance testing are in megohms (MΩ, millions of ohms).

Finally, never measure resistance in a circuit when power is applied. If the circuit has sufficient power, the meter can explode or burst into flames. Most modern meters are fuse and diode protected in order to prevent explosions, but would still be damaged by an overload of this magnitude.

## Low resistance ohmmeters (BS EN 61557-4)

These are used for checking protective conductor continuity, including main and supplementary equipotential bonding, ring final circuit continuity tests and polarity. Other applications include the checking of the integrity of welded joints, bolted lap joints on bus bars and verifying winding resistance in large motors and transformers. The test current may be d.c. or a.c. and should be derived from a source with a no-load voltage of not less than 4 V and no greater than 24 V and a short-circuit current of not less than 200 mA.

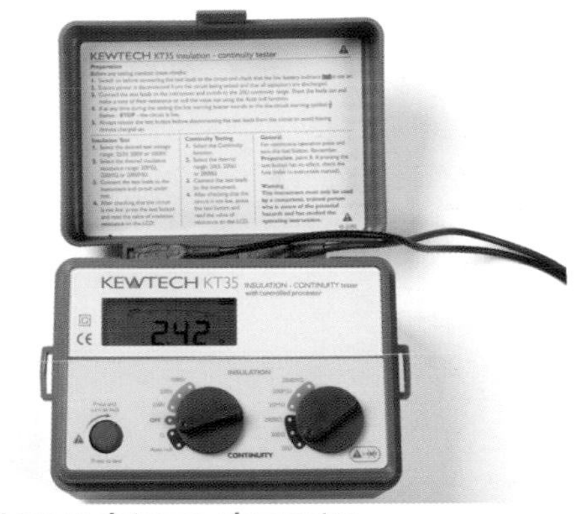
Low resistance ohmmeter

Continuity testing may be used to locate an open circuit conductor. To locate the conductor, the circuit may need to be broken into sections. To identify a lighting circuit that is not operating correctly, close the switch and test for continuity at the various light positions, operating the switches to confirm switch circuits are intact.

The easiest procedure to follow with any fault is to split the circuit into sections and systematically to test each section until the fault is identified. On a ring final circuit, the ring should be split at the distribution board and again at the approximate mid-point in the ring. You should then test each of the two sections before splitting each leg again and again, testing each length of cable between accessories until the fault is identified.

A continuity test can also be used to verify the continuity of each conductor including the circuit protective conductors. The test can ensure that there are no cross connections in a ring final circuit that might cause a shortened ring with several socket-outlets fed from spurs.

The test is a three-step process requiring:

- measurement of the end-to-end resistance of each conductor at the origin of the ring
- with the line and neutral conductors cross connected, measurement between line and neutral at each socket outlet (if the cross connections are made incorrectly, the readings obtained will tend to rise in value towards the mid-point of the ring and fall towards the origin)

- with the line and circuit protective conductors cross connected, measurement between line and cpc at each socket outlet (provided the cross connections are made correctly, all readings should be approximately equal to one quarter that of the first step, with the exception of spurs).

Locating a fault on a radial circuit for a single item of fixed equipment may be undertaken by disconnecting all of the conductors at the consumer unit or distribution board and linking them together in a connector block.

**ACTIVITY**

Describe how a half-test method could be applied to a ring final circuit.

Using a low resistance ohmmeter, test back to the consumer unit for continuity between line, neutral and the cpc. If there is a break in the circuit and no apparent branches, there could be a concealed junction box within the wiring. This junction box must be located and further tests undertaken to ensure continuity in both directions.

## Residual current tester (RCD) (BS EN 61557-6)

RCDs are sensitive devices, typically operating on earth fault currents as low as 30 mA and with response times as fast as 20 to 40 ms. BS 7671: 2008 provides for wider use of RCDs, for indoor circuits as well as for socket outlets intended to supply portable outdoor electrical equipment that is outside the zone of earthed equipotential bonding.

If an insulation resistance test has been carried on a suspected faulty circuit with a 30 mA RCD supplying an outside light and there is no low insulation resistance reading or other apparent fault, check the correct functioning of the RCD by injecting ½ × rated current and 1 × rated current. If the RCD trips at 15 mA, test its sensitivity by means of a ramp test.

Most modern RCD testers include a ramp test facility. The ramp test slowly increases the tripping current until the RCD trips.

The tester display is in milliamps (mA) and readings will identify if an RCD is susceptible to nuisance tripping or is out of specification.

The equipment can also be used to indicate which circuits have high leakage currents and can be used when there is evidence of frequent trips on a particular RCD.

The installation provides the source for RCD tests and, under certain circumstances, these tests can result in potentially dangerous voltages appearing on exposed and extraneous conductive parts within the installation. Therefore, suitable precautions must be taken to prevent contact with any such part.

Other users of the building should be made aware that tests are being carried out and warning notices should be posted, as necessary.

## Voltage testing

Within commercial, industrial and some domestic installations, different supply voltages can be encountered. It is important that suitable methods (such as an instrument in accordance with GS 38) are used to verify these voltages.

Voltages that are likely to be encountered are:

- 400 V – industrial and commercial equipment, such as motors and other fixed plant
- 230 V – industrial, commercial and domestic, such as lighting and power supplies
- 110 V – construction sites have reduced low voltage system for tools and lighting
- SELV – Class III (no greater than 50 V), such as portable lamps.

Instruments used solely for voltage verification fall into two categories:

- test lamps, which rely on an illuminated lamp to show if a voltage is present
- voltage detector meters, which give a visual readout (analogue or digital) of the voltage.

It is recommended that test lamps and voltage indicators are clearly marked with the maximum voltage that may be tested and any short-time rating, if applicable. Knowledge of the expected voltage enables the correct choice of instrument to be made and ensures its maximum voltage is not exceeded.

## Current testing

Clamp-meters, clip-on ammeters or 'tong-testers' are hand-held devices that can be used to determine the current in a current-carrying conductor, that is a circuit on-load. They can be used to identify whether a circuit is drawing too much current due to overloading by too many items of electrical equipment. Their operation relies on a current transformer (CT) that transfers a high rated current to a low rating meter and gives the reading on a calibrated scale. When using clamp-meters or clip-on ammeters it is important to ensure that the CT clamp is fully closed without any air gap as erroneous readings can be recorded if the clamp is not correctly closed.

Where current measurements are to be made using instruments other than insulated tong-testers, the connections should be made with the apparatus dead, and should be made secure before the power is switched on and rated, for both current and voltage.

Clamp-on ammeter

Voltage and current checks can be used to confirm the operation of newly installed equipment. If a new piece of equipment fails to function after installation, it should be confirmed that equipment is connected in accordance with the manufacturer's instructions, a voltage is present at the supply terminals and the equipment is drawing the correct current.

## ASSESSMENT GUIDANCE

You should be able to use your test instruments for the fault-finding exercise. If you need to use the centre instruments, take time to understand how to use them.

There are some simple rules that should be observed when using either a voltmeter or an ammeter.

- Familiarise yourself with the meter before use.
- Ensure it is safe to use – with no obvious damage to the meter or the meter leads.
- Ensure the test leads are in the correct sockets and the rotary switch is in the correct position for the desired measurement.
- Set the meter to the highest range available and reduce the range to obtain more accurate readings.
- If the reading is lower than the next available lower range on the meter, it is acceptable to set the meter to a lower range while the circuit is switched on.
- Keep fingers behind the finger guards on the test probes when making measurements.
- Always replace the battery as soon as the low-battery sign appears.

## ACTIVITY

State the reason for calibrating test instruments.

**Assessment criteria**

**4.4** Interpret testing data

## INTERPRET TESTING DATA

An inspector carrying out the inspection and testing of any electrical installation must have sufficient technical knowledge and experience to carry out the inspection and testing in such a way as to avoid danger to themselves and others. They must have knowledge of relevant technical standards, including BS 7671, and be fully conversant with the required inspection and testing procedures so that they are able to employ suitable test equipment during the test process. The inspector must also have sufficient experience to interpret the results obtained during the test process, being able to take a view and report on the condition of the installation.

It is a requirement that appropriate documentation is retained following testing and inspection, and reports can be produced in any durable medium such as written hard copy or by electronic means. The original copy of the report should be retained in a safe place and be made available to any person inspecting or undertaking work on the electrical installation in the future. If the property is vacated, the report will provide the new owner or occupier with details of the condition of the electrical installation at the time the report was issued.

Unless there are specific values that must be achieved for the installation to be deemed safe, readings such as insulation resistance should be considered relative. Readings obtained on one particular day for a piece of equipment, for example a motor, may not indicate a fault. However, the skill of the electrical technician is to determine what a trend in the readings may represent. For example, readings taken over time may show a trend that indicates failing insulation resistance and the need for some preventative maintenance. Periodic testing is, therefore, the best approach to preventive maintenance of electrical equipment.

## HOW TO DOCUMENT FAULT DIAGNOSIS

**Assessment criteria**

**4.5** Document fault diagnosis

BS 7671 requires that, following the inspection and testing of all new installations, alterations and additions to existing installations or periodic inspections, an Electrical Installation Certificate (EIC), together with a Schedule of Test Results, should be given to the person ordering the work; this is normally the client, duty holder or responsible person. Model forms for certification and reporting are contained in Appendix 6 of BS 7671.

There are two options available for the EIC, Form 1 or Form 2.

- *Form 1:* short form EIC, is to be used when one person is responsible for the design, construction, inspection and testing of an installation.
- *Form 2:* this EIC is to be used when more than one person is responsible for the design, construction, inspection and testing of an installation.

Whichever EIC is used, appropriate numbers of the following forms are required to accompany the certificate:

- Schedule of Inspections
- Schedule of Test Results.

When an addition to an electrical installation does not involve the installation of a new circuit a Minor Electrical Installation Works Certificate (MEIWC) may be used. This certificate is intended for use when work such as the addition of a socket outlet or lighting point to an existing circuit or a repair or modification to a circuit is undertaken.

EICs, EICRs and MEIWCs must be completed and signed by a competent person or persons in respect of the design, the construction and the inspection and test of the installation.

**KEY POINT**

An Electrical Installation Condition Report (EICR) will be provided when an inspection and test on an electrical installation has been undertaken in order to highlight any safety shortcomings, defects or deviations from the current version of the Requirements for Electrical Installations (BS 7671).

A competent person is defined in BS 7671 as: 'a person who possess sufficient technical knowledge, relevant practical skills and experience for the nature of the electrical work undertaken and is able at all times to prevent danger and, where appropriate, injury to him/herself and others'.

Therefore, competent persons must have a sound knowledge and relevant experience of the type of work being undertaken and of the technical requirements of BS 7671. They must also have a sound knowledge of the inspection and testing procedures contained in the Regulations and must use suitable testing equipment.

EICs and MEIWCs must identify who is responsible for the design, construction, inspection and testing, whether this is new work or an alteration or addition to an existing installation.

## Form 1 (short form) Electrical Installation Certificate

Form 1, the short form EIC is a shortened version of Form 2. It is used where one person is responsible for the design, construction, inspection and testing of a new installation or where a major alteration or addition to an existing installation is carried out and the installation, inspection and testing of that work is the responsibility of one person.

## Form 2 Electrical Installation Certificate (three signatory version)

Form 2, the EIC (three-signatory version) is used where more than one person is responsible for the design, construction, inspection and testing of a new installation or where a major alteration or addition to an existing installation is carried out. The Form 2 certificate also has space for the name, address, telephone number and signatures of two different design organisations. This can be used where different sections of the installation are designed by different individuals or companies (for example, the electrical installation may be designed by one consultant and the fire alarm system by another). In this situation the details and signatures of both parties are required. BS 7671 provides advice notes for the completion of each type of form and also guidance notes for the person receiving the certificate.

## Electrical Installation Condition Report

The EICR should only be used for reporting on the condition of an electrical installation. It is intended primarily for the person who is ordering the work to be undertaken (the client, duty holder or responsible person for the installation) and anyone subsequently involved in additional or remedial work or additional inspections, to confirm, so far as reasonably practicable, whether or not the electrical installation is in a satisfactory condition for continued service.

Each observation of a problem or concern relating to the safety of an installation should be given an appropriate classification code selected from standard classification codes as follows:

- *C1: danger present*. Risk of injury. Immediate remedial action required.
- *C2: potentially dangerous*. Urgent remedial action required.
- *C3: improvement recommended*.

## Electrical Installation Minor Works Certificate

The MEIWC is used for minor works, which are defined as an addition to an electrical installation that does not extend to the installation of a new circuit. This could cover, perhaps, the addition of a new socket outlet or of a lighting point to an existing circuit. This certificate includes space for the recording of essential test results but does not require the addition of a Schedule of Test Results.

Each certificate or report will have attached to it a set of guidance notes explaining its purpose.

When an electrical fault has been diagnosed, identified and then repaired the IET Wiring Regulations require that the circuit, system or individual piece of equipment be inspected, tested and functional tests carried out. These inspections and tests must be carried out in accordance with Part 6 of the Regulations and the results recorded in accordance with Part 6, Chapter 63. The requirements that apply to fault diagnosis and repair are:

- 631.2 – completion of an EICR
- 631.4 – the EICR shall be compiled and signed by a competent person or persons
- 631.5 – the report can be produced in any durable medium such as written hard copy or by electronic means.

Depending on the extent of the fault and subsequent report, either an EIC or a MEIWC will be issued to the person requesting the work.

The original certificate should be retained in a safe place and be shown to any person inspecting the installation or carrying out further work in the future. If the property is later vacated, this certificate will demonstrate to the new owner or occupier that the minor works covered by the certificate met the requirements of the Regulations at the time that the report was issued.

# ASSESSMENT CHECKLIST

## WHAT YOU NOW KNOW/CAN DO

| Learning outcome | Assessment criteria | Page number |
|---|---|---|
| 1 Understand how electrical fault diagnosis is reported | *The learner can:* | |
| | 1 Describe procedures for recording information on electrical fault diagnosis | 498 |
| | 2 Identify codes used in electrical condition report (BS 7671) for different faults | 503 |
| | 3 Explain implications of recorded information. | 506 |
| 2 Understand how electrical faults are diagnosed | *The learner can:* | |
| | 1 Explain safe working procedures that should be adopted for completion of fault diagnosis | 508 |
| | 2 Describe precautions that should be taken in relation to hazards of fault diagnosis | 513 |
| | 3 Explain the logical stages of fault diagnosis that should be followed | 522 |
| | 4 Specify the requirements for the safe and correct use of instruments to be used for fault diagnosis | 526 |
| | 5 Explain causes of electrical faults | 533 |
| | 6 Specify types of electrical faults found in different locations | 536 |
| | 7 Explain how faults are diagnosed by tests | 539 |
| | 8 Explain responsibilities where unsatisfactory results are obtained | 547 |
| | 9 Explain implications of fault diagnosed. | 549 |
| 3 Understand process for fault rectification | *The learner can:* | |
| | 1 Explain safe working procedures that should be adopted for completion of fault rectification | 550 |
| | 2 Describe precautions that should be taken in relation to hazards of fault rectification | 552 |
| | 3 Explain process of fault rectification | 558 |
| | 4 Explain factors which can affect fault rectification. | 563 |

| Learning outcome | Assessment criteria | Page number |
|---|---|---|
| 4 Be able to diagnose faults on electrical systems | *The learner can:* | |
| | 1 Implement safe system of work for fault diagnosis of electrical systems | 565 |
| | 2 Use logical approach for locating faults on electrical systems | 566 |
| | 3 Use testing instruments for completing fault diagnosis work | 568 |
| | 4 Interpret testing data | 574 |
| | 5 Document fault diagnosis. | |

## ASSESSMENT GUIDANCE

The assessment of this unit is divided into four parts:

- Part A Scenario-based questions (open-book exam)
- Part B Open-book questions
  You will need access to the internet and other research tools
- Part C Fault diagnosis and reporting
  This is a practical assessment. You will need a selection of tools and test equipment.
- Part D Common task – safe isolation
  This is common to 204, 303 and 304 and needs to be passed only once.

  You will need writing materials and a calculator for the open-book exams.

  You will need PPE for the fault-finding practical session.

## KNOWLEDGE CHECK

1 Identify three items of PPE that may be required when fault finding in a partially completed block of flats.

2 Information is to be provided to the client regarding activities involved in tracing a fault in a shopping centre. State:

a) to whom the information should be given

b) a means of recording the information for future reference.

3 Describe the most likely cause for each of the following faults.

a) three-phase motor running in reverse

b) switch-start fluorescent flickering and failing to start

c) immersion heater overheating

d) PIR-controlled lamp operating during daylight hours

4 State the precautions to be taken when carrying out:

a) I R test on lighting circuit containing dimmer switches

b) I R test on two-way lighting circuit

c) isolation of installation containing UPS systems

d) work on or near storage batteries.

5 Name three documents that could be referred to during the fault-finding process.

6 State the possible effect on the following groups during a loss of power in a shopping centre.

a) fellow workers

b) shoppers

c) security staff

d) shop assistants

7 Identify five factors that would affect the decision to repair or replace defective equipment.

8 State how the following materials should be disposed of at the end of a contract.

a) cardboard packaging

b) polystyrene packaging

c) scrap PVC/copper cable withdrawn during the contract

d) old transformer oil

9 Identify three outcome codes that could be inserted on an Electrical installation condition report. State the meaning of each code.

10 It is suspected that an RCD is suffering from nuisance tripping. Describe one method of determining the value (in mA) that it is tripping at.

# UNIT 308
# Career awareness in building services engineering

Career planning is an important part of life, regardless of age or qualifications. People who have not had the foresight to plan their career may find themselves drifting from one path to another, hoping to find their ideal job. This random method may work occasionally for some fortunate individuals. Instead of relying on luck, when there is an opportunity to take stock, think about your career so far and what you want to do in the future. Consider if and how you can reach your goals and think how you will know if you have reached your potential.

Looking for a new career or, particularly, looking for a new direction within an existing career, doesn't necessarily mean starting from absolute zero. It means looking at the skills and qualifications you have already gained and developed in previous employment, or even hobbies. Where appropriate, you can see what knowledge and skills are transferrable or provide a common grounding on which to develop the required new skills and qualifications.

## LEARNING OUTCOMES

There are two learning outcomes to this unit. The learner will understand:

1 how to plan for careers in building services engineering
2 the requirements to become a qualified operative in building services engineering.

This unit will be assessed by:

- a short written paper (closed book)
- a practical task that prepares you for making job applications.

OUTCOME 1

# Understand how to plan for careers in building services engineering

Assessment criteria

1.1 Identify resources to support career planning

SmartScreen Unit 308

Additional resources to support this unit are available on SmartScreen.

## ASSESSMENT GUIDANCE

You will be expected to complete a short assignment and a multiple-choice paper to achieve this unit. You will be given ample time to practise example questions and also spend some time at home attempting sample questions.

## RESOURCES FOR CAREER PLANNING

There are many publications that relate to career development, time management and personal improvement programmes. Reading them will involve substantial investment in books or spending a considerable amount of time in the library.

The internet is a logical and easy place to look for sources of information to support career planning. However, resources found on the internet should always be read carefully in the light of the original purpose of the website or blog and the actual benefit you can glean from them. For example, many websites relate just to overseas employment or specific sectors. Before relying on such information, make sure it is relevant to your needs.

The most useful sites for career planning are generally those provided by UK government departments or agencies sponsored to promote career awareness. The National Careers Service, at http://nationalcareersservice.direct.gov.uk, provides general advice and career planning information. (It is funded in association with the European Social Fund and is provided on behalf of the Department for Business Innovation and Skills.)

Once you have determined your career path, the National Careers Service website can provide useful information in a general context, including information on government support and guidance. Depending on circumstances, individuals may be able to seek government funding support for certain retraining.

Specific requirements and qualifications need to be researched from relevant trade organisations, competent person registration schemes, etc. There are many UK Accreditation Service (UKAS) accredited awarding bodies that provide support and guidance to trainees and to those already qualified and wishing to update themselves and maintain their own continuing professional development (CPD). If you are already qualified in a specific trade or profession, your professional institution will have a recognised development programme, criteria for meeting their requirements, a mapping process and access to **mentors** so you can complete the process.

**Mentor**

An advisor, counsellor or trainer who gives guidance and acts as a role model.

Organisations such as the Joint Industry Board (JIB) also provide support to member companies and their employees. Much of the JIB's work relates to practical issues such as pay and conditions, and relies on the employer accepting the JIB's terms. However, it gives employees of member organisations access to mentors, etc to ensure that they adopt an appropriate approach and undertake appropriate development in order to achieve ECS (Electrotechnical Certification Scheme) gold card status, once the training period has been completed.

**ACTIVITY**

Visit the IET website and investigate what technical assistance the institution offers to trainees or newly qualified operatives.

If you wish to develop other skills, such as estimating and design, the next step is to register with a professional institution such as the Institution of Engineering and Technology (IET). This provides a route to developing a career path in engineering or management. The IET provides information and guidance on the qualifications and experience you need to progress to the desired level. It also provides mentors to talk through application and development issues and make the process of gaining qualification and membership less painful.

## Recruitment

**KEY POINT**

A recruitment agency may try harder to place an individual in work if they have a greater chance of success, because the individual is registered with just the one agency.

Recruitment agencies operate in all employment sectors. These organisations will offer your *curriculum vitae* (CV) and services around their contacts. If an employer agrees to employ you, the agency will charge the employer a finder's fee.

Another common method of recruitment is to place advertisements in local or national newspapers or in professional journals. In this way, an employer can target a specific audience for an advertisement.

Other routes to employment include attending recruitment fairs and individual company recruitment events.

**ACTIVITY**

Obtain copies of trade magazines that contain job advertisements from employers and recruitment agencies. Note what qualifications and experience they are looking for.

**You will probably have several interviews during your career**

## Networking

Networking events are useful for finding out what opportunities are available in the market place and where they are being offered. It is also a useful method for targeting specific positions of work. It might be the slowest method for finding employment, but it has the potential to reap the best rewards.

**KEY POINT**

Some people find networking difficult. However, simply keeping in touch with colleagues can be an effective way of finding out what is happening within the job market.

**Assessment criteria**

1.2 Describe elements of career planning

# ELEMENTS OF CAREER PLANNING

When planning your career, the first thing to look at is what you can already do. Ask yourself: What am I already good at or do I have an aptitude for? It may be that you already have a number of skills and qualifications that are transferrable to a new career. A useful self-analysis tool is SWOT analysis.

## SWOT analysis

SWOT analysis is a simple tool for examining strengths, weaknesses, opportunities and threats. It is accredited to Albert Humphrey, who led a convention at the Stamford Research Institute in the 1960s. Although SWOT analysis is normally associated with examining the strategic position and aims of corporate organisations, it can also be applied to an individual's goals or objectives.

The SWOT analysis gives a measure of internal and external factors. It looks at the following factors:

- **Internal strengths** – What are you good at? For example, I can rewire a house more quickly than anyone else so I should focus on the domestic market where I excel over others.
- **Internal weaknesses** – What do you believe you could do better? This is normally something you don't like to do. For example, I don't fill in my inspection forms very neatly as my handwriting is poor.
- **Opportunities in the external environment** – What can you do that is different or how can you adapt to a changing environment to address new opportunities to be successful? For example, the surge in people wishing to train has created a demand for additional tutors, which gives me the opportunity to teach at evening classes.
- **Threats in the external environment** – The increase in general unemployment levels and the financial climate has created a shortage of work across all industries. However, with my qualifications I can look to work as an assessor for a Competent Person Scheme operator, which is less prone to the financial challenges faced by companies operating in the construction industry.

**KEY POINT**

SWOT stands for:

- **S**trengths
- **W**eaknesses
- **O**pportunities
- **T**hreats.

**ACTIVITY**

The days when a person left school at 15 or 16, found employment and was still in the same job 40 years later have long gone. Changes in technology have meant that many traditional jobs no longer exist. Can you think of two heavy industries that have declined in the last 30 years and two which have become large employment areas?

**ASSESSMENT GUIDANCE**

You may be asked about SWOT analysis. Try to find a way of committing it to memory.

These factors can be analysed for a particular venture or objective and represented in a matrix or diagram format, to allow easy interpretation of the results.

| | Positive factors | Negative factors |
|---|---|---|
| Internal | Strengths<br>• example<br>• example<br>• example<br>• example | Weaknesses<br>• example<br>• example<br>• example<br>• example |
| External | Opportunities<br>• example<br>• example<br>• example<br>• example | Threats<br>• example<br>• example<br>• example<br>• example |

**SWOT analysis diagram**

## ACTIVITY

Read the text below and then do a SWOT analysis for your own situation, using a diagram like the one opposite. Discuss your analysis with a friend/relative and then with a tutor. Think about your strengths, as seen by these people.

## Strengths

Strengths are the qualities that allow a business or individual to move towards their goal, assuming the *status quo* means that no changes are needed and they continue as they are.

## Weaknesses

These are the characteristics that hold back or impede a business or individual. If a business or an individual does not address these weaknesses, they might lose a particular job or opportunity to a competitor who has developed a strength in that area.

## Opportunities

Opportunities are the openings or changes in the external environment that can bring about new challenges or horizons. An example could be a newly qualified 20-year-old electrician who has shown an interest in inspection and testing. This type of work is normally given to more experienced workers but, due to an unusual level of sickness in the company, the 20-year-old is asked to carry out inspection and testing on a number of industrial projects. The company say that, subject to the work being completed to a satisfactory standard, any future work of this type will be given to the 20-year-old. When a permanent testing job is available, the 20-year-old will have the required experience.

## Threats

It is generally accepted that threats that influence a business can come from competition or changes in the market. They can come from anywhere and are usually more devastating when they come from an unanticipated source.

Therefore, when 'horizon scanning' for opportunities, it is necessary to check if there are also any threats, such as a competitor moving into a new area of business that overlaps yours. Every threat should be examined. Responses to threats bring about change, which will usually create opportunity.

## Knowing yourself

Having taken stock and reviewed your existing skills and qualifications, you will need to consider what you would like to do in the future. This should be done realistically; it may have been your life-long ambition to be an astronaut, but the likelihood of being successful in this career will depend on having specific skills and physical attributes, and accessing the right opportunities. Although you should never set your targets so low that you become quickly dissatisfied, you should also set realistic targets.

**ACTIVITY**

Stay with same firm or move around – what will suit you? Think about the pros and cons.

Once you have decided what you want to do next, you should determine what you need to do and learn in order to meet your goal. This will involve finding out what is really required to fulfil the job or career objective, including what training you will need.

**Assessment criteria**

**1.3** Describe documents to support career development

# DOCUMENTS TO SUPPORT CAREER DEVELOPMENT

## Career action plan

A career action plan is a useful document to use to plan career development. It allows you to plan your career and development in your existing job or to plan a different career path in the future. It assists you to look at the skills and qualifications you need, to reach your goal, and then monitor your progress. A typical career action plan can be developed by asking the following questions.

- Where am I now?
- What do I like about where I am?
- What do I want to change?
- How can I change what I don't like?
- What will I need to create that change?
- How do I get there?

After careful consideration, you can decide on a goal. An example might be to improve this year's turnover by aiming to have 70% returning clients. You should also set a target date for achieving your goal. In this example, it might be in 12 months' time.

In order to reach your goal, you will need to put a plan into action, similar to the one opposite.

| Objective | Action to achieve | Who is responsible or able to help | Target date | Success indicator or corrective action required | Actual result |
|---|---|---|---|---|---|
| Find out why customers do not always return | Write customer satisfaction feedback questionnaire. | Admin staff | 3 months' time | 20% of questionnaires returned with information | 17% of questionnaires returned |
| | Follow up returned questionnaires with telephone discussion as to how customer can be retained. | Customer service manager | 1 month after returns | Clear feedback from customers identifying individual areas of weakness | Three key areas where customers felt there were inconsistencies in service levels |
| | Address individual areas, eg training of customer service team | Customer service manager | 1 month after feedback | Customer service team receive additional training | Improved service levels by start of third quarter of year |
| | Example action 1/4 | | | | |
| Goal number 2 | Example action 2/1 | | | | |
| | Example action 2/2 | | | | |
| | Example action 2/3 | | | | |
| Goal number 3 | Example action 3/1 | | | | |
| | Example action 3/2 | | | | |
| | Example action 3/3 | | | | |

A typical action plan: actions and initiatives to meet goals

As you progress towards your goal, your action plan will make it clear when you are on track or going off track. It may even show that you need to change direction altogether, in which case you can re-plan and create a revised action plan.

## Plan–Do–Check–Act

Many organisations use the Plan–Do–Check–Act (PDCA) process in their professional development or career development programmes.

Also known as the Deming Cycle, the PDCA process is made up of these factors:

- **Plan** – establish objectives to deliver the required goals.
- **Do** – carry out the plan.
- **Check** – study the results to spot any differences between the actions planned and done.
- **Act/Adjust** – take corrective action to put the plan back on course.

Applied logically, this approach can be used for any type of development process, personal or professional.

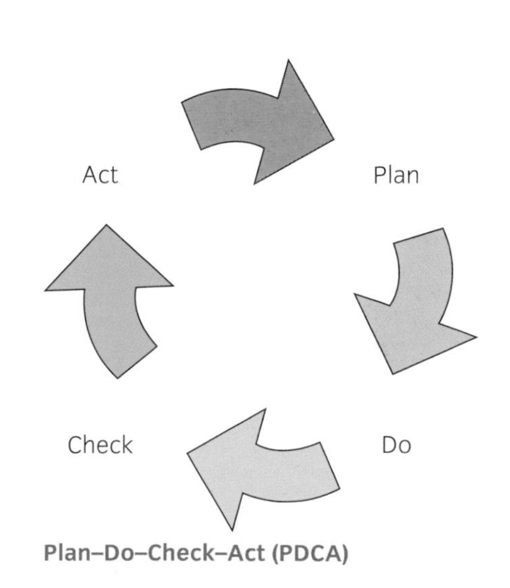


Plan–Do–Check–Act (PDCA)

## Portfolio

A portfolio is a collection of well-presented documents that demonstrates the type and nature of work an individual (usually a student or trainee) has carried out. It can be presented as a collection of evidence to show what the individual can do, either to the assessor of a course being undertaken or to a potential employer in place of a detailed *curriculum vitae* (CV).

## Curriculum vitae

A CV is an important résumé of your work and professional experience to date. Ideally CVs should be short and easily readable to keep the reader's interest. They should be supported by a covering letter and a statement of references should be available upon request.

**KEY POINT**

Present your CV in a clean and tidy fashion. No-one wants to be presented with a dog-eared document.

A good CV sets out all the details for a potential employer to decide whether you have the skills and experience required for the position you are applying for. The details must be informative, clearly presented and to the point. The aim is to get the employer to invite you for an interview.

Recipients usually scan CVs rather than read them thoroughly. Therefore, it is essential to limit your CV to two pages if possible, making sure that the important information is clearly visible.

## Organising and presenting your CV

There are many different ways of presenting a CV. However, information must be presented in a logical order.

### Initial information

This should include your personal details, address and contact information.

### Personal statement

**ASSESSMENT GUIDANCE**

The exam questions ask for short answers, not long ones. Be prepared to explain the major points of a CV, for instance.

The next element of the CV should be a personal statement outlining your key strengths and achievements. Include those you think will be important and relevant to the potential employer. If possible, tailor this statement to suit the needs of each potential employer, so that they can quickly check your skills and attributes against their requirements. Limit your personal statement to 200 words.

### Experience and qualifications

It is often useful to indicate your work experience next. It is essential to identify all your relevant experience, without including information that an employer does not need to know. It is usual to list work experience in reverse chronological order. This means presenting your current position first, with the date you were appointed, describing your role, responsibilities and any achievements or professional developments made.

Repeat this for each previous employer. Consider including less detail, as you go back in time, to help you keep your CV as brief as possible. There is little need for a chartered engineer of many years' experience, for example, to go into much detail about apprenticeship. It would be better to choose a few words to cover the overall experience and save the space for more important details elsewhere on the CV.

If you are still studying or if you have lots of experience in a large number of positions, you might choose to give your qualifications first. Then a prospective employer can see them more quickly. Present your education in reverse chronological order, indicating all the educational establishments you have attended or are still attending.

**ACTIVITY**

Personal and life skills can be as important as exam success. Can you list any specific skills you may have in this area?

### Other skills, hobbies and interests

Next include other skills that are relevant but not job specific. For example, you may have particular computer skills, fluency in a foreign language or be a qualified first-aider. A short section relating to your hobbies and interests may also be included if you feel they are particularly relevant to the position you are applying for.

### Referees

If requested, you should provide the contact details for your referees (see 'References' below). However, if your CV is speculative and not in response to a specific request, you simply say that referees' details are available on request so that you can keep control over other people's contact information.

**ACTIVITY**

Examine the example CV on pages 592–593. Does it follow the advice given? How could this CV be improved so that a prospective employer would be interested in interviewing the candidate?

## Covering letter

This letter should be professional and to the point, although it should say more than just 'please find the enclosed CV ...'. It provides a chance for you to show that you understand the employer's organisation and how your skills and experience can meet their requirements. It is a chance to stand out from the other applicants so that the reader will consider your application carefully.

## References

References are important in terms of verifying what sort of person you are and whether you are reliable, honest, trustworthy and, most importantly, employable.

Ideally, referees should know you personally. It is better to have a good appropriate reference from the supervisor who knows and works with you, rather than to have a standard employers' reference from the managing director, who may only exchange pleasantries with you.

If you have named someone as a referee, it is good manners (but often overlooked) to let them know so that they are prepared for the contact, which is sometimes initially by telephone. This avoids potential embarrassment and also ensures you have the referee's cooperation, which is useful in terms of receiving a positive reference.

# CURRICULUM VITAE: John Jones

## PERSONAL DETAILS

| | |
|---|---|
| Name | John Jones |
| Date of birth | 11 November 1991 |
| Address | 1 Market Street<br>Any Town<br>Any City<br>AB12 3CD |
| Mobile | 0799 321654 |
| E-mail | john.jones@interweb.co.uk |

### Personal statement

I am currently working as a personal care and support carer. I always worked hard at school and was head boy in my last year. I believe this shows I have maturity and a sense of responsibility. My school results show I am able to pass examinations. I believe I would be an excellent trainee electrician.

## SCHOOL EDUCATION

| | |
|---|---|
| 2003–2008 | Any Town High School<br>School Road<br>Any Town<br>AD12 3CE<br>Telephone: 01789 567123 |

### Qualifications obtained

| Subject | Level | Grade achieved | Date achieved |
|---|---|---|---|
| Maths | GCSE | A | July 2008 |
| English Language | GCSE | A | July 2008 |
| English Literature | GCSE | A | July 2008 |
| Science | GCSE | A | July 2008 |
| Additional Science | GCSE | C | July 2008 |
| Spanish | GCSE | B | July 2008 |
| Geography | GCSE | C | July 2008 |
| Media | GCSE | A | July 2008 |
| Drama | GCSE | B | July 2008 |

**FURTHER EDUCATION**

2009–2011 Any Town College
Any Town Road
Other Town
CT9 6AA
Telephone: 01907 232341

**Qualifications obtained**

| Subject | Level | Grade achieved | Date achieved |
|---|---|---|---|
| Performing arts | BTEC National Diploma | Pass | July 2011 |

**WORK EXPERIENCE**

**July 2004–2005: Premier Store**
I had a weekend job in the local paper shop, stacking shelves and working on the tills.

**July 2007–2008: Any Town Restaurant**
While studying for my GCSEs, I worked part time as a pot washer. I was then given the opportunity to work as a waiter.

**September 2007–2008: Local Hospice**
I worked here for two weeks on work experience. I enjoyed working with vulnerable people and ensuring that they were safe and comfortable. After work experience, I continued to work voluntarily for a year. I learned the skills of communication and dealing with difficult circumstances.

**July 2009 – January 2011: Any Town Nursing Home**
I undertook a full-time care-worker post, attending to people's personal needs and supporting them to live their daily lives. It was a very challenging role, as many patients had dementia, requiring special ways of communication. I became a strong, problem-solving member of staff. I undertook a course on dementia, which gave me deeper insight into the condition and provided me with different ways of dealing with it.

**July 2011 – present: Any City Homecare**
I wanted to expand my experience further, so undertook this post as a care and support worker. I provide personal/social/domestic care within people's own homes. Promoting independence and providing people with a sense of inclusion is important and I find that very enjoyable and rewarding. I have recently been given the title of Branch Trainer. I am delighted to have the opportunity to teach and have a positive impact on the way care workers deliver our service.

**HOBBIES**

- I participate in the local amateur dramatic society and have played a number of lead and supporting roles. I also work as part of the backstage crew, building sets.
- I used to play first horn in the school band.
- I love music and play a number of different instruments.

**REFERENCES**

Referee details are available upon request.

Assessment criteria

1.4 Explain the principles of goal setting

## PRINCIPLES OF GOAL SETTING

There is a proverb that states 'a journey of one thousand miles starts with a single step'. Setting goals is an important step towards a strategy for change that will lead you in the direction you want to go.

People often refer to setting goals and objectives without any thought to the difference between the two terms. However, it is important to understand the difference, so as to avoid using the wrong terminology.

**ACTIVITY**

You will probably need to draft out your goals and objectives. Use IT to help you do this.

### Goals

A goal is a strategic requirement or outcome that an individual or organisation is trying to reach. It is likely to be quite general and long range; it can look quite idealistic and doesn't include all the practical details of exactly what is required and when it is required by. An example of a goal could be to have a complete career change within the next five years. This may be not fully detailed but that goal is the starting point in the planning process. (See Assessment criteria 1.5 on page 596 for how to set goals.)

**Goal**

Where someone wants to be with their life and/or career in the future.

**ASSESSMENT GUIDANCE**

One goal you should aim for is to pace yourself through the exam and assignment. Do not write too much in one area and then have insufficient time to finish.

### Objectives

Once goals have been set, it is necessary to set objectives to meet the goals. Asking 'who, why, what, where and when?' questions will assist you in deciding on appropriate objectives.

**Objective**

A plan for how to achieve a specific goal.

Objectives are short-term tactical approaches that help to achieve a goal. They should be clearly defined and an effective objective needs to be SMART. A SMART objective is:

- **S**pecific (clearly defined so you know exactly what needs to be achieved)
- **M**easurable (concrete criteria for measuring progress and successful completion)
- **A**chievable (you know it is possible to achieve)
- **R**ealistic (you have the time and resources to achieve it)
- **T**imely (a deadline for completion).

**KEY POINT**

An effective objective needs to meet the SMART criteria:

- **S**pecific
- **M**easurable
- **A**chievable
- **R**ealistic
- **T**imely (time-bounded).

In setting each objective, check that it meets the SMART criteria and that the intended outcome will meet, or help to meet, the overall goal.

Progress with objectives should be reviewed regularly, at timely intervals that allow for specific stages to be realistically achieved. A review may show that the objective is working well towards the goal, needs changing slightly to achieve the same goal or needs changing altogether. It may also be that the overall goal needs to change, in which case new objectives will also be needed. If objectives need changing, you simply start the process again.

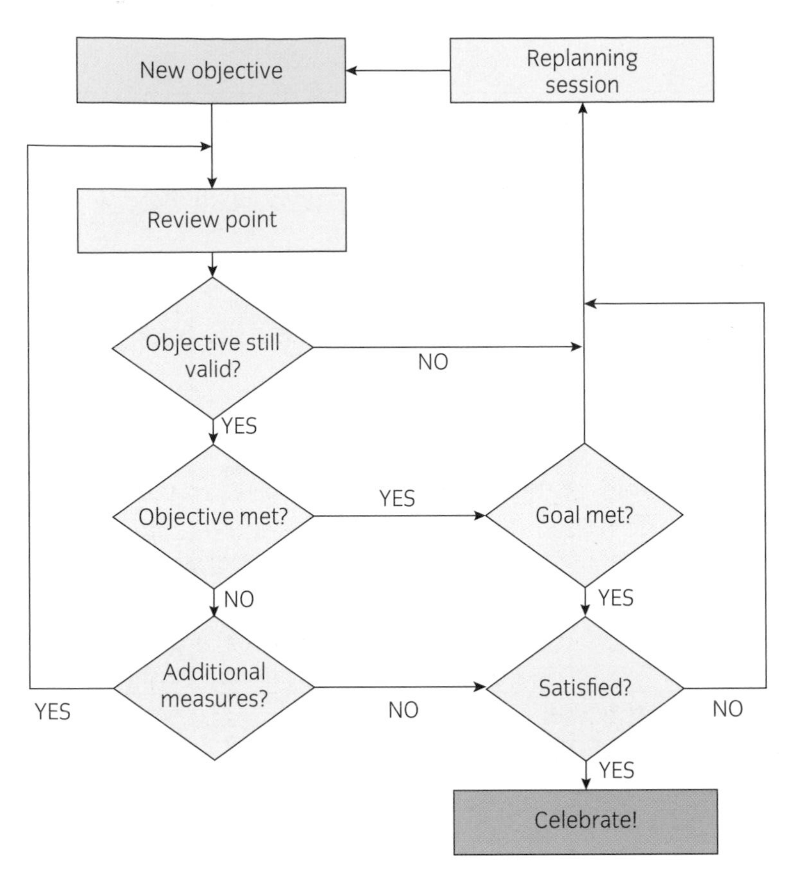


**Objectives review**

## Case study

After qualifying, an electrician should first consolidate their skills and then assess their own preferences, strengths and weaknesses. This will help them to decide what type of work they want to do and/or post they want to hold, how they want to get there and what they need to do next to get there.

Simon qualified as an electrician two years ago. He has now decided that he would like to work towards becoming an approved electrician. Having identified his goal, he set out the objectives to achieve his goal in the grid overleaf. He checked that the objectives were SMART.

### ACTIVITY

Check the action plan overleaf. Look at the objectives within the action plan and explain how they meet the SMART criteria.

| ACTION PLAN | | | | | |
|---|---|---|---|---|---|
| **GOAL:** to become a test and inspection electrician | | | **TARGET DATE:** 12 months from start date | | |
| **Objective** | **Action to achieve** | **Who is responsible or able to help** | **Target date** | **Success indicator or anticipated result/ corrective action required** | **Actual result** |
| Gain City & Guilds 2394 and City & Guilds 2395 by end of year | Locate local colleges training electricians for the City & Guilds.<br>Find out cost and entry requirements. | Employer | 4 weeks | Get enrolment details, costs and approval/ support from employer. | Obtain letter of support to pay fees. |
| | Enrol on City & Guilds 2394 and achieve the necessary pass for certification. | Employer/ college | 3 months | Attend 2394-01 course and sit exams. | Complete and pass exam.<br>Receive 2394-01 certificate. |
| | Enrol on City & Guilds 2395 and achieve the necessary pass for certification. | Employer/ college | 6 months | Attend 2395-01 course and sit exams. | Complete and pass exam.<br>Receive 2395-01 certificate. |
| Gain practical testing experience | Attend/assist with practical testing and inspection on behalf of company (new works). | Employer/ colleagues | 4 months | Carry out/assist inspectors on testing and inspection work. | Achieve competence in initial verification and testing. |
| | Attend/assist with practical testing and inspection on behalf of company (any type of work). | Employer/ colleagues | 6 months | Carry out/assist inspectors on testing and inspection work on different project types and age. | Achieve competence in initial verification and testing on any project of any age. |
| Become a competent test and inspection electrician | Work unassisted, including completing paperwork, etc. | Employer | 9–12 months | Complete all paperwork, understand the supporting information and be able to interpret results, etc. | Be competent in testing and inspection works and understand the duties of the qualifying supervisor. |

Simon's action plan

**Assessment criteria**

1.5 Describe how to set goals

## HOW TO SET GOALS

Goals are strategic requirements to meet a desired outcome. The goal will be quite general in its design, for example, to become a qualified electrician in the next few years.

Goals like this do not just appear out of nowhere. They are the result of considering significant issues such as what direction you want your career to take and where you want to be in the future. Lifestyle and career changes should not be embarked upon lightly, as they may involve considerable change and impact all those around you, so they should be part of a considered plan.

## Personal goal setting

Personal goal setting is an excellent way to think about your ideal future and to motivate yourself to turn your vision of this future into reality. Setting goals helps you choose where you want to go in life. It is generally acknowledged that goal setting motivates individuals to 'better themselves' or to achieve more.

Goals give an individual something to aim for. However, as with setting objectives, when setting a goal it is important that it is achievable. It is easy to be discouraged, go off target or even give up if your goal is not achievable. Impossible goals are merely pipe dreams, so aim high but keep it real.

Goals give you the opportunity to study the 'bigger picture', to take stock of your wishes and aspirations and to set down what you want to do and where you would like to go.

Initially, a brainstorming or concept-mapping session with a relative or mentor will allow you to record lots of different ideas. Goals can be selected in different categories or areas of life. In an ideal world they would be complementary to each other so that one supports or assists the other.

**ASSESSMENT GUIDANCE**

Set out your own goals as a practice for the assignment. Show the results to a friend or relation.

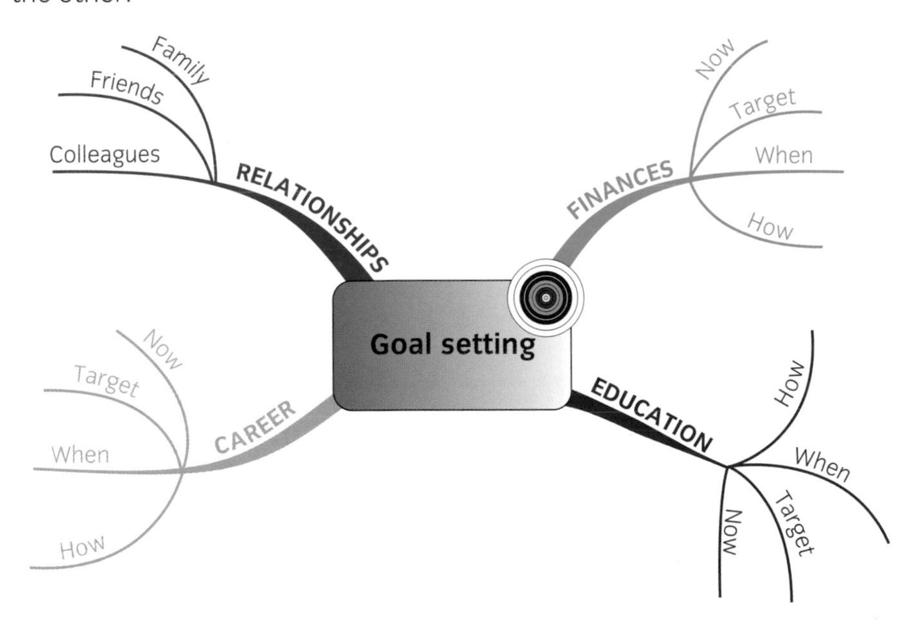


**You can use a diagram like this to map out your thoughts**

These areas might include lifetime goals such as:

- career
- education
- relationships
- finances.

**KEY POINT**

Keep the task simple and break large tasks down into small 'bite-sized' chunks.

The next stage after the brainstorming or concept-mapping session is to trim down or prioritise your goals, so that initially you don't have too many, as failure to achieve can often be a demotivator.

It is important that they are goals that you really want to achieve, not ones that others want you to achieve.

Setting these lifetime goals gives you perspective, allowing you to shape your decision-making processes for the future. In whatever area of life you set your goals, it is important that you write the goals down: this makes them tangible and real.

Once you have set lifetime or bigger picture goals, you need to break them down into smaller goals, as milestones on the way to achieving the overall goal. Milestone goals need to take a SMART approach in the same way as objectives (see page 594). If you never meet any of the smaller goals used as midway markers, they will put you off instead of providing motivation.

In summary, you should take these steps:

- Look at the bigger picture (assisted by brainstorming, mind-mapping, etc).
- Look at where you want to be (starting with only a few goals in small bite-sized chunks).
- Make sure the goals are your goals (not what someone else wants for you).
- Set milestone goals (to check you are on course and it is still what you want to do).
- Review and set short-term SMART objectives to ensure you meet your milestone goals.

**ACTIVITY**

Set some goals to take you forward from this point in your life. Discuss them with a tutor. Then decide how to break down your goals into small 'bite-sized' chunks. Decide on timely points to review your progress towards your goals.

## ROLES IN BUILDING SERVICES ENGINEERING

**Assessment criteria**

1.6 Define the different roles in building services engineering

There are a number of different trades within building services engineering. Although they may all be employed in the same construction team on a particular project, they are often provided by different contracting companies.

### Electrician

This trade contractor is responsible for the installation of all electrical wiring, including any containment systems for the wiring or specialist systems such as door-entry security or fire-alarm systems. On smaller projects, the electrician will install all services but may employ a specialist installer for systems such as the intruder alarm to meet any contractual or specification requirements.

On larger projects, the electrical contractor will usually be responsible for the low-voltage wiring to the building and provide containment and wiring systems for specialist installers working, for example, on fire and security systems. Such specialists will normally be sub-contractors of the electrical contractor, who is responsible for the management of these individuals as well as their own employees.

Electrician preparing cables for terminating

**ACTIVITY**

What would be the advantage to a main electrical contractor of employing sub-contracting electricians for a particular contract?

**ACTIVITY**

Find out what qualifications the other trades need and see how they compare. Are they at Level 2 or 3?

## Heating and ventilation (H&V) fitter/engineer

These engineers carry out pipework and heating installation works. They normally carry out installations in steel, whereas a plumber normally carries out installations in copper.

## Gas fitter

These specialist trade contractors are qualified to work on gas and combustion equipment installations such as boilers. They are registered with Gas Safe, which is a legal requirement for working on gas systems and boilers.

## Air-conditioning and refrigeration engineers

These trades are aligned closely to the electrical trades, although they undertake an amount of pipework installation. The refrigeration engineer must be aware of refrigeration legislation, as certain gases are banned in the UK.

## Installation electrician

This type of electrician generally works within or around the construction industry. They are trained in electrical installation techniques that include the testing and commissioning of electrotechnical equipment and systems. They are generally employed by organisations whose core business is electrical installation contracting.

**ACTIVITY**

Can you think of four areas where maintenance rather than contracting electricians might be employed?

## Maintenance electrician

Unlike the installation or contracting electrician, a maintenance electrician is usually employed by an organisation whose core business has nothing to do with electricity. For example, this could be a factory producing bars of soap, where a maintenance electrician is employed to keep production machines running or ensure that the safety of the installation is not compromised by the production work, by regular checking, testing and maintenance.

**KEY POINT**

The Engineering Council is the **regulatory body** for engineers and technicians in the UK.

**Regulatory body**

An organisation that regulates an industry or similar in a public protection role. Unlike professional bodies, it is not a membership organisation. The regulatory body is usually backed by a legal mandate or statutory requirement.

## Building services engineer

This is normally a professionally qualified individual who may or may not have been a tradesperson before taking advanced qualifications and training. Usually the engineer will have undergone a training programme that would satisfy the Engineering Council's requirements. They will have qualifications to carry out different tasks, such as estimating, contract management and design. Some specialist engineers are responsible for the overall safety of systems.

**Assessment criteria**

1.7 Explain opportunities for progression within building services engineering

# OPPORTUNITIES FOR PROGRESSION

Once qualified, a skilled operative should spend a period of time consolidating their skills and competency so that they are able to work without supervision. After that, they may start to think about the next stage of progression.

There are opportunities at different levels of commitment, responsibility and risk. Within an organisation, the next stage might be to become a highly skilled specialist or to take on a managerial role as a trade supervisor. These are not always two different positions. A trade supervisor will start to look at resources and delivery but will often also continue to be concerned with the craft and quality issues faced by the skilled operative.

Different positions suit different people. Not everyone has the aptitude, skills or tenacity to take on the role of specialist or trainer/assessor, or be the owner of a business.

A supervisor checking work carried out by an installer

## Supervisor/manager

Those who choose to take on managerial roles normally begin by becoming trade supervisors. They often progress to be contracts managers or even the owners of businesses.

These positions carry much more responsibility than those of skilled operative or leading hand/foreperson. People in these positions must have a good understanding of the trade skills and needs of the tradespeople they are leading. They must be able to lead others well, making appropriate decisions to allow their subordinates to work efficiently and effectively.

These positions also carry responsibility in law. Although all workers need to be aware of health and safety law, a supervisor or manager must take responsibility for the people who report to them, as well as for themselves. They also take responsibility for any liabilities that their actions or lack of action have for the company that employs them.

Supervisors and managers therefore have to undertake managerial training, which would usually include learning about employment law and health and safety law. Although good leadership cannot necessarily be taught, supervisors and managers can also be coached to adopt leadership styles to help them achieve the outcomes required.

**ACTIVITY**

Think about the best combination of trades for multi-skilling, perhaps where existing skills could be transferred to a new trade.

## Multi-skilling

Some tradespeople find that becoming multi-skilled, or retraining in an associated trade, is more rewarding from a financial or job satisfaction point of view. This might happen when a skilled operative finds that they have an aptitude for an area of construction they had not considered when they undertook their initial training. For example, an electrician may discover a real passion for transforming rooms and surfaces as a plasterer. This trade also fits very closely with rewiring and renovation works so multi-skilling would be very useful.

Tradespeople working within an organisation may also find that there is potential for staff who have undergone structured apprenticeships or have a thorough grounding in a particular trade to take on office and technical duties.

This level of practical experience, as well as academic qualifications, is often taken into account when an employer is looking to appoint a junior designer, estimator or surveyor. Academic qualifications for these roles can be gained through part-time study at local colleges, at university or through the Open University. Many designers, surveyors and estimators started their careers as tradespeople working on craft installations and there is considerable respect in the industry for people who take this career path through the technical grades.

## Trainer or assessor

Trainers and assessors are usually motivated by a desire to teach the next generation to practise their skills to the best of their ability. Such people usually have superior knowledge of the craft and care about technical excellence and improving standards of service.

An assessor may be a trainer or tutor, or someone who just carries out assessments. They know industry standards and are qualified to use a wide range of methods in assessing if someone possesses the skills required by industry.

## Designer

A number of opportunities occur in design. Usually a designer has to have achieved the status of engineering technician through to chartered engineer.

Designers need a good understanding in the core subject but also of the interfaces where electrical and other systems meet.

## Different types of employment

Within the building services sector there are different types of employment, ranging from casual labour through being employed by a company to offering expert advice as a consultant.

### Contract work

Contract work generally refers to a specific piece of work agreed by contract to be completed within a fixed period of time for a fixed sum of money. The person undertaking such work might be a self-employed **contractor** or the employee of a company.

**Contractor**

A person or organisation carrying out set pieces of work for fixed sums of money.

A self-employed contractor would be paid a higher rate than an employee, to allow for the fact that they have to pay the usual overheads of holiday pay, national insurance, etc. Agency fees also have to be paid for any workers found by an agency.

**ACTIVITY**

Self-employed electricians can run their own companies and may employ others, as work demands. Some work through an agency and others work for another contractor for the duration of the contract. See if you can find out how many registered electricians there are in the UK.

Contract work can often seem to favour the client. The contract may contain penalty clauses to compensate the client if the quality or timing of the work become an issue. On the other hand, issues over which the contractor has little or no control may arise. These may result in the contractor having to pay compensation for delays.

The secret to success is to overcome the disadvantages. Establishing trust between client and contractor, so that both parties are working towards a common goal, is very effective. It should avoid the necessity for constant referral to the contract and resorting to methods of punishing each other, which can be costly and bad for repeat business.

### Consultancy

A consultant offers specialist or professional advice and therefore requires a level of expertise and specialist skill. Clients rely on this expertise, so a consultant must hold professional indemnity insurance (PII) in case their advice results in damage to the client.

A consultant will normally be appointed by the client, under a form of contract called a consultancy agreement.

### Sub-contractor

A sub-contractor works for a contractor. There is very little difference between the two roles, although the sub-contractor has no contractual connection to the ultimate client and deals solely with the contractor. This means that all instructions and payments are given to the sub-contractor by the contractor. If the ultimate client defaults on payment, the sub-contractor should still be able to claim payment from the contractor, provided the sub-contractor is not part of the reason for the client withholding payment.

## Casual labour

Casual labour is provided by seasonal or non-permanent staff with a limited term of contract. It may include, for example, students working over holiday periods.

Casual workers do not have all the employment rights of permanent employees, but they are entitled to the national minimum wage, paid holidays, statutory sick pay, the right to work breaks and the right to make a **protected disclosure**. Other rights, such as the right to have time off to look after dependants, apply to all workers, regardless of length of service.

**Protected disclosure**

A disclosure of information about an employer honestly made in the interests of safety or the public interest for which the informant cannot be punished by the employer.

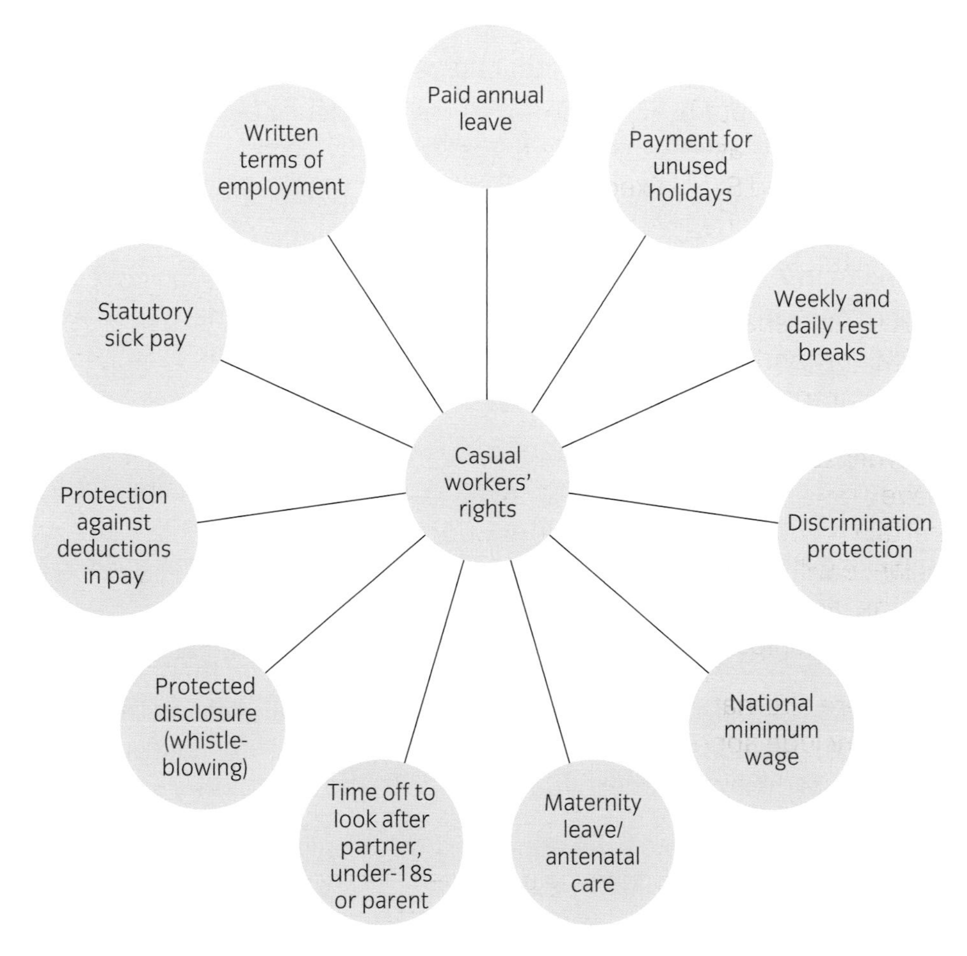


Casual workers' rights

### ACTIVITY

List three different types of job where employees do not have permanent year-round employment (eg holiday representatives).

# OUTCOME 2

# Understand the requirements to become a qualified operative in building services engineering

## REQUIREMENTS FOR BECOMING A QUALIFIED OPERATIVE

Assessment criteria

2.1 Describe specific requirements for career choices in building services engineering

To become a qualified operative involves not only passing the academic test, but being competent in the particular trade. Competence is assessed through qualifications, training and skill. The Health and Safey Executive (HSE) interpretation, given in HSR 25 Memorandum of Guidance to the Electricity at Works Regulations 1989, states:

> A person shall be regarded as competent ... where he has sufficient training and experience or knowledge and other qualities to enable him properly to assist in undertaking the measures.

**KEY POINT**

Competence is normally achieved through a mixture of experience and qualifications.

In order to become a qualified electrical operative, it is necessary to have a combination of nationally recognised qualifications, with a certain amount of experience to support the qualifications. This balance will normally be assessed by potential employers. Be prepared for the fact that the majority of employers worked through full apprenticeships and may view alternatives with less enthusiasm.

**ASSESSMENT GUIDANCE**

Do you know the roles of the ECA, NICEIC and JIB? Ask your tutor or use the internet to find out, so you can answer any questions in this area.

There are alternative routes to qualification as a qualified electrician, which include apprenticeships for older trainees through the Sector Skills Council (SSC). The SSC is licensed by the UK government through the UK Commission for Employment and Skills (UKCES).

The principal route to becoming a qualified electrician is through the apprenticeship route. You can now take this route as an adult trainee. This involves completing a number of qualifications within the apprenticeship framework. This framework is laid down by the SSC. The apprentice needs to show that they have carried out practical training with a registered scheme and achieved 2357 (NVQ Knowledge and Practice) and, if there was no prior learning such as GCSE:

- Maths Functional Skills Level 2
- English Functional Skills Level 2
- ICT Functional Skills Level 2
- Employer's Rights and Responsibilities (ERR)
- Personal Learning and Thinking Skills (PLTS).

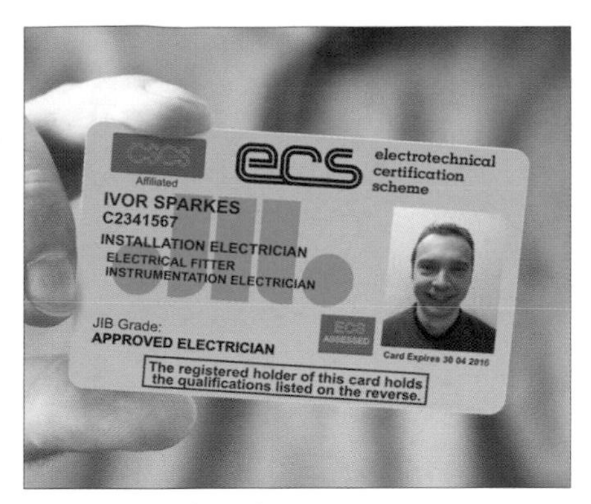


The ECS gold card

The alternative to the apprenticeship, which was Introduced in September 2012, is the City & Guilds 2365 Levels 2 and 3 or the older 2330 qualifications in conjunction with:

- City & Guilds 2357 Units 311–318
- 2357–399 Electrotechnical Occupational Competence (AM2).

These can be used as recognised prior learning (RPL), counting towards some of the units in the City & Guilds 2357 and hence allowing the trainee with employment that counts towards significant work experience to apply for an ECS (Electrotechnical Certification Scheme) gold card.

**Assessment criteria**

2.2 Identify the areas in building services which run competent person schemes

## AREAS THAT RUN COMPETENT PERSON SCHEMES

There are a number of areas in building services where a competent person scheme (CPS) operates, including:

- heating system installation regardless of boiler fuel
- removal or replacement of oil tanks
- installation of a new bathroom or kitchen if new plumbing is installed
- installation of additional radiators to an existing heating system
- new electrical installations in kitchens and bathrooms
- electrical work outside a house
- modification of circuits in kitchens and bathrooms (not like-for-like maintenance replacement of equipment)
- replacement of windows and door units
- replacement of flat/pitched roof coverings
- installation of fixed air-conditioning systems.

A contractor who is on a competent person scheme accreditation listing is allowed to self-certify that their work complies with Building Regulations in place of the normal Building Control notification required. Some of the areas listed, such as installation of additional radiators to an existing heating system, do not always require Building Control notification.

A group of five CPS operators has formed the Electrical Safety Register. This register allows potential customers to locate registered contractors within a pre-determined radius of their own postcode.

At the time of writing, the register only suggests contractors registered with two of the five operators.

### ACTIVITY

The Achievement Measurement 2 test is managed by NET. Find out where your nearest testing centre is located.

## COMPETENT PERSON SCHEMES

A competent person scheme (CPS) is a scheme that gives its members accreditation as being competent persons. Under the Electrotechnical Assessment Specification (EAS) Scheme, a competent person is as defined as:

> a person, considered by the Enterprise to possess the necessary technical knowledge, skill and experience to undertake assigned electrical installation work, and to prevent danger and where appropriate injury.

All CPSs are approved and administered by the Department for Communities and Local Government (DCLG). The UK Accreditation Service (UKAS) has been appointed as the accrediting body for each of the relevant CPS operators.

A CPS operator is a body offering a self-certification scheme covering the specific area of work and meeting the requirements of UKAS accreditation requirements and Schedule 3 of the Building Regulations 2010, as amended.

In order for a CPS to become accredited, the operator must meet and maintain the requirements of UKAS BS EN 45011: 1998 or the latest equivalent. Scheme operators must assess existing members and applicants as technically competent, against National Occupational Standards (NOS) under a Minimum Technical Competence (MTC) assessment procedure, where one is in place, for the relevant type(s) of work. They must also assess the competence of prospective and existing members of the scheme to deliver compliance with the Building Regulations. New applicants must be assessed before they can be registered with the scheme.

**Assessment criteria**

**2.3** Define the term 'competent person scheme' (CPS)

**ACTIVITY**

What is the advantage of belonging to a competent person scheme?

**ASSESSMENT GUIDANCE**

The books listed are all excellent in their own way. The IET Wiring Regulations and On-Site Guide explain how the legal requirements of the Electricity at Work Regulations may be implemented in practice.

## REQUIREMENTS FOR RENEWING CPS MEMBERSHIP

**Assessment criteria**

**2.4** Identify the renewal requirements for being part of competent persons schemes

### Visit to applicant's premises

To renew membership of a CPS, the applicant needs to satisfy certain requirements. The CPS operator will make an assessment at the applicant's premises to check that:

- supporting books and information are available, including:
  - current edition of BS 7671 (IET Wiring Regulations)
  - Memorandum of Guidance on the Electricity at Work Regulations 1989 (HSR 25)
  - approved Document P
  - IET On-Site Guide

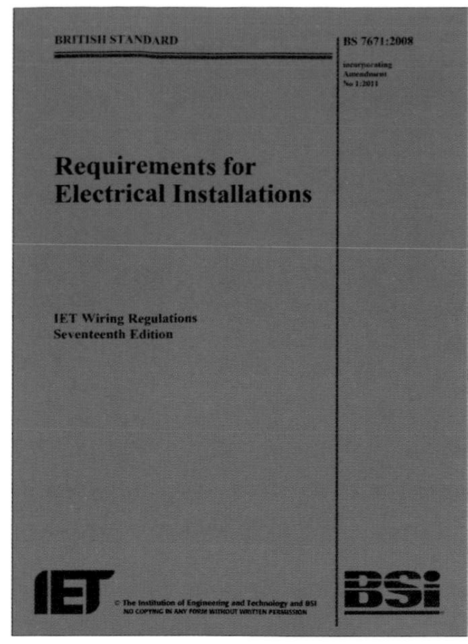

- instruments for the following are in working order and with calibration certificates:
  - voltage
  - continuity ($R_1 + R_2$)
  - insulation resistance
  - phase earth loop impedance ($Z_s$)
  - external earth loop impedance ($Z_e$)
  - RCD testing
- certificates – copies of all BS 7671 certificates issued
- public liability insurance – documentary proof of a current public liability insurance policy to an appropriate value
- complaints procedure – a review of complaints handling, checking the complaints log/register, details of any complaints received, the corrective and preventative actions taken to satisfy the complaint, and that all complaints are dealt with in a timely and effective manner.

If the applicant employs five or more people, the CPS operator will also check:

- **risk assessment** – that a written risk assessment procedure for the workplace is in place
- **health and safety policy** – that a signed and dated health and safety policy statement is in place; if there are fewer than five employees, the qualified supervisor should have a good working knowledge of safe working practices.

## Site visit

The CPS operator's assessor will need to make a site visit to see domestic installation work that is representative of the work typically undertaken.

## Assessment of competence

In order to assess competence, applicants are questioned about wiring regulations, inspection and testing.

## Additional renewal criteria

Once accepted onto a scheme, the registered enterprise is required to maintain the standards expected at registration. There will be surveillance inspections at intervals and, at each renewal of membership, the enterprise will be required to prove maintenance of the register of certificates issued and that each project requiring notification was notified to the CPS operator.

## ADVANTAGES OF CPS MEMBERSHIP

There is generally no requirement to join a CPS. The exception is the requirement to be on the Gas Safe Register for gas and boiler works.

However, for those who regularly undertake notifiable electrical installation work joining a CPS is recommended. Membership has the advantages of making the notification process simpler to administer and also indicates to the client that the enterprise is competent in the design, installation, testing and certification of domestic installations.

An enterprise that is not a member of a CPS is required to submit details of their work to Building Control and pay any necessary fees for the application to be checked and approved. Failure to notify Building Control can result in a fine of up to £5000.

The flow chart below indicates the process relating to Part P compliance for England.

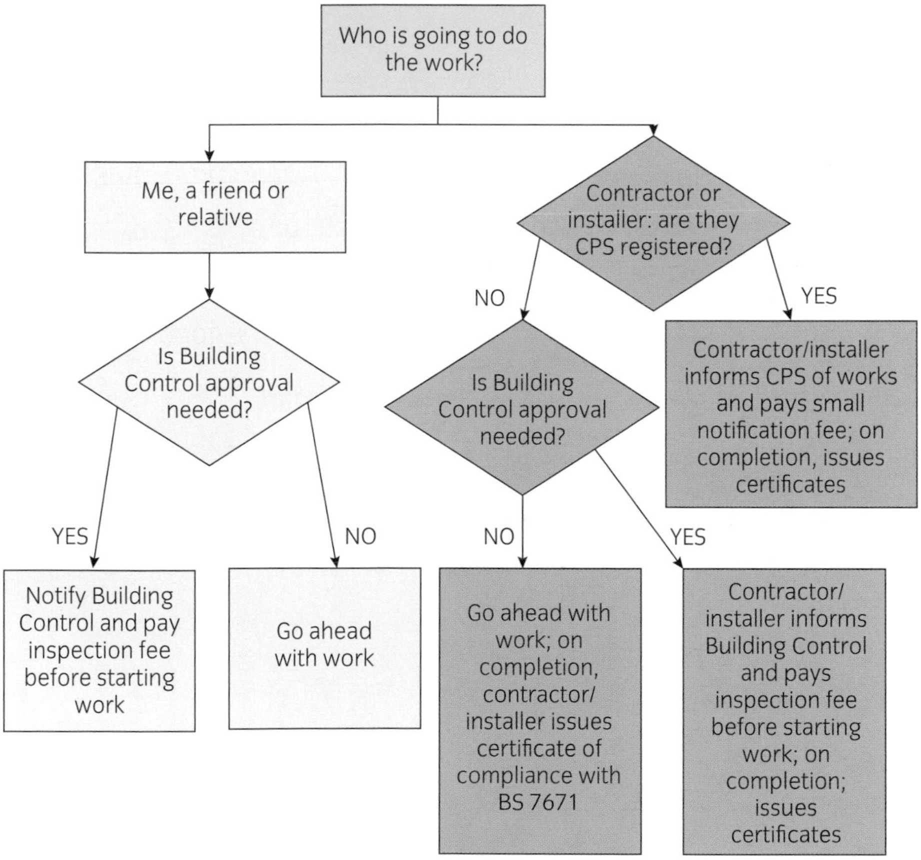

**Assessment criteria**

**2.5** Describe the consequences of not being part of the competent person scheme when working in building services engineering

**KEY POINT**

Joining a CPS makes good sense for a business that carries out domestic electrical installations, because it saves time and the cost of registering a job with Building Control.



**How to meet the requirements of the Building Regulations**

# ASSESSMENT CHECKLIST

## WHAT YOU NOW KNOW/CAN DO

| Learning outcome | Assessment criteria | Page number |
|---|---|---|
| 1 Understand how to plan for careers in building services engineering | *The learner can:* | |
| | 1 Identify resources to support career planning | 584 |
| | 2 Describe elements of career planning | 586 |
| | 3 Describe documents to support career development | 588 |
| | 4 Explain the principles of goal setting | 594 |
| | 5 Describe how to set goals | 596 |
| | 6 Define the different roles in building services engineering | 598 |
| | 7 Explain opportunities for progression within building services engineering. | 600 |
| 2 Understand the requirements to become a qualified operative in building services engineering | *The learner can:* | |
| | 1 Describe specific requirements for career choices in building services engineering | 605 |
| | 2 Identify the areas in building services which run competent person schemes | 606 |
| | 3 Define the term 'competent person scheme' (CPS) | 607 |
| | 4 Identify the renewal requirements for being part of competent person schemes | 607 |
| | 5 Describe the consequences of not being part of the competent person scheme when working in building services engineering. | 609 |

## ASSESSMENT GUIDANCE

For this unit, you will be assessed by

- A short written paper containing 11 question (closed book)
- A two-section task requiring you to produce documents about yourself that will help you make job applications. You will be required to do this using word processing tools on a computer.

Make sure you have suitable writing materials with you at all times.

Do not be afraid to ask questions at work if you are unsure of company policy

Read through any notes you are given each week.

Plan your work so that you do not have to rush at the end.

If you are unsure of anything before your assessment, ask.

## KNOWLEDGE CHECK

1 Identify one organisation that could provide information on each of the following issues.

   a) Working conditions and pay rates.

   b) Technical information regarding electrical Installations.

   c) Employment issues.

2 Identify five items which could be included in a CV.

3 In SWOT analysis what do the letters SWOT stand for?

4 State the minimum qualifications required to be a JIB qualified electrician.

5 Name four potential sources of employment openings in the electrical contracting industry.

# APPENDIX 1: UNIT 304 SMALL INDUSTRIAL UNIT FLOOR PLAN

## FLOOR PLAN OF A SMALL INDUSTRIAL UNIT SHOWING THE POSITIONS OF ALL PROPOSED ELECTRICAL EQUIPMENT

This floor plan relates to Unit 304 and allows you to appreciate the electrical layout of a small industrial unit. The positions of the incoming water and gas supplies are also highlighted. During your assessments, you will be required to read and understand this type of drawing. The following pages refer to this plan: 390–395 (inspection exercise); 402 (prescribed tests for a small industrial unit); 412–415 (continuity of protective conductors); 415–418 (continuity testing of a lighting circuit); 419–423 (continuity of a ring final circuit); 449–452 (calculation for earth fault loop impedance); 452 (earth fault loop impedance for radial socket outlets). This floor plan is also referred to in many activities. Please refer to Appendix 2 for supply characteristics and circuit details.

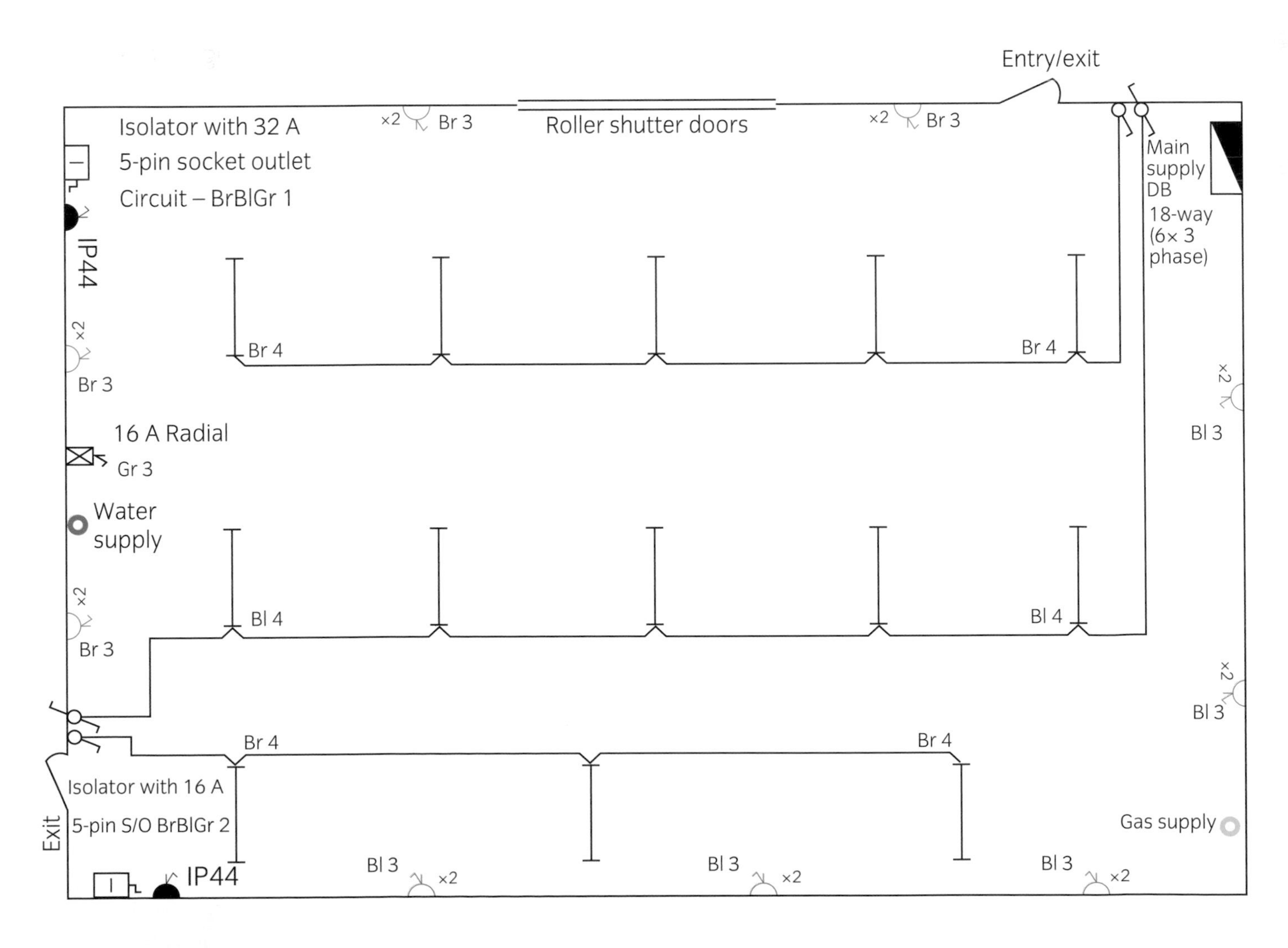

# APPENDIX 2: UNIT 304 DISTRIBUTION AND WIRING INFORMATION

Listed below are the supply characteristics and circuit details which relate to the small industrial floor plan shown in Appendix 1, relating to Unit 304.

These supply characteristics and circuit details are supplied to enable you to evaluate and calculate initial verification testing requirements in accordance with BS 7671:2008 (2011) and Guidance Note 3: Inspection and Testing, IET.

By using this information, you could fill in parts of the Electrical Installation Certificate and Generic Schedule of Test Results.

## Supply characteristics

Three-phase 230 / 400 V a.c. – 50 HZ

Assessed demand by designer = 48 kVA (approx. 70 A/phase)

Supply authority protective devices = 3 No. 100 A BS 88-3 type 1 fuses

TN-C-S (PME) $Z_e$ at the DB = 0.21 Ω PEFC = 1.09 kA

PSCC = 1.15 kA measured single phase (calculated three-phase [x2] = 2.3 kA)

## The circuits

Br/Bl/Gr 1 – C32 A three-phase isolator with socket outlet (5-pin)

Br/Bl/Gr 2 – C16 A three-phase isolator with socket outlet (5-pin)

Br 3 – B32 A ring final circuit for BS 1363 socket outlets

Bl 3 – B32 A radial final circuit for BS 1363 socket outlets

Gr 3 – B16 A radial final circuit for a 3 KW instantaneous water heater

Br 4 – B10 A lighting circuit

Bl 4 – B6 A lighting circuit

## Distribution board circuit details

| |
|---|
| Br/Bl/Gr 1 – C32 A BS EN 60898 3-phase + N Isolator with socket outlet (5-pin)<br>(all conductors 6 mm$^2$ including cpc) *measured distance of 15 m* |
| Br/Bl/Gr 2 – C16 A BS EN 60898 3-phase Isolator with socket outlet (5-pin)<br>(all conductors 4 mm$^2$ including cpc) *measured distance of 27 m* |
| Br 3 – B32 A BS EN 61009 (30 mA) ring final circuit for BS 1363 socket outlets<br>(2.5 mm$^2$ line and neutral conductors – cpc 1.5 mm$^2$) *48 m – total ring (no spurs)* |
| Bl 3 – B32 A BS EN 60898 radial final circuit for BS 1363 socket outlets<br>(4 mm$^2$ line and neutral conductors – cpc 1.5 mm$^2$) *22 m to final socket outlet* |
| Gr 3 – B16 A BS EN 60898 radial final circuit for a 3 KW instantaneous water heater – 13 A switch fused spur<br>(all conductors 2.5 mm$^2$ including cpc ) *measured distance 22 m* |
| Br 4 – B10 A BS EN 60898 lighting circuit<br>(all conductors 1.5 mm$^2$ including cpc) *measured distance 37 m* |
| Bl 4 – B6 A BS EN 60898 lighting circuit<br>(all conductors 1.5 mm$^2$ including cpc) *measured distance 28 m* |
| Gr 4 – spare |
| Br 5 – spare |
| Bl 5 – spare |
| Gr 5 – spare |
| Br 6 – spare |
| Bl 6 – spare |
| Gr 6 – spare |

## Cable and wiring methods

The wiring is in singles (single-core 70 °C – thermoplastic insulated) cables, installed in a combination of metal and PVC conduit and trunking (individual cpcs supplied for each circuit). All single-phase (single-pole) circuit breakers have a 6 kA rated short-circuit capacity.

ALL three-phase (triple-pole) circuit breakers have a 10 kA rated short-circuit capacity.

# GLOSSARY

## A

**Abrade** To scrape or wear away.

**Adiabatic** The adiabatic equation can be used in several ways. It can be used to determine the time taken for a given cable to exceed its final limiting temperature. Alternatively, it can be used to determine the minimum csa to be able to withstand a fault current for a given duration without exceeding the final limiting temperature.

**Audit** To conduct a systematic review to make sure standards and management systems are being followed.

**Azimuth** Ideally the modules should face due south, but any direction between east and west will give acceptable outputs. Azimuth refers to the angle that the panel direction diverges from facing due south.

## B

**Basic insulation** Insulation applied to live parts to provide basic protection. It does not necessarily include insulation used exclusively for functional purposes.

**Batter or slope** The angle in relation to the horizontal surface, of the trench walls of an excavation, to prevent the walls collapsing.

**Breaking capacity** The amount of current a protective device can safely disconnect.

**Breaking capacity** This is the maximum value of current that a protective device can safely interrupt.

**Busbar trunking system** This is a type-tested assembly in the form of an enclosed conductor system that comprises solid conductors separated by insulating material.

**Business opportunity** In this context, the opportunity to make profit from the work or contract.

## C

**Combustible** Able to catch fire and burn easily.

**Competent person** A person who possess sufficient knowledge, relevant practical skills and experience for the nature of the electrical work undertaken and is able at all times to prevent danger and, where appropriate, injury to him/herself and others.

**Compliance** The act of carrying out a command or requirement.

**Contact resistance** Sometimes, a meter may give a reading in which the value does not stabilise or which is quite different from the calculated value. In this situation, reverse the test meter lead connections and re-test. This may give more accurate values.

**Contamination** The introduction of a harmful substance to an area.

**Contractor** A person or organisation carrying out set pieces of work for fixed sums of money.

## D

**Duty holder** The person in control of the danger is the duty holder. This person must be competent by formal training and experience, and with sufficient knowledge to avoid danger. The level of competence will differ for different types of work.

## E

**Earth fault loop impedance** The impedance of the earth fault current loop starting and ending at the point of earth fault. This impedance is denoted by the symbol $Z_s$.

**Efficacy** The ratio of power in and power out, measured in two different units. For example, the ratio of light output in lumens to the electrical power measured in watts.

**Efficiency** The ratio between power in and power out, measured in the same unit.

**Enabling Act** An enabling Act allows the Secretary of State to make further laws (regulations) without the need to pass another Act of Parliament.

**Extent** The amount of inspection and testing. For example, in a third-floor flat with a single distribution board and eight circuits, the inspection will be visual only without removing covers and testing, and will involve sample tests at the final point of each circuit.

## F

**Fatality** Death.

**FELV** Functional extra-low voltage circuit (requirements can be found in Section 411.7 of BS 7671).

**Fine-wire cable** Cable with many strands, which gives it much more flexibility. The conductors may need preparing by solder or tinning before they are terminated.

## G

**Goal** Where someone wants to be with their life and/or career in the future.

**Granular soils** Gravel, sand or silt (coarse grained soil) with little or no clay content. Although some moist granular soils exhibit apparent cohesion (grains sticking together forming a solid), they have no cohesive strength.Granular soil cannot be moulded when moist and crumbles easily when dry.

## H

**Hazard** Anything with the potential to cause harm (eg chemicals, working at height, a fault on electrical equipment).

**Hazardous substance** Something that can cause ill health to people.

**Hot work** Work of any type that involves actual or potential sources of ignition and which is done in an area where there may be a risk of fire or explosion (for example welding, flame cutting and grinding).

## I

**IMD** Insulation monitoring device. An end-of-line resistor together with a fire detection system is an example.

**Impedance** Impedance (symbol *Z*) is a measure of the opposition that a piece of electrical equipment or cable makes to the flow of alternating electric current when a voltage is applied. Impedance is a term used for alternating current circuits.

**Inductive loads** Loads that involve magnetism, such as motors, discharge lighting and (although not a load) transformers.

## L

**Lead time** Lead time is the delay between the initiation and execution of a process. For example, the lead time in ordering, for instance, busbar trunking and delivery to site may be six weeks.

**Liability** A debt or other legal obligation in order to compensate for harm.

**Limitation** A part of the inspection and test process that cannot be done for operational reasons. For example, the main protective bonding connection to the water system, located in the basement, could not be inspected as a key for the room was not available.

**Load current** The amount of current an item of equipment or a circuit draws under full load conditions.

**Load** A device intended to absorb power, including active power and apparent power.

## M

**Manual handling** The movement of items by lifting, lowering, carrying, pushing or pulling by human effort alone.

**Mentor** An advisor, counsellor or trainer who gives guidance and acts as a role model.

## N

**Near miss** Any incident that could, but does not, result in an accident.

**Neutral conductor** A conductor connected to the neutral point of a system that contributes to the transmission of energy.

## O

**Objective** A plan for achieving a specific goal.

## P

**PELV** Protective extra-low voltage circuit.

**PEN** Protective earthed neutral. The PEN conductor is both a live conductor and a combined earthing conductor.

**Person in control of the premises** The person in control of the premises is likely to have physical possession of the premises or is the person who has responsibility for, and control over, the condition of the premises, the activities conducted on those premises and the people allowed to enter.

**Personal protective equipment (PPE)** All equipment, including clothing for weather protection, worn or held by a person at work, which protects that person from risks to health and safety.

**PME** Protective multiple earthing. This term is commonly used to describe a TN-C-S system.

**Protected disclosure** A disclosure of information about an employer honestly made in the interests of safety or the public interest for which the informant cannot be punished by the employer.

**Protective conductor (PE)** A conductor used for protection against electric shock; it is often called the 'earth wire' or 'earth conductor'.

## R

**RCM** Residual current monitor. This is much like an RCD but does not disconnect the circuit. Instead it gives a warning that a fault is present.

**Regulatory body** An organisation that regulates an industry or similar in a public protection role. Unlike professional bodies, it is not a membership organisation. The regulatory body is usually backed by a legal mandate or statutory requirement.

**Repose** The angle to the horizontal at which the material in the cut face is stable and does not fall away. Different materials have different angles of repose, for example, 90° for solid rock and 30° for sand.

**Responsible person** The person who is designated as the responsible person has delegated responsibility for certain aspects of a company's operational functions, such as fire safety, electrical operational safety or the day-to-day responsibility for controlling any identified risk such as Legionella bacteria.

**Risk** The chance (large or small) of harm actually being done when things go wrong (eg risk of electric shock from faulty equipment).

**Rotor speed** The speed at which the rotor rotates in revolutions/second (r/s)

## S

**Schedule of rates** This is a document that lists a labour time/cost for installing and connecting an individual item of equipment. Examples may be £2.50 to fix a ceiling rose securely to a plaster ceiling and £1.50 to connect the cable to it. Running the cable to the ceiling rose could be charged at a rate of £2.00 per metre.

**SELV** Separated extra-low voltage circuit.

**Slip** The difference between the synchronous speed and the rotor speed expressed as a percentage or per unit value.

**Statute** A major written law passed by Parliament.

**Sustainable design** Designing equipment, the built environment and services to comply with the principles of social, economic and ecological sustainability.

**Synchronous speed** The speed that the rotating field rotates around the field poles.

## T

**TN-C-S** T = source earth (terre).
N = neutral and protective conductor.
C = combined in the distribution system.
S = separated in the installation.

**TN-S** T = source earth (terre).
N = neutral and protective conductor.
S = separated in the installation.

**TT** T = source earth (terre).
T = installation earth (terre).

**Toxic** Poisonous.

# ANSWERS TO ACTIVITIES AND KNOWLEDGE CHECKS

Answers to activities and knowledge checks are given below. Where answers are not given it is because they reflect individual learner responses.

## UNIT 201 HEALTH AND SAFETY IN BUILDING SERVICES ENGINEERING

### Activity answers

*Page*

**6** Safety glasses, safety goggles or full-face shield.

**12** Such as Ribbmaster.

**13** Clear it away so that no one else can trip over it.

**16** Any four signs not included on page 16.

**18** One example would be smoke detectors.

**19** Mercury is a hazardous material. The lamps should be returned to the manufacturer or wholesaler, or taken to a suitable recycling facility, for safe disposal.

**20** Stop work immediately, leave the area and notify your supervisor.

**22** Gloves are commonly worn to protect against cuts and scratches. Cutting metal conduit or trunking would be a common application.

**25** The action required depends on the severity of the cut. Most injuries are slight and should be thoroughly cleaned and an antiseptic cream applied followed by a plaster. Cuts that require stitches should be referred to the local hospital.

**26** No.

**29** Protection by ADS plus 30 mA RCD for additional protection.

*Page*

**30** Light switches are unsuitable because: a) the off position is not reliably indicated; b) they cannot be restrained in the off position; c) there is insufficient contact gap to act as an isolator. They are not 'secure' and could be operated by unauthorised persons.

**33** For example, blue or yellow Artic flex; Butyl heat resistant flex.

**34** Yellow, blue, red.

**35** Multiple lock or Castell lock.

**38** Crimping relies on fewer components (flux, solder, expertise) to produce a more effective joint.

**38** Stop work immediately, turn off the gas and close the valve at the bottle, leave any hot equipment in a safe position.

**42** The water will spread the fire further.

**44** Broken (split) stiles or rungs, missing/uneven retaining cord, paint covering cracks and so on.

**46** IPAF MEWP Operator Licence.

**50** Check the route for any obstructions or obstacles.

### Outcome knowledge check answers

*Page 59*

| | | | | |
|---|---|---|---|---|
| **1** d) | **2** b) | **3** d) | **4** c) | **5** b) |
| **6** a) | **7** c) | **8** c) | **9** d) | **10** b) |
| **11** d) | **12** c) | **13** d) | **14** c) | **15** c) |
| **16** b) | **17** d) | **18** c) | **19** a) | **20** a) |

# UNIT 301 UNDERSTAND THE FUNDAMENTAL PRINCIPLES AND REQUIREMENTS OF ENVIRONMENTAL TECHNOLOGY SYSTEMS

## Activity answers

*Page*

**69** To circulate the primary fluid around the system.

**73** Due south.

**73** Wind over the roof can cause the panel to act like an aircraft wing, producing a force which can pull the panel off the roof.

**78** Both about 90% (British Gas, 2013).

**80** Rock or similar impervious material.

**82** Accumulator tank (used to store excess hot water).

**86** e.g. loft insulation, double glazing, cavity wall insulation, floor insulation.

**88** $CO_2$ is a greenhouse gas which contributes to global warming; it is vital to reduce $CO_2$ emissions into the atmosphere.

**91** e.g. access via narrow country lanes, narrow or weak bridges, sufficient storage in the event of road blockage due to snow or floods etc, provision of alternative heating system.

**96** There is a risk of electrocution from PV systems as they generate electricity in daylight conditions. There may be an output even in very dull conditions.

*Page*

**98** a) A roofing specialist or qualified PV installer,
b) a qualified electrician.

**99** To change the direct current produced by the PV array to alternating current.

**103** All test results are required.

**106** To light his holiday home.

**109** Birds.

**110** Money paid by the government to the homeowner for each unit of electricity produced.

**111** Migratory fish such as salmon and sea trout.

**116** 527.2 Sealing of wiring system penetrations. In particular, 527.2.1 – where wiring systems pass through the building structure, the openings must be sealed to the degree of fire protection required.

**120** Heat engine.

**125** Water butts (barrels).

**126** e.g. bore holes, reservoirs, rivers.

**129** No.

**133** It would be necessary to have two separate waste water systems – one discharging waste from toilets, etc. and the other for the greywater.

## Outcome knowledge check answers

*Page 143-146*

| | | | | |
|---|---|---|---|---|
| **1** c) | **2** d) | **3** a) | **4** b) | **5** b) |
| **6** c) | **7** d) | **8** b) | **9** b) | **10** b) |
| **11** a) | **12** c) | **13** a) | **14** b) | **15** b) |
| **16** d) | **17** b) | **18** c) | **19** b) | **20** d) |

# UNIT 302 PRINCIPLES OF ELECTRICAL SCIENCE

## Activity answers

*Page*

**151** Plot 0° = 0; 30° = 0.5 × 100 = 50 V; 60° = 0.866 × 100 = 86.6 V; 90° = 1 × 100 = 100 V; back to zero at 180° and then minus values back to 360°.

**156** 0.722 A.

**159** $I_s$ = 230/115 = 2 A; $V_{cap}$ = 2 × 200 = 400 V.

**162** It costs more than is saved; 0.95–0.97 is the required range.

**163** cos 1° = 0.9998, cos 10° = 0.98, cos 30° = 0.87, cos 45° = 0.71, cos 70° = 0.34, cos 90° = 1, $\cos^{-1}$0.75 = 41.41°

**166** Power factor from phasor is approx 0.7 leading, 38.46°

**167** Refrigerator, vacuum cleaner, washing machine, dishwasher, tumble dryer or any other motor-driven appliance.

**169** Individual, advantage: automatically switched on and off with load; individual, disadvantage: may need more capacitance than bulk correction. Bulk, advantage: less capacitance than individual capacitors; bulk, disadvantage: only corrects up to point of installation.

**172** a) 200 lux, b) 50 lux.

**187** a) 230 V, b) 240 V, c) 6.351 kV, d) 63.5 V.

**191** a) 40 A, b) 40 A.

**194** Copper loss in fields; copper loss in armature; windage (rotational) loss; bearing loss; brush voltage drop.

*Page*

**197** The field of a self-excited machine obtains its supply from the armature output. A separately excited machine obtains the field supply from an independent source.

**199** Given $N = f/p$, a) $N$ = 40/2 = 20 rev/s, b) $N$ = 50/2 = 25 rev/s, c) $N$ = 60/2 = 30 rev/s

**206** The electrolytic is used for starting; the paper capacitor stays in series with the run winding.

**209** One-third.

**218** Thermal and magnetic devices.

**222** BS 7671 states that the residual current rating must not exceed 30 mA.

**223** 30 A – red, 45 A – green.

**229** Various names are used but 'Off-peak tariff' and 'Economy 7' are typical; any tariff that provides cheaper electricity during the night.

**230** For short-term use; they heat the object (body) rather than the air.

**237** Zener diode connected across output; when the breakdown voltage is reached it conducts and causes voltage drop across the resistor; the capacitor charges and holds the output voltage steady (or similar description).

**238** It would vary the current flow through the device. Trigger at 90° and only the second half of the first half wave is conducted.

**241** Any from mica, paper, electrolytic, air, tantalum, film, ceramic, vacuum.

## Knowledge check answers

*Page 247–248*

1 a) Electrical resistance: property of a conductor due to which it opposes the flow of current. By Ohm's law an increase in resistance will cause a decrease in current flow.

b) By Lenz's law the back emf induced in the inductor opposes the force setting it up, limiting the current flow.

c) The voltage on the capacitor plates opposes the change in the applied voltage.

2 a) Lagging, b) leading, c) lagging, d) unity, e) lagging.

3 The supply current will go up. The lamp current will stay the same.

4 $I = \frac{P}{I \times \text{PF}} = \frac{5690}{230 \times 0.8} = 30.92\,\text{A}$

5 $\text{kW} = \frac{400 \times 20 \times 0.5}{1000} = 4\,\text{kW}$, $\text{kVA} = \frac{400 \times 20}{1000} = 8\,\text{kVA}$, $\text{kVA}_r = \sqrt{8^2 - 4^2} = \sqrt{48} = 6.93\,\text{kVA}_r$

6 In an unbalanced system the neutral carries the out-of-balance current back to the supply point. Without a neutral holding the star point to earth potential, the voltage would float and each phase would have a slightly different value.

7 a) Phase $\frac{400}{\sqrt{3}} = 230\,\text{V}$, Line = 400 V, b) Line = phase voltage = 400 V

8 a) Reverse armature or field connections, not both. b) Reverse armature or field connections, not both. c) Change over any two-phase connections. d) Reverse run or start winding connection. e) Not normally reversible unless specially designed.

9 $E = \frac{I}{d^2} = \frac{2000}{1.8^2} = 617\,\text{lux}$

$H = \sqrt{1.8^2 + 1^2} = 2.06$

$\text{Cos}\,\theta = \frac{\text{adj}}{\text{hyp}} = \frac{1.8}{2.06} = 0.8737$

$E = \frac{I\cos\theta}{h^2} = \frac{2000 \times 0.8737}{2.06^2} = 412\,\text{lux}$

10 i) Heat energy is radiated in direct lines from the source to the object, heating the object, not the air – radiant or infra-red heater. ii) The heater warms the air in the immediate vicinity of the unit. Warm air, being lighter than cold air, is pushed up by cold air flowing in below. This sets up a circulating air movement – block storage heater. iii) Direct contact between the element and medium to be heated – kettle or immersion heater element in water.

# UNIT 305 ELECTRICAL SYSTEMS DESIGN

## Activity answers

*Page*

**250** Green/yellow, brown, blue.

**251** Protective multiple earthing.

**253** Answer depends on individual system.

**254** Earth rods or pipes, earth tapes or wires, earth plates, underground structures or alternatives as Regulation 542.2.3.

**255** Suitable drawing showing fault, earth electrode, return path through ground, transformer electrode, winding and line supply.

*Page*

**256** No.

**260** $I = \frac{10000}{400 \times \sqrt{3} \times 0.8} = 18\,\text{A}$

**262** It reduces the length of cable run from the distribution board to the load and, therefore, the final circuit volt drop.

**266** Aerial terminal, down conductor, test point, system electrode.

**277** Fundamental principles.

**279** Zones 0, 1 and 2.

**287** a) 100 W lamp $I = \frac{100}{230} = 0.43\,\text{A}$,

b) 18 W lamp $I = \frac{18}{230} = 0.08\,\text{A}$

**290**

| $I_b$ load | $I_n$ BS EN 60898 | $I_t$ cable pvc/pvc |
|---|---|---|
| 15 | 16 | 16 |
| 25 | 32 | 37 |
| 3 | 6 | 16 |
| 60 | 63 | 64 |
| 41 | 45 | 47 |

**292** Ambient temperature, external heat sources, presence of water, presence of solid foreign bodies, presence of corrosive or polluting substance, impact, vibration and others as listed in Regulation 522.

**298** $I_{max}$ = 7000/230 = 30.43 A

First 10 A plus 30% of remainder:

10 A + 30% of 20.43 = 16.1 A + 5 A for socket = 21.1 A

$I_n$ = 25 A

Correction factors: Ca = 35 °C = 0.94;

$I_z = \frac{25}{0.94}$ = 26.6 A, 2.5 mm² = 27 A

Volt drop = $I_b$ × length × $V_d/am$

$$\frac{21.1 \times 15 \times 18\ \text{mv}}{1000} = 5.7\ \text{V}$$

**300** $t = \frac{115^2 \times 10^2}{2000^2} = 0.33\ \text{s}$

**302** Aluminium is a relatively soft metal and liable to break in the smaller sizes.

**307** $R = 230/20 = 11.5\,\Omega$, $P = \frac{V^2}{R} = \frac{220^2}{11.5} = 4208\ \text{W}$

or $\left(\frac{220}{230}\right)^2 \times 4600 = 4208\ \text{W}$

**310** Drill, jigsaw, angle grinder, chaser, vacuum cleaner etc.

**313** The following are approximate answers from the graphs: Type B = 20 s, Type C = 32 s; Type D = 38 s.

**314** 50 × (3.08 + 12.1) = 0.759 Ω

**318** Figures the same for 0.4 s and 5 s: 32 A Type C = 0.72 Ω, 100 A Type B = 0.46 Ω, 32 A Type D = 0.36 Ω

**319** As per Regulation 411.3.3.

**321** 16 A Type C – 160 A, 40 A Type B – 200 A; yes. 40 A Type B – 200 A, 63 A Type C – 630 A; yes. 50 A Type D – 1000 A, 100 A Type C – 500 A; no. Taken from Appendix 3, 0.4 to 5 s disconnection currents.

**324** You will still come across them in existing installations.

**327** By placing thermometers along the cable run.

**330** PME – 10 mm² Table 4.4(i); TT – 6 mm² Table 4.4(ii) note 2 (544.1.1)

**332** Any suitable device.

**340** If the body is wet it has a lower resistance and is more susceptible to shock.

**343** Damage to insulation from heat source. Risk of burns to operative from heat source.

**347** Loose connections, exposed conductors due to over-stripping insulation, terminal screw clamping on insulation, terminal screw over-tightened cutting conductors.

**351** RCD, PIR controls, boiler controls.

## Knowledge check answers

*Page 356–358*

1 a) Provides isolation from the supply for the whole installation, b) provides additional protection against electric shock, c) isolation of L–N from supply, d) isolation of individual circuit and overload/ earth fault protection.

2 The N–E forms a parallel path and if sufficient out-of-balance current occurs the RCBO will trip. Some of the neutral current flows to earth.

3 a) Resistance increases as length increases, b) as $Z_s = Z_e + (R_1 + R_2)$ then as $(R_1 + R_2)$ increases so does $Z_s$, c) as the impedance increases the current will decrease (Ohm's law), d) copper has a positive temperature coefficient of resistance, so as the temperature goes up so does the resistance.

4 a) Current-carrying capacity of conductor, b) tabulated current-carrying capacity of conductor, c) residual current setting/rating, d) correction factor for ambient temperature, e) correction factor for grouping.

5 The PEN conductor is common for neutral and earth on the supply system. Should this break, all exposed metal on the load side could become live. By earthing through electrodes at frequent intervals an alternative return path is provided.

6 Overload protection normally provided through thermal overloads or magnetic dashpots. Undervoltage protection provided by a no-volt coil which trips out below a predetermined voltage.

7 a) An emergency switch is used to disconnect all live conductors in the event of an emergency – a stop button in a control circuit, b) a switch used for day-to-day control equipment – A light switch.

8 S = cross sectional area of conductor, $I$ = fault current, $T$ = disconnection time, K = constant for cable composition (conductor/insulation).

9

| Device | Standard | Isolation | Emergency | Functional |
|---|---|---|---|---|
| Device with semi-conductor | BS EN 60669-2-1 | no | no | yes |
| Plug and socket | BS 1363 | yes | no | yes |
| Switched fused connection unit | BS 1363 | yes | yes | yes |
| Fuse | BS 1362 | yes | no | no |
| Cooker control unit | BS 4177 | yes | yes | yes |

10 Zone 0 is the area/enclosure containing water.

# UNIT 304 ELECTRICAL INSTALLATIONS: INSPECTION, TESTING AND COMMISSIONING

## Activity answers

*Page*

**361** Regulation 29 provides grounds for defence based on everything practical having been done to prevent such an incident.

**363** Make sure you check the casing, as well as the leads and probes.

*Page*

**364** There is always the remote risk of N and E being cross-connected or a voltage appearing on the earth due to a fault. Incorrect polarity or borrowed neutrals can also cause potentially dangerous voltages to appear on the neutral terminals. Although, in theory, these situations should not arise, if you speak to experienced electricians they will recount all kinds of oddities they have encountered. Check and be safe.

**365** Gather equipment available at your college or training centre and follow the procedure on pages 363–364, under supervision.

**366** If the Line is connected first, the other probe will be live until it is connected to the neutral or earth.

**367** 1) Shock, burns, fire, arcing and explosion risk.

2) The tester, the public, the client, the customer and other trades.

3) Loss of supply could be very dangerous. Essential supplies must be evaluated by competent persons on-site (usually the electrical engineer).

4) Ensure that the client backs-up the system and switches it off himself/herself.

**368** Dangers of working on or near live conductors include risk of electric shock by unintentional contact with conductors through normal body movements and blast or flashover through dropping items such as tools into live parts, or other suitable dangers.

**371** At a period determined by the last inspector, depending on the installation condition. The absolute maximum is 10 years; also on change of occupancy (5 years if rented).

**373** Design (designer), construction (installer) and test and inspection (test engineer). For small contracts this may be the same person.

**378** Someone with sufficient knowledge and experience to carry out the work. This is often interpreted as someone in possession of C&G 2391/2394/5 qualifications. Remember, qualifications are NOT competence.

**382** 1 a) The Health and Safety at Work etc. Act 1974, The Electricity at Work Regulations 1989, The Management of Health and Safety at Work Regulations 1999.

b) (IET) Wiring Regulations (BS 7671) 17th Edition, IET Guidance Note 3: Inspection and Testing, The IET On-Site Guide (OSG).

2 a) Electrical Installation Certificate, Schedule(s) of Inspections, Generic Schedule(s) of Test Results, Minor Electrical Installation Works Certificate

b) Electrical Installation Condition Report, Condition Report Inspection Schedule(s), Generic Schedule(s) of Test Results.

3 d) Competent.

**383** Alternating current can easily be transformed from one voltage to another by the use of transformers, which allows it to be transmitted over long distances with low losses. Direct current voltage is not easy to change and would result in heavy losses.

**384** The type and composition of each circuit, the number and size of conductors, additional protection may need to be considered under certain conditions, the isolating and switching devices, the equipment vulnerable to certain tests.

**386** $I_b \leqslant I_n \leqslant I_z$

**387** By visual inspection at the terminals when the circuit is isolated.

**388** IPXXB.

**390** IP4X or IPXXD.

**392** Check your answers for this activity on page 395.

**396** 25 mm$^2$.

**397** It allows the complete isolation of the consumer unit if it is necessary to work on internal connections. Without it some internal parts such as the incoming terminals will still be live.

**398** 1) Use table 54.7 to find the answer 50 mm$^2$ ÷ 2 = 25 mm$^2$,

2) 16 mm$^2$ tails – 16 mm$^2$,

3) 4 mm$^2$ in accordance with this regulation is the minimum,

4) see the equation on page 397.

**399** On and off position reliably indicated only when contacts fully open. Sufficient contact air gap as designed by the manufacturer.

**401** 1) Continuity of protective conductors (this test will be carried out during the process of ring final circuit testing), 2) Ring final circuit test, 3) Insulation resistance test, 4) Polarity test (4 is normally included in 2).

**402** They are carried out with the supply isolated – therefore the circuits are dead.

**403** Continuity of protective conductors, Insulation resistance, Polarity, $Z_s$, PFC (verification), Functional test.

**405** Low resistance ohmmeter, Insulation resistance tester, RCD tester, Earth loop impedance tester/prospective fault current tester.

**406** 1) Continuity testing is carried out by using a low resistance ohmmeter and the readings are in ohms (Ω).

2) Insulation resistance testing is carried out by an insulation resistance tester and the readings are in megaohms (MΩ).

**407** Check that it is not damaged in any way and the leads are in compliance with GS38.

**409** Appendix 3.

**410** Testing the conductor to verify its continuity.

**412** 1) Test from end to end and record result in milliohms.

2) Use Table B1 of Guidance Note 3 or the IET On-Site Guide Appendix I1, to find the resistance in milliohms per metre and divide into the tested value.

For example, tested value is 460 mΩ and the cable is 18.1 mΩ/m. 460 ÷ 18.1 = 25 m

**413** $V = I \times R,\ I = \frac{V}{R},\ R = \frac{V}{I}$

**414** Remove all paint, rust and so on with a file and polish to bright metal with wire wool all the way round the pipe.

**415** $R = \frac{13 \times 1.83}{1000} = 0.02\ \Omega$

**417** Personnel may trip over it.

**419** 1) $R_1 + R_2$ = 22 m of 4 mm²/1.5 mm² cable. From the table on page 417 this cable has a resistance of 16.71 mΩ/m:

$$R = \frac{22 \times 16.71}{1000} = 0.37\Omega$$

2) *Any* $R_1 + R_2$ = 27 m of cable 4 mm²/4 mm². From Table B1 of Guidance Note 3 or Table I1 of the IET On-Site Guide, this cable has a resistance of 9.22 mΩ/m:

$$R = \frac{27 \times 9.22}{1000} = 0.25\Omega$$

**420** If 2.5 mm² cpcs were used instead of a 1.5 mm² the overall resistance value would be lower.

**422** If the L–N and L–E loops are correct then the N–E must be correct.

**423** Various test results will apply.

**424** $R_t = \frac{16 \times 50}{16 + 50} = \frac{800}{66} = 12.12\text{M}\Omega$

**426** 1 ÷ 0.3695 = 2.70 MΩ (see the example on page 425)

**427** Switch feed and switch line must be connected together otherwise the wiring beyond the break will not be tested.

**428** At the output terminals of the meter or the supply company's double-pole switch if fitted.

**431** Ceiling rose.

**432** 1) Fuses, circuit breakers, switches, 2) circuits for socket outlets, fixed equipment, luminaires, 3) supply tails, distribution boards, consumer units.

**434** Use the simple diagram on page 434 and draw in a meter with the probes touching the L terminal and the PE terminal. The meter should read 0.8 Ω.

**437** Use the simple diagram on page 437 and draw in a meter with the probes touching the L terminal and the PE terminal. The meter should read 0.35 Ω.

**439** Use the simple diagram on page 439 and draw in a meter with the probes touching the L terminal and the PE terminal. The meter should ideally read 200 Ω or less (depending on the supply electrode resistance and the consumer electrode resistance).

**442** Memorise and redraw the diagram on page 441.

**444** 100 mA RCD = 500 Ω, 300 mA RCD = 167 Ω, 500 mA RCD = 100 Ω

**448** $Z_s = Z_e + (R_1 + R_2)$, $Z_e = 0.21\ \Omega$

$R_1 + R_2 = (14.82\ \text{m}\Omega \times 22\ \text{m}) \div 1000 = 0.33\ \Omega$ (at 20 °C)

Multiply by 1.2 (temperature correction) = 0.40 Ω

Therefore $Z_s = 0.21\ \Omega + 0.40\ \Omega = 0.61\ \Omega$

Maximum $Z_s$ value allowed for a B16 BSEN 60898 circuit breaker = 2.87 Ω

**454** Use Table 41.3 of BS 7671 for this task.

**459** Always read the manufacturer's instructions. Some meters can be set up to test between line voltages and other meters could be damaged.

**461** First press the test button to ensure operation (for new installations). Then the following tests apply. For each of the tests, readings should be taken on both positive (+ve) and negative (–ve) half cycles.

½ × rated tests are carried out to ensure that the RCD is not *too* sensitive.

1 × rated tests are carried out to ensure that the RCD is working to the manufactured standard.

5× rated must be used if the RCD is designed for additional protection.

**464** Generally a ramp test indicates how sensitive the RCD is to imbalance.

**465** At the isolator – it would be dangerous to do it at the motor.

**467** All switches, circuit breakers, socket outlets, isolators, luminaires (including two-way switching) to be checked for correct functioning.

**468** Answers are dependant on research. Draka and Amtech are two examples, and cable manufacturers also provide simple design packages.

**469** Power circuit will be 5% of 400 V = 20 V.

**471** 3% of 230 V = 6.9 V.

**472** The client would not necessarily know which documents need to handed over on completion. Therefore, it is up to you, as the qualified and competent person, to ensure that the following documents are handed over: full maintenance manual including manufacturer's instructions, Electrical Installation Certificate, Schedule of Inspections, Schedule of Test Results, any other safety information concerning the electrical installation.

**478** Provided the replacement is on a like-for-like basis, a Minor Electrical Installation Works Certificate.

**480** Electrical Installation Certificate, Schedule of Inspections, Schedule of Test Results.

## Knowledge check answers

### *Page 496*

1 50 V.

2 Low resistance ohmmeter.

3 3.33 MΩ.

4 a) sight, touch; b) sight; c) sight (plus appropriate test item); d) touch, sight.

5 a) The cpc has a smaller cross-sectional area, therefore a higher resistance,

b) $R = 0.8 + 0.8/4 = 0.4\ \Omega$, $R = 0.8 + 1.3/4 = 0.525\ \Omega$.

6 $V_d = 20 \times 11 \times 22/1000 = 4.84$ V.

7 Minor Electrical Installation Works Certificate.

8 1) Schedule of Inspections, 2) Schedule of Test Results.

9 Electrical Installation Certificate; Schedule of Inspections; 4 × Schedule of Test Results.

10 a) Long-lead method (method 2), b) $R_1 + R_2$ (method 1).

# UNIT 303 ELECTRICAL INSTALLATIONS: FAULT DIAGNOSIS AND RECTIFICATION

## Activity answers

*Page*

**502** Each part of the form must be filled in – sections must not be left empty.

**505** It is not compulsory to replace the rewireable fuses if they still provide the intended function. Suggest that the board is replaced by a board with circuit breakers/RCDs for added protection and convenience.

**508** Broken switch dolly, burnt contacts, loose terminal, flex pulled out of clamp, damage to flex, faulty thermostat, faulty thermal trip.

**512** Some organisations require PTWs for all tasks involving shut downs. PTWs may also be required for lone working or tasks in hazardous locations.

**514** The energy level in the system is so low that it cannot do any harm.

**516** Ensure that all materials used are conductive to prevent build up of static, see also BS EN 60601.

**522** Electric space heaters, furnaces, immersion heaters or suitable alternatives.

**526** Office staff, shopworkers, shoppers, security staff

**527** They generally do not have adequate short-circuit rating; they may not produce the required output; they are easily left on the wrong range.

**529** Return it to the manufacturer for checking and/or repair.

**530** Main earthing conductor.

*Page*

**534** Overload or short circuit causing fuse to blow on supply side.

**536** No.

**537** Fit RCBOs to individual circuits.

**540** The circuit is split into halves and each part tested separately. The section which is faulty is then divided again and each part is tested. This is repeated until the fault is found.

**544** If all circuit conductors were included, the overall magnetic flux would be zero – no reading.

**549** Dustsheet, brush and dustpan, broom, cleaning cloth, vacuum cleaner.

**551** The method statement will be based on the assumptions made. It should include details of how the task is carried out in a logical sequence.

**556** Sufficient ventilation, avoid spills, insulated tools.

**558** By smothering the fire with powder or gas.

**562** a) Return to store, b) send for recycling or disposal, c) package up for recycling, d) return to wholesaler or manufacturer for disposal.

**565** Voltage stick or similar.

**569** 1 mA.

**572** Divide the ring at approximately the mid-point and test each half, further dividing until the fault is found.

**574** To make sure readings are accurate.

## Knowledge check answers

*Page 581–582*

1 a) Overalls, b) hardhat, c) safety glasses or suitable alternatives.

2 a) The client, b) information about the work carried out and any observations about future problems should be recorded by permanent means such as paper forms.

3 a) Phase rotation is incorrect, change over any two phases at the motor terminals, b) faulty tube or starter, c) faulty thermostat and thermal cut-out (where fitted ), d) incorrectly set, or faulty sensor in PIR.

4 a) Disconnect dimmer and link out feed and return conductors (switch feed/sw wire), b) operate two-way switches to ensure strappers are properly tested, c) isolate UPS and test as this has a permanent live output, d) risk from discharge gases and short circuiting of terminals by uninsulated spanners etc, make sure room is well ventilated and all tools insulated.

5 a) Circuit charts, b) previous maintenance records, c) Manufacturers' data sheets (or suitable alternatives ).

6 a) Stop working, loss of income, delays in contract, b) closure of shops, loss of shopping facilities and tills, c) loss of closed circuit TV, d) loss of facilities such as tills, computer systems etc.

7 Age of equipment, availability of spare parts, availability of specialist staff, down-time of plant, loss of production or suitable alternatives.

8 a) Packaged up for recycling, b) can be difficult to recycle but if possible do so, c) packaged up and sent for recycling, d) returned to manufacturer for recycling.

9 C1 danger present, C2 potentially dangerous, C3 improvement recommended.

10 As a basic test the RCD could be tested at 0.5 $I_{\Delta n}$ to see if it trips. If this is not successful, set up a supply with a milliammeter, gradually increasing the current and recording the value at which the RCD trips.

## UNIT 308 CAREER AWARENESS IN BUILDING SERVICES ENGINEERING

### Activity answers

*Page*

**586** Decline: eg coal mining, ship building; new: eg IT technology, various service industries.

**591** Tailor the CV to the job application. For example, simple research into the company may show that it carries out a lot of work within hospitals – emphasising experience in care work may make this applicant stand out. Also, moving the science qualification higher up the order makes this relevant qualification stand out.

**595** Check the action plan to ensure that the objectives are realistic and achievable within the time allocated, considering any potential problems that could arise.

**599** It is possible to buy in specific skills to meet the contract requirements (or another suitable alternative answer).

**599** Electricians at Level 3, Plumbers at Level 2 and 3, various levels for other trades.

*Page*

**600** Banks, factories, hotels and shopping complexes, etc.

**602** Possible suggestions: plumber–gas fitter, electrician–heating engineer–refrigeration engineer, plasterer–tiler, etc.

**604** Examples may include seasonal workers: certain farm workers such as fruit pickers, ice cream vendors, people who work in UK holiday resorts (restaurant staff, tour guides and entertainers) and those whose work involves seasonal busy periods, such as delivery drivers around Christmas time.

**606** Go to the NET website and enter your post code.

**607** The person can self-certify their own work.

## Knowledge check answers

*Page*

1 a) JIB (Joint Industry Board), b) IET (Institute of Engineers and Technology), c) Unite (trades union), other suitable answers are acceptable.

2 Personal details (name, age, address, etc), academic qualifications (school, college, university), experience in electrical trade, if any, outside interests, hobbies, other experience, qualifications such as first aid, organisational skills.

3 S = strengths, W = weaknesses, O = opportunities, T = threats.

4 (1) Must have been a registered Apprentice or undergone some equivalent method of training and have had practical training in electrical installation work.

(2) Must have obtained a Level 3 NVQ in Electrotechnical Services (Electrical Installation – Buildings & Structures) or the Level 3 NVQ Diploma in Installing Electrotechnical systems and equipment (buildings, structures and the environment) (or approved equivalent) *(see Note 1)*.

(3) Must have obtained the Level 3 technical certificate relevant to the NVQ either as part of the NVQ Diploma or as a separate qualification (such as the City & Guilds 2360 Electrical Installation Theory and Practice Part 2 or Level 3 Certificate in Electrotechnical Technology (or approved equivalent)). Have obtained Achievement measurement 2 assessment.

5 Direct contact with possible employer, employment office, trade magazine, job fair, or other suitable alternatives.

# INDEX

## C

## D

## E

## F

## G

## J

## N



## O



## P

**Q**



**R**

**T**